S0-AZX-879

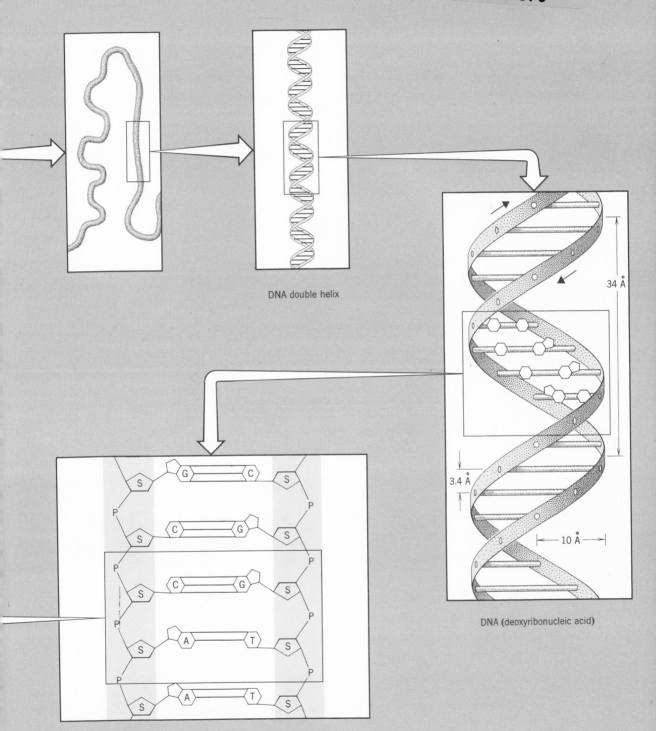

DNA double helix

DNA (deoxyribonucleic acid)

PRINCIPLES OF GENETICS

PRINCIPLES OF GENETICS

Eldon J. Gardner
Utah State University

5th Edition

John Wiley & Sons, Inc.
New York
London
Sydney
Toronto

This book was set in Trump by
Progressive Typographers, Inc. and
printed and bound by Quinn & Boden
Co., Inc. The text and cover were
designed by Jerome Wilke; the
drawings were designed and executed
by John Balbalis with the assistance
of the Wiley Illustration Department;
Stella Kupferberg was the picture
researcher; the editor was Eugene
Patti; Reiko Okamura supervised
production.

Copyright © 1960, 1964, 1968, 1972, 1975, by John Wiley &
Sons, Inc.

All rights reserved. Published simultaneously in Canada.

No part of this book may be reproduced by any means, nor
transmitted, nor translated into a machine language without the written permission of the publisher.

Library of Congress Cataloging in Publication Data:

Gardner, Eldon John, 1909–
    Principles of genetics.

    1. Genetics. I. Title. [DNLM: 1. Genetics.
QH431 G226p]
QH430.G37   1975   575.1   74-17433
ISBN 0-471-29131-5

Printed in the United States of America

10 9 8 7 6 5 4 3 2

PREFACE

This new edition is a continuing attempt to provide a general genetics text book that is current, readable, and challenging for students, and that can be taught in a one-quarter or one-semester course. It is a beginning-level text, focused on the basic principles of genetics, and the present edition has been reorganized and extensively rewritten. Historical, extrachromosomal, and detailed material in other areas has been deleted in order to include current experimental results in molecular, population, and human genetics. To make the book more concise, the aspects of basic genetics have been preserved and extended in depth.

Eighteen chapters of the previous edition have been condensed and integrated into sixteen chapters, which include a brief introductory chapter that gives a historical perspective and identifies the principles covered in the chapters that follow. The previous Chapter 1 has been integrated with parts of Chapter 5 to provide a unified statement of Mendelian genetics. Experimental evidence for the biochemical nature and action of the gene is introduced in Chapter 3, and cell mechanics is discussed in Chapter 4. These early statements of molecular and cell biology make it possible to deal with the aspects that follow in terms of DNA, RNA, and cell mechanics. Chapter 5 examines linkage and crossing over. Alleles and genetic structure are discussed in Chapter 6, aspects of gene regulation are considered in Chapter 7, and mutagenesis is studied in Chapter 8. Chromosomal structural changes and chromosomal numerical changes are examined in Chapters 9 and 10. Extrachromosomal inheritance (Chapter 11) is based on DNA in extranuclear organelles and extracellular symbionts and possible non-DNA inheritance. Chapters 12 to 14 are devoted to quantitative inheritance, population genetics, and systems of mating (including applications to plant and animal breeding). The current aspects of behavior genetics are outlined in Chapter 15. Examples from human genetics are given throughout the book, and the final summary (Chapter 16)

discusses basic genetics as applied to man. It comments on genetic engineering, prenatal diagnosis of genetic disorders and treatments for syndromes resulting from chromosome irregularities, metabolic disorders based on genetic defects, and population questions.

I thank the students, teachers, and colleagues who suggested improvements in the text and assisted in many ways. I especially thank D. Peter Snustad, J. R. Preer Jr. and John R. Simmons who read the revised manuscript and made valuable suggestions. Kandy Baumgardner, Bruce S. Shadbolt, and David M. Abbott read major parts of the manuscript and recommended improvements. Lois Cox gave most valuable editorial assistance. Credits for illustrations, tables, and quotations from other individuals or publications are appropriately acknowledged in the text.

ELDON J. GARDNER

Logan, Utah, 1974

CONTENTS

PRINCIPLES OF GENETICS

Versuche
über
Pflanzen-Hybriden

von

Gregor Mendel.
(Vorgelegt in den Sitzungen vom 8. Februar und 8. März 1865.)

Einleitende Bemerkungen

ONE

In a sense, genetics could be called a science of potentials since it deals with the transfer of information from parents to offspring and between generations. Geneticists are concerned with the why's and how's of these transfers, which are the basis for certain differences and similarities that we recognize in groups of living organisms. Not all the variation among living things is inherited, however. Environmental and developmental factors are also significant and therefore of interest to the geneticist.

Various basic concepts have been established by observation and experimentation as *principles of genetics*. Some of these principles are: (1) The gene is the unit of inheritance. (2) Genes are arranged in linear order on chromosomes. (3) Chromosomes are generally single units in reproductive cells (egg and sperm) but they are paired in fertilized eggs and in body cells that develop from fertilized eggs. (4) Members of a pair of genes and chromosomes segregate to different reproductive cells. (5) Members of different gene pairs are assorted independently with respect to those of other gene pairs

INTRODUCTION

Front page of Mendel's original paper and portrait of Mendel. (Courtesy of New York Public Library and American Museum of Natural History.)

1

in the formation of eggs and sperm. (6) Genes are units of deoxyribonucleic acid (DNA) and are capable of replication. They carry coded messages that can be transcribed and translated into polypeptides, which may be either enzymes or structural proteins. (7) Changes (mutations) occur in genes and in chromosomes. (8) Multiple genes control the inheritance of quantitative traits (e.g., size, pigmentation). (9) Genes in populations establish an equilibrium, the level of which can be changed by such factors as mutation, migration, and selection, phenomena that provide the basis for race and species formation. (10) Inheritance patterns are associated with different systems of mating (e.g., inbreeding or outbreeding).

In this book, these and other genetic principles are discussed in detail as they apply to microorganisms, plants, and animals, with strong emphasis on man. The behavior patterns that we see in animals (particularly in human beings) are also considered in terms of genetic principles.

THE BIRTH OF A SCIENCE

Gregor Mendel (1822–1884; Fig. 1.1) is appropriately called the "father of genetics." His precedent-setting experiments with garden peas (*Pisum sativum*) were conducted in a limited space in a monastery garden (Fig. 1.2) while he was also employed as a substitute teacher. The conclusions that he drew from his elegant investigations constitute the foundation of today's science of genetics.

Mendel was not the first to perform hybridization experiments, but he was one of the first to consider their results in terms of single traits. His predecessors had considered whole organisms, which incorporate a nebulous complex of traits; thus, they

Figure 1.1 Gregor Mendel, Austrian monk, whose experiments with garden peas laid the foundation for the science of genetics. (Courtesy of American Museum of Natural History.)

Figure 1.2 Garden at Altbrünn. Monastery garden where Mendel's experiments on garden peas were conducted. (Courtesy of Professor Dr. Ing. Jaroslav Kříženecký.)

could only observe that differences occurred among parents and progeny. Employing the scientific method, Mendel designed the necessary experiments, counted and classified the peas resulting from his crosses, compared the proportions with mathematical models, and formulated a hypothesis for these differences. Although Mendel devised a precise mathematical pattern for the transmission of hereditary units, he had no concept of the biological mechanism involved. Nevertheless, on the basis of his preliminary experiments and hypothesis, he predicted the subsequently verified results of later crosses.

In 1900, Mendel's paper was discovered simultaneously by three botanists: Hugo de Vries, in Holland, known for his mutation theory and studies on the evening primrose and maize; Carl Correns, in Germany, who investigated maize, peas, and beans; and Eric von Tschermak-Seysenegg, in Austria, who worked with several plants including garden peas. Each of these investigators obtained evidence for Mendel's principles from his own independent studies. They all found Mendel's report while searching literature for related work, and cited it in their own publications. William Bateson, an Englishman, gave this developing science the name **genetics** in 1905. He coined the term from a Greek word meaning "to generate."

CONCEPT OF THE GENE
In addition to naming the science, Bateson actively promoted Mendel's view of paired genes. Also during the early 1900s, a Frenchman, Lucien Cuénot showed that

genes controlled color in the mouse; an American, W. E. Castle related genes to sex and to fur color and pattern in mammals; and a Dane, W. L. Johannsen studied the influence of heredity and environment in plants. Johannsen began using the word **gene** from the last syllable of Darwin's term "pangene." The gene concept, however, had been implicit in Mendel's visualization of a physical element or factor (*anlage*) as the foundation for development of a trait. These men and their peers were able to build upon the basic principles of cytology, which were established between 1865 (when Mendel's work was completed) and 1900 (when it was discovered).

CHROMOSOME THEORY

Wilhelm Roux had postulated as early as 1883 that chromosomes within the nucleus of the cell were the bearers of hereditary factors. The only model he was able to devise that would account for his observed genetic results lined up objects in a row and duplicated them exactly. To explain the mechanics of gene transmission from cell to cell, he therefore suggested that nuclei must have beadlike structures that line up and duplicate themselves. The constituents of the nucleus that seemed best designed to carry the genes and fill these requirements were the **chromosomes.**

Experiments of T. Boveri and W. S. Sutton brought evidence in 1902 that a gene is part of a chromosome. The theory of the gene as a discrete unit of the chromosome was developed by T. H. Morgan and his associates from studies on the fruit fly, *Drosophila melanogaster.* H. J. Muller later promoted the merger of cytology and genetics as **cytogenetics.** A. H. Sturtevant and C. B. Bridges, along with Morgan, eventually indentified the genetic mechanisms of linkage and crossing over and prepared chromosome maps of *D. melanogaster.*

CHEMICAL NATURE OF THE GENE

In the 1930s G. W. Beadle, B. Ephrussi, and E. L. Tatum provided a basis for understanding the functional properties of genes and suggested extensions of the classical gene concept. The classical gene had been characterized as an indivisible unit of structure, a unit of mutation, and a unit of function with all three of these attributes considered equivalent. The classical gene could not be divided into smaller parts. C. P. Oliver reported, however, (in 1940) that the lozenge (*lz*) gene (for oval-shaped eyes) in *D. melanogaster* could be subdivided. Later data by M. M. and K. C. Green indicated that the *lz* gene could be divided into sites. This implied that all "genes" might be subject to subdivision.

Investigators then recalled that A. E. Garrod had demonstrated in 1909 that genes produce **enzymes.** The geneticists of the 1950s therefore sought an ideal experimental system with which they could investigate functional aspects of genes. The prokaryotes (organisms lacking well-defined nuclei and not undergoing meiosis, that is, bacteria and blue-green algae) were chosen as likely candidates. An immediate triumph was the identification by O. T. Avery et al. and by A. Hershey and M. Chase of macromolecules in these organisms that were carrying genetic information. The macromolecules proved to be deoxyribonucleic acid **(DNA)** and ribonucleic acid **(RNA).** The later Watson and Crick model of DNA as a double helix established the gene as a chemical unit. Genetics thus solved its central problem. DNA is the genetic material that replicates, carries coded specifications, and influences development. It thereby operates as the functional gene.

These developments prepared the way for **molecular genetics,** which blends chemistry (biochemistry) and physics (biophysics) with biology. Geneticists, in the past two decades, have been largely preoccupied with using the molecular aspects of biology. Much of this book is devoted to describing how both past and recent events generate today's progress, and to the implications of genetics for the future.

DNA carries the specifications for growth, differentiation, and functioning of the organism. The same genes are present in virtually all nucleated cells in the body. Different genes, however, are active at different times during development. At these times the information contained in the gene is **decoded** by the process of transcription and translation and proteins are produced. These proteins, called enzymes, catalyze cellular biochemical reactions. Contents of a single fertilized cell carry the information for development of an organism destined at one time to contain a million billion (10^{15}) cells. The DNA in the species population is the evolutionary storehouse carrying the information for that **species.**

REFERENCES

Dunn, L. C. 1965. *A short history of genetics.* McGraw-Hill, New York.

Iltis, H. 1932. *Life of Mendel.* (Trans. E. and C. Paul.) W. W. Norton, New York.

Stern, C. (ed.) 1950. "The birth of genetics." (Supplement of *Genetics.*) 35(5) part 2.

Stern, C., and E. R. Sherwood (eds.) 1966. *The origin of genetics.* A Mendel Source Book. W. H. Freeman, San Francisco.

Sturtevant, A. H. 1965. *A history of genetics.* Harper and Row, New York.

Tschermak-Seysenegg, E. von, 1951. "The rediscovery of Gregor Mendel's work." *J. Hered.,* 42, 163–171.

TWO

Mendel chose the garden pea as his experimental organism because it was an annual plant that had well-defined characteristics, and it could be grown and crossed easily. Moreover, garden peas have perfect flowers containing both female and male (pollen-producing) parts, and they are ordinarily self-fertilized. Pollen from another plant can be introduced to the stigma by an experimenter, but cross-pollination is rare without man's intervention.

Good fortune as well as wise judgment attended Mendel's choice because other properties of garden peas, unknown to Mendel, were important for his experiments. Through many generations of natural self-fertilization, garden peas had developed into **pure lines.** A single alteration in a trait was therefore demonstrated by a visible difference between varieties. Furthermore, in the seven pairs of contrasting traits Mendel chose to study, one form was dominant over a well-defined, contrasting alternative. Vines were either tall or dwarf; unripe pods were green or yellow and inflated or constricted between the seeds; flowers were either distributed along the stem (axial) or

MENDELIAN GENETICS

Garden peas growing on
a fence. (Grant Heilman.)

bunched at the top (terminal); nutritive parts of the ripe seeds were green or yellow; the outer surface of the seed was smooth or deeply wrinkled; and the seed coats were either white or gray. Flower color was positively correlated with this last trait: seeds with white seed coats were produced by plants that had white flowers, and those with gray seed coats came from plants that had violet flowers. Much of Mendel's success in his first experiments may be attributed to his good judgment in **making crosses,** as far as possible, between parents that differed in only one trait. When this was not feasible, he considered only **one trait at a time.**

MENDEL'S EXPERIMENTS

Crosses were made with great care when the peas were in blossom. To prevent self-fertilization in "test" flowers, the anthers were removed from those chosen to be seed parents before their pollen-receiving parts were fully mature. Pollen from the designated pollen parent was transferred at the appropriate time to the stigma of the seed-parent flower. Seeds were allowed to mature on the vines. With a trait such as seed color, classification could be made immediately; but before traits such as plant size could be classified, the seeds had to be planted in the next season and the plants raised to maturity. The hybridization experiments were carried through several generations, and backcrosses were made between the hybrids and the pure parent varieties. Mendel visualized clearly each problem to be solved and **designed** his crosses to that end. He observed that weather, soil, and moisture conditions affected the growth characteristics of the peas, but heredity was the main limiting factor under the conditions of his experiments. In a given environment, tall plants were 6 to 7 feet high, whereas dwarfs measured from 9 to 18 inches. No dwarfs ever turned into tall plants and no tall plants became dwarfs.

PRINCIPLE OF SEGREGATION

In one experiment, Mendel crossed tall and dwarf varieties of garden peas. All of the offspring in the first (F_1) generation (F symbolizes filial from the Latin, meaning progeny) were tall. The dwarf trait had disappeared in the F_1 progeny. When the tall hybrid plants were self-fertilized and the progeny (second, or F_2 generation) were classified, some were tall and some were dwarfs. Careful classification of the plants showed that when large numbers were considered, about three-fourths were tall and one-fourth were dwarfs. To be exact, an F_2 of 1064 consisted of 787 tall plants and 277 dwarfs, a near perfect **3/4 : 1/4 ratio.**

The experiment could have been concluded at this point, but to test his hypothesis that independent factors (genes) were responsible for the observed hereditary patterns, Mendel predicted what would occur in the F_3 generation and planted the F_2 seeds to test this prediction. On the basis of his hypothesis, he predicted that about one-third of the yellow F_2 seeds would produce only yellow seeds, whereas two-thirds would produce both yellow and green seeds. The green F_2 seeds were expected to produce all green. Of the 519 plants produced from yellow F_2 seeds, 353 had both yellow and green seeds in the proportion of about three yellow to one green, and 166 had only yellow seeds. As Mendel had anticipated, the green F_2 seeds gave only plants producing green seeds.

In other crosses, the remaining five of the originally selected seven pairs of contrasting traits were studied. One member of each pair dominated the other in the same way as tall dominated dwarf. This member Mendel identified as dominant in contrast to the other (recessive) member. Mendel's conclusions were based on his concept of **unit characters,** which was in marked contrast to the prevailing belief in a blending inheritance. On the basis of good experimental evidence, he visualized the **physical elements** as occurring in pairs or **alleles** (different forms of a given gene). In garden peas, for example, a gene for height has two alleles, one for tall and one for dwarf. The allele for tall behaves as a dominant whereas that for dwarf is recessive. Likewise, the gene for seed coat color has two alleles, a dominant for yellow and a recessive for green. During meiosis the members of each pair of alleles separate into different sex cells or gametes and thence into different offspring. Mendel called this separating or segregation process the "splitting of hybrids."

The most significant deduction from Mendel's results was the separation of pairs of determiners resulting in the "purity of gametes." This concept of segregation identified as Mendel's first principle is: **the separation of paired genes (allelic pairs) from one another and their distribution to different sex cells.**

SYMBOLS AND TERMINOLOGY

Mendel used letters of the alphabet as symbols for genes. Capitals signified the dominant, and lowercase the recessive member of a pair of alleles. He considered the factors as abstract units, any one of which could be symbolized by A or B or some other letter. Now many genes are known and several may affect the results of a single series of experiments. To avoid confusion as to which gene is indicated, appropriate letter symbols are chosen to represent particular genes. The **mutant** trait that deviates from the ancestral type is usually chosen as the basis for the symbol. This trait is usually produced by the recessive allele because most mutations occur as recessives. For example, the mutant vermilion is a recessive eye color in fruit flies and the gene symbol is v. Some mutant alleles such as the one for wrinkled wings in the fruit fly, however, are dominant, and are therefore represented by a capital letter (e.g., W).

If the mutant is to provide the symbol, the history of the organism under investigation must be well enough known to suggest which member of the contrasting pair (for example, tall or dwarf in peas) is the mutant trait. World collections of pea species show no dwarfs in natural populations. Dwarf peas occur only in certain cultivated stocks that have been developed by man. Since dwarf is probably the mutant, while tall is the "wild type," d is used to symbolize the allele for dwarf and D for tall. With the aid of these symbols, Mendel's experiment (Fig. 2.1) may be reconstructed in steps. The parents (P), each with two members of the two alleles (DD and dd), are symbolized as follows:[1]

$$\text{Tall parent} \quad \times \quad \text{Dwarf parent}$$
$$DD \qquad\qquad dd$$

---

[1] In diagrams of crosses, the female or seed parent is always written first.

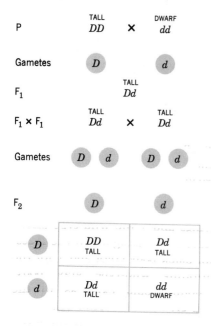

		TALL		DWARF
P		*DD*	×	*dd*

Gametes   *D*   *d*

F₁   TALL *Dd*

F₁ × F₁ TALL *Dd* × TALL *Dd*

Gametes *D* *d*   *D* *d*

F₂   *D*   *d*

	D	*d*
D	*DD* TALL	*Dd* TALL
d	*Dd* TALL	*dd* DWARF

Phenotypes	Genotypes	Genotypic Frequency	Phenotypic Ratio
Tall	*DD*	1	3
	Dd	2	
Dwarf	*dd*	1	1

Figure 2.1 Mendel's cross between tall and dwarf garden peas and summary of phenotypic and genotypic results.

Segregation, the separation of the pairs, occurs during the formation of mature reproductive cells or **gametes.** Each gamete produced by the tall parent carries only one *D* allele, and each gamete from the dwarf parent carries only one *d* allele. Therefore, the fertilized egg **(or zygote),** which results from the fusion of the male and female gametes, must have one allele of each kind (*Dd*). Because the *D* was always present and dominant, the F₁ plants (first generation progeny) were tall. When the F₁ tall (*Dd*) plants were self-fertilized, half of the gametes carried the *D* allele and half the *d* allele. The results of selfing the F₁ indicated to Mendel that the alleles were entirely separate from each other.

When the F₁ (*Dd*) plants from Mendel's experiments were **crossed back** to the dwarf (*dd*) variety, half of the progeny were tall and half dwarf, as illustrated in Fig. 2.2. This demonstrated further the principle of segregation, but the separation of alleles could be detected only in the parent (*Dd*) that produced two kinds of gametes (*D*) and (*d*). The dwarf parent (*dd*) could only produce one kind of gamete. A cross between a heterozygous individual or one of unknown genotype and an individual homozygous for the recessive gene in question is called a **test cross.** It is an extremely important device which has many applications in genetics.

Gene symbols represented in pairs designate zygotes and the individual plants or animals that have arisen from zygotes. Members of the pairs of alleles are represented separately to designate mature germ

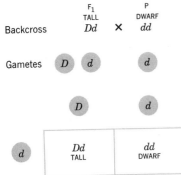

	F₁ TALL	P DWARF
Backcross	Dd	×	dd

Gametes D d d

D d

	D	d
d	Dd TALL	dd DWARF

Summary: 1/2 tall, 1/2 dwarf

Figure 2.2 Backcross between F₁ tall garden pea and the dwarf parent variety from the cross illustrated in Fig. 2.1.

cells or gametes, either eggs or sperm. Circles or brackets placed around gamete symbols indicate mature germ cells, as distinguished from plants or animals. A female and a male gamete combine in fertilization to produce a zygote. Zygotes or individual organisms carrying two units of one allele (such as DD or dd) are **homozygous,** and those with two alleles (such as Dd) are **heterozygous.** Two other useful terms, **phenotype** and **genotype,** refer to the visible expression or trait, and the actual gene constitution, respectively. Letter symbols are used to represent genotypes.

As the genetics of a particular species advanced, the 26 letters in the alphabet were soon associated with genes and more symbols were needed. Drosophila geneticists met this limitation by adding a second letter and a third and fourth, when necessary, taken from the name of the mutant phenotype. Another technical advance from Drosophila geneticists came in the use of + to symbolize wild-type genes known to be alleles of recognized mutant genes. For example, Cy symbolized the dominant gene for curly wings and + the wild-type allele. The lowercase b was assigned to the recessive gene for black body color and + for the allele controlling wild-type gray body. Whenever doubt could exist as to the meaning of a given +, the letter symbol of the mutant was added and either the letter symbol or the + was used as a superscript. Thus, w^+ became the symbol for the wild-type allele of w, (white eye in fruit flies). Another useful device initiated by Drosophila geneticists was the separation of alleles by a crossbar or slash mark. A heterozygous pair of alleles at the w locus position occupied by the w gene in a chromosome, for example, was symbolized w^+/w.

Hybrids are offspring from a cross between two genetically unlike individuals (e.g., $AA \times aa \rightarrow Aa$). **A monohybrid** is heterozygous for one pair of alleles (e.g., Aa). By extension, crosses (e.g., $AA \times aa$) involving parents that differ with respect to one pair of alleles are called "monohybrid crosses."

MONOHYBRID CROSSES

Monohybrid crosses are basic to Mendelian genetics. Pertinent information about genetic segregation as it occurs in monohybrid combinations is summarized in Table 2.1 and discussed in the following pages. Such crosses may occur in all major groups

TABLE 2.1

Crosses Involving One Gene Pair. Expected Gametes, Genotypic Frequency of Progeny, and Phenotypic Ratios for Dominance and Intermediate Inheritance Are Given for the Different Combinations

Mating Combinations	Gametes		Progeny		
	First Parent	Second Parent	Genotypic Frequency	Phenotypic Ratio When Dominance Is	
				Complete	Intermediate
$AA \times aa$	A	a	all Aa	all dom.	all int.
$Aa \times Aa$	A a	A a	$\frac{1}{4}AA, \frac{1}{2}Aa, \frac{1}{4}aa$	3:1	1:2:1
$Aa \times AA$	A a	A	$\frac{1}{2}AA, \frac{1}{2}Aa$	all dom.	1:1
$Aa \times aa$	A a	a	$\frac{1}{2}Aa, \frac{1}{2}aa$	1:1	1:1

of sexually reproducing organisms. Dominance is the major form of interaction between alleles.

DOMINANT ALLELES

If allele A is completely dominant, AA and Aa individuals are alike phenotypically. In the heterozygous (Aa) condition, allele a is completely masked and is called a recessive gene. The phenotypic evidence of dominance may be influenced by factors in the internal and external environment and therefore is not caused by a single gene alone. In practice, however, phenotypes attributable to single allele substitutions are called dominants, and those requiring homozygous combinations for expression are called recessives. Dominants are easier to detect than recessives because they are expressed when paired with either kind of allele. Criteria for identifying dominant, defect-transmitting alleles from human pedigree studies are summarized as follows. (1) The trait is transmitted by a parent to about half of his or her children. (This assumes that the parent is heterozygous, which is usually the case because such

genes are almost always rare, hence heterozygous.) If each family includes three or four children, the trait usually occurs in every generation. (2) Persons who do not express the trait do not carry the gene and therefore do not transmit it to their children.

Dominant inheritance can be illustrated by the results of a study of the occurrence of brown teeth in a family group (Fig. 2.3). Among the descendants of II-I, 16 people have brown teeth. In the families of those 16, 15 individuals have white teeth. Thus, about half of the children who have one parent with brown teeth express the trait. This is the expected result from matings between heterozygous people with brown teeth and people who are homozygous for the recessive allele for white teeth. Among the descendants of white-toothed family members, no brown teeth have occurred. Thus, the two criteria for identifying dominant genes were met. X-rays of the brown teeth (Fig. 2.4) showed that the central pulp cavity of each tooth was filled with dentine and no nerves entered the teeth. Histological studies confirmed that dentine, which is completely covered in

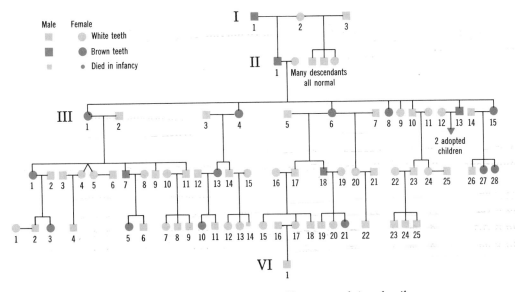

Figure 2.3 Pedigree chart showing the distribution of brown teeth in a family group. (From Gardner, Journal of Heredity, 42, 289–290, 1951.)

normal teeth, could be seen through the deficient enamel and this gave the teeth their brown appearance.

From the pattern of dominant inheritance and the probability factors involved, it might be predicted that in future genera- tions those who have brown teeth (presum- ably heterozygous) and marry individuals with white teeth might expect about half of their children to have brown teeth. In- dividuals who do not express the trait will not transmit it.

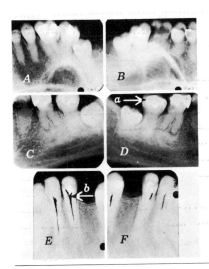

Figure 2.4 Roentgenograms of soft, brown teeth lacking enamel, compared with white teeth in which the enamel developed normally. Plates *A* to *D* show all teeth remaining in mouth of a woman 21 years of age. Pulp cavities were closed and fillings were large and numerous. A filling is identified by arrow a. Plates *E* and *F* show normal teeth of a brother of the woman. Normal pulp cavity is identified by arrow *b*. (Reprinted with permission from the *Journal of Heredity*.)

CODOMINANT ALLELES

When both alleles of a pair are fully expressed in a heterozygote they are called **codominants.** Such alleles act in distinctive ways and yield different products that can be traced in the heterozygous condition. For example, in man the allele A for A-type blood is codominant with allele A^B for B-type blood. The heterozygote (A/A^B) expresses the characteristics of both A and B type. Since the two alleles control different protein products, a mating between a homozygous A-type person (AA) and a homozygous B-type person $(A^B A^B)$ would result in heterozygous AA^B offspring. Matings between heterozygotes $(AA^B \times AA^B)$ would result in a ratio of $1\ AA:2\ AA^B:1\ A^B A^B$. A ratio of **1:2:1** thus replaced the 3:1 ratio because the alleles are codominant.

RECESSIVE ALLELES

Recessive alleles produce a different pattern of inheritance because the phenotype is expressed only in homozygous (aa) individuals. A population of crossbreeding organisms usually includes all three genotypes $(AA, Aa, \text{and } aa)$ but has more heterozygous (Aa) **carriers** than homozygous (aa) individuals who express the trait. Carriers may not be detectable phenotypically because they carry a dominant allele, but they can be identified experimentally with a test cross.

In human family groups, the influence of recessive alleles can be detected from pedigree studies on the basis of the following criteria: (1) the trait usually first appears in the family group only in a sib (offspring of the same parents) but not in parents, or other relatives; (2) on the average, one-fourth of the sibs of parents, both of whom are known to be carriers of the recessive allele, are affected; and (3) males and females are equally likely to express the trait unless the recessive gene is sex linked (see Chapter 5). The gene associated with albinism, for example, is a relatively rare recessive that is not sex linked. Albino people are characterized by a marked deficiency or complete absence of pigment in the skin, hair, and iris of the eye. Homozygous (cc) individuals are albino, whereas CC and Cc are within normal limits of pigmentation. In one Caucasian family of six, for example, two albinos have occurred. Parents are within the normal range of pigment for Caucasian people. Presumably, both parents are carriers (Cc) and a chance combination of recessives (cc) occurred in the two albino sons. The cross is represented diagrammatically in Fig. 2.5. On the average, three

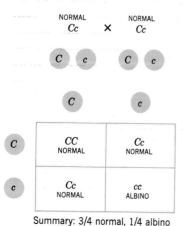

Summary: 3/4 normal, 1/4 albino

Figure 2.5 Cross between two normally pigmented people, both of whom were carriers for the gene (c) for albinism.

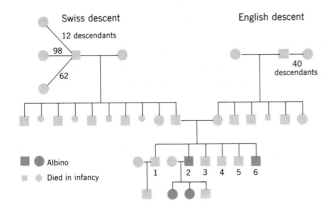

Figure 2.6 Pedigree of a family group in which albinism has occurred.

normal children to one albino would be expected. Son No. III-4 on the pedigree chart (Fig. 2.6), an albino, married a normally pigmented woman. The couple had two albino daughters and one normal son. Presumably, the wife was a carrier (*Cc*) of the recessive allele (*c*). For an infrequent allele, this occurrence would be rare if the parents (III-3 and III-4) were not related. This example represents an exception to the first criterion for detecting recessives from pedigrees. The cross is the backcross type and is similar to the one reconstructed in Fig. 2.2 for a different trait. About half of the children of son No. III-4 would be expected to express the albino trait whereas the other half would be normal phenotypically but carriers (*Cc*) of the allele for albinism.

LETHALS

Genes may affect viability as well as visible traits of an organism. Appropriate experiments have shown that animals carrying certain genes are disadvantaged through impaired biochemical as well as physical functioning. White-eyed and vestigial-winged Drosophila, for example, have lower viabilities than the wild type. Detrimental physiological effects are apparently associated with the genes involved (*w* and *vg*, respectively). Some other genes have no effect on the appearance of the fly, but do influence viability in some way. Some genes have such serious effects that the organism is unable to live. These are called **lethal genes.** They represent an important class of altered genes or mutations. Obviously, if the lethal effect is dominant and immediate in expression, all individuals carrying the gene will die and the gene will be lost. Some dominant lethal genes, however, have a delayed effect so that the organism lives for a time. Recessive lethals carried in heterozygous condition may come to expression when matings between carriers occur.

The dominant gene (*C*) in chickens, for example, is responsible for profound developmental changes that result in aberrant forms called "creepers," and the homozygous genotype (*CC*) is lethal. These birds have short, crooked legs and are of little value except as novelties. When two creepers were mated, a ratio of **2 creepers to 1 normal** instead of 3:1 appeared, as illustrated in Fig. 2.7. This is a characteristic ratio for all crosses involving lethals. In this particular case the *CC* class is missing. All creepers that lived could be shown by testcrosses to be heterozygous (*Cc*). When a creeper was mated with a normal chicken the expected backcross result of 1 creeper to 1 normal was obtained (Fig. 2.8).

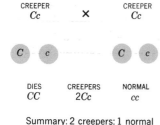

CREEPER
Cc × CREEPER
Cc

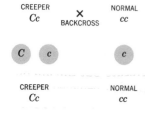

DIES
CC CREEPERS
2*Cc* NORMAL
cc

Summary: 2 creepers: 1 normal

Figure 2.7 Cross between two creeper chickens. The 2:1 ratio replaces the 3:1 because the homozygous (*CC*) embryos die.

CREEPER
Cc ×
BACKCROSS NORMAL
cc

C   c   c

CREEPER
Cc NORMAL
cc

Summary: 1 creeper: 1 normal

Figure 2.8 Cross between a creeper and a normal chicken. Expected ratio is 1 creeper:1 not-creeper (normal).

Walter Landauer compared the development of creepers with that of normal chickens. He found that the mutant gene causes a general retardation of growth in the early embryo at the time when the limb buds are forming. The effects are widespread in the body. Characteristic abnormalities occur in the head, eyes, and other body structures of heterozygotes (*Cc*) but the leg abnormality is the most conspicuous. Defects in growth can be observed on the second and third days of incubation and the homozygous (*CC*) embryos die on about the fourth day.

At least 27 recessive lethal genes are known to occur in cattle. Some of these have been spread widely through artificial insemination. Thirteen of the 27 abnormalities involve bone formation. In some, or perhaps all of the 13, the bones fail to develop properly in the embryo. The same phenotypes have been identified in different breeds, indicating either that different mutations have occurred independently or that the genes came from common ancestors. So-called "bulldog" calves (Fig. 2.9) are the best known of all the abnormalities resulting from lethal

Figure 2.9 Bulldog calf. This condition is caused by a single recessive gene in homozygous condition, that causes the calves to be aborted at six to eight months from the beginning of development.

genes in cattle. They were first described in Germany in 1860. The name came from the abnormal head which resembles that of a bulldog with a short face and short upper jaw. Other conspicuous features are short legs and cleft palate. Usually bulldog calves are aborted at about six to eight months. The Dexter breed, in which this trait is prevalent, is no longer used to any great extent, mainly because of the serious losses from lethals.

PRINCIPLE OF INDEPENDENT ASSORTMENT

Mendel also crossed plants that differed in two pairs of alleles (Fig. 2.10). In this cross, designed to clarify the relation of genes in different pairs of alleles, he crossed plants having round, yellow seeds with plants having wrinkled, green seeds. F_1 progeny from such a cross between homozygous parents are hybrids (heterozygotes) for two gene pairs. The F_1 progeny ($Gg\ Ww$) are dihybrids and by extension the $GGWW \times ggww$ cross is a *dihybrid cross*. Alleles for both round and yellow were known from previous studies to be dominant over their respective alleles producing wrinkled and green seeds.

All the F_1 seeds resulting from the cross were round and yellow as expected. When the F_1 hybrids were allowed to self-fertilize, four F_2 phenotypes were observed in a definite pattern. From a total of 556 seeds, the following distribution was obtained: 315 round, yellow; 108 round, green; 101 wrinkled, yellow; and 32 wrinkled, green. When reduced to lowest terms (i.e., $\frac{315}{556} = \frac{9}{16}$, $\frac{108}{556} = \frac{3}{16}$, $\frac{101}{556} = \frac{3}{16}$, $\frac{32}{556} = \frac{1}{16}$), these results closely fit a ratio of $9:3:3:1$. Mendel recognized this as the result of two monohybrid crosses, each expected to result in a $3:1$ ratio, operating together. The product of the two **monohybrid** ratios $(3:1)^2$ or $(3 + 1)^2$ was equal to the **dihybrid** ratio $(3 + 1)^2 = (9 + 3 + 3 + 1)$, thus conforming to the law of probability which states: the chance of two or more **independent events occurring together** is the **product** of the **chances of their separate occurrences.**

The results were those expected from the assortment of two independent pairs of alleles, each showing dominance of one member. Not only did the members of each pair of alleles segregate but the allelic pairs of different genes also behaved independently with respect to each other. Mendel therefore drew another conclusion: members of different pairs of alleles (included in his study) **assort independently into gametes.** This concept, involving the independent combinations of different pairs of alleles, is designated as his second principle. Mendel's two principles were set forth in a paper entitled, "Experiments in Plant Hybridization," which was read before the Brünn Natural History Society in 1865 and published in the proceedings of that society in 1866.

Mendel's principle of independent combinations has a practical application in plant and animal breeding. Desirable traits carried in different varieties can be combined and maintained in a single type. For example, a variety of barley resistant to rust was needed in a rust-infested area in the United States. The best available rust-resistant variety, however, like most barley varieties, had hulls on the seeds and did not thresh well. Another variety had no hulls and threshed out clean, like wheat, but had poor rust resistance. These two varieties were combined by appropriate crosses, and a valuable new strain with rust resistance and no hulls was obtained.

DIHYBRID RATIOS

The basic mechanics of genetics were postulated and later established from particular ratios such as 3:1 and 9:3:3:1. Ratios of this kind merely represent the grouping expected when particular conditions are met. Common patterns such as the 9:3:3:1 ratio may serve as models for an-alyzing results of experiments. When such a ratio is obtained from a cross in which the parental genotypes are not known, the geneticist may postulate that **two independent pairs of alleles** are involved; and that one member of each pair behaves as a dominant over its allele. Mendel's dihybrid cross between plants with round, yellow

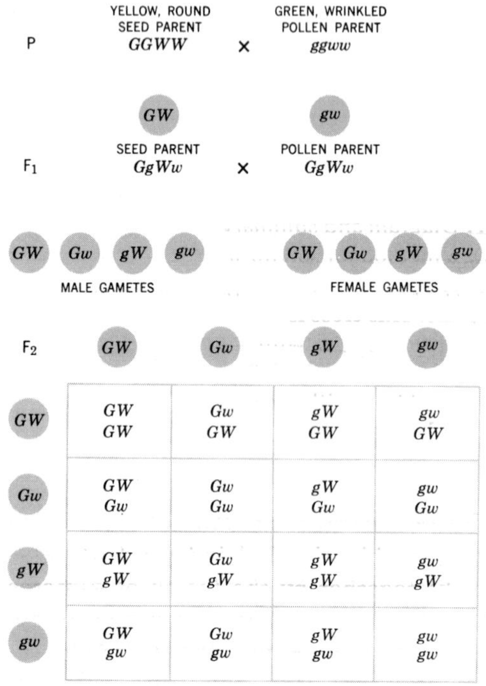

Phenotypes	Genotypes	Genotypic Frequency	Phenotypic Ratio
Yellow, round	$GGWW$	1	9
	$GGWw$	2	
	$GgWW$	2	
	$GgWw$	4	
Yellow, wrinkled	$GGww$	1	3
	$Ggww$	2	
Green, round	$ggWW$	1	3
	$ggWw$	2	
Green, wrinkled	$ggww$	1	1

Figure 2.10 Diagram and summary of a cross between a variety of garden peas with yellow, round seeds and a variety with green, wrinkled seeds. The $F_1 \times F_1$ represented illustrates a dihybrid cross.

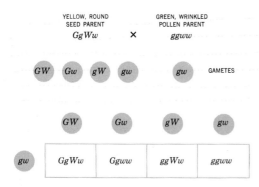

Phenotypes	Genotypes	Genotypic Frequency	Phenotypic Ratio
Yellow, round	$GgWw$	1	1
Yellow, wrinkled	$Ggww$	1	1
Green, round	$ggWw$	1	1
Green, wrinkled	$ggww$	1	1

Figure 2.11 Diagram and summary illustrating a method for solving a backcross type problem involving two gene pairs. This cross is between an F_1 garden pea with yellow, round seeds and the fully recessive parent type with green, wrinkled seeds.

seeds and those with wrinkled, green seeds is represented diagrammatically in Fig. 2.10 as a pattern applicable in analyzing other crosses.

When the F_1 plants were selfed (that is, pollen and eggs from the same plant were united), four kinds of gametes were produced by the male parts and four by the female parts of the F_1. At the top of the checkerboard (Punnett square) (Fig. 2.10) the four kinds of gametes from the seed parent are shown. The four possible gametes from the pollen parent are represented in a vertical row at the left. This **Punnett square** is merely a geometrical device that helps in visualizing all the possible combinations of male and female gametes. It is valuable as a learning exercise, but will be replaced presently with less time-consuming methods. Letter symbols

in the 16 squares of the Punnett square represent combinations of independent genes brought together by the fusion of gametes. When these are collected according to the phenotypes represented, the **9:3:3:1 ratio** becomes apparent. The completed summary chart illustrates the F_2 results of the cross in tabular form.

The **1:1:1:1** ratio is expected from a **dihybrid backcross** to the recessive parent; that is, a cross between an F_1 that carries two heterozygous pairs of alleles and a parent type with the full recessive combination for these two genes. This cross is illustrated in Fig. 2.11. A cross of this type (called a test cross) is used in practical breeding programs to determine the genotype of an individual that may carry recessive alleles, the expression of which could be obscured by dominant alleles.

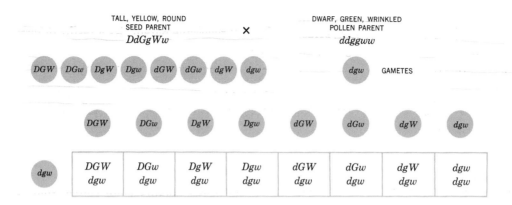

Phenotypes	Genotypes	Genotypic Frequency	Phenotypic Ratio
Tall, yellow, round	*DdGgWw*	1	1
Tall, yellow, wrinkled	*DdGgww*	1	1
Tall, green, round	*DdggWw*	1	1
Tall, green, wrinkled	*Ddggww*	1	1
Dwarf, yellow round	*ddGgWw*	1	1
Dwarf, yellow wrinkled	*ddGgww*	1	1
Dwarf, green, round	*ddggWw*	1	1
Dwarf, green, wrinkled	*ddggww*	1	1

Figure 2.12 Method for solving backcross type problems involving three gene pairs. This cross is between an F₁ garden pea with tall vines, yellow and round seeds, and the fully recessive parental type with dwarf vines, green and wrinkled seeds.

TRIHYBRID RATIOS

Virtually all cross-fertilizing plants or animals differ in more than one or two pairs of alleles. Therefore, matings in natural breeding populations usually produce new combinations of many genes. Genetic analysis of such crosses may be prohibitively complicated. In many cases, however, complex combinations can be simplified by resolving them into monohybrid crosses, or by using formulas devised to handle several factors in the same problem. A cross between homozygous parents that differ in three gene pairs (i.e., producing **trihybrids**) are combinations of three single pair crosses operating together. Thus ($AA \times aa$) ($BB \times bb$) ($CC \times cc$) could be combined in the same cross as $AABBCC \times aabbcc$.

What results might be expected (1) in the F_1, (2) in the backcross to the fully recessive parent, and (3) in the F_2 from a cross between two varieties of garden peas differing in three traits? A diagram of a cross in which the seed parent is homozygous for the genes producing a tall vine and yellow, round seeds ($DDGGWW$), and the pollen parent has a dwarf vine and green, wrinkled seeds ($ddggww$) can answer that question. The three traits represented in the seed parent are known from previous experiments to depend on dominant genes. The first-generation cross may be illustrated as follows:

$$DDGGWW \quad \times \quad ddggww \quad \text{P}$$
$$DGW \qquad\qquad dgw \quad \text{gametes}$$
$$DdGgWw \qquad\qquad \text{F}_1$$

When the F_1 plants are crossed with the full recessive type, $DdGgWw \times ddggww$, eight kinds of gametes (DGW, DGw, DgW, Dgw, dGW, dGw, dgW, dgw) are produced by the F_1 parent and only one kind, dgw, by the full recessive parent. As a result of fertilization, eight kinds of peas are expected in equal proportion. Thus the trihybrid backcross ratio of $1:1:1:1:1:1:1:1$ is explained by the fertilization of eight different kinds of gametes from F_1 by the one kind of gamete from the fully recessive parents. The sequence involved in this backcross and the summarized results are illustrated in Fig. 2.12.

When the F_1 plants were selfed (that is, $DdGgWw \times DdGgWw$), eight kinds of gametes, DGW, DGw, DgW, Dgw, dGW, dGw, dgW, dgw, were produced from both the male and female parts. These gametes represent all combinations. If the $F_1 \times F_1$ cross were represented by a Punnett square, $64 = (8^2)$ squares would be required with a ratio of $27:9:9:9:3:3:3:1$ as a result. A less time-consuming method for determining the results of complex combinations is explained in the next section.

FORKED-LINE METHOD FOR GENETIC PROBLEMS

A method for bringing the combinations of a trihybrid cross together may be illustrated as follows: first, visualize the trihybrid

$$DdGgWw \times DdGgWw$$

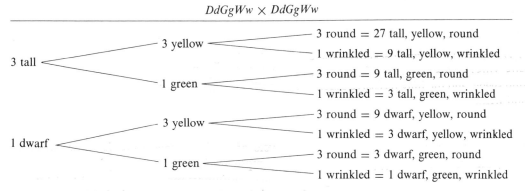

Figure 2.13 Forked-line method for solving genetic problems in which independent assortment is involved. Genotypes of parents are given, but only phenotypes of progeny are listed in this example.

cross as three monohybrid crosses; that is, $Dd \times Dd$, $Gg \times Gg$, and $Ww \times Ww$, operating together. If one member of each pair is dominant, a 3:1 ratio would be predicted from each monohybrid cross. Since the three pairs are independent, each mono-hybrid segregant may occur with any combination possible from any other pair of alleles. The combinations, therefore, can be systematically arranged together. The 3:1 ratio from $Dd \times Dd$ may be combined with the 3:1 ratios from each of the other two

$DdGgWw \times DdGgWw$

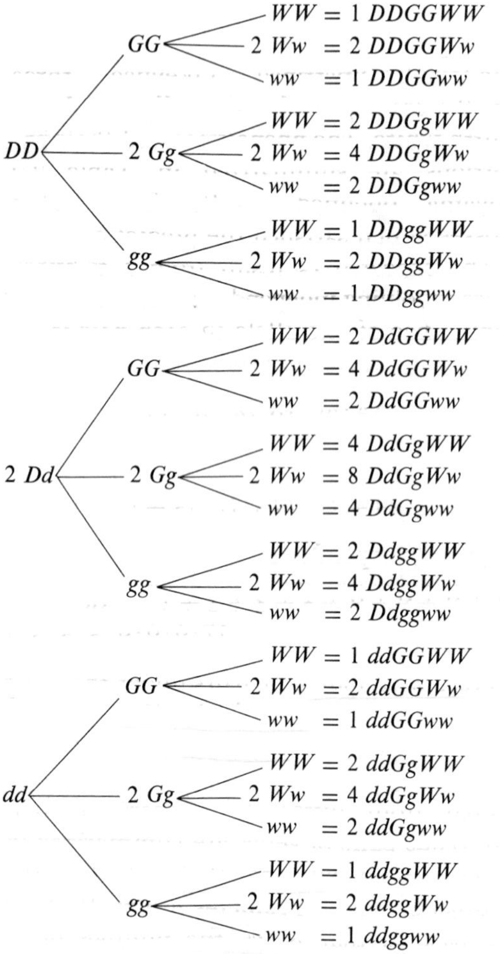

```
            ┌ WW = 1 DDGGWW
      GG ───┼ 2 Ww = 2 DDGGWw
            └ ww = 1 DDGGww
            ┌ WW = 2 DDGgWW
DD ─ 2 Gg ──┼ 2 Ww = 4 DDGgWw
            └ ww = 2 DDGgww
            ┌ WW = 1 DDggWW
      gg ───┼ 2 Ww = 2 DDggWw
            └ ww = 1 DDggww

            ┌ WW = 2 DdGGWW
      GG ───┼ 2 Ww = 4 DdGGWw
            └ ww = 2 DdGGww
            ┌ WW = 4 DdGgWW
2 Dd ─ 2 Gg─┼ 2 Ww = 8 DdGgWw
            └ ww = 4 DdGgww
            ┌ WW = 2 DdggWW
      gg ───┼ 2 Ww = 4 DdggWw
            └ ww = 2 Ddggww

            ┌ WW = 1 ddGGWW
      GG ───┼ 2 Ww = 2 ddGGWw
            └ ww = 1 ddGGww
            ┌ WW = 2 ddGgWW
dd ─ 2 Gg ──┼ 2 Ww = 4 ddGgWw
            └ ww = 2 ddGgww
            ┌ WW = 1 ddggWW
      gg ───┼ 2 Ww = 2 ddggWw
            └ ww = 1 ddggww
```

(a)

Phenotypes	Genotypes	Genotypic Frequency	Phenotypic Ratio
Tall, yellow, round	DDGGWW	1	27
	DDGGWw	2	
	DDGgWW	2	
	DDGgWw	4	
	DdGGWW	2	
	DdGGWw	4	
	DdGgWW	4	
	DdGgWw	8	
Tall, yellow, wrinkled	DDGGww	1	9
	DDGgww	2	
	DdGGww	2	
	DdGgww	4	
Tall, green, round	DDggWW	1	9
	DDggWw	2	
	DdggWW	2	
	DdggWw	4	
Tall, green, wrinkled	DDggww	1	3
	Ddggww	2	
Dwarf, yellow, round	ddGGWW	1	9
	ddGGWw	2	
	ddGgWW	2	
	ddGgWw	4	
Dwarf, yellow, wrinkled	ddGGww	1	3
	ddGgww	2	
Dwarf, green, round	ddggWW	1	3
	ddggWw	2	
Dwarf, green, wrinkled	ddggww	1	1

(b)

Figure 2.14 (a) cross between two F_1 garden peas of the genotype $DdGgWw$. The forked-line method is employed and the genotypes are illustrated. These results represent the F_2 of a cross similar to those obtained from the *Punnett square* method involving 64 squares. (b) summary of F_2 from trihybrid cross resulting in a 27:9:9:9:3:3:3:1 phenotypic ratio.

TABLE 2.2

Expectations and Results from Crosses Between Striped, Yellow, and Unstriped, White Moths (Data from Toyama)

Phenotypes	Proportions	F_2 Expectations	Observed
Striped, yellow	$\frac{3}{4} \times \frac{3}{4}$	$\frac{9}{16}$ 6368.4	6385
Striped, white	$\frac{3}{4} \times \frac{1}{4}$	$\frac{3}{16}$ 2122.8	2147
Unstriped, yellow	$\frac{1}{4} \times \frac{3}{4}$	$\frac{3}{16}$ 2122.8	2099
Unstriped, white	$\frac{1}{4} \times \frac{1}{4}$	$\frac{1}{16}$ 707.6	691

monohybrid crosses, $Gg \times Gg$ and $Ww \times Ww$, as shown in Fig. 2.13.

Usually the genotypes as well as the phenotypes are necessary for the complete solution of such a problem. The same forked-line system may be employed to represent and combine **genotypes** expected from monohybrid crosses. From each monohybrid cross in the example, a genotypic frequency of 1:2:1 may be predicted. The three monohybrid units may be combined as shown in Fig. 2.14. The forked-line system is merely another device for analyzing crosses.

MATHEMATICAL METHOD FOR GENETIC PROBLEMS

This introduction to combinations suggests a third way to anticipate the results from crosses involving independent pairs of alleles. A mathematical manipulation provides a way to arrive at the **product** of the combinations without drawing them out mechanically. For example, consider the crosses made by Toyama between two varieties of the silk moth *Bombyx mori*. In one variety, the caterpillars were striped and the cocoons were yellow, and in the other variety the caterpillars were unstriped and the cocoons were white. From previous crosses, striped was known to be dominant over unstriped and yellow over white. What proportions might be expected in the F_2?

If the striped and unstriped alleles are considered separately, three-fourths of the

F_2 progeny are expected to be striped and one-fourth unstriped. Likewise, three-fourths are expected to be yellow and one-fourth white. The phenotypes and their proportions are summarized in Table 2.2. Toyama obtained results from actual crosses which satisfied the predictions.

The expected F_2 result from a trihybrid cross involving independent assortment and dominance of one allele in each pair is the product of three pairs such as

$$(A:a) \quad (B:b) \quad (C:c)$$

Written algebraically

$$(3+1) \quad (3+1) \quad (3+)$$

or $(3+1)^3$ expands to

$$27+9+9+9+3+3+3+1 \quad \text{or}$$
$$27:9:9:9:3:3:3:1$$

A cross with four gene pairs under the same conditions would result in $(3+1)^4$, five gene pairs $(3+1)^5$, and so on. Numbers of gametes, genotypes, and phenotypes expected from different numbers of heterozygous pairs of genes are summarized in Table 2.3. It will be observed that the number of kinds of **gametes** is a multiple of the base 2, that is 2^n; the number of F_2 **genotypes is a** multiple of the base 3, that is 3^n; and the number of **phenotypes is** 2^n when dominance is present. This pattern forms the basis for predicting results when any number of independent pairs of alleles is involved in production of hybrids.

TABLE 2.3
Relations Among Pairs of Independent Alleles, Gametes, F$_2$ Genotypes, and F$_2$ Phenotypes When Dominance Is Present

Number of Heterozygous Pairs	Number of Kinds of Gametes	Number of F$_2$ Genotypes	Number of F$_2$ Phenotypes
1	2	3	2
2	4	9	4
3	8	27	8
4	16	81	16
10	1024	59049	1024
n	2^n	3^n	2^n

SEMIDOMINANCE IN DIHYBRIDS

In the absence of complete dominance, every genotype may have a distinguishable phenotype. Semidominant alleles may produce the same genetic product but in unequal quantity. In heterozygous condition the total product is the sum of the separate quantities of the two alleles. **Semidominant alleles** therefore act **additively** so that the expressions of **heterozygotes are intermediate** between those of the two homozygotes. The ratio for the monohybrid cross then becomes $1:2:1$ instead of $3:1$ whereas the dihybrid ratio, when the two pairs of alleles are semidominant, is $1:2:1:2:4:2:1:2:1$.

In snapdragons, for example, intermediate-size leaves are produced by a heterozygous gene combination (BB'). Plants with broad leaves and narrow leaves have the homozygous gene arrangements, BB and $B'B'$, respectively. Likewise, pink flowers are controlled by a heterozygous (RR') pair, with red (RR) and white ($R'R'$) attributable to the corresponding homozygotes. A cross between plants with broad leaves and red flowers ($BBRR$) and those

with narrow leaves and white flowers ($B'B'R'R'$) produced plants with intermediate leaves and pink flowers. The F$_2$ were classified into nine phenotypic classes corresponding with the genotype combinations as illustrated in Fig. 2.15. The $9:3:3:1$ dihybrid ratio was thus replaced by the $1:2:1:2:4:2:1:2:1$ ratio, as expected for semidominance involving two pairs of alleles.

EPISTASIS

Epistasis (Greek, standing upon) is a widespread type of gene interaction. Any gene or gene pair that **masks** the expression of another, **nonallelic** gene is **epistatic** to that gene. The gene suppressed is said to be **hypostatic.** Epistatic genes are inactive genes that produce defective enzymes or no enzymes at all. They block particular reactions and mask the effects of normal or hypostatic genes.

This form of suppression should be distinguished from simple dominance, which is associated with members of allelic pairs.

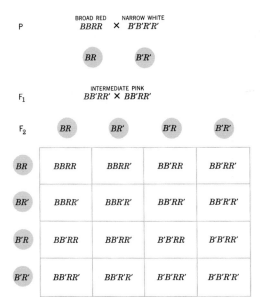

P	BROAD RED *BBRR*	×	NARROW WHITE *B'B'R'R'*

BR   B'R'

| F₁ | INTERMEDIATE PINK
BB'RR' × *BB'RR'* |

F₂

	BR	BR'	B'R	B'R'
BR	BBRR	BBRR'	BB'RR	BB'RR'
BR'	BBRR'	BBR'R'	BB'RR'	BB'R'R'
B'R	BB'RR	BB'RR'	B'B'RR	B'B'RR'
B'R'	BB'RR'	BB'R'R'	B'B'RR'	B'B'R'R'

Phenotypes	Genotypes	Genotypic Frequency	Phenotypic Ratio
Broad red	*BBRR*	1	1
Broad pink	*BBRR'*	2	2
Broad white	*BBR'R'*	1	1
Int. red	*BB'RR*	2	2
Int. pink	*BB'RR'*	4	4
Int. white	*BB'R'R'*	2	2
Narrow red	*B'B'RR*	1	1
Narrow pink	*B'B'RR'*	2	2
Narrow white	*B'B'R'R'*	1	1

Figure 2.15 Cross between snapdragons with broad leaves and red flowers and those with narrow leaves and white flowers, illustrating semidominant inheritance.

A classical example of epistasis based on the results of crosses between different varieties of chickens was reported in the early part of the present century by William Bateson and his associate, R. C. Punnett, (after whom the Punnett square was named). Bateson began to confirm and extend Mendel's work immediately after its discovery in 1900 and he became a pioneer in transmission genetics. He had chickens in his research coops and sweet peas in his garden for immediate use in genetic investigations.

Domestic breeds of chickens have different comb shapes (Fig. 2.16). Wyandottes have a characteristic type of comb called "rose," whereas Brahmas have a "pea" comb. Leghorns have "single" combs. The investigators crossed Wyandottes and Brahmas, and all the F₁ chickens had walnut combs, a phenotype not expressed in either parent. When the F₁ chickens were mated among themselves and large F₂ populations were produced, a familar dihybrid ratio, 9:3:3:1, was recognized, but the phenotypes representing two of the four classes

Figure 2.16 Comb types characteristic of different breeds of chickens: (*a*) rose, Wyandottes; (*b*) pea, Brahmas,; (*c*) walnut, hybrid from cross between chickens with rose and pea combs; and (*d*) single, Leghorns. (Courtesy of Ralph G. Somes, Jr.)

were different from those expressed in the parents. About $\frac{9}{16}$ of the F$_2$ birds were walnut, $\frac{3}{16}$ rose, $\frac{3}{16}$ pea, and $\frac{1}{16}$ had single combs. Neither single comb nor walnut was expressed in the original parental lines. These two phenotypes were explained as the result of **gene product interaction.** The results, based on a total of 16, indicated that two different allelic pairs were involved; one pair was introduced by the rose-comb parent and one by the pea-comb parent. A gene for rose and a gene for pea

would interact and produce walnut, as in the F$_1$.

Analysis of the F$_2$ results and appropriate test crosses indicated that the $\frac{9}{16}$ class, with the two dominant genes (*R-P-*), were walnut, like the F$_1$ chickens. The $\frac{1}{16}$ class representing the full recessive combination (*rrpp*) was characterized by single combs. The two $\frac{3}{16}$ (rose and pea) classes were *R-pp* and *rrP-*. It was then determined that the homozygous genotype of the rose-comb parent (Wyandotte) was *RRpp* and of the pea-combed parent (Brahma), *rrPP*. Although the usual 9:3:3:1 ratio was obtained, the result from this cross was unusual in two important respects: (1) the F_1 progeny differed from those of either parent, that is, none were rose or pea, but all were walnut; and (2) two phenotypes (walnut and single), not expressed in the original parents, appeared in the F_2.

Genes *R* and *P* were nonallelic but each was dominant over its allele (i.e., *R* over *r* and *P* over *p*). When *R* and *P* were together as in the F_1 (*RrPp*), the two different products interacted to produce walnut comb. The two nonallelic genes R and P acted independently in different ways, similar to the ways in which codominant alleles act. **Nonallelic genes** behaving in this way are called **coepistatic genes.**

Another early classical study, also conducted by Bateson and Punnett, demonstrated gene interaction in plant material resulting in a 9:7 ratio. When two white-flowered varieties of sweet peas, *Lathyrus odoratus*, were crossed, the F_1 progeny all had colored flowers. When the F_2 were clas-

sified, $\frac{9}{16}$ were purple and $\frac{7}{16}$ were white. The base number of 16 was again recognized as that associated with a dihybrid cross, but there were only two classes instead of the usual four. This suggested a modification of the 9:3:3:1 ratio in which the $\frac{3}{16}$, $\frac{3}{16}$, and $\frac{1}{16}$ classes were indistinguishable and therefore were grouped together when the plants were classified phenotypically. The fact that the proportion was based on a total of 16 rather than 4, 64, or some other number was good evidence that two pairs of genes were segregating.

The hypothesis formulated by the investigators was that **two nonallelic gene pairs** were necessary to produce the purple color. Thus the flowers of the original parents could be white for different genetic reasons; that is, in the absence of either one (or both) of two complementary genes. It has since been shown that, in the presence of *C* and *P* genes, anthocyanins (pigments) are formed in the flowers. In the absence of either of the complementary genes (*CCpp, Ccpp, ccPP, ccPp* or *ccpp*), the flowers are white. Nine of sixteen F_2 progeny carried *C-P-*, but

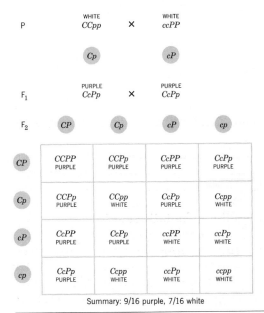

Summary: 9/16 purple, 7/16 white

Figure 2.17 Cross between two white varieties of sweet peas from which an F_2 ratio of 9:7 was obtained.

TABLE 2.4

Epistatic Interactions Among Genes, Resulting in Modifications of the 9:3:3:1 Mendelian Ratios from Crosses, *AaBb* × *AaBb*, Each Pair Assorting Independently. Genotypes Expected from This Cross Are as Follows:

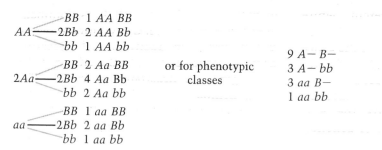

$$
AA
\begin{cases}
BB & 1\ AA\ BB \\
2Bb & 2\ AA\ Bb \\
bb & 1\ AA\ bb
\end{cases}
$$

$$
2Aa
\begin{cases}
BB & 2\ Aa\ BB \\
2Bb & 4\ Aa\ Bb \\
bb & 2\ Aa\ bb
\end{cases}
\quad
\begin{array}{l}
\text{or for phenotypic} \\
\text{classes}
\end{array}
\quad
\begin{array}{l}
9\ A-\ B- \\
3\ A-\ bb \\
3\ aa\ B- \\
1\ aa\ bb
\end{array}
$$

$$
aa
\begin{cases}
BB & 1\ aa\ BB \\
2Bb & 2\ aa\ Bb \\
bb & 1\ aa\ bb
\end{cases}
$$

Interactions		Genotypes										
		AABB	AABb	AaBB	AaBb	AAbb	Aabb	aaBB	aaBb	aabb		
Epistasis	Dominance	Phenotypic Classes									Example	Phenotypes
(classical ratio)	complete for *A,B*	9				3		3		1	Garden peas seed color and surface	9 yellow round, 3 yellow wrinkled, 3 green round, 1 green wrinkled.
aa epistatic to *B,b*	complete for *A,B*	9				3		— 4 —			mice coat pattern and color	9 agouti, 3 colored 4 white
A epistatic to *B,b*	complete for *A,B*	— 12 —						3		1	summer squash fruit color	12 white, 3 yellow 1 green
A epistatic to *B,b; bb* epistatic to *A,a*	complete for *A,B*	⌐ 13						3		①	chickens color	13 white, 3 colored
aa epistatic to *B,b; bb* epistatic to *A,a*	complete for *A,B*	9				— 7 —					yellow daisy color	9 purple, 7 yellow center
A epistatic to *B,b; B* epistatic to *A,a*	complete for *A,B*	— 15 —								1	shepherd's purse seed capsules	15 triangular 1 ovoid
aa epistatic to *B,b; bb* epistatic to *A,a; bb* epistatic to *aa*	complete for *A* partial for *B*	×	−	×	−	−	×	⌐ 4	3	①	flour beetle color	6 sooty (−) 3 red (×) 3 jet, 4 black

AABB	AABb	AaBB	AaBb	AAbb	Aabb	aaBB	aaBb	aabb

$\frac{7}{16}$ lacked one or both. The original white parent varieties were *CCpp* and *ccPP*. The cross is reconstructed diagrammatically in Fig. 2.17. In later experiments, colorless extracts made from the different white varieties were combined in a test tube and produced color. The interaction of the genes was thus associated with simple indicator type chemicals that the genes produced.

Historically, the discoveries of different modified F_2 ratios, particularly the **9:7**, were landmarks in the science of genetics. They required a broader foundation of Mendelian heredity than that based on a strict interpretation of Mendel's principles. Now that the principle of interaction is well established, modified ratios have been encountered in many examples including both plants and animals.

The cross between two different varieties of sweet peas both with white flowers illustrated again both dominance and epistasis. In both gene pairs, one allele is dominant over a recessive allele. *C* is dominant over *c* and *P* over *p*. In addition, each recessive homozygote is epistatic to the effects of the other gene. As illustrated in the F_2 (Fig. 2.17), homozygous allele *cc* prevents the expression of purple color by *P* and thus is epistatic to *P*. Likewise, the homozygous allele *pp* prevents the expression of color by *C* and is epistatic to *c*. Both *C* and *P* are hypostatic, *C* to *pp* and *P* to *cc*. Other **modified F_2 ratios** have been identified in the results of dihybrid crosses, (Table 2.4).

ATAVISM

Through epistatic effects and other gene interactions, traits may remain hidden for generations. Occasionally a "throwback" occurs in a strain of domestic animals or plants; that is, **an ancestral trait is expressed** unexpectedly. Reappearence of traits after several generations was first discussed scientifically by Charles Darwin and was called **atavism.** It is now explained by the chance combination of genes that allows a long-suppressed, hidden characteristic to finally come to expression.

Occasionally, throwbacks occur in highly selected strains of domestic pigeons that have been bred for show birds. Progeny that are "wild type" in appearance, resembling the rock pigeon, *Columba livia,* are produced when a chance combination of genes removes the effect of epistatic genes that, for generations, have suppressed the expression of certain ancestral traits in the show birds. Intermatings between Indian fantails (Fig. 2.18a), for example, sometimes result in pigeons resembling wild rock pigeons (Fig. 2.18b). Without constant attention from the breeders, different types of pigeons would mate indiscriminately, and their progeny would tend to revert to wild (or so-called "mongrel") birds. This phenomenon was one of the lines of evidence presented by Charles Darwin in explaining the ancestry of the domestic pigeon. Atavism was also one of Darwin's supporting arguments for broader aspects of evolution.

The variations possible in any species are explained on the basis of numerous **genes segregating** in the population and of **interactions** among the products of genes and the environment. Expressions of genes may by suppressed, enhanced, or altered by gene interactions. Since the phenotype rather than the genotype is directly affected by the environment, gene interactions producing unique expressions are of interest in studies of evolution as well as in man's selection of animals and plants.

(a)   (b)

Figure 2.18 Pigeons illustrating atavism. (*a*) Indian fantail pigeon; (*b*), wild rock pigeon. Intermatings between highly selected varieties, such as the fantail sometimes produce progeny resembling the wild rock pigeon (*a*) Grant Heilman; (*b*) photo by H. P. Macklin, courtesy of Wendell Levi).

PROBABILITY IN MENDELIAN INHERITANCE

Mendel had studied probability in his mathematics courses, and he recognized the 3 : 1 ratio as a particular mathematical relation which suggested to him a model for the mechanism of segregation. If two alleles, one dominant and one recessive, could segregate freely many times in succession, or if several similar pairs were behaving in this way at one time, the expected summation would be about 3 of the dominant expressions to 1 recessive. The analysis, therefore, was based on a mathematical relation with which Mendel was familiar, but its application and the concept of **segregation** were Mendel's own contribution.

The laws of probability apply to the genetic mechanism as well as to other processes in which uncertainty exists. In the F_2 results of Mendel's cross between tall and dwarf garden peas, $\frac{1}{4}$ were dwarf, $\frac{1}{4}$ homozygous tall, and $\frac{1}{2}$ heterozygous tall. A similar result might be obtained from a simple experiment in tossing coins. A coin that is tossed freely is equally likely to fall heads or tails. If one coin is tossed 100 times, it would be expected to fall heads about 50 times and tails about 50 times. When two coins are tossed together, each behaves independently and falls either heads or tails. From 100 trials, about 25 heads-heads, 50 heads-tails, and 25 tails-tails (1 : 2 : 1) would be expected. This result, which parallels Mendelian segregation is merely the chance occurrence of independent events. The experimenter would not always obtain exactly 25 heads-heads, 50 heads-tails, and 25 tails-tails. It would be surprising if precisely those results were obtained very often. The ratio represents only an average of expected results when independent events occur.

How frequently can the various combinations be expected to occur in succession? The law of probability applied for the solution of this problem is stated as follows: **if two or more events are independent, the chance that they will occur together is the product of their separate probabilities.** When a single coin is tossed repeatedly, the chance of heads occurring twice in succession would be $\frac{1}{2} \times \frac{1}{2} = \frac{1}{4}$. The chance of three such occurrences would be $(\frac{1}{2})^3$ or $\frac{1}{8}$, and of four, $(\frac{1}{2})^4$ or $\frac{1}{16}$. When two coins are tossed together the tails-tails combination is expected in one-fourth of the trials $(\frac{1}{2} \times \frac{1}{2} = \frac{1}{4})$. The chance of occurrence of two such tails-tails combinations for two coins in succession would be $(\frac{1}{4})^2$ or $\frac{1}{16}$.

FITTING RESULTS FROM CROSSES TO HYPOTHESES

The "goodness of fit" of the numerical result obtained from an actual cross or other experiment, relative to predicted results based on a particular mechanism, and a perfect genetic segregation, is of vital concern. The geneticist must know how much the experimental result can differ from the hypothetical or calculated figure and still be regarded as statistically close to expectation. In evaluating the results of crosses and determining which modes of inheritance are involved, how much deviation is permissible without casting some doubt as to whether the data agree with a given hypothesis? Too much deviation would surely make the investigator **question his hypothesis** or discard it entirely. Where shall he draw the line? Unfortunately, there is no precise answer to this question. The best the geneticist can do is to determine the likelihood of the deviation occurring by chance, and use his statistical inferrence to decide whether a particular result supports a given hypothesis. These numerical data are his only means of **evaluating goodness of fit** of an experimental result as compared with a particular expectation.

CHI-SQUARE

The chi-square (χ^2) test is a valuable tool that aids the investigator in determining goodness of fit. The test takes into account the **size of the sample** and the **deviations** from the expected ratio. It not only can be used for samples of different sizes but it can also be adapted to ratios with different numbers of classes such as those of monohybrid crosses with two classes and those of dihybrid crosses with four classes. Essentially, the chi-suqare test is a mechanism by which deviations from a hypothetical ratio are reduced to a single value based on the size of the sample. This allows the investigator to determine the probability that a given sum of deviations will occur by chance. Expected values are obtained from the total size of the sample. If the hypothesis is that a $1:1$ ratio should result from a cross, the total is divided into two equal parts. For any other expected ratio the total is divided into appropriate proportions.

A formula for χ^2, designed for a sample consisting of two classes (that is, $1:1$ or $3:1$ ratios), is symbolized as follows:

$$\chi^2 = \frac{(O_1 - e_1)^2}{e_1} + \frac{(O_2 - e_2)^2}{e_2}$$

where O_1 is the experimentally observed number for the first class and e_1 is the expected number for the same class derived from the ratio; O_2 is the observed for the second class and e_2 the expected. This formula can be simplified by representing as a single figure the difference between each observed and expected value. For example, $(O_1 - e_1)$ or $(O_2 - e_2)$ may be represented by a single deviation (d). When each of these deviations is squared (d^2) and divided by the expected value (e) for that class, the resulting fractions can be added (Σ) to give a single χ^2 value. The simplified formula can be symbolized as follows:

$$\chi^2 = \Sigma \frac{d^2}{e}$$

where d is the deviation between each observed and expected class value, e the expected value in the respective class, and the Greek letter Σ is the summation sign. If the deviations of expected from observed are small, χ^2 approaches 0; if the deviations are large, χ^2 increases and the fit becomes poorer. As an example, calculate χ^2 for the two arbitrary samples, $15:35$ and $240:260$, (each having the same actual deviation) on the basis of a $1:1$ hypothesis. This example will illustrate how the χ^2 relates the size of the deviation to the size of the sample. For the $15:35$ result, with a total of 50, the expected (e) value for each class is $15 + 35 \div 2$, or 25. The deviations (d) on either side of e are 10; that is, $25 - 15 = 10$ and $25 - 35 = -10$. For the larger sample, $240:260$, the expected (e) for each class is 250. The deviations (d) are $250 - 240 = 10$, and $250 - 260 = -10$.

(1) $\quad \chi^2 = \Sigma \dfrac{d^2}{e} = \dfrac{(10)^2}{25} + \dfrac{(-10)^2}{25}$

$\qquad = \dfrac{200}{25} = 8$

(2) $\quad \chi^2 = \Sigma \dfrac{d^2}{e} = \dfrac{(10)^2}{250} + \dfrac{(-10)^2}{250}$

$\qquad = \dfrac{200}{250} = 0.8$

The χ^2 value of 8 for the smaller sample is considerably greater than that of 0.8 for the larger sample, even though the actual deviations in the two examples are the same.

When more than two groups are classified from the sample (for example $1:2:1$ or $9:3:3:1$ ratios), each class is included in the summation, which is χ^2. It should be emphasized that the χ^2 formula is based on actual numerical frequencies and not on percentages. When data are reduced to percentages, the total automatically becomes 100, thus eliminating the important factor of sample size in the evaluation.

The next step is to interpret the χ^2 value in terms of probability. In any experimental procedure dealing with quantitative data, some variation, called **experimental error** can be attributed to chance. It is important to determine whether observed deviations from a hypothesis are significantly different from the experimental error. For interpreting χ^2 values, the number of classes on which a χ^2 is based must be considered. It is therefore, necessary to include the **number of classes** contributing to a given χ^2 in evaluating the "goodness of fit." The effect of the number of independent classes is included in the mathematical concept as **degrees of freedom.** For example, a person may have two gloves for his two hands, but in placing the gloves he has only one degree of freedom. If he places a glove on one hand, the other glove must go on the other hand. When the total number of objects or classes is fixed, and all except one have been placed, the one remaining is not free but must fill a particular niche. In general, therefore, the number of degrees of freedom is one less that the number of classes. Ratios of $1:1$ or $3:1$ for example, have one degree of freedom; $1:2:1$ and other 3-class ratios have 2; $9:3:3:1$, 3 degrees of freedom, and so on.

When χ^2 and the degrees of freedom have been determined, Table 2.5 may be consulted for the probability (P) value. Locate

TABLE 2.5
Table of Chi-square (See also Appendix A)

Degrees of Freedom	$P = 0.99$	0.95	0.80	0.50	0.20	0.05	0.01
1	0.000157	0.00393	0.0642	0.455	1.642	3.841	6.635
2	0.020	0.103	0.446	1.386	3.219	5.991	9.210
3	0.115	0.352	1.005	2.366	4.642	7.815	11.345
4	0.297	0.711	1.649	3.357	5.989	9.488	13.277
5	0.554	1.145	2.343	4.351	7.289	11.070	15.086
6	0.872	1.635	3.070	5.348	8.558	12.592	16.812
7	1.239	2.167	3.822	6.346	9.803	14.067	18.475
8	1.646	2.733	4.594	7.344	11.030	15.507	20.090
9	2.088	3.325	5.380	8.343	12.242	16.919	21.666
10	2.558	3.940	6.179	9.342	13.442	18.307	23.209
15	5.229	7.261	10.307	14.339	19.311	24.996	30.578
20	8.260	10.851	14.578	19.337	25.038	31.410	37.566
25	11.524	14.611	18.940	24.337	30.675	37.652	44.314
30	14.953	18.493	23.364	29.336	36.250	43.773	50.892

Taken from Table 3, R. A. Fisher, *Statistical Methods for Research Workers,*
14th edition, copyright © 1972 by Hafner Press.

the figure representing the degrees of freedom at the left, read across horizontally, and find the figures nearest the χ^2 value in the body of the table; then read the P values directly above on the top line. The χ^2 of 8, calculated for the first example, is not on the table. The highest value on the line for one degree of freedom is 6.635, which has a P of 0.01. This indicates that the probability for obtaining, by chance, deviations as great as or greater than those of $\chi^2 = 8$ would be less than one percent. Experimental error is expected but the difference here is much greater than that expected by chance. Therefore, the fit of these results to a 1:1 ratio is not good. Another hypothesis (other than the 1:1 ratio) might be considered for these data.

In the second example, $\chi^2 = 0.8$ falls between 0.455 and 1.642, or between the P values of 0.50 and 0.20. The probability of obtaining a deviation due to *chance* as great as

or greater than $\chi^2 = 0.8$ is between 0.20 and 0.50. This probability value indicates that if numerous independent repetitions of an ideal experiment involving two independent events were conducted, chance deviations as large as or larger than those observed here (± 10 corresponding to $\chi^2 = 0.8$) would be expected to occur in 20 to 50 percent of the trials. Such a deviation could be explained readily by chance. The data fit the 1:1 ratio hypothesis very well.

A hypothesis is never proved solely by a P value. Neither is it disproved by a P value, but a P value can make it highly unlikely. Results of an experiment are evaluated by the investigator as **acceptable** or **unacceptable** with respect to the hypothesis. The 5 percent point (0.05) on the table is usually chosen as an arbitrary standard for determining the significance or goodness of fit. Probability at this point is one in twenty that a true hypothesis will be rejected.

Sometimes the 1 percent point (0.01) is used as a level of significance. At this level there is a smaller probability (0.01) that a true hypothesis will be rejected, but a correspondingly greater chance that a false hypothesis will be accepted.

It should be emphasized that these are arbitrary points, and judgment is required in making interpretations. In any event, the P value represents the probability that a deviation as great as or greater than that obtained from the experiment will occur by chance alone. If the P is small, it is concluded that the deviations are not due entirely to chance, and the hypothesis is rejected. If the P is greater than the predetermined level (for example, 0.05), the data conform well enough to the hypothesis and the hypothesis is accepted. Curves prepared by James. F. Crow showing P values at different confidence levels are given in Appendix A. These provide a dynamic picture of probability and may be consulted for interpolation between values in Table 2.5 and for an extended view of probability.

INDEPENDENT ASSORTMENT AND PROBABILITY

Probability must be considered in explaining the Mendelian principle of independent assortment as well as that of segregation. It was through Mendel's understanding of the mathematical laws of combinations that he was able to recognize and interpret the dihybrid ratio of $9:3:3:1$ as a multiple of the $3:1$ monohybrid ratio. If, for example, the $3:1$ ratio is changed to an algebraic expression, $3+1$, the product of the expected results of two monohybrid crosses is: $(3+1)^2 = 9 + 3 + 3 + 1$. Because the terms represent separate classes, they are not grouped together, but the product can be converted back to a ratio: $9:3:3:1$. This is an example of the **binomial expansion** of $(a + b)^n$, in this case $(A + a)^2$, where $A = 3$ and $a = 1$.

Using the F_2 results of the cross between peas with round, yellow seeds and those with wrinkled, green seeds, Mendel tested the mathematical relation between the monohybrid and dihybrid cross. He observed that about three-fourths ($\frac{423}{556}$) of the F_2 seeds were round and one-fourth ($\frac{133}{556}$) were wrinkled. Likewise, seeds from about three-fourths ($\frac{416}{556}$) were yellow, and those from one-fourth ($\frac{140}{556}$) were green. This observed proportion provided a cross-check for the hypothesis of **independence.** When the two characters were considered together, the results conformed to the mathematical model expected for two independent events occurring together.

On the basis of the law of probability, Mendel predicted that nine-sixteenths ($\frac{3}{4} \times \frac{3}{4}$) of the F_2 would be round, yellow; three-sixteenths ($\frac{3}{4} \times \frac{1}{4}$) round, green; three-sixteenths wrinkled, yellow; and one-sixteenth ($\frac{1}{4} \times \frac{1}{4}$) wrinkled, green. The results that Mendel actually obtained ($315:108:101:32$) resembled very closely the calculated ratio of $9:3:3:1$, based on the hypothesis of complete independence of the genes influencing the shape and color of the seeds. When the χ^2 test for goodness of fit between the actual and the expected result is applied to these figures, the probability (P) of finding deviations as great as or greater than those obtained by Mendel is between 0.80 and 0.95.

$$\chi^2 = \sum \frac{d^2}{e} = \frac{(2.25)^2}{312.75} + \frac{(3.75)^2}{104.25}$$

$$+ \frac{(-3.25)^2}{104.25} + \frac{(-2.75)^2}{34.75}$$

$$= 0.016 + 0.135 + 0.101 + 0.218$$

$$= 0.470$$

$$P = 0.80 - 0.95$$

The actual results fit very closely those expected on the basis of the hypothesis of independent assortment. In fact, it is so close that the probability is less than 0.20 that further data would fit the hypothesis this well. Investigators are as concerned about results that are very close to a calculated expectation as about those with large deviations. A very close fit may be obtained occasionally by chance. When the observed data fit "too well," the investigator may have intentionally or unintentionally biased his results. Mendel presented other results that had poor agreement with expectation. In one experiment from which he expected a $3:1$ ratio, he obtained 43 round and only 2 wrinkled. From another experiment with the same expectation be obtained 32 yellow and 1 green. He explained these results on the basis of fluctuations due to chance, and in the second case there was also some difficulty in distinguishing between yellow and green seeds.

From his results, Mendel was able to predict the numbers of genotypes to be expected when more than two pairs of alleles were involved in the cross.

EXPANSION OF A BINOMIAL

Many applications can be made of binomial expansion. This method is different from the simple multiplication of separate probabilities because it includes all possible combinations of alternative events. When probabilities, for example, are concerned with two or more events, that are mutually exclusive, or in some way mutually dependent, they follow the addition theorem. This theorem or rule is stated as follows: **the probability of occurrence of one or the other of two (or more) mutually exclusive events is the sum of their separate probabilities.** If boys and girls occur in equal proportion in a family and they are mutually exclusive, the probability that one or the other will be obtained from a particular single birth pregnancy is the sum of the separate probabilities or $\frac{1}{2} + \frac{1}{2} = 1$ or certainty. The probability that a baby will be of one sex (male or female) is $\frac{1}{2}$. In families of four, the probability that all will be of one sex, either boys or girls, is $\frac{1}{2} + \frac{1}{2} + \frac{1}{2} + \frac{1}{2} = \frac{1}{8}$.

Various combinations in groups of a given size representing a particular ratio can be calculated by the binomial expansion of $(p + q)^n$, where p and q represent the probabilities of occurrence of two alternative events (for example, $p = $ boys, $q = $ girls) and n is the size of the groups involved. How many boys and how many girls would be expected in randomly selected families of 2, 3, 4, 5, or more? Combinations for families of a given size may be calculated by the binomial expansion of $(p + q)^n$. Thus,

TABLE 2.6
Distribution of Boys and Girls in Families

Number of Children in Family	$(p + q)^n$	Distribution
1	$(\frac{1}{2} + \frac{1}{2})^1$	$\frac{1}{2}(1 \; \male) + \frac{1}{2}(1 \; \female)$
2	$(\frac{1}{2} + \frac{1}{2})^2$	$\frac{1}{4}(2 \; \male) + \frac{1}{2}(1 \; \male : 1 \; \female) + \frac{1}{4}(2 \; \female)$
3	$(\frac{1}{2} + \frac{1}{2})^3$	$\frac{1}{8}(3 \; \male) + \frac{3}{8}(2 \; \male : 1 \; \female) + \frac{3}{8}(1 \; \male : 2 \; \female) + \frac{1}{8}(3 \; \female)$
4	$(\frac{1}{2} + \frac{1}{2})^4$	$\frac{1}{16}(4 \; \male) + \frac{4}{16}(3 \; \male : 1 \; \female) + \frac{6}{16}(2 \; \male : 2 \; \female) + \frac{4}{16}(1 \; \male : 3 \; \female) + \frac{1}{16}(4 \; \female)$
5	$(\frac{1}{2} + \frac{1}{2})^5$	$\frac{1}{32}(5 \; \male) + \frac{5}{32}(4 \; \male : 1 \; \female) + \frac{10}{32}(3 \; \male : 2 \; \female) + \frac{10}{32}(2 \; \male : 3 \; \female) + \frac{5}{32}(1 \; \male : 4 \; \female) + \frac{1}{32}(5 \; \female)$

for two-child families,

$$(p + q)^2 = p^2 + 2pq + q^2$$
$$p = q = \tfrac{1}{2}$$
$$p^2 = \text{families of 2 boys} = \tfrac{1}{4}$$
$$2pq = \text{families of 1 boy and 1 girl} = \tfrac{1}{2}$$
$$q^2 = \text{families of 2 girls} = \tfrac{1}{4}$$

Among families of two children, $\tfrac{1}{4}$ would be expected to be composed of all boys, $\tfrac{1}{2}$ of the one boy and one girl, and $\tfrac{1}{4}$ of all girls. For three-child families,

$$(p + q)^3 = p^3 + 3p^2q + 3pq^2 + q^3$$
$$p^3 = \text{families of 3 } \male = (\tfrac{1}{2})^3 = \tfrac{1}{8}$$
$$3p^2q = \text{families of 2 } \male, 1 \female = 3(\tfrac{1}{2})^2 (\tfrac{1}{2}) = \tfrac{3}{8}$$
$$3pq^2 = \text{families of 1 } \male, 2 \female = 3(\tfrac{1}{2}) (\tfrac{1}{2})^2 = \tfrac{3}{8}$$
$$q^3 = \text{families of 3 } \female = (\tfrac{1}{2})^3 = \tfrac{1}{8}$$

($\male$ is the male symbol, shield and spear, of Mars, Roman war god; $\female$ is the female symbol, mirror of Venus, Roman goddess of love). The binomial expansion includes all possible combinations of the two events. In three-child families, for example, there are eight combinations.

If the chance of a particular birth order is included in the problem, the probability of each sequence is $\tfrac{1}{8}$. If, on the other hand, only the total number of boys and girls is considered, the probability of all boys is $\tfrac{1}{8}$, two boys and one girl $\tfrac{3}{8}$, one boy and two girls $\tfrac{3}{8}$, and all girls $\tfrac{1}{8}$. The expected distribu-

tions of males and females in families of one to five children are summarized in Table 2.6.

The probability for each combination can be determined from the **binomial coefficient** for this combination compared with all possible combinations. In sibships of five, for example, the following proportions of boys and girls would be expected: $\tfrac{1}{32}$ all boys, $\tfrac{5}{32}$ four boys and one girl, $\tfrac{10}{32}$ three boys and two girls, $\tfrac{10}{32}$ two boys and three girls, $\tfrac{5}{32}$ one boy and four girls, and $\tfrac{1}{32}$ all girls. The coeffecients in a given bionomial expansion can be found by referring to Pascals pyramid, Fig. 2.19.

When the probability of only a certain combination in a given size group is required, factorials may be employed. These are **products of factors** derived from functions by successively increasing or decreasing by a constant, usually one. For example, factorial 4 (4!) is the product of $4 \times 3 \times 2 \times 1$ (or $4! = 4 \times 3 \times 2 \times 1 = 24$). (Factorial 0 (0!) = 1 and 1! = 0! by definition). The probability for a particular combination may be calculated from the following formula: (the $x \times 1th$ term in the binomial expansion of $(p + q)^n$:

$$P = \frac{n!}{x! \, (n-x)!} \, p^x q^{(n-x)}$$

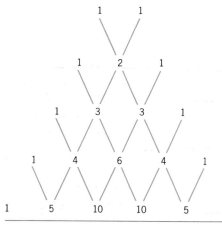

Figure 2.19 The Pascal pyramid, which can be used to obtain the coefficients of the terms in binomial equations. Each number after the first is obtained by adding the numbers above it. The second number in each line indicates the power of the expansion. Thus, the last line has 5 as its second number and is used in expanding $(a + b)^5$.

where $n!$ is the product of the integers making up the total size of the group (n = the total number in the group); $x!$, the product of the integers for one class (p); and ($n - x$)! the product of the integers for one class (q). The symbol p represents the probability for one occurrence (e.g., boys) and q is the probability for the other (e.g., girls). If, for example, six babies are born in a given hospital on the same day, what is the probability that two will be boys and four will be girls? For this problem, assume that $p = \frac{1}{2}$ and $q = \frac{1}{2}$. Substituting

$$P = \frac{n!}{x! \ (n-x)!} \ p^x q^{(n-x)}$$

$$= \frac{6!}{2!4!} \left(\frac{1}{2}\right)^2 \left(\frac{1}{2}\right)^4$$

$$= \frac{6 \times 5 \times 4 \times 3 \times 2 \times 1}{2 \times 1 \ (4 \times 3 \times 2 \times 1)} \left(\frac{1}{4}\right) \left(\frac{1}{16}\right)$$

$$= 15 \times \frac{1}{4} \times \frac{1}{16}$$

$$= \frac{15}{64}$$

The probability of two boys and four girls in groups of six is $\frac{15}{64}$.

For examples of boys and girls in families of different sizes, the probability values p and q were equal ($p = q = \frac{1}{2}$). The binomial distribution can be applied for other values of p and q. If, for example, the trait being considered is albinism in a human family, and the parents are known to be heterozygous (Cc), the probability for a normally pigmented child (p) would be $\frac{3}{4}$, and the probability for an albino child (q) would be $\frac{1}{4}$. In families of four, what is the probability that two will be normally pigmented and two will be albino? Substituting

$$P = \frac{n!}{x! \ (n-x)!} \ p^x q^{(n-x)}$$

$$= \frac{4!}{2! \ (2)!} \left(\frac{3}{4}\right)^2 \left(\frac{1}{4}\right)^2$$

$$= \frac{4 \times 3 \times 2 \times 1}{2 \times 1 \ (2 \times 1)} \left(\frac{3}{4}\right)^2 \left(\frac{1}{4}\right)^2$$

$$= \frac{54}{256} = \frac{27}{128}$$

The probability of two normally pigmented and two albino children in families of four children from heterozygous (Cc) parents would be $\frac{27}{128}$. Other values could be substituted into the binomial expansion for families of given size and the probability for various combinations could be calculated.

PROBABILITY IN PEDIGREE ANALYSIS

Practical applications of the Mendelian principles and the laws of probability are made by human geneticists and, in some instances, by animal breeders in analyzing pedigrees. Traits with a simple pattern of inheritance may sometimes be traced accurately enough to justify predictions concerning the likelihood of their expression in future children, if related individuals or those with a family history of such traits, marry. The first step in such an analysis is to determine whether the trait in question is behaving as a dominant or a recessive. Although most human traits are affected by many genes, a few have been associated with the differential action of certain specific genes and have been identified as dominants or recessives in family groups. A confusing feature of this type of analysis is that some phenotypes (e.g., deafness) may behave as dominant in some families and recessive in others. Obviously, several gene substitutions can result in deafness.

Recessive genes are difficult to keep track

of because they may **remain hidden** by their dominant alleles generation after generation. Carriers in the population usually cannot be identified until an expression occurs. Traits dependent on recessive genes sometimes appear unexpectedly in families having no previous history of such a trait. Recessives are expressed more frequently in families in which the father and mother are closely related than they are in the general population. The likelihood of similar genes being present is enhanced when the parents have descended from a common ancestor.

In the absence of data to indicate which individuals are carriers, the geneticist may resort to probability as the best available tool for determining the likelihood of expression of a given recessive gene in a certain family. If no expression has occurred in the history of the family, an estimate indicating the frequency of the gene in the general population may be used as a basis of probability. If the trait has appeared in the family, more precise calculations are sometimes possible. Probability is then based on the family history, which may be recorded in a pedigree chart.

The use of probability in human pedigree analysis is illustrated in Fig. 2.20. The trait, adherent ear lobes (Fig. 2.21), dependent on a recessive gene, appeared only once in the known history of the family, as indicated by the single darkened circle. No information other than that shown on the chart is available. Unless there is evidence to the contrary, it may be assumed (to avoid dealing with small probabilities) that those individuals who have married into the family are homozygous for the dominant genes and do not carry the gene in question.

The first step is to identify the genotypes of as many individual family members as possible from the information given. The woman (II-2) in whom the trait is expressed must be homozygous (aa) for the recessive gene. Each of her parents (I-1 and I-2) who did not express the trait but contributed an a gene to their daughter (II-2) must carry the heterozygous genotype (Aa). The sister (II-1) and brother (II-3) must be AA or Aa.

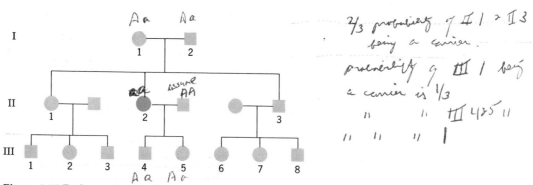

Figure 2.20 Pedigree chart to illustrate probability in pedigree analysis. In this family group a trait (adherent ear lobes), dependent on a recessive gene, has appeared in one individual (identified by the darkened circle). Three basic steps may be followed in all pedigree analyses: (1) what is the chance that one parent is a *carrier* of the gene in question? (2) What is the chance that the other parent is a *carrier* for the same gene? (3) What is the chance that a child of these two parents (genotypes) could *express* the trait involved? The *product* of the separate probabilities is the chance that a particular future child will express the trait.

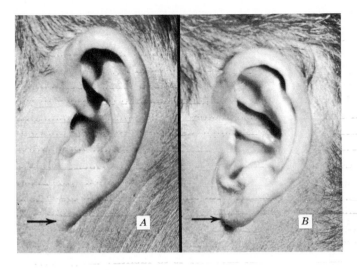

Figure 2.21 Adherent ear lobes compared with free ear lobes. (*a*) adherent; (*b*) free.

Obviously, they are not *aa* because they do not express the trait. There is no way to determine whether each of these individuals is *AA* or *Aa*. Therefore, the probability that each individual is a carrier (*Aa*) represents the best information available. From the parent cross (*Aa* × *Aa*), the probability of the occurrence of *Aa* in any child with free ear lobes is $\frac{2}{3}$ and the probability for the occurrence of *AA* is $\frac{1}{3}$. In the absence of more definite information, II-1 and II-3 may be considered *Aa* with $\frac{2}{3}$ probability. The children of II-1 and II-3 have a $\frac{1}{2}$ chance of being carriers for the gene (*a*) if their parent is a carrier. Therefore, the probability that III-1, III-2, III-3, III-6, III-7, or III-8 is a carrier is $\frac{2}{3} \times \frac{1}{2} = \frac{1}{3}$. The children of II-2 (III-4 and III-5) must be carriers (probability = 1).

The problem may be carried a step further by calculating the likelihood for an expression (*aa*) of the trait in the first child resulting from a marriage between two of the cousins represented in generation III. The mating III-1 × III-5 will serve for the example. The probability that III-1 is a carrier (*Aa*) is $\frac{2}{3} \times \frac{1}{2}$, and the comparable probability for III-5 is 1. Both could be carriers and yet avoid an expression of the trait in their family. Therefore, another probability must be included, that of parents with genotypes *Aa* having an *aa* child, (*Aa* × *Aa* = 1*AA*, 2*Aa*, 1 *aa*), which is $\frac{1}{4}$. Probability for an expression of the trait in the child of the individuals indicated is

$$\frac{2}{3} \times \frac{1}{2} \times 1 \times \frac{1}{4} = \frac{1}{12}$$

The chance of each future child expressing the trait is one in twelve.

If the first child should express the trait, the probability that a second child would also express the trait would be $\frac{1}{4}$ because evidence would then be available to indicate that the genotypes of III-1 and III-5 were both *Aa*. Two elements of uncertainty or probability would thus be eliminated. At best, probability is a poor substitute for certainty. It is employed in analyses only when definite information is not available.

MODERN EVALUATIONS OF MENDEL'S CONCLUSIONS

New interpretations are inevitable in scientific disciplines as additional data accumulate. Mendel considered a single gene to be responsible for a single trait. It is now known that many genes are involved in the production of some traits, although single gene substitutions can influence basic biochemical reactions and thus be responsible for alternative end products. Furthermore, it is the **genes and not the traits that are inherited.** Genes behave as separate units, whereas traits may result from complex interactions involving many genes.

Complete dominance was indicated in all seven allelic pairs that Mendel reported. It was natural, therefore, for him to consider dominance as an inherent property of genes. When sweet peas and snapdragons were studied, shortly after the discovery of Mendel's paper, **intermediate traits** were observed in hybrids (as we saw in Fig. 2.15). Crosses between homozygous snapdragons with red flowers and those with white flowers resulted in F_1 progeny with pink flowers. Heterozygotes could thus be distinguished phenotypically from both parents. Dominance has now been shown to be influenced by factors in the external, internal (hormonal), and genetic environment. Therefore, Mendel's view of dominance as a fundamental inherent property of the gene itself is no longer tenable for all cases. Dominance of some genes may eventually be explained on the basis of modifier genes that are present in the genetic environment. In other cases, dominance may depend on the quantity or activity of enzymes that are gene-controlled.

The most important concepts that Mendel inferred from his experiments were: (1) **segregation,** the process through which alleles separate and produce pure gametes; and (2) **independent assortment** of different pairs of alleles. These principles are the basic foundation of Mendelian heredity.

REFERENCES

Bateson, W. 1909. *Mendel's principles of heredity.* The University Press, Cambridge.

Carlson, E. A. 1973. *The gene: a critical history,* 2nd ed. W. B. Saunders, Philadelphia.

Fisher, R. A. 1946. *Statistical methods for research workers.,* 10th ed. Oliver and Boyd, London.

Kříženecký, J. (ed) 1965. *Fundamenta genetica: the revised edition of Mendel's classic paper with a collection of twenty-seven original papers.* Oosterhaut, The Netherlands.

Mendel, G. 1866. "Versuche über pflanzen-hybriden." (Available in the original German in *J. Hered.,* 42, 1–47). English translation under the title "Experiments in plant hybridization." Harvard University Press, Cambridge, Mass.

Morgan, T. H. 1926. *The theory of the gene.* Yale University Press, New Haven, Conn.

PROBLEMS AND QUESTIONS

2.1 On the basis of Mendel's hypothesis and observations, predict the results from the following crosses in garden peas: (a) a tall (dominant and homozgous) variety crossed with a dwarf variety; (b) the progeny of (a) above selfed; (c) the progeny from (a) crossed with the original tall parent; (d) the progeny from (a) crossed with the original dwarf parent variety.

2.2 Mendel crossed pea plants producing round seeds with those producing wrinkled seeds. From a total of 7324 F_2 seeds, 5474 were round and 1850 were wrinkled. Using the symbols W and w for genes, (a) symbolize the original P cross; (b) the gametes; and (c) F_1 progeny. (d) Represent a cross between two F_1 plants (or one selfed); (e) symbolize the gametes; and (f) summarize the expected F_2 results under the headings phenotypes, genotypes, genotypic frequency, and phenotypic ratio.

2.3 The French biologist Cuénot crossed wild, gray-colored mice with white (albino) mice. In the first generation, all were gray. From many litters he obtained in the F_2, 198 gray, and 72 white mice. (a) Propose a hypothesis to explain these results. (b) On the basis of the hypothesis, diagram the cross and compare the observed results with those expected.

2.4 A woman has a rare abnormality of the eyelids called ptosis, which makes it impossible for her to open her eyes completely. The condition has been found to depend on a single dominant gene (P). The woman's father had ptosis, but her mother had normal eyelids. Her father's mother had normal eyelids. (a) What are the probable genotypes of the woman, her father and mother? (b) What proportion of her children will be expected to have ptosis if she marries a man with normal eyelids?

2.5 What phenotypic ratio would be expected from a test cross of F_1 and a full recessive if the F_2 ratios were as follows:
(a) 13:3
(b) 15:1
(c) 9:3:4
(d) 12:3:1
(e) 1:2:1:2:4:2:1:2:1

2.6 In pigeons, the checkered pattern is dependent on a dominant gene C and plain on the recessive allele c. Red color is controlled by a dominant gene B and brown by the recessive allele b. Diagram completely a cross between homozygous checkered, red; and plain brown birds. Summarize the expected F_2 results.

2.7 In mice, the gene (C) for colored fur is dominant over its allele (c) for white. The gene (V) for normal behavior is dominant over (v) for waltzing. Give the probable genotypes of the parent mice (each pair was mated repeatedly and produced the following results): (a) colored, normal, mated with white, normal, produced 29 colored, normal, and 10 colored, waltzers; (b) colored, normal mated with colored normal produced 38 colored, normal; 15 colored waltzers; 11 white, normal; and 4 white, waltzers; (c) colored, normal mated with white, waltzer, produced 8 colored, normal; 7 colored, waltzers; 9 white, normal; 6 white, waltzers.

2.8 In rabbits, black fur is dependent on a dominant gene (B) and brown on the recessive allele (b). Normal length fur is determined by a dominant gene (R) and short (rex) by the recessive allele (r). (a) Diagram and summarize the F_1 and F_2 results of a cross between a homozygous black rabbit with normal length fur and a brown, rex rabbit. (b) What proportion of the normal, black F_2 rabbits from the above cross may be expected to be homozygous for both gene pairs? (c) Diagram and summarize a test cross between the F_1 and the fully recessive brown, rex parent.

2.9 In shorthorn cattle, the gene (R) for red coat is not dominant over that for white (R'). The heterozygous combination (RR') produces roan. A breeder has white, red, and roan cows and bulls. What phenotypes might be expected from the following matings and in what proportions:

(a) red × red   (d) roan × roan
(b) red × roan   (e) roan × white
(c) red × white   (f) white × white

(g) Would it be easier to establish a true-breeding (homogeneous for color) herd of red or a true-breeding herd of roan shorthorns? Explain.

2.10 Albinism in man is controlled by a recessive gene (c). From marriages between normally pigmented people known to be carriers (Cc), and albinos (cc): (a) what proportion of the children would be expected to be albinos? (b) What is the chance that any pregnancy would result in an albino child? (c) What is the chance in families of three that one would be normal and two albino?

2.11 If both partners were known to be carriers (Cc) for albinism, what is the chance of the following combinations in families of four: (a) all four normal? (b) three normal and one albino? (c) two normal and two albino? (d) one normal and three albino?

2.12 In Drosophila, a dominant gene (D) for a phenotype called "dichaete" alters the bristles and also makes the wings remain extended from the body while the fly is at rest. It is homozygous lethal. (a) Diagram a cross between two dichaete (Dd) flies and summarize the expected results. (b) Diagram a cross between dichaete and wild type and summarize the expected results.

2.13 In man, two abnormal conditions, cataracts in the eyes and excessive fragility in the bones, seem to depend on separate dominant genes located in different chromosomes. A man with cataracts and normal bones, whose father had normal eyes, married a woman free from cataracts but with fragile bones. Her father had normal bones. What is the probability that their first child will (a) be free from both abnormalities; (b) have cataracts but not fragile bones; (c) have fragile bones but not cataracts; (d) have both cataracts and fragile bones?

2.14 The inheritance pattern represented by colored squares and circles (symbolizing the same trait in different families) may be assumed to depend on a single autosomal dominant or a single autosomal recessive gene. (a) Indicate which is the most likely mode of inheritance for the trait. (b) Based on your answer to (a), symbolize the probable genotype for each individual in each of the four pedigrees.

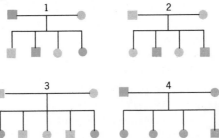

X **2.15** In garden peas, the genes for tall vine (D), yellow seed (G), and round seed (W) are dominant over their respective alleles for dwarf (d), green (g), and wrinkled (w). (a) Symbolize a cross between a homozygous, tall, yellow, round plant and a dwarf, green, wrinkled plant. Represent the gametes possible from each parent and the F_1. (b) Symbolize a cross between two F_1 plants. Complete this cross by making use of the forked-line method and summarize the expected phenotypes. (c) Using the forked-line method, diagram a cross between the F_1 and the dwarf, green, wrinkled parent. Summarize the results for phenotypes, genotypes, genotypic frequency, and phenotypic ratio.

X **2.16** How many F_1 gametes; F_2 genotypes; and F_2 phenotypes would be expected from: (a) $AA \times aa$; (b) $AABB \times aabb$; (c) $AABBCC \times aabbcc$? (d) What general formula can be applied for F_1 gametes, F_2 genotypes, and F_2 phenotypes?

2.17 The shape and color of radishes are controlled by two independent pairs of alleles that show no dominance; each genotype is distinguishable phenotypically. The color may be red (RR), purple ($R'R$), or white ($R'R'$), and the shape may be long (LL), oval (LL'), or round ($L'L'$). Using the Punnett square method, diagram a cross between red, long ($RRLL$) and white, round ($R'R'L'L'$) radishes and summarize the F_2 results under the headings phenotypes, genotypes, genotypic frequency, and phenotypic ratio.

2.18 In poultry, the genes for rose comb (R) and pea comb (P) together produce walnut comb. The alleles of both in a homozygous condition (that is, $rrpp$) produce single comb. From information concerning interactions of these genes given in the chapter, determine the phenotypes and proportions expected from the following crosses: (a) $RRPp \times rrPp$; (b) $rrPP \times RrPp$; (c) $RrPp \times Rrpp$; (d) $Rrpp \times rrpp$.

2.19 Rose-comb chickens mated with walnut-comb chickens produce 15 walnut, 14 rose, 5 pea, and 6 single-comb chicks. Determine the probable genotypes of the parents.

2.20 White-fruit color in summer squash is dependent on a dominant gene (W), and colored fruit on the recessive allele (w). In the presence of ww and a dominant gene (G), the color is yellow, but when G is not present (that is, gg), the color is green. Give the F_2 phenotypes and proportions expected from crossing a white-fruited $(WWGG)$ with a green-fruited $(wwgg)$ plant.

2.21 The White Leghorn breed of chickens is known to carry in homozygous conditions a color gene (C) and a dominant inhibitor (I) which prevents the action of C. The White Wyandotte $(iicc)$ has neither the inhibitor nor the color gene. Give the F_2 phenotypes and proportions expected from crossing a White Leghorn $(IICC)$ with a White Wyandotte $(iicc)$.

2.22 In the F_2 generation of a certain tomato experiment, 3629 fruits were red and 1175 were yellow. A $3:1$ ratio was expected. (a) Are the discrepancies between the observed and expected ratios significant? (b) In the same experiment 671 plants with green foliage and 569 with yellow were counted. This was a backcross and the hypothetical ratio was $1:1$. Test with χ^2 and explain.

2.23 The following are some of Mendel's results with the hypotheses to which they were fitted. Test each for goodness of fit and indicate whether each is significantly different from the hypothesis.

Cross	Results	Hypothesis
(a) Round × wrinkled seed (F_2)	5474:1850	$3:1$
(b) Violet × white flower (F_2)	705:224	$3:1$
(c) Green × yellow pod (F_2)	428:152	$3:1$
(d) Round yellow (F_1) ×		
wrinkled green	31:26:27:26	$1:1:1:1$
(e) Round yellow (F_1) ×		
wrinkled green	24:25:22:27	$1:1:1:1$

2.24 When four coins are tossed together in a series, (a) what proportion of the total results will be in the class of four heads? (b) four tails? (c) three heads and one tail? (d) three tails and one head? and (e) two heads and two tails?

2.25 If four babies are born at a given hospital on the same day: (a) what is the chance that two will be boys and two girls? (b) What is the chance that all four will be girls? (c) What combination of the boys and girls among four babies is most likely to occur? Why? (d) If a certain family has four girls, what is the chance that the fifth child will be a girl?

2.26 What is the probability in families of six of: (a) one boy and five girls? (b) three boys and three girls? (c) all six girls?

2.27 The trait represented by colored squares and circles in the following pedigree chart is inherited through a single dominant gene. Calculate the probability of the trait appearing in the off-spring if the following cousins should marry: (a) 1 × 3; (b) 2 × 4.

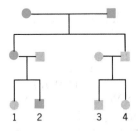

DO

2.28 In the family pedigree shown in the following chart, an abnormal trait is inherited as a simple recessive. Unless there is evidence to the contrary, assume that the individuals who have married into this family do not carry the recessive gene for the trait. Colored squares and circles represent expressions of the trait. Calculate the probability of the trait appearing in a given offspring if the following cousins and second cousins should marry: (a) 1 × 10; (b) 4 × 12; (c) 6 × 11; (d) 16 × 17.

a) $1 \times \frac{1}{4} \times \frac{1}{2} = \frac{1}{8}$

b) $\frac{1}{2} \times \frac{1}{2} \times \frac{1}{4} = \frac{1}{16}$

c) $1 \times \frac{1}{4} \times \frac{2}{3} = \frac{1}{6}$

d) $\frac{1}{2} \times \frac{2}{3} \times \frac{1}{2} \times \frac{1}{4} = \frac{1}{24}$

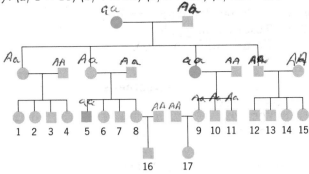

2.29 Phenylketonuria in man is caused by a recessive allele *p*. If both partners were known to be carriers (*Pp*), what is the chance in the following combinations with five children: (a) all normal? (b) four normal, one affected? (c) three normal, two affected? (d) two normal, three affected? (e) one normal, four affected? (f) all affected?

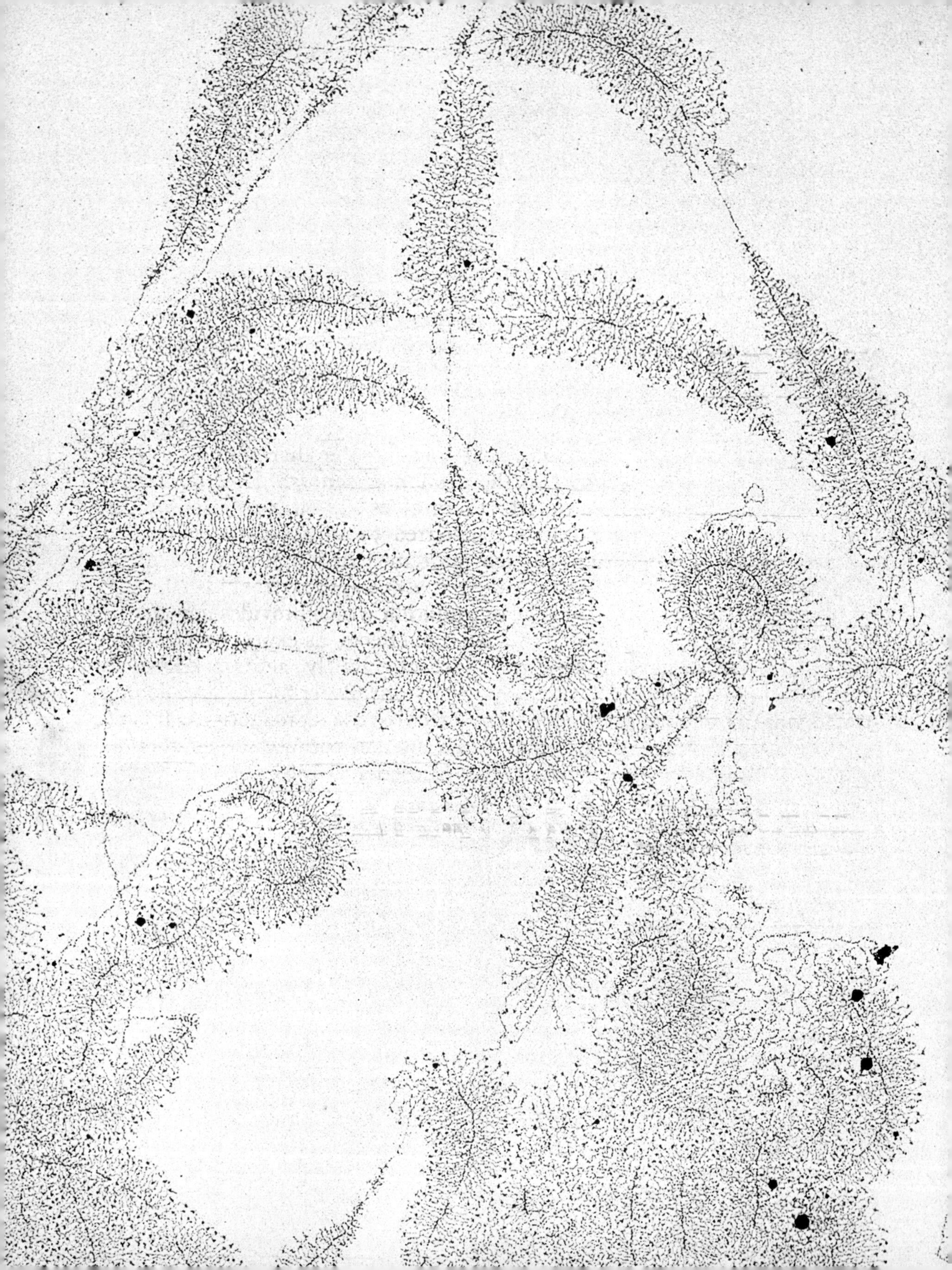

THREE

As Mendel's experiments and many others showed, genes carry from generation to generation the information that specifies the characteristics of the plant or animal. They accomplish their function through: (1) a **replication** process that produces more units like themselves, (2) a **transcription** process through which the information is transferred to an appropriate place for translation, and (3) **translation,** whereby proteins that function in the metabolism of the cell are synthesized. Although genes are extremely stable, they are susceptible to occasional change or **mutation** which provides new alleles. Mendel first postulated genes from their end effects, as expressed in altered phenotypes. Genes have now been defined chemically, and are known for what they do in directing the synthesis of proteins. Experiments have demonstrated that the nucleic acid component (DNA) of the reproductive cell is the genetic material. Nucleic acid, first called nuclein because it was obtained from the cell nucleus, was isolated by F. Miescher in 1869.

GENETIC MATERIAL

Fibrils of RNA attached to central DNA strands in gene transcription. The small spheres on DNA strands at the base of each RNA fibril are molecules of RNA polymerase. (25,000×). (O. L. Miller, Jr. and Barbara R. Beatty, Visualization of nucleolar Genes," *Science,* 164:955–957, May 23, 1969. © 1969 by the American Association for the Advancement of Science.)

DNA, THE GENETIC MATERIAL

Evidence that **DNA** is genetic material, that is, the chemical of which genes are composed, came from investigations on pneumococcus (*Diplococcus pneumoniae*) bacteria beginning in the 1920s. Many strains or types of pneumococcus can be distinguished by serological characteristics (that is, by antigen and antibody interactions). Type specificity is a genetically controlled property of the organism.

Genetic strains are identified with Roman numerals (i.e., Type I, Type II, etc.) The most obvious point of difference between colonies of all strains is the carbohydrate capsule, which determines the appearance of each colony when cultures are grown on blood agar. Some colonies are smooth because the cells have polysaccharide capsules, and some are rough, with no capsules. Capsules also function in antigenic specificity and virulence (Table 3.1). The bacterial colonies that have a rough appearance and no capsules are not virulent. When mice are inoculated with such avirulent cells, no ill effects occur. By contrast, when virulent cells are introduced into mice or other susceptible mammals, a severe illness usually follows. Infrequently, virulent cells lose their capsules through spontaneous mutation and become avirulent. Likewise, avirulent cells may mutate to virulent, capsulated cocci.

Frederick Griffith in England used this background information in studying the effects of virulent and avirulent pneumococci. In a series of experiments (in 1928), he introduced heat-killed virulent cocci into one group of mice, avirulent cocci into another group, and avirulent cocci mixed with heat-killed virulent cocci into a third group. The first two groups were not affected by the injections, but mice in the third group became ill, as illustrated in Fig. 3.1. Griffith thus made the remarkable discovery that a fatal case of pneumonia could be produced if mice were given a mixture of live, avirulent Type II organisms, and virulent but killed Type III organisms. The change could not have arisen by mutation because the avirulent bacteria were from Type II. A mutant for a single trait would be expected to represent the same genetic strain. The disease-generating organisms were of Type III. Something in the dead cocci was apparently transferred to the live cocci. **Avirulent Type II had been transformed into virulent cocci.** This phenomenon was at first called the "Griffith effect" and later became known as **transformation.** It was a forerunner in the establishment of the modern foundation for molecular genetics. Griffith's work provided indirect evidence that a chemical is the genetic material. These experiments demonstrated an important effect but did not identify the chemical involved.

TABLE 3.1
Characteristics of Smooth and Rough Colonies of Pneumococcus *Diplococcus Pneumoniae* **Raised on Blood Agar**

Appearance	Size	Virulence	Capsule
Smooth	Large	Virulent	Present
Rough	Small	Avirulent	Absent

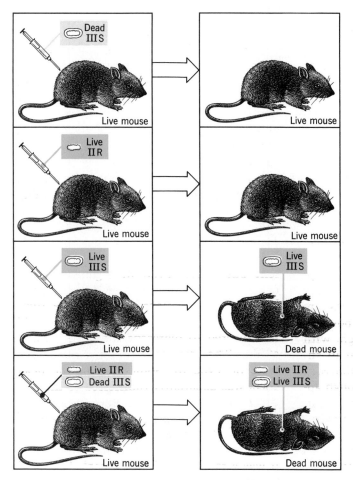

Figure 3.1 The Griffith experiment that demonstrated transformation. When smooth, Type III pneumococci were killed and introduced into mice no ill effect occurred. Likewise, when rough Type II living cells were introduced the mice were not affected. Living Type III cocci killed the mice. When killed Type III material was introduced with living Type II (avirulent) cocci the mice were killed. A dead substance had been incorporated into a living bacterium and transferred to the mice.

TRANSFORMATION

Techniques were refined and experiments were designed by several investigators to test the transforming principle more critically. When cells of the rough form of pneumococcus Type II, which had arisen by spontaneous mutation from Type II smooth, were combined with heat-killed cells of Type III smooth, a few cells could be shown to have the serological properties as well as the virulence of Type III smooth. Transformation in this experiment went toward the type from which the heat-killed cells were derived, demonstrating that a transfer had been made from the killed cells to the live cells. Later experiments were carried out *in vitro*, that is, in a test tube rather than in the animal. These showed that debris, or an extract from the donor cells, could carry the transforming principle, which removed an explanation based on the earlier investigation that the live avirulent cells somehow restored the dead cells to viability. *In vitro* studies led to the biochemical definition of the transferred material.

In 1944, O. T. Avery, C. M. MacLeod, and M. McCarty published the results of their extensive investigations, which had been conducted over a period of ten years. Their

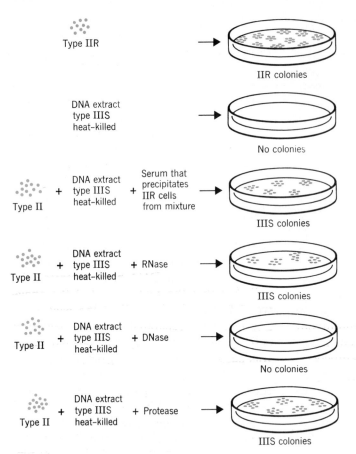

Figure 3.2 Experiments demonstrating transformation in pnuemococcus by Avery, MacLoed, and McCarty. DNA was shown to be the transforming substance. Purity of the DNA extract was verified by digestion experiments. DNA was disintegrated by DNase but not by ribonuclease (RNase) or protease (trypsin).

long and tedious experiments had eliminated every other possibility and identified the transforming substance as deoxyribonucleic acid (DNA). When DNA extracted from one strain of pneumococcus was highly purified and allowed to penetrate the cells of another strain, the recipient organisms developed certain characteristics of the donor strain. The chemically (DNA) transmitted traits were continued generation after generation, indicating that the **genetic material** as well as the phenotype of recipient organisms **had been changed.** Evidently the DNA (Fig. 3.2), which was purified, was incorporated into the genotype of the recipient bacterium.

When the extracted DNA was found to be sensitive to DNase, an enzyme that degrades DNA, but not sensitive to RNase or protease, which degrade RNA and protein respectively, all doubt concerning its purity was eliminated.

When the transformed cells were cultured, they remained encapsulated, and extracts prepared from them had the same transforming ability as those of the original encapsulated strain. It was thus demonstrated that *in vitro* transformation resulted in a hereditary change similar to that which Griffith had demonstrated *in vivo*; that is, in the living mice. This now classical experiment of Avery and his associates (Fig.

3.2) resulted in two important conclusions: (1) **DNA is the genetic material** in pneumococci, and (2) **DNA acts as an agent** that specifies an end product or trait (for example, a characteristic of the polysaccharide capsule), but DNA itself is not the end product.

Transformation is now known to also occur in *Hemophilus influenzae* (a minute rod-shaped pathogen), *Bacillus subtilis* (a common soil bacterium), *Shigella paradysenteriae* (the causative agent for a diarrhea resembling mild dysentery), and several other microorganisms. The transformation principle, therefore, is not unique to pneumococcus, but has general significance, at least in several bacteria. In some instances it has been possible, through transformation, to transfer genetic material from one bacterial species to another. Streptomycin resistance, for example, has been transferred from pneumococci to streptococci. This suggests that bacteria causing one disease might transfer resistance for antibiotics to other types of bacteria, thus increasing the problems of disease control.

DNA ACTIVITY IN A VIRUS

By using **radioactive tracers** A. D. Hershey, a 1969 Nobel prize winner, and M. Chase provided (in 1952) further direct proof that DNA is the genetic material in certain bacterial viruses. These investigators were studying the bacteriophages that attack the bacterium, *Escherichia coli*. They prepared a chemically defined medium that limited phosphate (as orthophosphate) and sulfur (as $MgSO_4$). Known quantities of radioactive isotopes of phosphorus (^{32}P) and sulfur (^{35}S) were added to the medium. The phage particles that infected the radioactive bacteria then incorporated the P^{32} into their DNA. Phage proteins do not contain appreciable amounts of phosphorus, thus only the DNA was labeled with ^{32}P.

Similarly, the protein envelope around the phage was selectively labeled with ^{35}S. By this method it was possible to differentially label the phage protein (Fig. 3.3).

After phage growth, the virus particles were separated from the host cells by centrifugation. The radioactive viruses were next introduced to nonradioactive bacterial cultures where they attacked the bacteria. Subsequently, the adsorbed viruses were separated from the host cells by agitation and the contents of ^{32}P and ^{35}S in host and parasite were determined. The phosphorus label was found to be associated with the bacterial cells and the sulfur label was in the protein coats left in the medium. This indicated that the DNA had penetrated the cells but that the **protein coat of the phage was left outside** the wall of the bacterium and had been separated from the bacterial cell by agitation. Only the labeled **DNA was passed on to the next generation.**

Significantly, it was concluded that when the DNA of a virus particle entered a host cell, essentially all of its protein remained outside and, therefore, it did not represent the genetic material. Only the DNA of the virus was in the bacterium during viral replication. After the DNA part of the virus was reproduced within the bacterial cell, new protein was synthesized and became associated with the DNA units. New infective virus particles were thus formed. When the host (bacteria) lysed, numerous infectious virus particles emerged, ready to enter other bacterial cells and repeat the cycle. The interpretation of the Hershey and Chase experiments has been fully confirmed by further studies including electron micrographs of the infection process. These experiments, taken together, demonstrated beyond doubt that **DNA is genetic material.**

In some viruses, RNA substitutes for DNA in storing and transmitting genetic information. In the tobacco mosaic virus (TMV), proof that RNA and not protein is

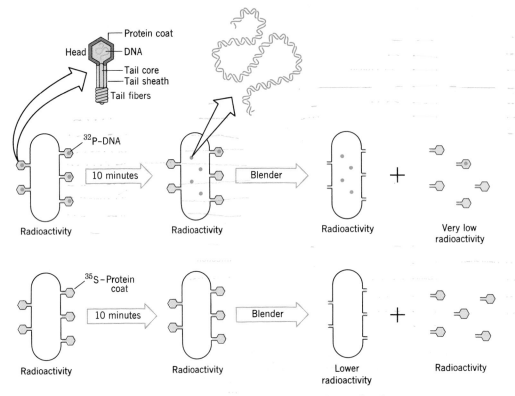

Figure 3.3 Hershey and Chase experiment demonstrating that only phage DNA enters the bacterial host cell after infection. Cells were infected with ^{32}P labeled phage, and after being allowed time for infection, they were agitated in a blender which sheared off the phage coats. Radioactivity was measured at each step in the procedure. Very little of the original ^{32}P radioactivity was lost from the cells. Thus, the phage DNA had been incorporated into the host cells. When phage were labeled with ^{35}S in their proteins, and the same experiment performed, the results were very different. Most of the radioactivity was found in the supernatant with the phage protein; very little entered the cells. (Based on Ruth Sager and F. J. Ryan, *Cell Heredity*, John Wiley and Sons, 1961.)

the genetic material has been provided through the simple but elegant reconstitution experiment of H. Fraenkel-Conrat and B. Singer. The TMV consists of an RNA coil surrounded by protein that aids in infection of plant cells and protects the RNA from degradation by such enzymes as ribonuclease (RNase). The particular kind of protein envelope is specified by the TMV strain.

When the investigators separated the protein by biochemical methods from the RNA, it was noninfective, but when reconstituted with isolated RNA the particles became infective and produced new virus. RNA and protein from the same strain could be reconstituted by rubbing pure RNA on live tobacco leaves, thus causing synthesis of new TMV virus like the parental strain. It was also possible to re-

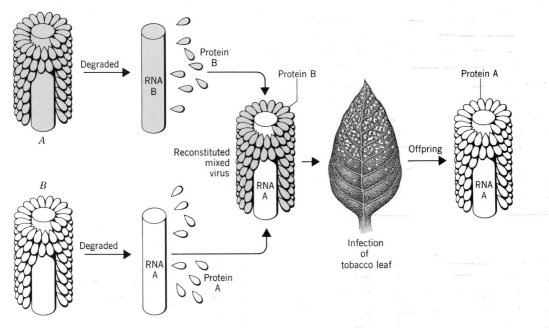

Figure 3.4 Reconstitution experiment demonstrating that RNA and not protein is the genetic material in tobacco mosaic virus. RNA from one strain (A) was reconstituted with protein from strain (B). When the reconstituted virus was rubbed on a live tobacco leaf the offspring virus was like strain A from which the RNA had been obtained and not strain B from which the protein was obtained. (After H. Fraenkel-Conrat and B. Singer.)

constitute TMV of one strain with protein from another strain (Fig. 3.4). When this virus was used to infect live tobacco leaves, the **protein of the new virus** was always de- **termined by the parental RNA** and not the parental protein. Synthesis of new TMV could be prevented by RNase which degrades RNA.

WATSON AND CRICK MODEL

While the investigations that chemically identified DNA as genetic material were in progress, and especially following the report of Avery and his associates (1944), attempts were made to explain the physical and chemical nature of DNA. As early as 1938, **X-ray diffraction** pictures of nucleic acid had indicated that the main chemical components in the molecule (organic bases) were stacked on each other. Preliminary models were prepared to illustrate the data being accumulated. Some of these early models represented the bases in ascending steps resulting in a ladderlike configuration. All of this information was available to J. D. Watson (Fig. 3.5) and F. H. C. Crick when they began, in the early 1950s, to develop a model that would account for all the data bearing on the chemical and physical nature of DNA.

Figure 3.5 J. D. Watson, American investigator in biochemical genetics. Along with the British investigators F. H. C. Crick and M. H. F. Wilkins, Dr. Watson won the Nobel Prize in physiology and medicine in 1962. The prize was awarded for the contribution furthering our knowledge and understanding of the chemical nature of the gene. Their progress up to 1953 was summarized by Watson and Crick in a model for the structure of the DNA molecule. (Courtesy of Harvard University News Office.)

Watson and Crick eventually proposed a model for DNA that could not, at that time, be fully supported from experimental data, but which has since gained strong support. Their model was based on **linear sequences of nucleotides,** each composed of a pentose sugar, a phosphate, and an organic base (Figs. 3.6, 3.9). Four kinds of nucleotides were recognized, each including a different nitrogenous base: **adenine, thymine, cytosine,** or **guanine** (Fig. 3.7). These investigators received a Nobel prize in 1962 for their 1953 model and hypothesis which had been substantiated in the intervening years.

The physical criteria used in creating the model were derived from X-ray diffraction pictures produced by M. H. F. Wilkins (also a 1962 Nobel prize recipient) and his co-workers, using isolated DNA. Their X-ray diffraction pictures were crude compared with those subsequently developed, but they did show that the DNA molecule was in the shape of a helix (Fig. 3.8). The early pictures also were suitable for making measurements of the spacing between the bases. Examinations of isolated nuclei and sperm heads proved that the structure found in the isolated DNA also exists in the cell. These pictures indicated that DNA molecules were composed of two or three strands forming helices.

Chemical analyses used by Watson and Crick were those of Chargaff and his associates, which showed a **1:1 relation between adenine (a purine) and thymine (a pyrimidine), and between cytosine (a pyrimidine) and guanine (a purine).** This relation did not exist between the two purines or the two pyrimidines but the total amount of purine equalled the total amount of pyrimidine. These investigators found the same basic chemical pattern in DNA

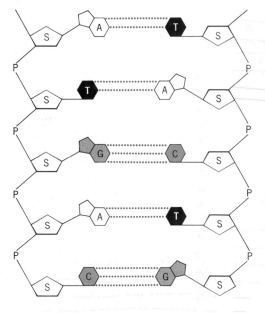

Figure 3.6 Linear sequence of nucleotides as proposed in the DNA model of Watson and Crick. Each nucleotide is composed of a phosphate (P), a sugar (S), and an organic base: adenine (A), thymine (T), guanine (G), or cytosine (C). The phosphate-sugar rims in the configuration are held together firmly by chemical bonds but the two organic bases in each cross link are connected less firmly with each other by hydrogen bonds. (After Crick.)

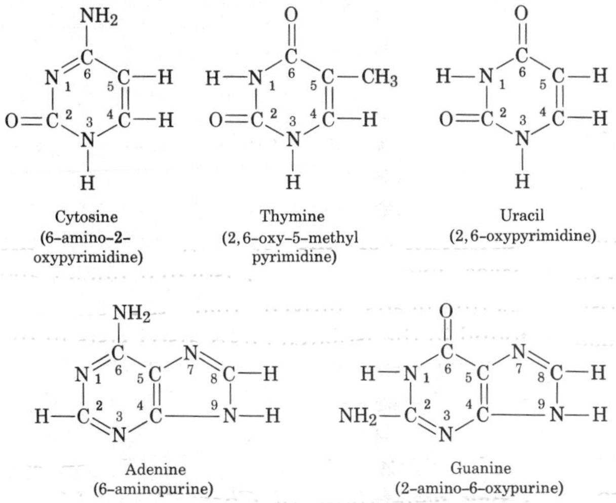

Cytosine
(6–amino-2–
oxypyrimidine)

Thymine
(2,6–oxy-5–methyl
pyrimidine)

Uracil
(2,6–oxypyrimidine)

Adenine
(6–aminopurine)

Guanine
(2–amino–6–oxypurine)

Figure 3.7 Structural formulas for organic bases in DNA and RNA. Four bases—cytosine, thymine, adenine, and guanine—are in DNA. In RNA thymine is replaced by uracil.

from many sources. The Watson and Crick model is illustrated schematically in Fig. 3.9. The relatively rigid long strands that form the spirals around an axis are made up of phosphates and pentose sugars. Cross-wise, the strands are less rigidly connected by the organic bases through hydrogen (H) bonds. Adenine and thymine are connected by two H bonds, whereas cytosine and guanine are connected by three. In each cross link, the bases are arranged in such a way that a certain purine is bound to a certain pyrimidine, adenine-thymine and cytosine-guanine. **Pairs of organic bases may**

Figure 3.8 One of the first X-ray diffraction photographs of DNA prepared by Wilkins and his associates showing repeating patterns of symmetrical molecules. An expert in the field can recognize from the patterns that the structure is a helix. (M. H. F. Wilkins, ''Molecular configuration of nucleic acids,'' *Science*, 140, 941–950, 1963. Copyright 1963 by the American Association for the Advancement of Science)

be arranged in any order. The length of one spiral (that is, within 34 Å), contains 10 base pairs.

Although Watson and Crick relied heavily on the work of others, they made enormous contributions themselves. They put together the X-ray diffraction, chemical, and physical data and created a model that incorporated the functional requirements of the genetic material. Watson and Crick were the first scientists to propose a **double helix** held together by **hydrogen bonds** between specific pairs of bases (thymine to adenine and cytosine to guanine) and with each strand complementary to its partner strand in terms of base sequence.

The Watson and Crick initial publication was a short paper published in *Nature* under the heading, ''Molecular Structure of Nucleic Acids.'' Another paper in the same issue was the one by Wilkins and associates presenting the X-ray diffraction data used by Watson and Crick (see the references at the end of this chapter).

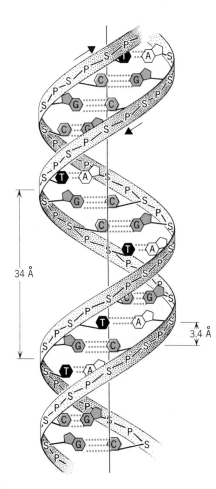

Figure 3.9 Watson and Crick model for DNA showing the two-stranded double helical structure. Symbols are the usual chemical symbols and those for nucleic acid bases, with S for sugar and P for phosphate residues.

34 Å

3.4 Å

REPLICATION OF GENETIC MATERIAL

Living things perpetuate their kind through duplication or reproduction. Duplication of the organism is preceded by replication of the DNA molecules. Each DNA molecule uses its own structure as a template and takes from its immediate environmental substratum the necessary materials for replication.

In the Watson and Crick model, each of the two original strands is the complement of the other. When duplication occurs, the hydrogen bonds between the bases break and the strands replicate as they unwind, as illustrated in Fig. 3.10. Each strand acts as a model or **template** for the formation of a **new complementary chain.** Two pairs of chains thus appear where before only one pair existed. Furthermore, each chain is a complement of the one from which it was specified, carrying genetic qualities deter-

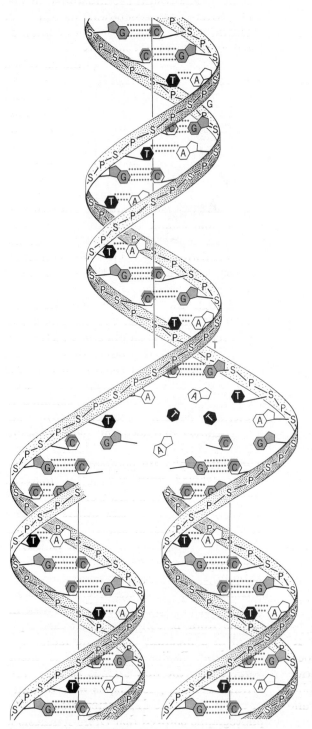

Figure 3.10 Replication of DNA. As the double-stranded parent DNA unwinds, the separated strands serve as templates for the alignment of nucleotides, which combine to form new strands complementary to the parental strands.

mined by the original structure. This complementarity is maintained because of the pairing relations that exist between bases.

Strong support for the Watson and Crick explanation for DNA replication came from a series of experiments by M. S. Meselson and F. W. Stahl. From the same experiments these investigators answered another significant question: whether replication results in original DNA chains in one molecule and new chains in another (conservative replication) or an original and a new chain together in a molecule (semiconservative replication). These investigators grew bacteria (*E. coli*) for many generations in a simple medium in which the only nitrogen source was ammonium chloride (NH₄CL), which contained only the heavy isotope of nitrogen, ^{15}N. During the reproductive process, the DNA of the bacteria became labeled with heavy nitrogen. An excess of the ordinary isotope of nitrogen, ^{14}N, was then added to the medium. At intervals, cells were removed from the culture and the DNA was extracted from them. The density of the individual molecules in each sample was then examined by the method of equilibrium density-gradient centrifugation to determine the relative content of ^{15}N and ^{14}N.

The equilibrium density-gradient method is a procedure by which the DNA being examined is placed in a concentrated salt solution such as cesium chloride (CsCl) in water and centrifuged in an ultracentrifuge. After several hours of centrifugation, Meselson and Stahl found that the distribution of CsCl became essentially stable resulting in a density gradient. Eventually each molecule reached the layer in the density gradient with a buoyant density equal to its own. The ^{15}N and ^{14}N DNA became localized in different layers. The concentration of the DNA could be measured photographically and the proportion of ^{15}N and ^{14}N in the daughter molecules could be de-

termined. The results are illustrated in Fig. 3.11. After one generation in the ^{14}N medium, each DNA molecule contained equal amounts of ^{15}N and ^{14}N. Each molecule was "hybrid" in density and each had one "light" and one "heavy" strand. After the second generation, there were equal numbers of totally "light" ^{14}N containing molecules and hybrid molecules with one "light" (^{14}N) and one "heavy" (^{15}N) strand. In each replication the new strand incorporated the ^{14}N in the medium but the ^{15}N label was retained in the DNA strands originally labeled. These studies indicated that the original strand was maintained during each replication as shown in Fig. 3.11*b*. and not as shown in Fig. 3.11*a*. The replication process was, therefore, **semiconservative** and not conservative.

Although the Meselson and Stahl experiments showed that the net product of DNA replication is an old DNA strand and a new synthesized strand, they did not show the sequence of steps by which this kind of replication is accomplished. Furthermore, the experiments were confined to bacteria. Further studies with autoradiography extended the data on semiconservative replication to higher forms. Autoradiographs are photographs of a substance labeled with radioactive material and developed over a period of time. By this method, John Cairns and others supplied additional data about sequential events in **chromosome replication** while J. H. Taylor and others have extended the conclusions to include higher organisms.

Cairns in 1963, for example, obtained autoradiographs (Fig. 3.12) at different time intervals from *E. coli* DNA that had replicated in a medium containing radioactive thymidine. Radioactive label was incorporated into the new strands as they were synthesized and photographs were obtained of the steps in the duplication process. The photographs showed that DNA synthesis is

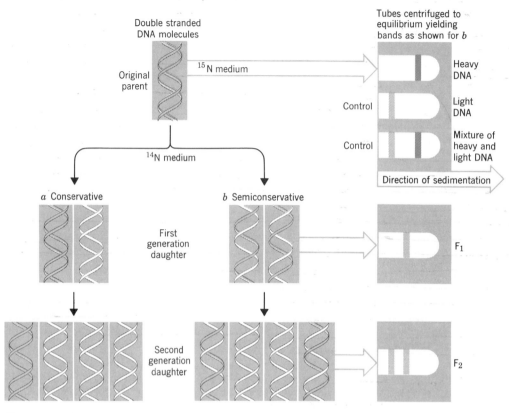

Figure 3.11 The Meselson and Stahl experiment in 1957–58 was designed to determine whether DNA replication is conservative or secmiconservative. *E. coli* cells were grown for many generations in a medium containing only ^{15}N and the DNA become labeled with ^{15}N. ^{14}N was then added to the medium. Cells were removed, DNA was extracted and individual molecules were examined by equilibrium density—gradient centrifugation to determine the proportion of ^{15}N and ^{14}N. Each molecule reached the layer with a buoyant density equal to its own and the ^{15}N and ^{14}N molecules became localized in different layers as shown at right. Heavy DNA formed a single band near the end of the tube, light DNA and a mixture of heavy and light, introduced as controls, yielded separate bands some distance apart, F_1 DNA yielded a single band intermediate between the heavy and light, and F_2 DNA yielded two bands, one intermediate like the F_1 and one similar to that of the light DNA. The results showed semiconservative replication, illustrated in b and not conservative replication as shown in a to be the mechanism for replication of DNA. (After Meselson and Stahl: Proc. Nat'l., Acad. Sci., 44, 671–682, 1958.)

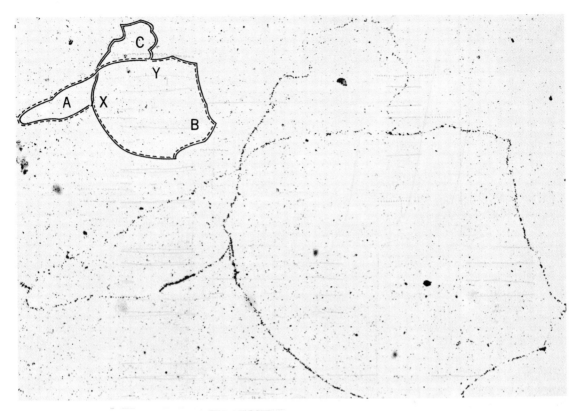

Figure 3.12 Autoradiograph of circular *E. coli* chromosome that is in the process of duplication. The DNA was labeled with ³H thymidine. (John Cairns, Imperial Cancer Research Fund, London, England.)

closely associated with the separation of the two original strands of the double helix. Each original strand gave rise to a new strand in semiconservative fashion, as indicated by the Meselson and Stahl data and postulated in the Watson and Crick model (Fig. 3.9). Studies of the green algae *Chlamydomonas reinhardi* have also shown evidence for semiconservative replication.

Taylor obtained end results similar to those of Meselson and Stahl but he investigated chromosomes of higher organisms and worked with autoradiography rather than equilibrium density-gradient centrifugation. When chromosomes of dividing cells of the broad bean *Vicia faba* replicated in a radioactively labeled medium, the new chromosomes appeared as labeled because they consisted of a labeled parental strand and a newly synthesized radioactive strand (Fig. 3.13). When an additional replication was allowed to occur in an unlabeled medium, only the previously radioactive strand of each chromosome was labeled. When the strands separated, the parental strand retained its label but the new strand synthesized in the unlabeled medium did not incorporate the label. The results of Taylor and others, like those of Meselson and Stahl, showed that DNA replication requires the separation of the strands in the double helix and that replication is semiconservative.

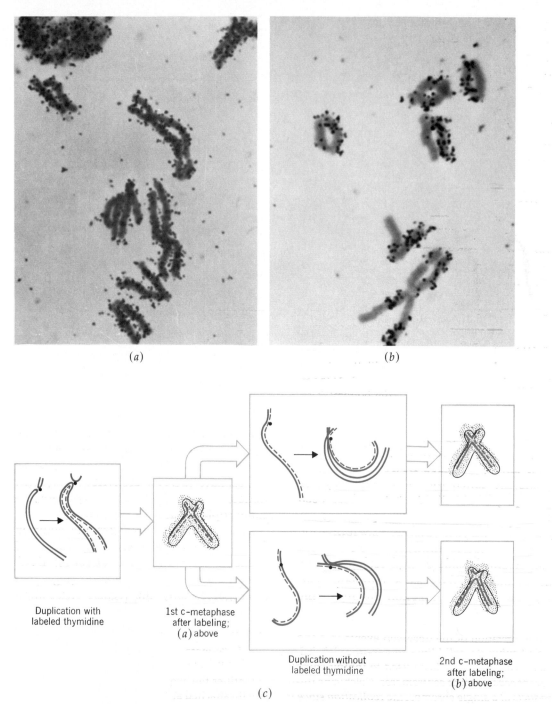

(a)

(b)

Duplication with
labeled thymidine

1st c-metaphase
after labeling;
(a) above

Duplication without
labeled thymidine

2nd c-metaphase
after labeling;
(b) above

(c)

Figure 3.13 (a) Radioautograph of *Vicia faba* chromosomes after one replication in radioactively labeled medium. (b) Appearance of such labeled chromosomes after an additional replication in unlabeled medium.

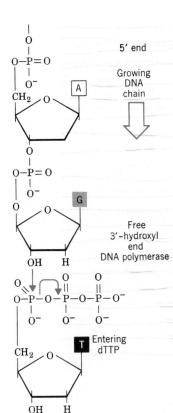

5' end

Growing
DNA
chain

Free
3'-hydroxyl
end
DNA polymerase

Entering
dTTP

Figure 3.14 DNA chain extension by DNA polymerase. From the 5^1 end (at the top) dATP and dGTP have already entered the chain (with the release of 2 phosphate groups). At the free 3^1 end dTTP (at the bottom) is entering the chain.

Arthur Kornberg, Nobel prize winner in 1959, studied the chemistry of DNA synthesis and synthesized DNA *in vitro* with an enzyme called the "Kornberg enzyme" (DNA polymerase I) which has since been shown not to be the true replicating enzyme. This enzyme catalyzes the formation of internucleotide linkages of DNA from deoxyribonucleoside 5'-triphosphates using one strand of DNA as a template. Participants in the reaction include the 5'-triphosphates of each of the four deoxyribonucleosides: deoxyribose adenine triphosphate (dATP), deoxyribose guanine triphosphate (dGTP), deoxyribose thymine triphosphate (dTTP), and deoxyribose cytosine triphosphate (dCTP), and Mg^{++} ions. DNA polymerase I catalyzes the addition of mononucleotides to the 3'-hydroxyl end of the DNA chain thus directing DNA synthesis in the 5' → 3' direction (Fig. 3.14). It recognizes only the regular sugar-phos-

(c) Interpretation of the observed events in *a* and *b*. Broken lines represent labeled subunits, solid lines represent unlabeled subunits. In "c-metaphase" colchicine has been used to inhibit spindle fibers and thereby prevent separation of centromeres. Colchicine treatment identifies the two products of a single chromosome replication since they remain attached at the centromere. (From J. H. Taylor, 1963. The replication and organization of DNA in chromosomes. *Molecular Genetics,* Part 1. J. H. Taylor (ed.) Academic Press, New York.)

phate portion of the nucleotide precursor and does not determine sequence specificity. **DNA is the direct template for its own formation.**

These studies did much to clarify the nature of the biochemical processes involved in DNA replication, but several questions remain to be answered. A critical one is how do the *in vitro* observations relative to DNA replication relate to the formation of genes in living cells? The DNA strands are at the molecular level and in a different order of magnitude as compared with chromosomes and other structures seen in the cells of higher organisms with the light microscope. Most of the data accumulated by Kornberg and others concerning the replication of DNA have come from studies on bacteria and viruses; but extensive evidence indicates that DNA replication occurs by the same or very similar processes in higher organisms.

Based on the Watson and Crick model, many different chemical arrangements could be established and replicated. If, for example, each of the four bases, adenine (A), cytosine (C), guanine (G), and thymine (T), is paired with a complementary base, many combinations may be illustrated as in the three series below, each representing a characteristic double helix.

A-T	G-C	T-A
C-G	C-G	A-T
G-C	T-A	C-G
T-A	A-T	G-C
T-A	C-G	T-A
G-C	G-C	C-G
C-G	C-G	G-C
T-A	G-C	A-T
A-T	T-A	C-G
G-C	A-T	T-A
. .	. .	. .
. .	. .	. .
. .	. .	. .

The great number of bases present in any given molecule permits a **tremendous number of arrangements** (4^n where n is the number of base pairs). DNA in one virus particle, for example, contains some 200,000 base pairs. Higher forms of life contain many times more in each cell. In physical dimensions, a bacterial cell has about 2mm of DNA whereas a human cell has about 6 feet.

Since Kornberg's initial discovery of a DNA polymerase, several enzymes have been found to be intimately involved in DNA replication. The first indication that Kornberg's enzyme (polymerase I) is not "the" replicating enzyme came with the discovery of an *E. coli* mutant (pol A$^-$) that lacks DNA polymerase I but, nevertheless, replicates its DNA at the normal rate. Since this mutant has increased sensitivity to ultraviolet rays, it was concluded that it must lack one of the enzymes involved in repair processes. Another enzyme, **DNA polymerase III,** is now generally accepted as the replicating enzyme.

Moreover, the specificity of base pairing may be enzymatically determined. DNA polymerases seem to be involved in base selection, to fit the DNA template, as well as connecting the bases together in strands. Hydrogen bonding alone seems to be too weak and inefficient to account for the speed and accuracy of chromosome replication. The site of polymerase activity for the bacterial chromosome is apparently in the region where the chromosome is attached to the cell membrane. Membrane-bound polymerase may be responsible for chromosome replication whereas free polymerase may be involved with DNA repair (Chapter 8).

Other enzymes have been found to fulfill further requirements for DNA synthesis. **Polynucleotide ligase,** isolated from *E. coli* by M. Gellert (in 1967) catalyzes the formation of a covalent linkage between the 3' end of one polynucleotide chain and the 5'

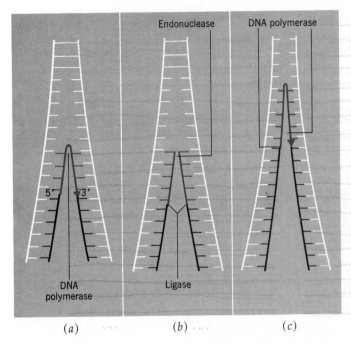

Figure 3.16 Knife and fork model for DNA replication. (*a*) DNA polymerase has copied the segment shown in red at left, jumped to the right side of the fork and is completing the corresponding segment of the second strand. Replication has already been completed in daughter strands shown in gray (lower left and right). (*b*) Endonuclease ("the knife") has cut the base of the fork and ligase has sealed the copied segments. (*c*) New DNA polymerase has copied from one side of another segment, jumped to the other side of the fork and is completing another daughter segment. (After R. Barzilai and C. Thomas.)

chains which are later **linked together.** Polynucleotide ligase provides the required capacity to link short chains of DNA produced by discontinuous synthesis at the replication fork.

Replication of the bacterial chromosome is followed by segregation of the two daughter chromosomes, each with one old and one new strand, into the two daughter cells at the time of cell division (Chapter 4).

REPLICATION IN A PHAGE

Chromosome replication has been more readily studied in bacteriophage ΦX174 than in bacteria. This is a DNA phage containing a small, single-stranded, circular chromosome. When ΦX174 infects an *E. coli* cell it replicates immediately, producing a complementary strand. The injected strand is called "**positive**" and the newly synthesized strand "**negative.**" The two strands remain together for several replications, in a closed circular structure

called the parental replicative form (RF). After replicating several times, these forms shift to production of positive, single-stranded rings, synthesize a protein coat, and are released when the bacterial host cell ruptures (lyses).

The positive strands can be copied *in vitro* by the Kornberg enzyme (DNA polymerase I) to produce **complementary negative strands.** If these strands are sealed into rings by ligase and isolated, they can **produce** *in vitro* positive strands with biological activity. The **rolling circle model** has been proposed to explain (1) how RF replication occurs in the host cell and (2) how RF production of phage with single positive strands occur.

A parental RF (Fig. 3.17*a*) is nicked by an endonuclease (E) at a particular point in its positive strand (Fig. 3.17*b*). The 5' end of the positive strand unravels and is peeled back while the negative strand remains closed forming a "rolling circle intermediate" (Fig. 3.17*c*). Host DNA poly-

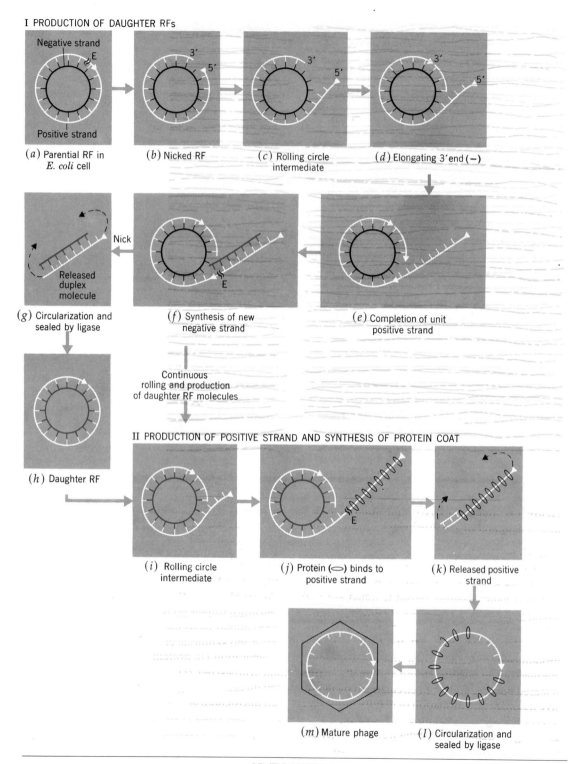

I PRODUCTION OF DAUGHTER RFs

Negative strand
E
Positive strand

(a) Parental RF in
E. coli cell

(b) Nicked RF

3'
5'

(c) Rolling circle
intermediate

3'
5'

(d) Elongating 3'end (−)

Nick

Released
duplex
molecule

(g) Circularization and
sealed by ligase

E

(f) Synthesis of new
negative strand

(e) Completion of unit
positive strand

(h) Daughter RF

Continuous
rolling and production
of daughter RF molecules

II PRODUCTION OF POSITIVE STRAND AND SYNTHESIS OF PROTEIN COAT

(i) Rolling circle
intermediate

E

(j) Protein (⊂⊃) binds to
positive strand

I

(k) Released positive
strand

(m) Mature phage

(l) Circularization and
sealed by ligase

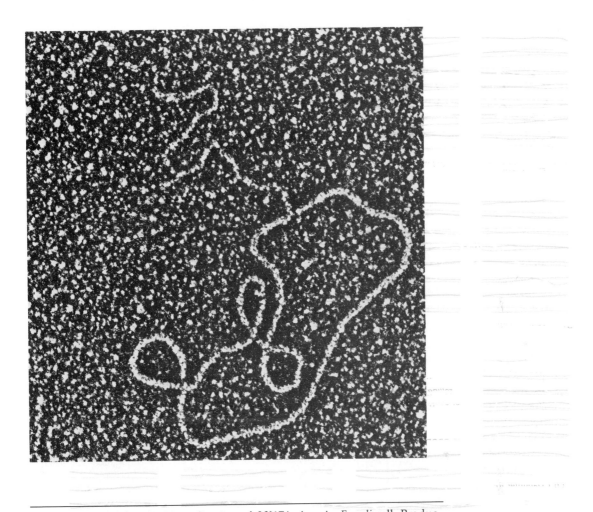

Figure 3.17 Model for DNA replication of ΦX174 virus in *E. coli* cell. Production of daughter RFs. Arrowhead identifies a specific nucleotide sequence on the positive strand which is recognized by a specific endonuclease (E). (*a*) Parental RF with positive strand (outside) and negative strand (inside). (*b*) RF nicked, (*c*) positive strand peeled back and (*d*) DNA replication of positive strand begins by elongating 3′ end. (*e*) New positive strand is completed as "old" positive strand is rolled out to the right. (*f*) New negative strand is synthesized from the old positive strand and the linear duplex (*g*) is released by an endonuclease inflicted nick. (*h*) Circularized duplex molecule with ends sealed by ligase. II. Production of positive strand and synthesis of protein coat. (*i*) Rolling circle intermediate with DNA replication (*j*), protein synthesized on positive strand and small negative strand synthesized, (*k*) released positive strand, (*l*) circularized by ligase and (*m*) digestion of negative fragment and formation of mature virus. (*n*) Electron micrograph of ΦX174 rolling circle. Double-stranded replicating intermediate circle (below) and single-stranded tail (above), similar to (*e*) in diagram. (From D. Dressler, Proc. Nat'l. Acad. Sci. 67:1934–1942. 1970.)

merase adds new nucleotides in a $5' \rightarrow 3'$ direction to the 3' end of the positive strand (Fig. 3.17d). The direction of synthesis is always $5' \rightarrow 3'$. When the growing point has proceeded around the circle (Fig. 3.17e), the positive strand can be copied in reverse order (but in $5' \rightarrow 3'$ direction) by a new DNA polymerase to form a new negative strand (Fig. 3.17f). When the duplex is nicked, it disengages from the parental RF (Fig. 3.17g) and forms a double-stranded circle by the overlapping of complementary base sequences at the two ends. If these ends are sealed by ligase, a daughter RF will be formed (Fig. 3.17h). The negative circle continues to "roll" and to form new copies.

To affect the change from production of RF to that of positive strands, phage protein is postulated with high affinity for the base sequence of the positive strand (Fig. 3.17i). As this protein is synthesized in the cell, (Fig. 3.17j), it is expected to associate with positive strands as they are generated from the rolling circle (Fig. 3.17k). This prevents the strands from serving as templates for negative strand synthesis and RF formation (Fig. 3.17l). The positive strand is sealed by ligase to form a circle, synthesizes a protein coat, and becomes a **mature phage** (Fig. 3.17m).

The rolling circle model describes well the replication of the ΦX174 chromosome and represents a plausible mechanism for DNA replication of other viruses. Replication of most other viral and bacterial chromosomes, however, does not require the addition of new DNA to one of the old parental strands as does this model. Although the rolling circle model is well established for ΦX174, it does not explain the mechanism of duplex molecules. Great diversity apparently exists in the ways different chromosomes are copied and transmitted from parent to offspring. Two constants are established: (1) DNA synthesis proceeds in a $5' \rightarrow 3'$ direction and (2) **replication of duplex DNA is semiconservative.**

FUNCTIONS OF RIBONUCLEIC ACID

Ribonucleic acid (RNA) has important functions in protein synthesis. In eukaryotic cells, it is located in the nucleus and also in the cytoplasm particularly in the cytoplasmic structures called **ribosomes.** Only RNA is present in some viruses and in these particles it fulfills the genetic function that is ordinarily restricted to DNA.

RNA also participates actively in cell metabolism. The turnover rate changes with the metabolic activity of the cell. Usually, the information that DNA carries is transmitted through RNA, which directs protein synthesis in the ribosomes within the cytoplasm of the cell.

RNA molecules in most organisms are

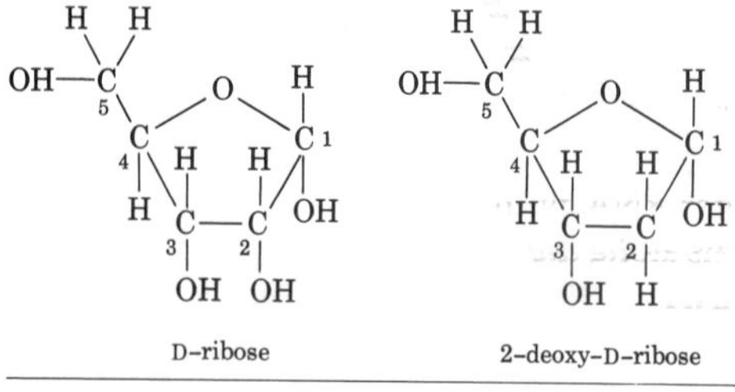

Figure 3.18 Structural formulas for sugars found in RNA and DNA. D-ribose is a part of the RNA molecule and 2-deoxy-D-ribose with one less atom of oxygen (see position 2) is in DNA.

D–ribose

2-deoxy-D-ribose

single stranded rather than double stranded. RNA contains ribose (Fig. 3.18) instead of deoxyribose sugar and the organic base **uracil** instead of thymine. Like DNA, RNA gives a maximum absorption of ultraviolet light with a wavelength of about 260 millimicrons. Also like DNA, **RNA stains with basic dyes.** RNA can be distinguished from DNA with cytochemical tests. When, for example, **RNase,** an enzyme that **degrades** or hydrolyzes **RNA** to mononucleotides is introduced, it breaks down the cytoplasmic basophilia (acidic contents that stain with basic dyes). At least three kinds of RNA have an important function in the transcription and translation of genetic information: (1) **messenger RNA,** (2) **ribosomal RNA** and (3) **transfer RNA.**

TRANSCRIPTION OF GENETIC INFORMATION

Transcription is the formation of messenger RNA (mRNA) from a DNA template, with the aid of an enzyme DNA-dependent polymerase, abbreviated as **RNA polymerase.** Transcription is much like DNA replication but differs in several respects: (1) the base uracil in RNA replaces thymine in DNA, (2) RNA transcripts do not remain bonded with the DNA template but "peel off" and participate in protein synthesis, (3) DNA replication has one or a few initiation points and conserves the entire genome, whereas synthesis of RNA molecules is initiated at close intervals producing copies of individual genes for immediate and **short-term use in protein synthesis.**

The attachment of RNA polymerase to a DNA molecule (Fig. 3.19) opens up a section of the double helix, thereby allowing free bases on one of the DNA strands to specify complementary bases and thus to provide a transcription of the base sequence. For example, thymine (T) of DNA is represented by adenine (A), on the

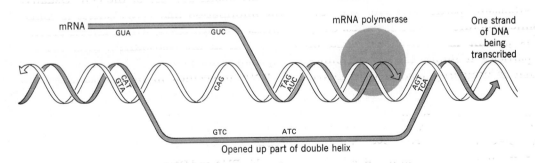

Figure 3.19 Transcription of DNA specifications on messenger RNA with the aid of mRNA polymerase. A few bases are shown at intervals along the DNA and mRNA strands to illustrate DNA replication and mRNA transcription.

new RNA strand, cytosine (C) is represented by guanine (G), guanine by cytosine and adenine by uracil (U). As the RNA polymerase moves along its DNA template, the newly assembled mRNA bases are joined in tandem by an enzymatic reaction. Due to repeated transcription, the quantity of the mRNA product resulting from this synthesis greatly exceeds that of the DNA. The coded message is complementary to only one of the two DNA strands at a particular point or gene (asymmetric transcription).

Evidence for the specificity of mRNA in carrying DNA's information came first from studies on bacteriophages. When *E. coli* was infected with virus T2, a T2 mRNA was synthesized that represented the base complement of T2 DNA. Investigations with virus T4 that used radioactive tracers have identified an mRNA in the ribosomes of infected *E. coli* that was specified by a particular viral DNA. These experiments and many others have demonstrated that **ribosomes are manufacturing centers** for protein, but mRNA must supply them with the proper information from the DNA before a particular protein can be synthesized.

The mRNA transcript peels off from the DNA template, passes from the nucleus to the cytoplasm (in eukaryotes) and becomes attached to a ribosome. Following removal of the mRNA transcript from the nuclear DNA template, hydrogen bonds that were separated during transcription reform between complementary DNA strands and DNA is again doublestranded.

RNA polymerases are remarkable enzymes. They **recognize** individual bases in DNA, **select** the appropriate complementary ribonucleotides, **catalyze** bond formation between ribonucleotides, pick the correct strand to be transcribed and **initiate transcription** at appropriate **promotor sites** along the chromosome. Isolated RNA polymerase from *E. coli* is a large protein including the "core polymerase" and five subunits. One subunit called the **sigma factor** initiates transcription in the appropriate place (Fig. 3.20). Sigma may become dissociated from the core polymerase in the cell. When this occurs the core polymerase

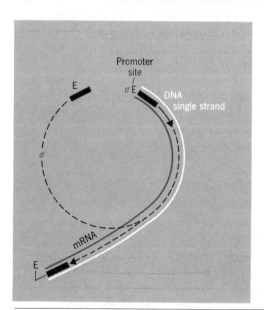

Figure 3.20 The sigma cycle. When σ is associated with E (polymerase) at promoter site (top of circle), mRNA transcription from DNA template is initiated; when transcription is in progress, σ is released to find a new core polymerase (bottom of circle) and transcription continues.

is capable of copying DNA but it begins at any place along the template and transcribes from both strands. When the sigma factor is again associated with the core polymerase, transcription begins only at the true promotor sites and on the correct strand. Sigma apparently initiates transcription by recognizing a promoter region and **binding a core polymerase to the promoter.** When the core enzyme has become bound and begun transcription, sigma may be **released** to become associated with another core polymerase. The sigma factor is thought to be the most important regulator in the process of RNA transcription.

SYNTHESIS OF PROTEINS

Chemically, proteins are giant organic molecules that have different shapes, sizes, and functions. Molecular weights vary widely, from 6000 for insulin, 13,500 for ribonuclease and 66,200 for hemoglobin, to some 500,000 for certain proteins. Each molecule varies from 51 amino acids in insulin, 124 in ribonuclease, and 574 in hemoglobin, to several thousand in the larger protein molecules. Although the subunits may number into the hundreds of thousands, only 20 of about 80 kinds of known amino acids are commonly encountered in proteins. Amino acids contain one or more amino groups (NH_2) and one or more carboxyl groups (COOH). The basic structure of all the common amino acids (except proline) is shown in Fig. 3.21.

In the protein molecule, the amino acids are held together by **covalent peptide bonds.** Bonding involves elimination of a hydroxyl from the carboxyl group of one amino acid and a hydrogen from the amino group of another (Fig. 3.22). The twenty common amino acids found in proteins (Fig. 3.23) may occur in numerous combinations. Therefore, the number of potential kinds of proteins is enormous. Only a comparatively small proportion of the enormous number of possible proteins have been identified in nature.

Two major aspects in the mechanics of protein synthesis will be considered here: (1) locations in the cell where the process is accomplished, and (2) how the parts (amino acids) required for building proteins are assembled and incorporated into particular proteins.

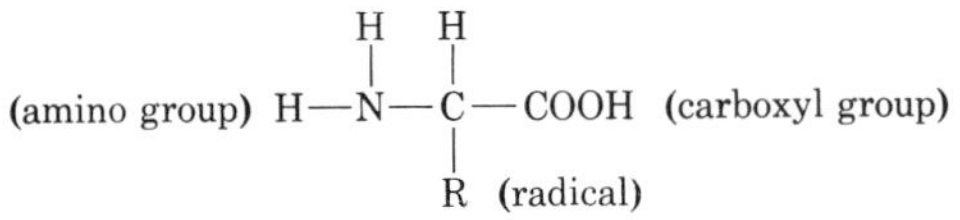

$$\text{(amino group)} \quad H-\underset{\underset{R \;\; \text{(radical)}}{|}}{\overset{\overset{H \quad H}{|\quad\;|}}{N-C}}-COOH \quad \text{(carboxyl group)}$$

Figure 3.21 Amino acid, basic structure on which proteins are built.

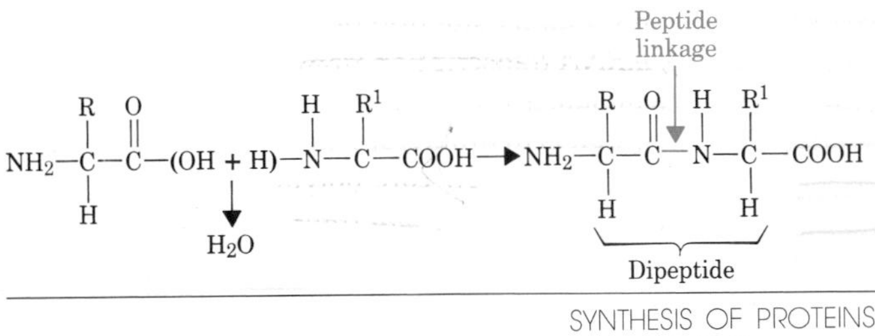

Peptide linkage

$$NH_2-\underset{\underset{H}{|}}{\overset{\overset{R}{|}}{C}}-\overset{\overset{O}{||}}{C}-(OH + H)-\underset{}{\overset{\overset{H}{|}}{N}}-\overset{\overset{R^1}{|}}{C}-COOH \longrightarrow NH_2-\underset{\underset{H}{|}}{\overset{\overset{R}{|}}{C}}-\overset{\overset{O}{||}}{C}-\underset{\underset{H}{|}}{\overset{\overset{H}{|}}{N}}-\overset{\overset{R^1}{|}}{C}-COOH$$

$$H_2O$$

Dipeptide

Figure 3.22 Peptide linkage joining amino acids in formation of proteins. Synthetase enzymes catalyze the joining process.

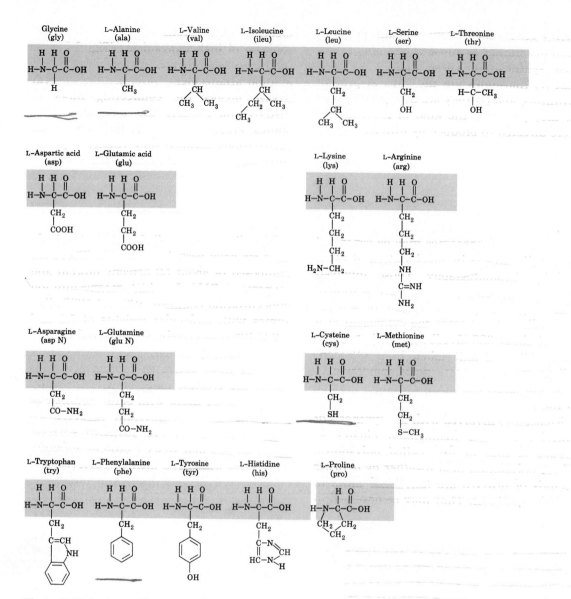

Figure 3.23 Amino acids commonly found in proteins.

RIBOSOMES, CENTERS OF PROTEIN SYNTHESIS

In early *in vitro* studies of protein synthesis, A. E. Mirsky and others showed that while a small amount of protein is synthesized in the nucleus, most of it is manufactured in the cytoplasm. Continuing these studies, Zamecnik and others incorporated radioactive amino acids into mammalian cell systems that were accumulating protein and showed that the cytoplasmic ribosomes became radioactive. Later, ribosomes from *E. coli* were brought together in a test tube with mRNA from a phage, an energy-generating system, and amino acids. Proteins similar to those of the phage coat were synthesized. These studies indicated that (1) **ribosomes** are the sites of protein synthesis, and (2) they are **nonspecific**; that is, they produced the kind of protein they were directed to produce by mRNA.

STRUCTURE AND ACTIVITY OF RIBOSOMES

Ribosomes are cytoplasmic bodies composed of protein and ribosomal RNA (rRNA). Single ribosomes may be separated from cell extracts by precipitation at pH 5 and by chromatography. Those involved with hemoglobin synthesis have a molecular weight or particle weight of 4×10^6 and sediment out from the cytoplasm at a rate of about 80S (Svedberg units). A Svedberg unit is a unit taken from ultracentrifuge measurements and is indicative of size. The larger the value, the larger the size of the particles.

Ribosomes involved with the synthesis of proteins such as hemoglobin cluster in groups of four or five, called **polysomes.** These polysomes, held together by delicate strands of mRNA, are functional units in protein synthesis. They may be observed in electron micrographs of the cellular cytoplasm of higher organisms (Fig. 3.24). In liver cells of adult rats the polysomes have a spiral arrangement. Brief treatment with **RNase breaks the connecting links** of mRNA and releases the individual ribosomes. Polysomes are not observed in all preparations because they readily separate into single ribosomes during the usual laboratory methods or preparation for observation. They can be obtained intact after very gentle lysis and sucrose-density gradient centrifugation.

The ribosomes studied most extensively thus far have been from *E. coli.* These are composed of about 60 percent protein and 40 percent rRNA. In a high magnesium ion concentration (1.0mM), 70S and 100S ribosomes with molecular weights of 2.8 and 5.9×10^6, respectively, are observed in electron micrographs. The *E. coli* ribosomes that are most active in protein synthesis under the greatest range of environmental conditions are those with **sedimentation of 70S.** When the magnesium is decreased below 1.0mM, the 70S particles separate into 50S and 30S RNA + protein subunits (Fig. 3.25), which can then be observed separately. The RNA units, while intact, are most efficient in performing their two im-

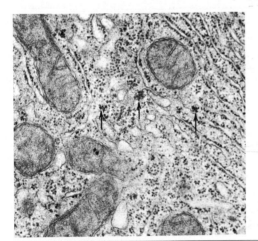

Figure 3.24 Electron micrograph of a rat liver cell showing clusters of ribosomes or polysomes (see arrows) in cytoplasm. Magnification ×35,000 (Courtesy of R. L. Wood.)

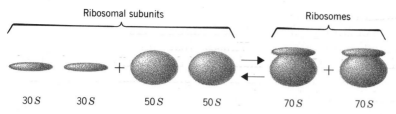

Ribosomal subunits   Ribosomes

30 *S*   30 *S*   50 *S*   50 *S*   70 *S*   70 *S*

Figure 3.25 Appearance of *E. coli* ribosomes based on electron microscope studies. The sedimentation coefficient is given below each particle. The dimensions of the discs (30S particles) are approximately 150 to 180 angstroms long, the 50S particles are approximately 140 to 160 angstroms in diameter, and the 70 S particles have dimensions of about 160 × 130 angstroms, being somewhat compressed in the vertical dimension. (After studies by Tissières et al.; Hall and Slayter; Huxley and Zubay.)

portant functions: (1) **reading the mRNA transcription** "tape" carrying the coded message from DNA; and (2) **combining with transfer RNA,** which brings amino acids to the ribosomal assembly plant.

POSITIONS AND ORIGIN OF RIBOSOMES IN CELLS

The *E. coli* ribosomes that are active in protein synthesis are dispersed widely throughout the entire cell. This is also true of other microbial systems that have been studied. Higher organisms that have a distinct separation between nucleus and cytoplasm have ribosomes that are mostly **cytoplasmic,** even though the ribosomes originate in the nucleus. In mammalian reticulocytes, which are involved in the synthesis of hemoglobin, for example, ribosomes are abundant in the cytoplasm. These ribosomes are composed of nearly

half RNA; the remainder is protein. Three species of RNA and about 30 ribosomal proteins have been identified in ribosomes. These make up the structural framework of the ribosome and function as enzymes. Ribonuclease is one of the ribosomal enzymes. Ribosomes have also been identified in the cytoplasm of many other higher animals and several higher plants such as pea seedlings, maize root tips, and ripening pears. With electron micrographs, ribosomal bodies may be observed in the **endoplasmic reticulum** of some cells. When the endoplasmic reticulum is ruptured, the ribosomes are suspended freely in the cytoplasm.

The origin of ribosomes in the cell is under **control of the genes.** About 0.3 percent of the DNA of the several different organisms that have been studied has been shown to carry the information required for the synthesis of ribosomal RNA. F. M. Ritossa and S. Spiegelman and their associates found that the **nucleolar organizing regions** on the X and Y chromosomes of *Drosophila melanogaster* included the DNA areas responsible for **rRNA synthesis.** The rRNA synthesis region is flanked on either side by "heterochromatin," a state of chromatin (as compared with "euchromatin") which stains differently and is not usually resolved into conventional genes. About 85 percent of the total cellular RNA is rRNA.

Nucleolar organizer regions and the nucleoli have a significant role in the life of the organism. The nucleolus may be the initial repository for newly synthesized rRNA in the cell. Data on this subject have come from studies of a mutation in the clawed toad *Xenopus laevis.* Normally, Xenopus cell nuclei have two nucleoli, but in the presence of a mutation (anucleolate) in heterozygous condition the nuclei carry only one nucleolus. Cells of the homozygote have no nucleoli and the toad dies in the tadpole stage. No ribosomal RNA is present in homozygous anucleolate toad embryos whereas the other kinds of RNA are in normal amounts. Homozygotes contain only the original rRNA of the egg. Synthesis of rRNA, which usually begins in the gastrula stage, is absent and the organism dies when the supply of rRNA from the egg is depleted.

VIRUSES AND HOST RIBOSOMES

Viruses that invade other organisms make use of the **host ribosomes** for synthesis of their protein coats. In one investigation, a virus that infected *E. coli* was found to have a unique component. The host DNA was composed of adenine, thymine, guanine, and cytosine, but the DNA of the virus was composed of adenine, thymine, guanine, and hydroxymethylcytosine. The viral DNA could thus be distinguished readily from the *E. coli* DNA, and it provided a tool for studying the process by which new viruses were synthesized in bacterial cells. In subsequent experiments, bacteria were infected with viral DNA. The viral DNA was replicated by using the facilities of the host bacteria in the regular replication process. Viral mRNA, which was required to specify the enzymes necessary to make hydroxymethylcytosine, was also produced. It moved to the host-cell ribosomes where it provided the information for enzyme synthesis. At least 30 different enzymes were involved in the reactions. **Viral proteins** were produced in the host under the **direction of viral DNA acting through viral mRNA.** Each amino acid was incorporated at the proper time into the protein-synthesizing process.

TRANSLATION OF GENETIC INFORMATION

Translation is the formation of a protein directed by a specific mRNA molecule. The information from nuclear DNA is translated on the ribosomes during protein synthesis. A crucial step in protein synthesis is that of **bringing to the ribosomes** the necessary **amino acids,** which are dispersed in the cell. In man, some of these amino acids are synthesized in the cell, but others ("essential" amino acids synthesized in the bodies of other animals or plants) must be supplied in the diet. The transfer of amino acids to ribosomes is accomplished through a kind of RNA called **transfer RNA** (tRNA).

As predicted by Crick in 1958, each amino acid is carried to the mRNA template by an adapter molecule containing nucleotides that fit the mRNA template. An amino acid to be incorporated into a polypeptide is brought to its position in the protein assembly line on a ribosome by a specific tRNA molecule (Fig. 3.26). The mol-

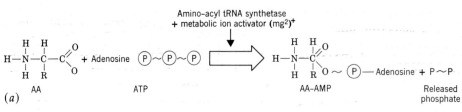

Amino–acyl tRNA synthetase
+ metabolic ion activator (mg²)⁺

(a) AA   ATP   AA–AMP   Released
                                                            phosphate

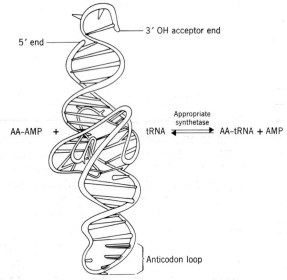

3′ OH acceptor end

5′ end

AA–AMP + ____ tRNA ⇌ AA–tRNA + AMP

Appropriate
synthetase

Anticodon loop

(b)

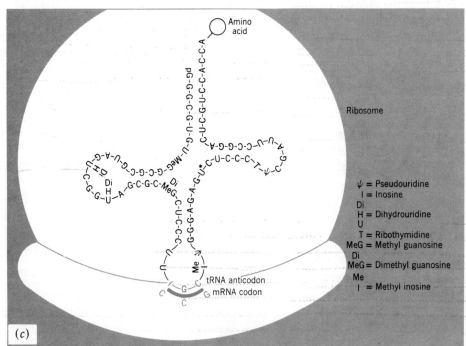

Amino
acid

Ribosome

ψ = Pseudouridine
I = Inosine
Di
H = Dihydrouridine
U
T = Ribothymidine
MeG = Methyl guanosine
Di
MeG = Dimethyl guanosine
Me
I = Methyl inosine

tRNA anticodon
mRNA codon

(c)

ecule is single stranded, like mRNA, but much smaller. It is a selecting and transporting agent. A specific tRNA exists for each amino acid. Molecules of tRNA are synthesized in the cell on the DNA template. They are transcripts of tRNA genes adapted for specific and immediate functions which are: (1) that of **seeking** out **particular amino acid** molecules from the intracellular amino acid pool, and (2) that of **attaching** themselves to a **ribosome** in a position specified by the mRNA template.

Activation of amino acids for attachment to tRNA is accomplished by a reaction that combines an amino acid (AA) with adenosine triphosphate (ATP) (Fig. 3.26a). ATP is an energy-rich molecule that furnishes energy for many reactions by losing one or two of its three phosphate groups. The reaction is catalyzed by a specific enzyme that combines a particular amino acid with ATP to form AA-AMP (adenosine monophosphate). In a separate, second step, this enzyme then transfers AA-AMP to a tRNA forming AA-tRNA and free AMP. There are more than 20 amino-acyl-tRNA synthetases that perform these functions. A metallic ion activator, for example, magnesium (Mg^{++}), is also involved in the reaction. One or more types of amino-acyl-tRNA synthetases are available in the cell for each of the amino acids commonly found in proteins; that is, some are specific for phenylalanine, others for leucine and so on. A free end of a particular tRNA molecule attaches to a specific activated amino acid (Fig. 3.26b). The amino acid complex remains bound to the amino-acyl-tRNA synthetase, which transfers it to the tRNA. These enzymes have **two different binding sites** and are thus specific for both the **amino acid** and the **tRNA amino-acid receptor site.**

Another receptor, the **anticodon**, located on the tRNA molecule binds to a unit of three bases called a **codon** on the mRNA strand (Fig. 3.26c). The tRNA molecules bring particular amino acids to the assembly site on the ribosome-mRNA complex, where each is connected by a peptide bond to the nascent polypeptide or protein. The amino acid corresponding to the beginning of the mRNA "message" is inserted along with its tRNA adapter into an opening in the 50S part of the ribosome-mRNA complex, called the binding site (Fig. 3.27). The place where the amino acids are assembled into polypeptides is called the growing site of the ribosome. When a tRNA unit has **released its amino acid** the ribosome moves along on the mRNA to the next base triplet where another tRNA with its amino acid is already attached. From its initial attachment at the binding site, the tRNA-amino acid unit is moved to the growing site. The previous tRNA leaves the ribosome and is free in the cytoplasm **to become attached to another activated amino acid** of the same kind.

The second amino acid is then joined

Figure 3.26 Illustration of activation of amino acid, attachment of activated amino acid to tRNA and the base sequence of tRNA with attachment regions for amino acid and mRNA codon. (a) Activation of an amino acid by ATP, (b) the attachment of the activated amino acid to the tRNA with an appropriate synthetase (enzyme), and (c) the transfer of the activated amino acid by the tRNA anticodon to the appropriate mRNA codon strand being held on a ribosome. (Adapted from R. W. Holley et al, *Science* 147: 1462–1465, 1965. Copyright 1965 by the American Association for the Advancement of Science.)

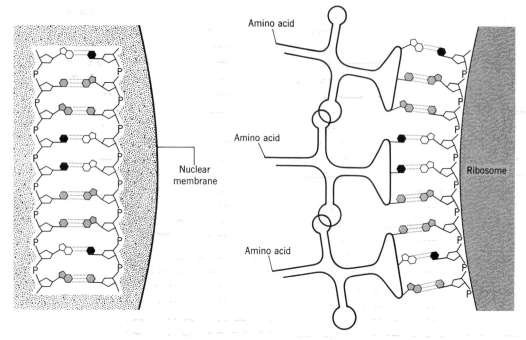

Figure 3.27 Scheme for protein synthesis in a eukaryote cell. A single strand of DNA (far left) forms a template for mRNA. The mRNA then passes through the nuclear membrane (curved black line) and attaches to one or more ribosomes (far right) in the cytoplasm. This mRNA serves as a template for the assembling of a chain of amino acids. To do this, tRNA, reads the code in the mRNA and attaches to the appropriate amino acid. The tRNA's, with their amino acids, pair with the codons of the mRNA. The amino acids will bond together to form a chain that will eventually become a protein molecule. The tRNA will then become detached from the amino acids.

with the one preceding by a peptide bond and the tRNA unit is released from the ribosome as a new tRNA-amino acid complex is attached to the binding site. Two tRNA units are thus attached to a particular ribosome at the same time, one in the growing site and the other at the binding site. When the first is released, the second is moved into the growing site and another tRNA with its activated amino acid moves into the binding site (Fig. 3.28).

Translation of genetic information into protein synthesis is accomplished when a tRNA picks up an amino acid and places it in the **position specified** by the mRNA codon in an mRNA-ribosome system. Stepwise growth of the polypeptide chain is paralleled by the concomitant movement of the translation apparatus along the mRNA. In effect, the ribosome and its attendant protein function as a computer that **translates the nucleotide** sequences of DNA into **protein.**

It should be emphasized that the sequence in which the amino acids must be assembled to form a particular protein is specified by the triplet bases of mRNA and not the ribosomes or the tRNA. One dem-

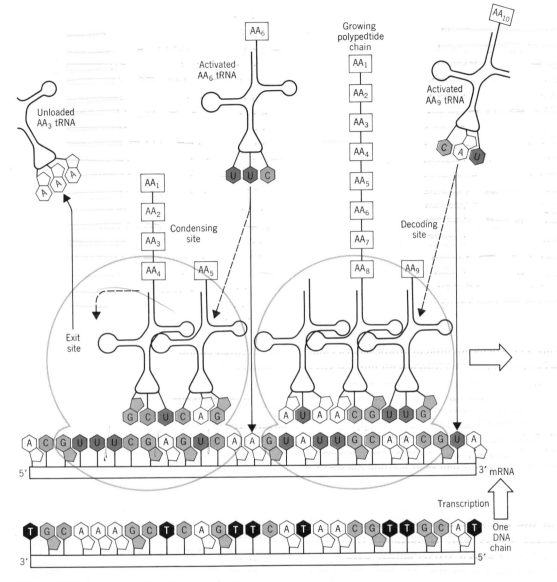

Figure 3.28 A scheme of protein synthesis.

onstration that the specificity is in the mRNA was made by John Knowland when he injected mRNA from tobacco mosaic virus into amphibian oöcytes and showed that the message of the mRNA donor was translated. Tobacco mosaic virus protein was translated in an amphibian cell.

NUCLEOTIDE SEQUENCE

Transfer RNA molecules appear superficially to be double helices somewhat similar to the double helix of DNA. X-ray diffraction studies indicate that part of the molecule is helical. **Only one strand is present,** however, and the apparent doubling occurs as a result of folding. Complete nucleotide sequences have been determined for several tRNA's. An alanine tRNA isolated from yeast and analyzed by 1968 Nobel prize winner, R. W. Holley and his associates, has short, folded, and therefore double-stranded regions. The molecule is considerably smaller than that of either mRNA or DNA, containing only 77 nucleotides compared with several hundred to many thousands in mRNA and DNA. One part of the tRNA molecule binds with an amino acid while another part contains an anticodon for pairing with a particular part of the mRNA transcript.

Following Holley's successful analysis of the nucleotide sequence of alanine tRNA, D. R. Mills, F. R. Kramer and S. Spiegelman determined the complete nucleotide sequence of a replicating RNA molecule (MDV-1). This is a very small virus containing only 218 nucleotides. A molecule with completely known sequence may be used to: (1) determine the way in which an enzyme (RNA polymerase) selects molecules for replication, (2) investigate the details of replication mechanism, and (3) determine what base changes occur when mutations change one replicating form to another.

H. G. Khorana (1968 Nobel prize winner), and his associates have **synthesized** *in vitro* the **DNA nucleotide sequence** (the gene) for the alanine tRNA in yeast. When Khorana began this project the base sequence for tRNA had recently been established by R. W. Holley. Since the sequences of the DNA gene and the tRNA are complementary, knowledge of one provides a pattern for the other. Khorana started with the four nucleotide bases, A, T, G, and C, which he chemically joined in proper sequence to form single-stranded segments of the molecule. These segments were enzymatically joined to make the complete 77 nucleotide molecule. To do this, Khorana lined up

single-stranded segments opposite each other. Of these, one segment was always several nucleotides longer than its counterpart so that at the end of each paired structure an unpaired end would protrude. This end would have complementary bases on a single-stranded terminal section corresponding to the next segment. The **ends were then joined** by the enzyme **DNA ligase.** By this procedure of joining segments together, the three sections (*a, b,* and *c*) of the molecule were constructed and these were finally joined together as shown in Fig. 3.29.

It was possible to check the sequence of the nucleotides and to demonstrate that all were joined together in the correct order. The artificially synthesized alanine tRNA gene, however, while structurally correct, was nonfunctional both in cells and in test tube experiments because it did not contain the start and stop signals that initiate and regulate the synthesis of alanine tRNA. It has been impossible to test this gene further for biological activity because yeast cells already produce tRNA.

The second artificial gene synthesized by Khorana and his associates is a 126 nucleotide DNA fragment that codes for the production of tyrosine tRNA in *E. coli.* The investigators synthesized two complementary 126 unit polynucleotides to form a **two-stranded helical gene.** This was accomplished by first synthesizing short segments (10 to 14 nucleotides) of each strand. When complementary segments of unequal length were placed in solution, they formed a two stranded complex with the longer strand extending from one end. This short segment of single-stranded DNA then acts as a template for binding the adjacent segment of the complementary strand, which can be linked to the first segment with a DNA ligase. Part of the newly added segment then extends as a single strand to serve as a template for the addition of an-

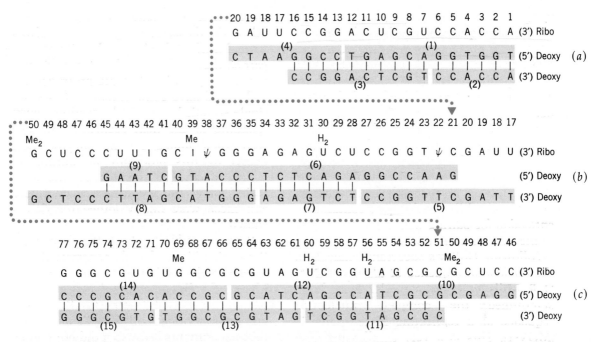

Figure 3.29 The full 77 nucleotide sequence of the gene which Khorana synthesized is shown under the line of letters representing transfer RNA bases. The gene was synthesized in three separate double-stranded segments which were later joined end to end. (After Khorana.)

other segment. In this way four major subunits were joined together (as in the earlier synthesis of the yeast alanine tRNA gene) to produce the tyrosine tRNA gene. An *E. coli* mutant was obtained with a defect in the stop signal of the tyrosine tRNA gene. This mutant acts as a suppressor of tRNA which can be overcome by introducing the active gene through the agency of a bacteriophage (Chapter 6). The mutant strain thus provides a means of determining whether the artificial gene is functional. This is a readymade test for biological activity. **Start and stop sequences** are being synthesized so this test can be accomplished.

The translation of genetic information is accomplished when appropriate amino acids have been assembled in proper order and joined by peptide bonds to form peptides. These are then united into proteins, which make up enzymes. In turn, **enzymes control** virtually all of the numerous chemical processes carried on in living systems. Genes therefore **act by synthesizing proteins,** and proteins represent the all-important chemicals through which inherited traits are expressed.

THE GENETIC CODE

The Watson and Crick model of DNA, the transcription mechanism from DNA to mRNA, and the synthesis of proteins in the ribosomes indicated that **nucleotides corresponded with amino acids in a linear arrangement.** It was postulated that the linear sequence of amino acids in a protein was specified by a code composed of a linear sequence of nucleotides in a gene.

The first step in understanding the code was to determine the kind of "language" in which the information was coded. How can four organic bases carry the information required for the synthesis of particular proteins in living systems? The "letters" in the "language" were presumed from the earlier studies to be the bases, the "words" were groups of bases or codons, each specifying one amino acid–tRNA complex.

DECIPHERING THE CODE

M. W. Nirenberg (1968 Nobel prize recipient), S. Ochoa (1959 Nobel prize winner) and others were involved in an exciting investigation designed to decipher the genetic code. Experiments with *E. coli* demonstrated an enzyme source that would promote the synthesis of proteins in a test tube. In another significant technical advance, different kinds of mRNA-like polymers were synthesized. The synthetic **mRNA** could then be **tested to see what kind of polypeptide it would specify.**

Nirenberg and J. H. Matthaei reported the production of a synthetic mRNA, polyuridylic acid (poly-U), consisting of molecules with only one base, uracil. This was accomplished with the aid of an enzyme, polynucleotide phosphorylase, that links ribonucleotides together. The investigators added the synthesized mRNA to a cell-free extract containing protein-synthesizing enzymes and ribosomes from *E. coli* together with a mixture of the 20 common amino acids.

To make such a cell-free extract, the investigators allowed the bacteria to grow rapidly by providing a suitable nutrient medium and optimal temperature conditions. When large numbers of bacteria had been produced, they were harvested by cen-

trifugation. Bacterial cells were then broken by gently grinding them in a mortar with a pestle. Cell contents containing all the enzymes necessary for protein synthesis, energy sources, and other necessary factors were prepared in a test tube.

When poly-U was added to such a cell-free extract, the only molecules synthesized were polyphenylalanine. The incorporation process was traced by repeated experiments, in each of which a **single amino acid** was labeled with radioactive carbon (C^{14}) as a tracer. When the mixtures from the different experiments were checked for the fate of the labeled amino acids, **only** radioactive **phenylalanine was incorporated** into macromolecular form. All other radioactive amino acids remained as free amino acids. Peptide linkages had joined units of this amino acid, in sequence, making polyphenylalanine. The base sequence, uracil-uracil-uracil (UUU), in mRNA base sequence was a transcription of the complementary DNA sequence, adenine-adenine-adenine (triplet AAA). Thus, the **DNA codon** for one amino acid (phenylalanine) had been **discovered.** It consisted of three bases (AAA) and was transcribed in the mRNA by three complementary bases (UUU).

The next obvious step was to try to similarly determine the code symbols for other amino acids. Progress was slow at first because it was necessary to prepare a fresh cell-free extract for each experiment and to synthesize fresh mRNA. Later, methods were devised by which the enzyme extracts could be stored for long periods of time without loss of activity, and soon agencies were established to prepare synthetic mRNAs for investigators which speeded up the experimental procedures.

In other experiments, polycytidylic acid (poly-C) was introduced into a cell-free system, and the amino acid, proline, was incorporated into a proteinlike substance. Synthesis was discontinued in the system

TABLE 3.2

The genetic code as represented by sixty-four mRNA codons, sixty-one of which code for amino acids and three for chain termination. Three of those coding for amino acids are also identified as initiators.

Second letter

First letter	U	C	A	G	Third letter
U	UUU ⎫ 　　　⎬ Phe UUC ⎭ UUA ⎫ 　　　⎬ Leu UUG ⎭	UCU ⎫ UCC 　　　⎬ Ser UCA UCG ⎭	UAU ⎫ 　　　⎬ Tyr UAC ⎭ UAA Ochre 　　　(terminator) UAG Amber 　　　(terminator)	UGU ⎫ 　　　⎬ Cys UGC ⎭ UGA (terminator) UGG Tryp	U C A G
C	CUU ⎫ CUC 　　　⎬ Leu CUA CUG ⎭	CCU ⎫ CCC 　　　⎬ Pro CCA CCG ⎭	CAU ⎫ 　　　⎬ His CAC ⎭ CAA ⎫ 　　　⎬ GluN CAG ⎭	CGU ⎫ CGC 　　　⎬ Arg CGA CGG ⎭	U C A G
A	AUU ⎫ AUC ⎬ Ileu AUA ⎭ AUG Met 　　　(initiator)	ACU ⎫ ACC 　　　⎬ Thr ACA ACG ⎭	AAU ⎫ 　　　⎬ AspN AAC ⎭ AAA ⎫ 　　　⎬ Lys AAG ⎭	AGU ⎫ 　　　⎬ Ser AGC ⎭ AGA ⎫ 　　　⎬ Arg AGG ⎭	U C A G
G	GUU ⎫ GUC 　　　⎬ Val GUA (initiator) GUG (initiator)	GCU ⎫ GCC 　　　⎬ Ala GCA GCG ⎭	GAU ⎫ 　　　⎬ Asp GAC ⎭ GAA ⎫ 　　　⎬ Glu GAG ⎭	GGU ⎫ GGC 　　　⎬ Gly GGA GGG ⎭	U C A G

when the mRNA was depleted (no DNA was present as a template for making more mRNA). When an appropriate mRNA was added to the system, protein synthesis was resumed. This provided a **stop-and-go system for** testing more **synthetic mRNAs.**

By following the same procedures with improved experimental designs and different "code words," Nirenberg and others determined the amino acid message for some 50 nucleotide sequences. These studies and others demonstrated significant properties of the code. (1) **Codons are linear** but do not overlap, that is, any one base is a part of only one codon. (2) The **message is continuous** ("commaless"), without breaks or pauses between codons. (3) The **three-letter code** is always read from the beginning of a gene. (4) The **code sequence corresponds with the amino acid sequence,** the polypeptide chain being synthesized from the free amino (NH$_2$) end rather than from the carboxyl (COOH) end of the peptide chain. Eventually other types of experiments were designed that gave information about the base order in the codons. Of the 64 possible combinations of the four DNA bases in groups of three (codons), 61 specify amino acids (Table 3.2).

PUNCTUATION FOR CODED MESSAGE

Base sequences, UAA, UAG, and UGA that do not code for any amino acid were at first considered to be "nonsense" triplets, with no function in protein synthesis. Further study has shown that they are termination codons providing stop signals or spacers at the cistron level of the coded message. Spacers do not occur between codons. The message at that level is continuous from triplet to triplet.

In one early experiment, proteins from *E. coli* were found to contain characteristic NH$_2$-terminal amino acids, predominately methionine and valine. This suggested that the codons for methionine or valine might be **initiation codons** for protein synthesis. Methionine was also found to be the NH$_2$-terminal amino acid for the *E. coli* host-specific bacteriophages. Formylmethionine was found to be the amino-terminal group of the coat proteins of f$_2$ and R17 bacteriophages produced in cell-free systems. For example, an mRNA triplet, **AUG,** which specifies the incorporation of N-formylmethionyl-tRNA has been identified as the **start signal** for synthesis of a virus-coat protein. In this chain initiation, the amino group of methionine is blocked by formylation and cannot be involved in peptide bond formation. GUG and GUA may serve as initiator codons as well as specifying valine.

REDUNDANCY OF THE DNA CODE AND ITS SIGNIFICANCE

The above experiments not only showed how each code word is "spelled" and how the message is punctuated but also demonstrated **redundancy.** More than one code triplet was found to code for most of the amino acids. If the code were not redundant (repeated) and each of the 20 amino acids was specified by a single unique codon, the remaining 44 base combinations would not specify amino acids. In such a situation, most one-step mutations would lead to triplets that would not specify amino acids. As a result, many unfinished polypeptide chains would be expressed in the organisms as defects.

A redundant code, on the other hand, with one to six codons specifying each amino acid and only a minimum number of codons not specifying amino acids would **minimize** mutant phenotypes. If triplets corresponding to the same amino acid shared at least one base in common, one-step mutations would change the specified amino acid as infrequently as possible. This is the existing pattern (Table 3.2). Leucine, for example, can be coded by any one of six different codons. A high proportion of codons for the same amino acid (synonyms) differ in only one base. Mutations that result in no amino acid substitution (silent mutations) may occur in synonyms with shared doublets. If, for example, a C, G, or A in the third position of three codons for threonine should mutate to U, the result would be ACU, another codon for threonine.

Redundancy, as used here, does not imply lack of specificity in protein synthesis. It merely means that a particular amino acid can be directed to its place in the peptide chain by more than one base triplet. Redundancy has apparently become established through natural selection as a mechanism to help stabilize phenotypes by lessening the effects of random mutation. It also minimizes potential consequences of base-pairing errors occurring in the transcription and translation of the DNA information. A code that can buffer the effects of mutations and noninherited errors (for instance, those produced by drugs) increases the reliability of the entire system for gene expression and thus has a **selective advantage.**

More recent investigations by Nirenberg and others have provided further evidence for the triplet nature of the code and have shown that the sequence of the three bases in a codon is a significant factor in specifying amino acids. **The first and second bases in a triplet are more important than the third** in distinguishing amino acids. In one series of experiments, the sequence of bases within the codons for phenylalanine, serine, and proline was determined. This was accomplished by following the attachment of C^{14} labeled amino acid-tRNA complex with chemically defined trinucleotides to the ribosomes. Both UUC and UUU codons were found to specify phenylalanine, UCC and UCU serine, and CCC and

CCU proline. A general pattern of redundancy was proposed in which identical bases in the first and second positions and U or C in the third position serve as codons for the same amino acid. The same relation has been indicated for A and G in the third position; that is, identical bases in the first two positions specify the same amino acid when either A or G is in the third position.

UNIVERSAL GENETIC CODE

A code would be universal if the same base triplet codes the same amino acid in all organisms. Transformation experiments indicated a genetic material in pneumococcus and other bacteria. The Hershey and Chase, as well as other experiments showed that DNA specifies hereditary traits in certain viruses. Other studies and logical extrapolations have indicated that DNA has widespread and nearly **universal significance** as the genetic material. DNA is a constant part of the eukaryote cell and it has the properties that are expected to characterize genes. By means of a specific staining technique (Feulgen reaction), DNA was located in the chromosomes in interphase cells as well as in those undergoing the division process. Furthermore, it eventually was shown that DNA is confined to the chromosomes and certain cytoplasmic bodies, that is, chloroplastids and mitochondria (Chapter 11). The concentration of DNA in any given eukaryote cell is comparatively uniform and constant. This was demonstrated when radioactive phosphorus (P^{32}) was incorporated into the DNA of dividing cells. The cells with the labeled DNA were then maintained as nondividing cells and checked periodically for changes in their amounts of DNA. Very little change was detected. This indicated that the amount of DNA remains constant and does not respond to the metabolic activities of the cell.

Figure 3.30 Electron micrograph of tobacco mosaic virus particles. Magnification ×18,200 (Courtesy of Carl Zeiss Company.)

The genetic code carried by RNA has been investigated in the tobacco mosaic virus (TMV) by H. G. Wittmann, H. Fraenkel-Conrat, and several others. These investigators have shown that the **same code** that has been discovered for bacteriophage, bacteria, and mammalian hemoglobin, also applies to TMV (Fig. 3.30). The RNA of the TMV particle also acts as mRNA and thus directs the protein synthesis of progeny viruses.

Several investigators have studied the amino acid replacements that result from single base changes in TMV RNA. The experiments showed that TMV protein was a complex of about 2200 single peptide chains, each containing 158 amino acids.

Fourteen different amino acids were identified. The TMV studies supported those conducted with DNA organisms and they reiterated that single base changes can cause amino acid substitutions. The TMV studies thus provide further evidence for a universal code. Samples from widely different groups of living organisms tend to indicate such universality.

COLINEARITY OF GENE AND PROTEIN STRUCTURE

The discovery that genes are units of DNA organized in linear sequence in chromosomes logically led to questions about the linear relation between nucleotide sequence and the sequence of corresponding amino acids in the protein being synthesized under specifications from the DNA units. Studies conducted soon after the discovery that DNA was genetic material indicated that specific steps could be identified in the synthesis of proteins. Crick hypothesized that DNA determines the sequence of amino acids in a polypeptide and that, once the sequence is determined, the structure of the protein to be synthesized by the DNA unit is determined. Several investigators tested this hypothesis, which requires linearity of both the sequences of nucleotides and corresponding amino acids in a polypeptide (protein). If the hypothesis is correct, a particular codon should determine a specific amino acid, and a mutational change in a codon should produce a change in a corresponding amino acid. That is, the nucleotide and amino acid sequences must be colinear. The sequence of nucleotide changes should correspond with amino acid substitutions specified by the mutational changes.

Results of a series of studies on the protein, tryptophan synthetase of *E. coli*, supported the colinearity hypothesis. Mutant strains of *E. coli* with alterations in the gene coding for the tryptophan synthetase A protein were accumulated. The positions of the mutations were determined by genetic mapping (Chapters 5 and 6). Polypeptide chain A includes 280 amino acids and the corresponding polynucleotide double helix includes 1680 nucleotides.

Yanofsky and his colleagues obtained 16 mutants in the polynucleotide chain that were related to amino acid substitutions in protein A. Each mutant strain could produce a particular abnormal tryptophan synthetase molecule. Furthermore, each abnormal synthetase molecule had an amino acid substitution in a particular position. Positions of altered codons (A-I) compared with positions of amino acids in the poly-

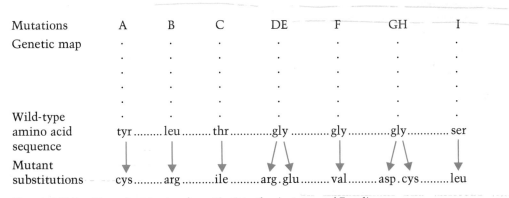

Figure 3.31 Position of mutant codons (A–I) in the A cistron of *E. coli* tryptophan synthetase and corresponding positions of amino acid substitutions in the A polypeptide chain. DE and GH are sites in the same gene (glycine in each case) but mutations result in arg or glu in DE site and asp or cys in GH site. The genetic map at the top represents the order and relative (linkage) position of the codons in the polynucleotide chain of DNA. The wild-type (normal) sequence of amino acids corresponding with codon positions on the genetic map is given below the genetic map. The amino-acid substitutions associated with the mutants are given at the bottom. The exact point-by-point relationship was demonstrated by the results of these investigations. (After C. Yanofsky et al.)

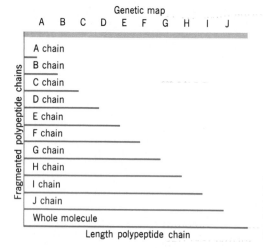

Figure 3.32 Position of mutant codon on genetic map related to length of polypeptide chain in head protein produced by different T4 amber mutants. (Data from A. Sarabhai et al.)

peptide chain A are illustrated diagrammatically in Fig. 3.31. **Amino acids corresponded in the same relative linear order with the mutationally altered nucleotides.** It was shown further that mutations with alterations extremely close to one another in the genetic material resulted in amino acid substitutions very close to one another in the protein. Some mutations which map very close together but at distinct (recombinable) sites actually are changes involving different nucleotides within the same codon and therefore result in amino acid substitutions at the same position in the polypeptide. On the other hand, distantly separated mutants produced amino acid substitutions that were widely separated from one another.

Colinearity of genetic and protein structure was also demonstrated by A. Sarabhai and associates on the head protein of bacteriophage T4. Certain phage T4 (amber) mutants grow on *E. coli* strains that have particular suppressors. They produce only partial polypeptide chains. Action of the suppressor permits short chains to be completed. Different mutants, therefore, produce polypeptide chains of different lengths. When the order of codons (A-J) was compared with termination points on the polypeptide chain (Fig. 3.32), colinearity was demonstrated.

REFERENCES

Avery, O. T., C. M. MacLeod, and M. McCarty. 1944. "Studies on the chemical nature of the substance inducing transformation in pneumococcal types." *J. Expl. Med.*, 79, 137–158.

Chargaff, E., and J. N. Davidson. 1955. *The nucleic acids.* Vol II. Academic Press, New York.

Hartman, P. E., and S. R. Suskind. 1969. *Gene action*, 2nd ed. Prentice-Hall, Englewood Cliffs, N.J.

Holley, R. W. 1966. "The nucleotide sequence of a nucleic acid." *Sci. Amer.*, 214, 30–39.

Ingram, V. M. 1965. *The biosynthesis of macromolecules.* W. A. Benjamin, New York.

Kornberg, A. 1960. "Biologic synthesis of deoxyribonucleic acid." *Science,* 131, 1503–1508.

Meselson, M. S. and F. W. Stahl. 1958. "The replication of DNA in *Escherichia coli.*" *Proc. Nat'l. Acad. Sci.* (U.S.). 44, 671–682.

Mills, D. R., F. R. Kramer, and S. Spiegelman. 1973. "Complete nucleotide sequence of a replicating RNA molecule." *Science,* 180, 916–927.

Nirenberg, M. W., and J. H. Matthaei. 1961. "The dependence of cellfree protein synthesis in *E. coli* upon naturally occurring or synthetic polyribonucleotides." *Proc. Nat'l. Acad. Sci.* (U.S.)., 47, 1588–1602.

Spirin, A. S. and L. P. Gavrilova. 1969. "The ribosome." *Molecular biology, biochemistry and biophysics.* Vol. 4. Springer-Verlag, New York.

Watson, J. D. 1970. *Molecular biology of the gene,* 2nd ed. W. A. Benjamin, New York.

Watson, J. D., and F. H. C. Crick. 1953. "A structure for deoxyribose nucleic acid." *Nature,* 171, 737–738.

Watson, J. D., and F. H. C. Crick. 1953. "Genetical implications of the structure of the deoxyribonucleic acid." *Nature,* 171, 964–967.

Wilkins, M. H. F., 1963. "Molecular configuration of nucleic acids." *Science,* 140, 941–950.

Wittmann, H. G. (ed.) 1968. *Molecular genetics.* Springer-Verlag, New York.

Woese, C. R. 1967. *The genetic code, the molecular basis for genetic expression.* Harper and Row, New York.

Yanofsky, C. 1967. "Gene structure and protein structure." *Sci. Amer.,* 216(5), 80–82.

Yčas, M. 1969. *Biological code, frontiers.* North-Holland Publ. Co., Amsterdam.

PROBLEMS AND QUESTIONS

3.1 (a) How did the transformation experiments of Griffith differ from those of Avery and associates? (b) What was the significant contribution of each? (c) Why was Griffith's work not direct proof for DNA as the genetic material whereas Avery et al. provided direct proof?

3.2 (a) How did the phenomenon of transformation support the hypothesis that DNA is the genetic material? (b) How widespread is the occurrence of transformation?

3.3 How could it be demonstrated that the mixing of heat-killed Type III pneumococcus with live Type II resulted in a transfer of genetic material from Type III to Type II rather than a restoration of viability to Type III by Type II?

3.4 How could it be demonstrated that transformation in bacteria is the result of a transfer of genetic material rather than a direct interaction of one chemical with another?

3.5 (a) What was the objective of the experiment cited, by Hershey and Chase? (b) How was the objective accomplished? (c) What is the significance of this experiment?

3.6 (a) What background material did Watson and Crick have available for developing a model of DNA? (b) What was their contribution to the building of the model?

3.7 (a) Why was a double helix chosen for the basic pattern of the molecule? (b) Why were hydrogen bonds placed in the model to connect the bases?

3.8 (a) If a virus particle contains double-stranded DNA with 200,000 base pairs, how many nucleotides would be present? (b) How many complete spirals would occur on each strand? (c) How many atoms of phosphorus would be present? (d) What would be the length of the DNA configuration in the virus?

3.9 *E. coli* cells grown on a medium with only the heavy isotope of nitrogen, ^{15}N, are labeled only with ^{15}N (a) If a twenty-fold excess of the ordinary isotope, ^{14}N is added to the medium, what relative contents of ^{15}N and ^{14}N and what arrangement in DNA strands would be expected after one generation assuming (1) conservative and (2) semiconservative replication? (b) If the F_1 bacteria were maintained in the ^{14}N medium, what would be expected in the F_2 assuming (1) conservative and (2) semiconservative replication?

3.10 If one strand or helix on the Watson-Crick model should have bases in the order GTCATGAC, what would be the order of the bases on the complementary DNA strand?

3.11 Distinguish between DNA and RNA (a) chemically, (b) functionally, and (c) locationally in the cell.

3.12 What bases on the mRNA transcript would represent the following DNA sequence: TGCAGACA?

3.13 What bases on the DNA strand would transcribe to the following mRNA strand: CUGAU?

3.14 From what evidence was the messenger RNA hypothesis established?

3.15 In a general way, describe the molecular organization of proteins and distinguish proteins from DNA, chemically and functionally. Why is the synthesis of proteins of particular concern to the geneticist?

3.16 (a) At what different locations in the cell may protein synthesis occur? (b) Evaluate the relative importance of these locations.

3.17 Characterize ribosomes in general as to size, location, function, and chemical makeup.

3.18 What methods are available for study and comparison of ribosomes?

3.19 (a) Where in the cells of higher organisms do ribosomes originate? (b) Where in the cells are ribosomes most active in protein synthesis?

3.20 Identify three different major kinds of RNA and give principal locations, characteristics, and functions of each in the living cell.

3.21 (a) How is messenger RNA related to polysome formation? (b) How does rRNA differ from mRNA and tRNA in specificity? (c) How does the tRNA molecule differ from that of DNA and mRNA in size and helical arrangement?

3.22 Outline the process of activation of amino acids.

3.23 (a) How was the genetic code first decoded? (b) What refinements have since been incorporated in the technique?

3.24 In what sense and to what extent is the genetic code (a) redundant, and (b) universal?

3.25 How was the colinearity hypothesis for nucleotides in a cistron, and amino acids in a polypeptide, supported?

3.26 Why is colinearity between codons and polypeptides significant?

3.27 What requirements are needed for DNA synthesis? What role does each requirement play?

3.28 Draw an analogy between the processes of transcription and translation with building a house.

3.29 What is the significance of Khorana's gene?

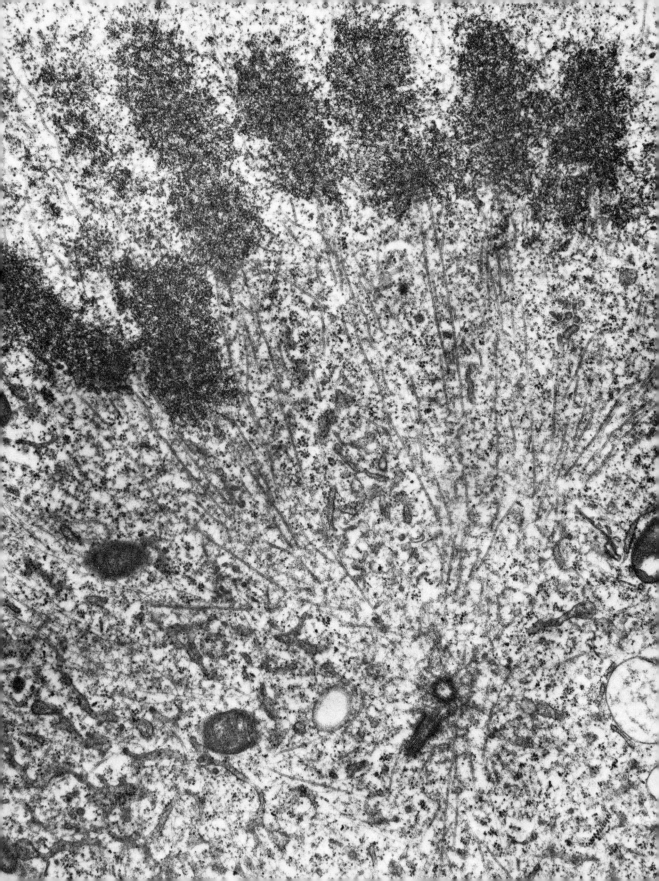

FOUR

The structural characteristics of a generalized cell are illustrated by the photomicrograph in Fig. 4.1. Although cells within a single plant or animal can vary widely in structure, shape, and function, they all represent units of living material and have some important properties in common. For example, both plant and animal cells are essentially alike in terms of having genes, chromosomes, and other factors related to inheritance. The chromosomes and their DNA are involved in the division cycles of individual cells as well as in the basic reproduction processes of the entire organism (whether animal or plant).

Cell division is the process through which cells reproduce themselves and multicellular organisms grow. When cells divide, each resultant part is a complete, although at first relatively small, cell. Immediately following division, the newly formed daughter cells grow rapidly, soon reaching the size of the original cell. Cell division is really **duplication** or multiplication rather than division in the usual sense. In unicellular animals, however, cell

CELL MECHANICS, SEX DETERMINATION, AND DIFFERENTIATION

Dividing cell in culture, mid-anaphase. (24,900×). (EM by Richard Wilkinson, tissue prepared by Harry Danforth.)

Figure 4.1 Electron photomicrograph of a liver parenchymal cell showing typical cell organelles; p, plasma membrane; microv., microvilli; m, mitochondrion; n, nucleus; nucleo., nucleolus; n.e., nuclear envelope; h, heterochromatin; g.e.r., granular endoplasmic reticulum; g, glycogen. Magnification ×4800. (Courtesy Dr. Robert L. Wood.)

duplication equates with reproduction, since two new individuals are formed from the original parent.

In higher organisms, **growth** occurs through cell division with subsequent enlargement and differentiation of the cells produced. Each human being, for example, began as a single fertilized cell or **zygote**, which developed eventually into more than a million billion cells. Most organisms consisting of more than a single cell grow in this way. However, a few organisms (e.g., rotifers) have a constant number of cells throughout their adult life. Some tissues within certain other animals, such as the salivary glands of the larval fruit fly, grow by increase in the size of individual cells. In most animals, however, continued growth of individual cells is rare. Cell duplication is the basic growth mechanism in almost all higher organisms.

The complex body of a multicellular organism eventually contains a variety of specialized cells. Epithelial cells have a relatively short life span; replacements must be made continuously. Those in the lining of the respiratory, digestive, and urinary tracts, for example, are replaced within a few days. Some gland cells have a life span of only a few hours. Cells in particular parts of the nervous system, on the other hand, once established do not divide, and the number remains fairly constant through the mature life of the individual.

CELL DIVISION

The details of cell reproduction were elucidated in animal cells in the latter part of the nineteenth century by Walther Flemming, and in plant cells by Edward Strasburger and several other investigators. Two interrelated processes were found to be involved: (1) **mitosis, the nuclear division;** and (2) **cytokinesis, the changes in the cytoplasm that include division of the cell proper.** Figure 4.2 illustrates these processes as they occur in plant cells.

The most fundamental part of cell division, the **replication of DNA,** apparently occurs before any changes related to the early stages of mitosis can be observed. This has been demonstrated by autoradiography in the broad bean, domestic fowl, and other plant and animal materials (see Chapter 3). Cells removed from a radioactive medium (^{3}H-thymidine) can be covered with photographic film that (on development) shows the extent of thymidine incorporation into the chromosomes. These studies have shown that DNA replication for both mitosis and meiosis occurs during the interphase before cell division begins. The essential genetic complex of each duplicated chromosome and each daughter cell is thus prepared before the cell begins its sequence of visible division stages.

The names interphase (between divisions), prophase, metaphase, anaphase, and telophase have been associated with different stages of the continuous mitotic cycle for convenience in describing the changes that occur. The prophase and telophase stages of mitosis are usually long and involved, whereas metaphase and anaphase are commonly brief.

During interphase, chromosomes appear to be thin, uncoiled, and filamentous, but in the beginning of prophase they become increasingly coiled, shortened, and more distinct as the mitotic process progresses. The pronounced shortening is accomplished as the number of coils decreases with a concomitant increase in the diameter of each coil. Eventually, in fixed and stained preparations, each whole chromosome appears to be solid and oval or rod shaped and double stranded. Each strand is a chromatid and was produced during interphase. The chromatids can be observed in late prophase in the many plant and animal materials that are favorable for chromosome observation. At late prophase, the two chromatids of each chromosome are held together at a clear constricted area called the **centromere** (or kinomere or kinetochore). This is the point of spindle fiber attachment on the

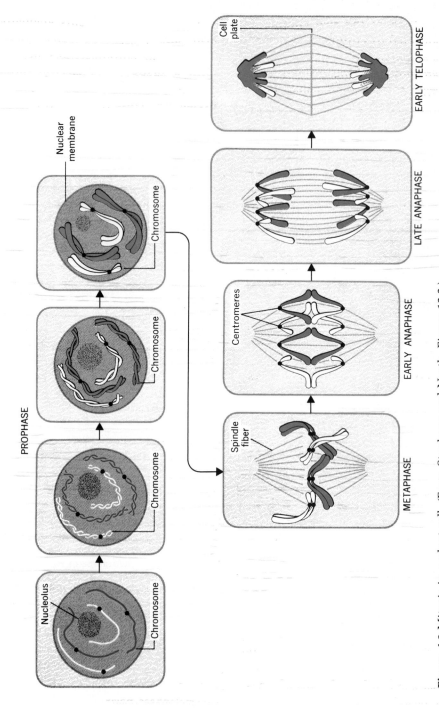

Figure 4.2 Mitosis in a plant cell. (From Stephens and North, Figure 12.3.)

chromosome. Each of the **two chromatids** becomes connected to a different pole of the **spindle** by a single fiber.

During late prophase the discrete chromosomes begin to take their places in the center or equatorial plane, the nuclear membrane and nucleolus gradually disappear and a spindle-shaped structure is formed. The chromosomes in the center form a figure called an equatorial plate. This part of the cell division cycle is called metaphase. During this stage, the chromo-

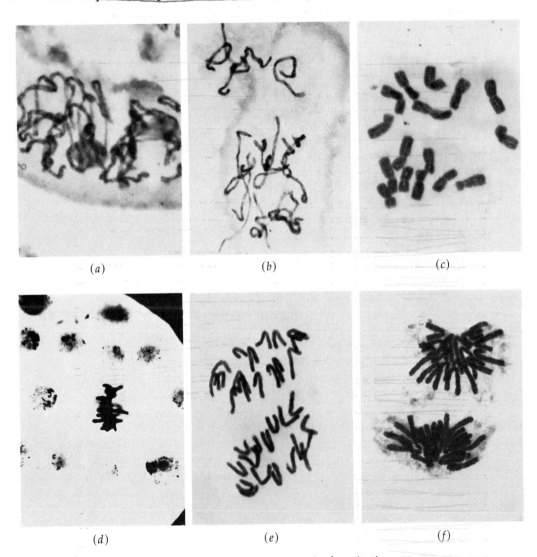

(a) (b) (c)

(d) (e) (f)

Figure 4.3 Photomicrographs representing major stages in the mitotic sequence of the onion, *Allium cepa,* root tip. This species has 16 chromosomes. (*a*) early prophase; (*b*) middle prophase; (*c*) metaphase, polar view; (*d*) metaphase, side view; (*e*) anaphase showing separate chromosomes; (*f*) telophase. All six photographs are made to the same scale ×800. (Courtesy of W. S. Boyle.)

somes are particularly discrete and tightly coiled, thus facilitating chromosome counts and gross structural comparisons.

The chromatids separate first at the centromere and ultimately along their entire length. This separation marks the beginning of anaphase. Each unit now has its own centromere and thus becomes a chromosome. The chromosomes elongate by changes in their coiling pattern and move to the respective poles of the spindle. The duplication and separation of chromatids fulfill the requirements of Roux's models (see Chapter 1).

During telophase, a nuclear membrane is reconstructed around each daughter nucleus and the nucleoli reappear. In the final stage (cytokinesis) of cell division, the cytoplasmic part of the cell divides. Animal cells, with their flexible outer layers, accomplish this by a constriction that converges from the two sides and eventually separates the two daughter cells. The surface around the equator pushes in toward the center and pinches the cell into two parts. Plant cells, with their rigid walls, form a partition or **cell plate** between the daughter cells. After the middle lamella (cell plate) is formed, walls of cellulose are deposited on either side.

In living materials, each cell division, mitosis plus cytokinesis, is a **continuous process** from the time a cell first shows evidence of beginning to divide until the two daughter cells are completely formed. The entire procedure ordinarily requires a few hours to several days, with variations dependent on the type or organism and environmental conditions. The mitotic (karyokinesis), and cytokinetic (cytoplasm division) phases are distinct but coordinated processes. The actual mitotic sequence is illustrated in Fig. 4.3.

GAMETE FORMATION IN ANIMALS

Virtually all normal cells can reproduce themselves. The sex or germ cells, however, can initiate reproduction of the **entire organism.** A special sequence (gametogenesis, the formation of haploid female and male gametes) results in the development of sex cells. **Gametogenesis** includes **meiosis** (from Greek, to reduce). During meiosis the chromosome number is changed from the **diploid or 2n** number, characteristic of body cells and premature germ cells, to the **haploid or n** number that is characteristic of mature germ cells. Gametogenesis also includes **differentiation of eggs and sperm** — a process necessary for their functioning. Eggs of animals usually accumulate nutrient materials that sustain the developing embryo for a brief period; sperm of most animal species develop a flagellum for independent motility.

SPERMATOGENESIS

Sperm originate in the male reproductive organs or **testes** through a sequence called **spermatogenesis** (Fig. 4.4). Considered grossly, the process consists of growth in cell size, **two successive cell divisions,** and a metamorphosis of the resulting cells from spherical static bodies to elongated motile sperm. Spermatogenesis is initiated in diploid (2n) or unreduced germ cells. These cells, called **spermatogonia,** enlarge and become **primary spermatocytes.** The spermatocytes then undergo the first meiotic division, with each producing two **secondary**

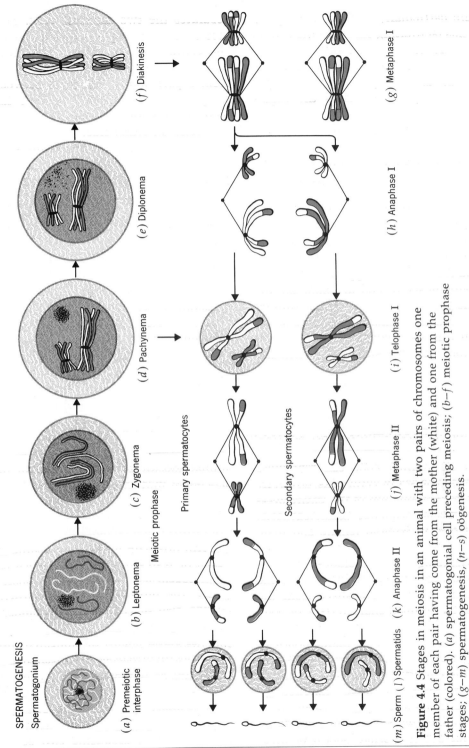

Figure 4.4 Stages in meiosis in an animal with two pairs of chromosomes one member of each pair having come from the mother (white) and one from the father (colored). (*a*) spermatogonial cell preceding meiosis; (*b*–*f*) meiotic prophase stages; (*g*–*m*) spermatogenesis; (*n*–*s*) oögenesis.

SPERMATOGENESIS
Spermatogonium

(*a*) Premeiotic interphase

(*b*) Leptonema

(*c*) Zygonema

(*d*) Pachynema

(*e*) Diplonema

(*f*) Diakinesis

Meiotic prophase

(*g*) Metaphase I

(*h*) Anaphase I

(*i*) Telophase I

Primary spermatocytes

Secondary spermatocytes

(*j*) Metaphase II

(*k*) Anaphase II

(*l*) Spermatids

(*m*) Sperm

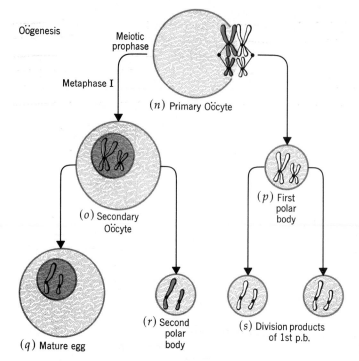

Oögenesis

Meiotic prophase

Metaphase I

(n) Primary Oöcyte

(o) Secondary Oöcyte

(p) First polar body

(q) Mature egg

(r) Second polar body

(s) Division products of 1st p.b.

Figure 4.4 (*Continued*)

spermatocytes. In turn, each secondary spermatocyte undergoes the second meiotic division to become two **spermatids.** Each spermatid then changes shape, develops a motile organelle, and becomes a **sperm.** While the cells are dividing **twice,** the chromosomes undergo intricate and significant processes, but duplicate themselves only **once.** Reduction in chromosome number is thus accomplished.

Cells of the salamander *Amphiuma means tridactylum* Cuvier (the three-toed Amphiuma) are unusually large and the chromosomes are conducive to studies of mitosis and meiosis. A series of photographs (Fig. 4.5) showing chromosomes from this salamander illustrate the actual appearance of stages in the meiotic sequence of spermatogenesis.

The chromosomes, which appear as single threads in the early prophase of the first meiotic division (*a*), represent the maternal and paternal chromosomes received by the individual (male) from the gametes of his parents. The chromosome number $(2n = 28)$ is characteristic of the species to which the three-toed Amphiuma belongs.

The sequence of developmental changes in the meiotic prophase, which precedes the first meiotic cell division, is usually long and involved. These changes occur during the spermatogonium stage. Five major **prophase stages** are distinguishable in the transition from prophase to metaphase of the first meiotic division. The preliminary stages of meiosis, in which the chromosomes appear as single thin filaments, is called the **leptotene** stage (the noun form, leptonema, identifies the visible configuration at this stage). Figure 4.5*a* shows some single leptotene threads and some that have entered the **zygotene** stage (zygonema) that follows.

The pairing process, called **synapsis** (from Greek, meaning conjunction or union), brings together maternal and paternal members of the same pair of chromo-somes (Fig. 4.5b). Pairing is accomplished while the cell (primary spermatocyte) is en-larging. Corresponding segments of particu-lar maternal and paternal chromosomes come together along their length in zip-perlike fashion. When synapsis is finished the apparent number of threads is half that in leptonema and the visible bodies in the nucleus are now bivalents rather than single chromosomes.

Following zygonema, homologous chro-mosomes can be observed side by side. This is the **pachytene** stage (pachynema) (Fig. 4.5c). The chromosomes continue to shorten, thicken, and become more distinct so that the four chromatids are apparent. Each group of four chromatids is called a tet-rad. This is the **diplotene** stage (diplonema) in which four chromatids are held together at the centromere. Paired chromosomes ap-pear to repel each other, causing the strands to separate longitudinally in some areas and to form loops (d, e).

In diplonema, the centromere in each chromosome is not split and the longi-tudinal separation of the chromosomes is incomplete. Tetrads are held together at various places along their length because of interchanges between chromatids (chias-mata). From one to several **chiasmata** may be observed in favorable preparations, de-pending on the length of the tetrad. Each chiasma observable at this stage apparently represents an exchange between nonsister chromatids (those that underwent synapsis in the zygotene stage). The point where the chiasma appears, however, is not necessar-ily where the chromatid exchange actually occurred, because chiasmata tend to slip toward the ends of the **bivalents** and thus to become terminalized as the meiotic pro-phase continues. Genetic implications of such exchanges or genetic crossovers are discussed in Chapter 5.

Shortening of the tetrads continues through the next stage (f), **diakinesis.** The result is discrete units which, in favorable preparations, can be counted and found to represent half the 2n chromosome number of the salamander or 14, as expected fol-lowing the pairing process. As the meiotic prophase is completed, the **tetrads** become angular or oval in appearance, and take their places in the equatorial plane, forming the **equatorial plate** of metaphase I (g and h).

The first of the two cellular divisions in the meiotic sequence separates the homol-ogous chromosomes (nonsister chromatids) that paired during the zygotene stage. In the anaphase of this so-called reduction divi-sion, the original maternal and paternal chromosomes (each composed of one cen-tromere and two chromatids) separate. Thus the number of chromosomes in each resultant cell is **reduced** from the original 2n to the n number. These are shown in prophase II (i) and **metaphase II** (j). If ex-changes between chromatids have occurred, both maternal and paternal chromosome parts will be present in each member of the pair.

During the secondary spermatocyte divi-sion, the centromere of each bipartite chro-mosome divides, providing each new chro-mosome with its own centromere (k). Each chromosome then moves to a pole of the spindle. The chromosome number in a sper-matid is haploid (n), the same as it is in the secondary spermatocyte. However, the chromosomes of the spermatids are unipar-tite (l) whereas those of the secondary sper-matocyte are bipartite, being composed of two chromatids. In other words, **each sper-matid** nucleus has a **single set of dissimilar chromosomes.**

The second division is a mitotic type di-vision, (called the equational division) be-cause it separates the duplicated (sister) chromatids. The division is different from the reductional division in which the ho-

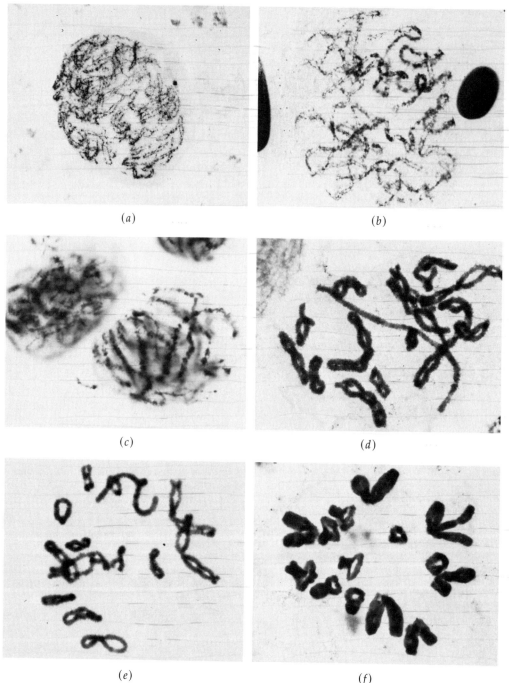

(a)

(b)

(c)

(d)

(e)

(f)

Figure 4.5 Stages in the meiotic sequence of spermatogenesis in *Amphiuma means tridactylum* Cuvier. (*a*) Leptotene-zygotene in spermatocyte with single filaments and some pairing visible. Feulgen stained. ×400. (*b*) Detail of zygotene spermatocyte showing paired strands. Feulgen stained. ×710. (*c*) Early pachytene stage showing some thin filaments which may represent asynaptic regions. Feulgen stained. ×1320. (*d*) Diplotene spermatocyte showing long arms of bivalent (in center) and chiasmata in other bivalents. ×740. (*e*) Late diplotene with 14 bivalents. ×840. (*f*) Diakinesis stage

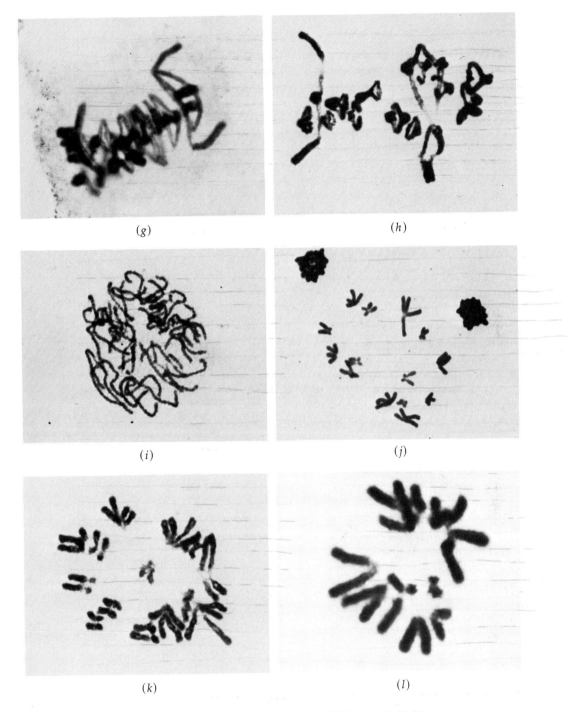

(g)

(h)

(i)

(j)

(k)

(l)

showing 14 distinct bivalents and a lack of chiasmata. ×780. (g) and (h) Metaphase I. ×880. (i) Prophase II with identifiable reduced (n) chromosomes each with 2 chromatids. ×1070. (j) Metaphase II. ×680. (k) Anaphase II showing early separation of daughter chromosomes. ×1070. (l) The haploid complement of 14 chromosomes after anaphase II separation. ×1700. (Courtesy of Grace M. Donnelly, Arnold H. Sparrow, and Robert F. Smith. Brookhaven National Laboratory.)

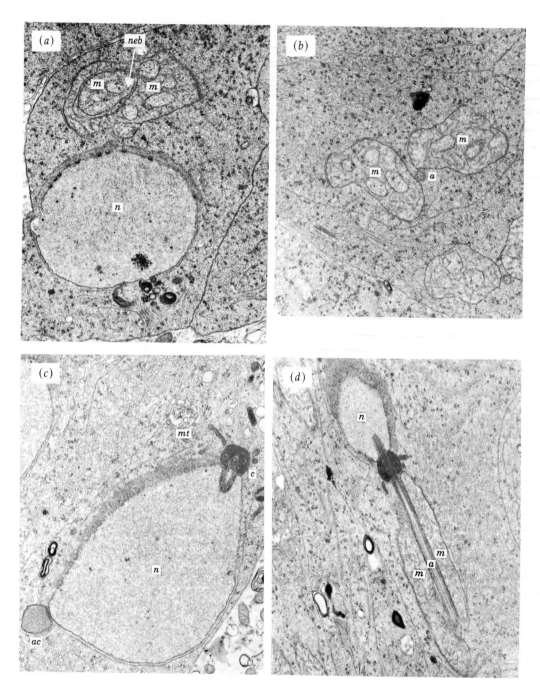

Figure 4.6 Normal sequence in spermiogenesis of *D. melanogaster*; *n*, nucleus; *neb*, nebenkern; *m*, mitochondrial derivative; *a*, axoneme; *mt*, microtubule; *c*, centriole, *ac*, acrosome. (*a*) Early spermatid prior to elongation. Nucleus is spherical and mitochondrial derivatives are established in nebenkern. ×11,800. (*b*) Cross-section through mitochondrial derivatives at the same

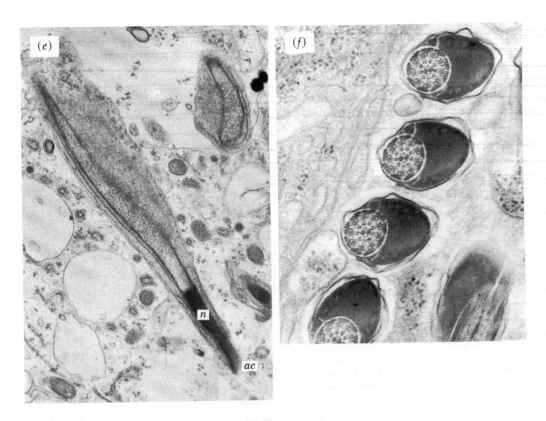

stage as *a*. The axoneme with its 9 + 2 tubule pattern is located between the two mitochondrial derivatives. ×10,500. (*c*) Spermatid beginning elongation. Acrosome is attached to the tip of the nucleus and the centriole, embedded in dense material, is attached to the nucleus. Microtubules extend anteriorly and posteriorly from the dense material. They are especially evident along the lower border of the nucleus in the region of nuclear pores. ×13,400. (*d*) Another spermatid at the same stage as the spermatid in *c*. The axoneme cut in longitudinal section lies between the two mitochondrial derivatives. ×11,800. (*e*) A later stage in elongation showing the acrosome attached to the nucleus. The chromatin within the nucleus has begun to condence. ×16,100. (*f*) Cross section through the tails of mature spermatozoa. Large primary mitochondrial derivative now filled with the paracrystalline material. The secondary derivative remains as a small dense body. In the axoneme the central and outer tubules are now cored. Nine peripheral doublets are seen with spokes directed from them toward the central pair of tubules. Cell membrane surrounds each spermatozoön. ×43,000. (Courtesy of Lynn J. Romrell.)

mologous chromosomes that came together in synapsis separate. With respect to the distribution of genes, however, neither of the two divisions can be considered to be completely reductional or completely equational, because of crossing over (see Chapter 5). If **exchanges** have occurred between chromatids in the tetrad the parts of the chromatids near the centromere will be dividing **reductionally** in the same division when parts beyond the point of exchange will be dividing **equationally.**

The terminal process in spermatogenesis is a complicated differentiation called **spermiogenesis** (Fig. 4.6). Through a progressive sequence of changes, each of the comparatively large, spherical, nonmotile spermatids is metamorphosed into a small, elongated, **motile sperm** composed typically of three parts: head, middle piece, and tail.

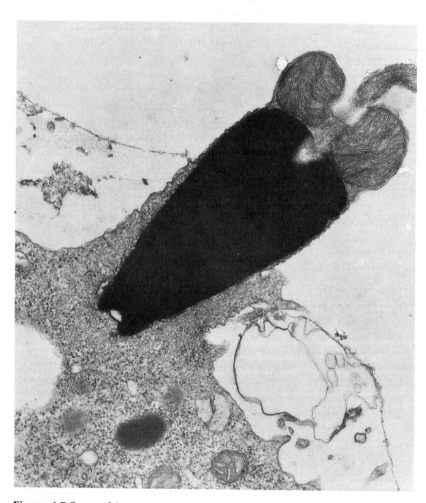

Figure 4.7 Sea urchin sperm entering egg. Magnification ×24,000. The egg cytoplasm bulges up around the sperm head forming a fertilization cone through which the sperm enters. (Courtesy of Everett Anderson, Harvard Medical School, Boston. From *J. Cell Biol.* 37:514, 1968.)

In most animal species spermatids begin differentiation by secretion of the apical body or acrosome, division of the centriole into two, and production of the flagellum by one centriole. Sloughing off of cytoplasm diminishes the overall size, as the developing sperm changes from a spherical to an oval and finally to an elongate shape. The **nucleus** moves to one edge of the cell, becoming elongated and increasingly compact. The **acrosome,** which is produced by the Golgi apparatus, takes its place around the anterior end of the sperm **head.** It contains lytic enzymes that presumably have a dissolving action that facilitates **sperm entry into ova** (Fig. 4.7).

The **middle piece** contains the centriole, which lies next to the nucleus, and the proximal part of the axial filament, which continues in the **tail.** The mitochondria of the spermatid become concentrated around the axial filament in the middle piece. The tail of the sperm is composed of two parts: the outer sheath, which is cytoplasmic in origin, and the axial filament inside the sheath, which extends from the base of the head to the posterior end of the tail. Much of the cytoplasm of the spermatid is not used in formation of the sperm and is reabsorbed, (taken up by Sertoli cells that presumably have nutritive and endocrine functions).

In most sexually mature male animals, spermatogenesis is constantly or periodically occurring in the testes and many millions of sperm are produced. Insects generally require only a few days to complete their cycle of spermatogenesis, but in mammals the cycle extends over weeks or months. In mature human males, spermatogenesis occurs in the seminiferous tubules of the testes. Spermatogonia undergoing mitotic division and continuing the population of stem cells, can be observed in cross sections of tubules at the periphery. The spermatogonia that appear to be in the innermost cell layer of the periphery enlarge and form primary spermatocytes with 23 bivalents. These spermatocytes undergo the two mechanically interwoven meiotic divisions in rapid succession and produce spermatids, which develop into sperm with single sets of 23 chromosomes. It should be noted that the sperm cells that develop in the testes of the male descend from original or primordial cells that migrated to the testes during early embryogenesis. The developmental time from primitive spermatogonia to mature human sperm is about 74 days.

OÖGENESIS

The process of gamete formation in the female (oögenesis, the origin of the egg) is also illustrated in Fig. 4.4. **Oögenesis** is essentially the same as spermatogenesis as far as meiosis is concerned, but other aspects of the process are quite different. Much more **nutrient** material is accumulated during oögenesis than during spermatogenesis. This is particularly true of oviparous animals—those that lay eggs that hatch outside the body of the mother. These animals must provide yolk material for the nourishment of the developing embryo outside of the mother's body. Even in viviparous animals, which retain and nourish the young inside the body of the mother, a considerable amount of nutrient material accumulates. Because of the accumulated nutrient materials, an egg is usually considerably larger than a sperm of the same species.

In addition, the cells that result from divisions in oögenesis are of unequal size. Nutrient material in the primary oöcyte is not divided equally into four cells that result from the meiotic sequence. One large cell in each division (Fig. 4.4) retains essentially all of the yolk, while the other cells, called **polar bodies**, get very little. First and second polar bodies, however, receive the same chromosome complements as the secondary oöcytes and ova from the respective divisions, but they do not become functional sex cells. During differentiation, special egg membranes are formed and the nucleus is generally reduced in size. The cytoplasm is filled with **ribosomes** and **RNA.** All mRNA required for early cell activity and differentiation is produced and stored in the cytoplasm. This is important in the expression of maternal inheritance (see p. 133).

In some animal species, oögenesis proceeds rapidly and continuously in sexually mature females and numerous eggs are produced. Usually, these eggs complete the second meiotic division and become mature before encountering sperm. In many other animals, including the mammals, the meiotic divisions are not completed until after sperm entry.

In the human female, for example, oögenesis begins before birth. Oögonia located in the follicles of cortical tissue in the fetal ovary begin to differentiate into primary oöcytes at about the third month of intrauterine development. At the time of birth of the female infant, the primary oöcytes are in the prophase of the first meiotic

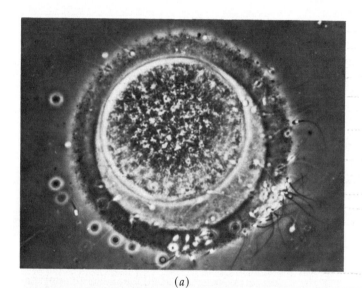

(a)

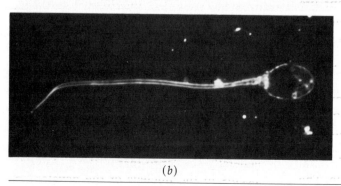

(b)

Figure 4.8 Human egg and sperm. (a) Egg with its surrounding layer of supporting cells, being penetrated by sperm. Several sperm are shown at lower right but only one will penetrate the egg. (b) Single human sperm greatly enlarged. (Photographs by Lester V. Bergman and Associates, Cold Spring, New York)

division. They remain in **suspended prophase** for many years until **sexual maturity** is reached. Then, as the ovarian follicles mature, the meiotic prophase is resumed. The first meiotic division for each developing egg is completed shortly before the time of ovulation for that egg. One cell becomes a secondary oöcyte and the other a polar body.

The second meiotic division is in progress when the developing egg is extruded from the ovary and passes into a Fallopian tube. This division is not completed, however, until after penetration by the sperm, which usually occurs in the tube (Fig. 4.8). Sperm entry is a random process in that any available sperm may fuse with any mature egg. If penetration by a sperm is accomplished, the **secondary oöcyte divides** and forms a **mature ovum** with a pronucleus containing a single set of 23 maternal chromosomes. The other cell resulting from this division is a second polar body not capable of further development. The sperm head forms a pronucleus with 23 paternal chromosomes. After the two **pronuclei fuse,** the resultant zygote, with (2n) 46 chromosomes, begins **mitotic division** or first cleavage, which produces the two-cell stage of a beginning **embryo.**

SPORE FORMATIONS IN PLANTS

Gamete formation in plants, like that in animals, requires a reduction in chromosome number from $2n$ to n. The meiotic process itself is similar to that in animals, but the life cycle of plants is somewhat more complicated (Fig. 4.9). Gamete formation usually does not follow meiosis directly. Sporogenesis in plants involves the formation of spores rather than sex cells. Spores produce gametophytes, and gametes come from gametophytes in which the chromosomes number is already reduced.

An **alternation of generations** between a haploid and diploid phase of the plant life cycle can separate the reduction of chromosome number from fertilization in plant reproduction. This two-phase system is a characteristic of virtually all plants. Diploid plants, called **sporophytes,** undergo **meiosis** to produce spores with the reduced chromosome number. Spores grow into haploid gametophytes which ultimately produce **gametes** that are capable of **fertilization.** Zygotes resulting from fertilization develop into sporophytes, completing the cycle. The sporophyte and gametophyte phases vary in length and importance in different plants.

MEIOSIS IN A PLANT
The details of microsporogenesis, the development of the microspore, are illustrated (Fig. 4.10) with some actual stages from living rye, *Secale cereale.* The zygotene stage of the meiotic prophase is shown in Fig. 4.10a.

Synapsis can be observed in progress in various parts of the chromosome configuration (see arrow). The diplotene stage is illustrated in Fig. 4.10b. In diakinesis and metaphase I (c to e), seven bivalents are visible. Later stages in the first and second divisions are shown (f to h) with the resulting four haploid sets of chromosomes. Following division of the cytoplasm, four independent microspores are formed. A pollen grain develops from each microspore.

The meiotic sequence in the female part is generally similar to that in the male part. It begins in the nucleus of the developing

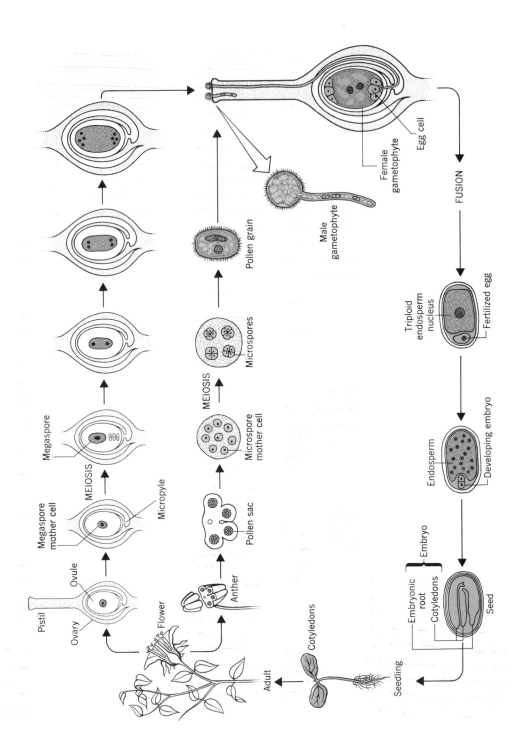

Pistil

Ovary
Ovule

Megaspore mother cell

MEIOSIS

Megaspore

Female gametophyte

Egg cell

FUSION

Micropyle

Flower

Anther

Pollen sac

Microspore mother cell

MEIOSIS

Microspores

Pollen grain

Male gametophyte

Triploid endosperm nucleus

Fertilized egg

Developing embryo

Endosperm

Embryo

Embryonic root

Cotyledons

Seed

Seedling

Cotyledons

Adult

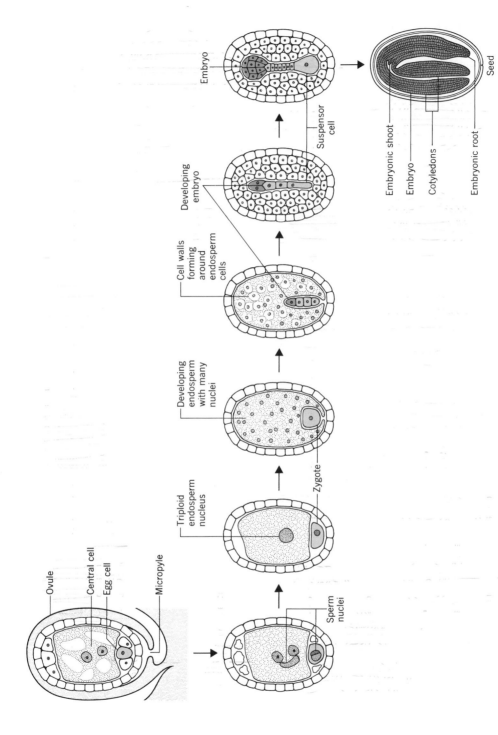

Figure 4.9 The life cycle of a seed plant. These enlarged diagrams show fertilization and seed formation.

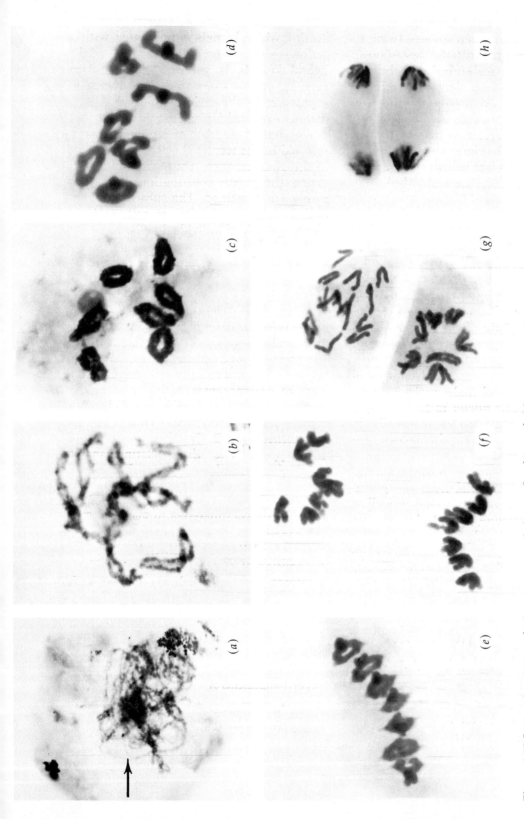

Figure 4.10 Successive stages of microsporogenesis in rye, *Secale cereale* (*a*) Zygotene stage, arrow shows synapsis in progress; (*b*) diplotene stage; (*c*) diakinesis, showing seven bivalents and nucleolus; (*d*) (*e*) metaphase I; (*f*) telophase I; (*g*) (lower figure) metaphase II and (upper figure) anaphase II; and (*h*) telophase II. All eight photographs are made to the same scale ×900. (Courtesy of W. S. Boyle.)

embryo-sac mother cell. After the meiotic prophase, the mother cell divides twice to form four embryo-sac initials (megaspores). Three of these are reabsorbed and the one functional megaspore undergoes a series of mitotic divisions, eventually forming the embryo sac (Fig. 4.9). One of the eight nuclei becomes the **egg,** two fuse to form a larger 2n nucleus that later gives rise to the endosperm nucleus (see below) and, in most seed plants, five nuclei are reabsorbed.

PLANT FERTILIZATION

The **pollen tube** contains three nuclei, the tube nucleus and two generative nuclei (Fig. 4.9). Generative nuclei are carried through the micropyle into the embryo sac and accomplish the **double fertilization** process characteristic of the higher plants. One male nucleus fuses with the egg nucleus and gives rise to the 2n **zygote,** which divides repeatedly to form the **embryo** of the seed. The second male nucleus unites with the two polar nuclei to form a triple fusion nucleus that divides repeatedly in typical seed plants to form the nutrient **endosperm** of the seed.

The process of double fertilization introduces genetic material from the pollen parent into the endosperm tissue, as well as the embryo. Therefore, it might be expected that both maternal and paternal inheritance would be represented. This hereditary influence of the pollen parent genes on the endosperm is called **xenia.** When, for example, maize from a variety normally bearing white kernels is pollinated with a yellow-kernel variety, the endosperm of the hybrid kernels is yellow. The dominant gene for yellow from the pollen comes to expression in the endosperm in the same manner expected for embryonic tissue.

The diploid number of chromosomes is restored in the fertilized cell that gives rise to the plant embryo. Thus, through fertilization, the genetic contributions from each parent are combined. The subsequent continuous mitotic division of cells and, in plants, growth of individual cells, produces a new individual, representative of the species to which the parent belongs.

GENES AND CHROMOSOMES

The physical bases of independent assortment and of segregation are inherent in the **chromosome mechanism at meiosis.** By observing microscopically distinguishable pairs of chromosomes during a reduction division, the maternal and paternal chromosomes can be followed. This was accomplished by E. E. Carothers in studies of grasshoppers. The three distinguishable pairs of maternal and paternal chromosomes are symbolized AA'BB'CC', and the various observed arrangements of the division spindle are illustrated in Fig. 4.11. The independence of the reduction division explains Mendel's second principle. A chance distribution of maternal and paternal chromosomes occurs in such a way that each

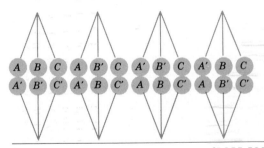

Figure 4.11 Arrangements of maternal and paternal chromosomes that could be identified in experiments with grasshoppers.

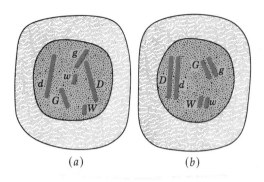

Figure 4.12 Cells showing three chromosome pairs (*a*) before synapsis has occurred, (*b*) after synapsis has occurred.

(*a*) (*b*)

has **equal chance** of facing one pole or the other on the equatorial plate. When the chromosomes become established on the equatorial plate, each goes to the pole it was facing and eventually becomes a part of the daughter cell formed around that pole of the spindle.

To illustrate this concept, consider the phenotypes vine height, seed color, and shape observed in F_1 peas (Chapter 2). These are located on separate homologous chromosome pairs. Gene *D* in the example is located on the maternal member of one pair whereas its allele *d* is in the corresponding position of the homologous (paternal) chromosome. Figure 4.12*a* shows, diagrammatically, six chromosomes representing three homologous pairs in unpaired condition, as would be expected during a mitotic division. Maternal and paternal members of each pair carrying their respective alleles are present in the same nucleus. At synapsis in meiosis, each chromosome finds its mate, and alleles come together in corresponding positions, as illustrated in Fig. 4.12*b*.

A particular maternal chromosome is equally likely to face one or the other pole of the spindle. Eight different meiotic metaphase arrangements are possible (two ways for each pair) and equally probable (independent of each other) in the example (Fig. 4.12*b*). These are illustrated in Fig. 4.13. Once the bivalents take their places on the metaphase plate, their positions are fixed. In the anaphase, each chromosome moves to the **nearest pole.**

Genes occurring in haploid gametes are nonallelic because meiotic division separates homologous members of allelic pairs of genes into different gametes. **Nonalleles** are symbolized by different letter symbols such as *A, B, C,* and those in a given chromosome are bound together in **linear order.** Those included in all the chromosomes of a gametic set are called a **genome.** Nonallelic genes that are in the same chromosome are **linked,** while those in different chromosomes are **not linked.**

Zygotes resulting from fusion of haploid (*n*) gametes are diploid (*2n*) and therefore carry **pairs** of alleles. Alleles are alternate

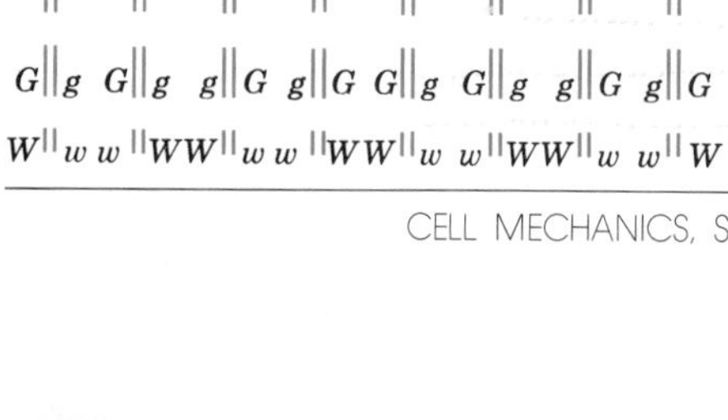

$D\|d\ D\|d\ D\|d\ D\|d\ d\|D\ d\|D\ d\|D\ d\|D$

$G\|g\ G\|g\ g\|G\ g\|G\ G\|g\ G\|g\ g\|G\ g\|G$

$W\|w\ w\|W W\|w\ w\|W W\|w\ w\|W W\|w\ w\|W$

Figure 4.13 Eight possible arrangements of three independent chromosomes in metaphase I. The eight different kinds of gametes are illustrated twice in the diagram to show the right and left positions which each member of each pair might take in the equatorial plate.

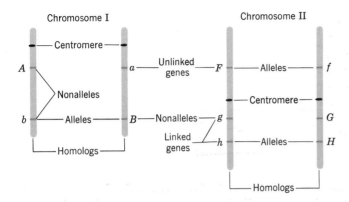

Figure 4.14 Alleles and nonalleles and linked and unlinked genes.

forms of a given gene located at the same chromosome **locus,** which may substitute for one another. Body cells of higher organisms originate from the zygote through continued cell duplication and differentiation. Like the zygote, body cells arising from the zygote carry two genomes included in two sets of homologous chromosomes. Two pairs of homologous chromosomes illustrating alleles and linked and unlinked nonalleles are shown in Fig. 4.14. The zygote from which these two pairs were taken would carry the diploid number characteristic of the species.

In animals and plants with a large number of chromosomes, an almost infinite number of **possible combinations** of chromosomes may be expected. For example, in

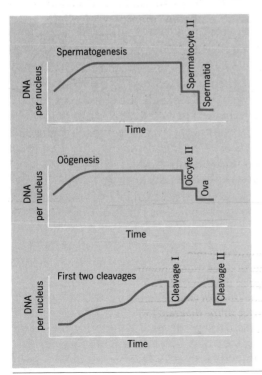

Figure 4.15 Relative concentrations of DNA in cells of grasshopper during stages of meiosis and mitosis. (After H. Swift in *International Review of Cytology,* Vol. 2, Academic Press, Inc., New York.)

organisms with 23 pairs of chromosomes, the probability that a gamete produced by an individual in the population will have any specific combination of chromosomes is $(\frac{1}{2})^{23}$, which is in the order of one in eight million. This calculation is an underestimate resulting from the possibility of any crossing over, which is another source of variability, discussed in Chapter 5. Further, increased numbers of gene combinations are possible in zygotes that result from **random fertilization.** Much of the variation observed in natural populations can therefore be explained on the basis of the **recombination** of chromosomes and genes already present in the breeding population.

DNA is the stable material in the chromosome that preserves genetic information and carries it from cell to cell and from generation to generation. New DNA is synthesized when chromosomes duplicate themselves. The proportional amount of DNA per cell is related to the number of sets of chromosomes in that cell. Haploid eggs and sperm, for example, contain half the quantity of DNA that is present in diploid somatic cells of the same species. During mitosis and meiosis the DNA content builds up in cell predivision periods and drops abruptly with each separation of the daughter chromosomes, as illustrated in Fig. 4.15.

SEX DETERMINATION

The first investigations relating chromosomes to sex determination were carried out at the turn of the century. H. Henking, a German biologist, discovered, in 1891, that a particular nuclear structure could be traced throughout spermatogenesis of certain insects. Half the sperm received this structure and half did not. Henking did not speculate on the significance of this body, but merely identified it as the "X body" and showed that sperm differed because of its presence or absence. In 1902 these observations were verified and extended by C. E. McClung, who made cytological observations on many different species of grasshoppers and demonstrated that the somatic cells in the female grasshopper carry a different chromosome number than do corresponding cells in the male. He followed the X body in spermatogenesis but did not succeed in tracing the oögenesis of the female grasshopper. McClung associated the X body with sex determination, but erroneously considered it to be peculiar to males. Had he been able to follow oögenesis, his interpretation would undoubtedly have been different.

Contributions to basic knowledge about sex determination were made in the early part of the century by E. B. Wilson and his associates. Wilson reported extensive cytological investigations on several different insects, notably from the genus Protenor, an uncommon group of insects closely related to the boxelder bug. In these insects, different numbers of chromosomes were observed in the germ cells of the two sexes. He succeeded in following oögenesis as well as spermatogenesis and found that the unreduced cells of the male carried 13 chromosomes, and those of the female carried 14. Some male gametes were found to carry 6 chromosomes, whereas others from the same individual carried 7. The female gametes all had 7. Eggs fertilized with 6-chromosome sperm produced males and those fertilized with 7-chromosome sperm produced females.

MECHANISMS OF SEX DETERMINATION

The "X" body of Henking thus was found to be a chromosome that influenced sex determination. It was identified in several insects and became known as the sex or X chromosome. All eggs of these insects carried an X chromosome, but it was included in only half of the sperm (Fig. 4.16). All sperm, however, had the usual complement of other chromosomes (autosomes). Eggs fertilized by sperm containing the X chromosome produced zygotes with two X chromosomes, which became females. Eggs receiving sperm without an X chromosome produced zygotes with one X, which became males.

Wilson observed another chromosome arrangement in the milkweed bug, *Lygaeus turcicus*. In this insect, the same number of chromosomes was present in the cells of both sexes. The one identified as the mate

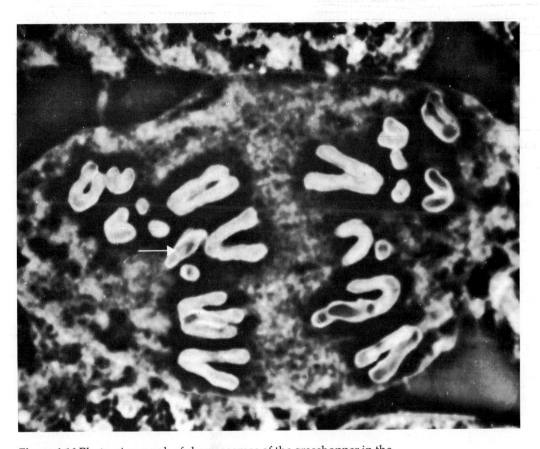

Figure 4.16 Photomicrograph of chromosomes of the grasshopper in the anaphase of the first division of spermatogenesis. Twelve chromosomes are at the left of the equatorial plane and 11 are at the right. The difference is the single X chromosome (arrow). Zygotes receiving X chromosomes from the sperm will become females and those receiving sperm with no X chromosomes will become males. (Courtesy A. M. Winchester, University of Northern Colorado.)

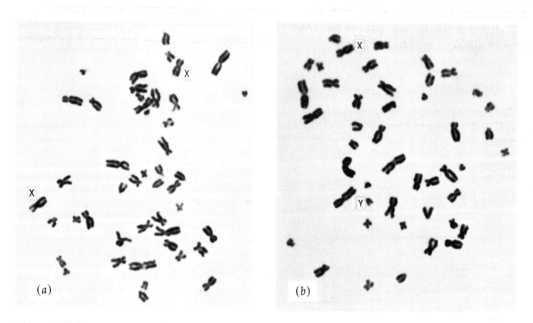

Figure 4.17 Human X and Y chromosomes exhibiting "end-to-end" pairing at meiosis.

Y   X

to the **X**, however, was distinctly smaller and was called the **Y** chromosome. Sex determination based on equal chromosome numbers in the two sexes, but with different kinds of chromosomes making up one pair, was called the **XY** type. As evidence was accumulated from a wider variety of animals, the XY mechanism was found to be more prevalent than the XO. The XY type is now considered characteristic in most of the higher animals and occurs in at least some plants (for example, *Melandrium album*).

Man also follows the XY pattern; the human X chromosome is considerably longer than the Y, as shown in Fig. 4.17. The total complement of human chromo-somes includes 44 autosomes, XX in the female, and XY in the male (Fig. 4.18). Eggs produced by the female in oogenesis have the usual complement of autosomes (22) plus an X chromosome. Sperm from the male have the same autosomal number and either an X or a Y. Eggs fertilized with Y chromosome sperm result in zygotes that develop into males; those fertilized with X chromosome sperm develop into females. Segregation of the XY pair and random fertilization thus explain, at least superficially, why some individuals develop into females and some into males, and why about half of the members of each population of higher animals are males and half are females.

Historically, the association of the most

Figure 4.18 Chromosomes of man. (*a*) Normal female cell metaphase with two X chromosomes and 44 autosomes. (*b*) Normal male cell metaphase with one X, a smaller Y chromosome, and 44 autosomes. (Courtesy of J. H. Tjio and T. T. Puck, Department of Biophysics, University of Colorado Medical Center, Denver.)

conspicuous phenotype (that is, sex) with a particular chromosome greatly strengthened the hypothesis that genes are in chromosomes. This idea originally had been postulated largely because of the parallel observed between the separation of chromosomes in the meiotic process and genetic segregation. Research-substantiated evidence that sex determination was controlled by a particular chromosome provided tangible support for a fundamental premise that **genes are in chromosomes.**

Experiments with insects formed a basis for speculation and experimentation concerning the sex-determining mechanism in higher forms. Because invertebrate hormones are not functionally comparable with the steroid hormones in birds and mammals, however, such animals as chickens and mice were mainly employed for the experimental work on the secondary sex characteristics (those characteristics that distinguish the two sexes, but that have no direct role in reproduction) and hormonal influences on phenotypes.

BALANCE CONCEPT OF SEX DETERMINATION

Soon after sex chromosomes were identified, it became obvious that sex determination was more complicated than preliminary observations had indicated. A more intricate mechanism than the segregation of a single pair of chromosomes was in evidence. The most fundamental contributions to the definition of this mechanism came from Bridges' investigations on Drosophila, which showed that **female determiners** were located in the X chromosome and male **determiners** were in the **autosomes.** More than one gene, perhaps a great many (in the X chromosome), were found to influence femaleness. Bridges also demonstrated that genes for maleness were not located in the Y chromosome of Drosophila, but were distributed widely among the autosomes. No specific loci have been identified, and the present evidence suggests that many chromosome areas are involved. Thus, it was shown that sex-determining genes are carried in certain

TABLE 4.1
Ratio of X Chromosomes to Autosomes and Corresponding Sex Type in *Drosophila melanogaster* **(After Bridges)**

X Chromosomes (X) and Sets of Autosomes (A)	Ratio X/A	Sex
1X 2A	0.5	Male
2X 2A	1.0	Female
3X 2A	1.5	Metafemale
4X 3A	1.33	Metafemale
4X 4A	1.0	Tetraploid female
3X 3A	1.0	Triploid female
3X 4A	0.75	Intersex
2X 3A	0.67	Intersex
2X 4A	0.5	Male
1X 3A	0.33	Metamale

chromosomes in Drosophila, and that all individuals carry genes for both sexes. The genetic **balance theory of sex determination** was devised as a more detailed explanation of the mechanics of sex determination.

The XO or XY chromosome segregation was interpreted as a means of tipping the balance between maleness and femaleness, whereas more deepseated processes were involved in the actual determination. Bridges experimentally produced various combinations of X chromosomes and autosomes in Drosophila and deduced from comparisons of the results that one X chromosome (X)

and two sets of autosomes (A) had a **ratio** in terms of **sex determining capacity** of $\frac{1}{2}$ or 0.5. This combination of 1X and 2A resulted in a male (Table 4.1). Two X and 2A produced a female.

The first irregular chromosome arrangement from Bridges' experiments resulted from **nondisjunction**, the failure of paired chromosomes to separate in anaphase of the reduction division. X chromosomes, which ordinarily came together in pairs in meiotic prophase of oögenesis and separated to the poles in anaphase, remained together and went to the same pole. As a result, some

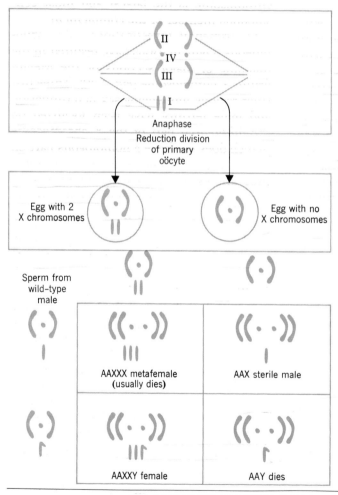

Figure 4.19 Nondisjunction in Drosophila, and zygotes resulting from fertilization by wild-type males (AAXY). The trisomic females (AAXXY) and monosomic males (AAX) were the exceptional flies in Bridge's experiment. In the primary oöcyte, autosomes II and III are represented by pairs of bent rods, autosomes IV which are small and take their places in the center of the equatorial plate are represented by a pair of dots and the X chromosomes (I) are symbolized as short rods. The Y chromosome introduced by the sperm is illustrated by a rod with a hook or short arm. (After Bridges)

female gametes received two X chromosomes and some received no X chromosomes (Fig. 4.19). Following fertilization by sperm from wild-type males (AAXY), all zygotes had $2n$ autosomes (2A) but some received two X from the mother and an X from the father (3X). The ratio of $\frac{3}{2}$ resulted in flies, called metafemales that were highly inviable. The XXY flies (2X/2A) from the same mating were normal females in appearance; XO (1X/2A) males were sterile while those with a Y but no X chromosomes did not survive. These results indicated that, in Drosophila, the Y chromosome is not involved in sex determination but that it does control **male fertility.**

Flies produced experimentally with 4X/3A were also metafemales. Those with 4X/4A and also those with 3X/3A both with an X/A ratio = 1 were females. The combinations 3X/4A = 0.75, and 2X/3A = 0.67, produced experimentally were intermediate in characteristics between males and females and were called **intersexes.** Combinations of 2X/4A = 0.5 were males and those of 1X/3A = 0.33 were **metamales.**

No other animals or plants have been investigated with equal thoroughness, but indirect evidence suggests that some such balance is involved in many organisms. Intersexes can be produced experimentally in some animals by upsetting this balance during the developmental stages. In nature, a margin of safety makes intermediates between the two sexes uncommon.

Y CHROMOSOMES IN SEX DETERMINATION

In Drosophila, the Y chromosome was shown by Bridges to have no influence on sex determination since all sex determiners are carried in the X chromosome and the autosomes. This seems to be true in some other organisms now investigated. Exceptions, however, include the plant genus Me-

landrium, a part of the pink family, and the amphibian genus Axolotl, in which **sex determination depends on the Y chromosome.** The Y chromosome also carries determiners for maleness in mice and in man.

In *Melandrium album*, which follows the XY mode of sex determination, H. E. Warmke, M. Westergaard, and others have shown that sex is determined by a balance between male-determining genes in the Y and female-determining genes in the X and in the autosomes. In this plant, which is normally unisexual, XY individuals are staminate (that is, pollen bearing), and XX plants are pistillate (egg bearing). The Y chromosome is the largest and most conspicuous member of the complement.

Experimental investigations using spontaneous fragmentations have resulted in the mapping of major sections of the Melandrium Y chromosome (Fig. 4.20). Three distinct regions influencing sex determination and male fertility have been localized on the differential part of the Y chromosome (which does not have a homologous part on the X). Region I suppresses femaleness and thus allows maleness to be expressed. In the absence of this region, plants are bisexual; that is, they express both male and female characteristics. Region II promotes male development. When this region (with or without Region I) is missing, a female plant is produced. Region III carries male fertility genes; loss of this region results in male sterility. A part of the Y chromosome is homologus with a part of the X, but the major part of the X is differential with no structural counterpart in the Y.

Although most female-determining genetic material is in the X chromosome, Westergaard found that the autosomes were also involved in female determination. Female plants of this species, like those of many other species of plants and animals, have a potentiality for maleness.

In man the Y chromosome plays an im-

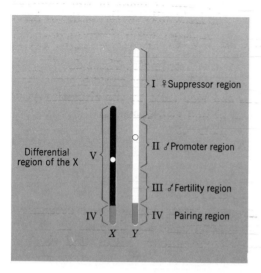

Figure 4.20 Sex chromosomes in Melandrium. Regions I, II, and III of the Y chromosome do not have homologous segments in the X, and thus they make up the differential portion of the Y. Regions IV are homologous in the X and Y, and are pairing regions at meiosis. V is the differential portion of the X chromosome. When I is lost from a Y chromosome, a bisexual plant is produced. When II is lost, a female plant is produced. If III is absent, male-sterile plants with abortive anthers appear. (After Westergaard, *Hereditas*, 34:257–279, 1948.)

portant part in sex determination. The XX chromosome arrangement is associated with the female sex and the XY arrangement with the male sex. When the chromosome constitution is irregular (Chapter 10) X chromosomes alone in any number are usually associated with females. The **Y** chromosome induces development of the undifferentiated gonadal **medulla** into a testis whereas an **XX** chromosomal complement induces the undifferentiated gonadal **cortex** to develop **ovaries**. In the presence of three or more X chromosomes in abnormal arrangements, the presence of a single Y chromosome is usually sufficient to produce testes and male characteristics.

MALE HAPLOIDY IN HYMENOPTERA

More involved mechanisms for sex determination have been described in the insect order, Hymenoptera, which includes ants, bees, wasps, and sawflies. In several species of Hymenoptera, males arise parthenogenetically, that is, without fertilization, and have a haploid chromosome number (16 in the drone honey bee). The queen honey bee and workers, which arise from fertilized eggs, carry the diploid chromosome number (32). Something associated with the **haploid-diploid** chromosome arrangement is involved in sex determination in bees. Parthenogenesis also occurs in a genus of parasitic wasps, Habrobracon, in which females are diploid with 20 chromosomes, and males are haploid with 10 chromosomes. **Females** originate from **fertilized eggs,** but **males** ordinarily come from **unfertilized eggs.**

Some Habrobracon males produced experimentally by Whiting came from fertilized eggs and were diploid, whereas others came from unfertilized eggs and were haploid. All females were diploid. Results of experiments showed that the homozygous or heterozygous status of certain chromosome segments controlled sex determination. Stated more precisely, haploid males have segments Xa, Xb, or Xc: diploid males were XaXb, XaXc, or XbXc. A complementary action of different alleles was postulated for the production of femaleness. Any allele, whether present in single or double condition, that had no complement with which to interact, would produce maleness.

SINGLE GENES AND SEX DETERMINATION

Sex determination in some organisms is influenced by the differential action of single genes. Maize, for example, is monoecious, (both sexes in the same plant) having staminate flowers in the tassel and pistillate flowers in the ear. A substitution of two single gene pairs makes the difference between monoecious and dioecious (separate sexes) plants. The gene for barren plant (*ba*), when homozygous, makes the stalk staminate by eliminating the silks and ears. On the other hand, the gene for tassel seed (*ts*), when homozygous, transforms the tassel into a pistillate structure that produces no pollen. A plant of the genotype *ba/ba ts/ts* lacks silks on the stalk but has a transformed tassel and is therefore only pistillate (female). A plant with *ba/ba ts$^+$/ts$^+$* is only staminate (male). These data suggest how **monoecious plants could become dioecious and vice versa by the alteration** (mutation) **of just two genes:** in this case, *ba$^+$* to *ba* and *ts$^+$* to *ts*.

EXTERNAL ENVIRONMENT AND SEX DETERMINATION

In some lower animals, sex determination is nongenetic and depends on the external environment. Males and females

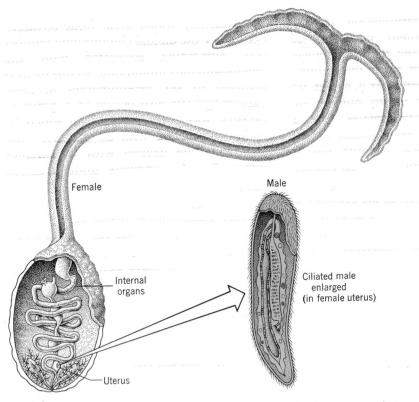

Female — Internal organs — Uterus

Male — Ciliated male enlarged (in female uterus)

Figure 4.21 Female and male of the marine worm *Bonellia viridis*. The male is shown in the uterus of the female and greatly enlarged at the right to show details of internal structure. (Redrawn from Dobzhansky, *Evolution, Genetics and Man*, 1955, John Wiley & Sons, Inc.)

have similar genotypes, but a stimulus from an environmental source initiates development toward one sex or the other. For example, the male of the marine worm Bonellia is small and degenerate and lives within the reproductive tract of the larger female (Fig. 4.21). The male is conveniently located for fertilization of the eggs, but little else can be said for his situation. All the organs of his body are degenerate except those making up the reproductive system. As excellent material for the study of an elementary type of sex determination, this worm has been investigated extensively.

F. Baltzer found that any young worm reared from a single isolated egg became a female. If he released newly hatched worms in water containing mature females, however, some of the young worms were attracted to the females and became attached to the female proboscis. These were transformed into males and eventually migrated to the female reproductive tract where they became parasitic. Genetic determiners for both sexes are apparently present in young worms. **Extracts** made from the **female** proboscis will influence young worms toward **maleness.**

GYNANDROMORPHS

In some animals such as insects, a typical chromosomal behavior can result in **sexual mosaics** called **gynandromorphs.** Some parts of the animal express female characteristics while other parts express those of the male. Some gynandromorphs in Drosophila are bilateral intersexes (Fig. 4.22), with male color pattern, body shape, and sex comb on one half of the body and female characteristics on the other half. Both male and female gonads and genitalia are sometimes present.

Bilateral gynandromorphs have been explained on the basis of an irregularity in mitosis at the first cleavage of the zygote. Infrequently, a chromosome lags in division and does not arrive at the pole in time to be included in the reconstructed nucleus of the daughter cell. If one of the **X chromosomes** of an XX (female) zygote **lags** in the center of the spindle, one daughter cell would get only one X chromosome, while the other received XX, as illustrated in Fig. 4.23. The basis for a mosaic pattern would thus be established. One cell in the two-cell stage would be XX (female) and one would be XO (male). In Drosophila, the right and

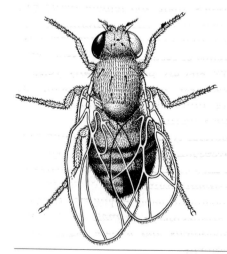

Figure 4.22 Bilateral gynandromorph in Drosophila. (After Morgan.)

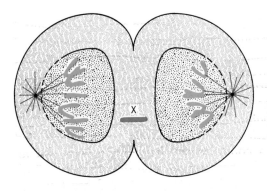

Figure 4.23 A lagging X chromosome in the first cleavage of Drosophila illustrating the origin of a bilateral gynandromorph. (After Morgan and Bridges.)

left halves of the body are determined at the first cleavage. One cell gives rise to all the cells making up the right half of the adult body and the other gives rise to the left half. If the same chromosome loss occurred at a later cell division, a smaller proportion of the adult body would be included in the male segment. The position and size of the mosaic sector are determined, therefore, by the place and time of the division abnormality.

Gynandromorphs were described in Drosophila by Sturtevant, Morgan, and Bridges beginning in 1919. Following the original description, a few conspicuous examples were reported in flies, but the condition was considered extremely rare. More extensive observations have since shown that a gynandromorph of some kind is produced in every 2000 to 3000 flies; many of these represent small sections of tissue involving only a few cells. Spencer Brown and Aloha Hannah-Alava devised a technique that increased the frequency of gynandromorphs by using a ring X chromosome (first discovered by L. V. Morgan). Structural modifications fuse the two ends of each of these chromosomes, forming rings. These aberrant chromosomes are frequently eliminated by a natural process from older eggs

thus causing a greater frequency of chromosome difference between the two blastomeres. When virgin females are not mated until they are 8 to 17 days old, the proportion of gynandromorphs is greatly increased, and when excess yeast is added to the culture, even more of the progeny are "gynanders." Investigators can thus **partially control** the natural process of chromosome removal and produce more gynandromorphs than would ordinarily occur. In treated cultures, up to 20 percent of the flies are gynanders. By placing marker genes such as y for yellow body color in the X chromosomes, male and female tissue can be identified. When one X chromosome is eliminated from a cell of y/y^+ heterozygote, some patches of resultant male tissue carry the y gene and are phenotypically yellow. These methods have proved useful in studies of differentiation of bristles and other body structures.

The bilateral type of gynandromorph typical of Drosophila is not the only type observed in insects. In Habrobracon, a parasite wasp, gynandromorphs may occur in the anterior-posterior plane, giving rise to such peculiar arrangements as male heads with female abdomens and female heads with male abdomens.

SEX DIFFERENTIATION

Cell diversity in development arises from the unity of the fertilized ovum. The myriad of cell types composing an individual all result from successive divisions of a single zygote. DNA contains the biochemical information specifying the primary structures of such subsidiary molecules as RNA, proteins, enzymes, and antigens and ultimately of the organism itself. Under unusual circumstances these molecules may be influenced or even overridden by local factors such as hormones, viruses, or drugs, thus altering the differentiated state of cells.

During normal development of the mammalian **sperm, differentiation results in a continual restriction of RNA transcription.** The mature sperm has no nuclear RNA synthesis. The mature **ovum** on the other hand, is not differentiated and **has extensive RNA synthesis.** When the nontranscribing sperm fuses with the mature ovum during fertilization the genome carried by the sperm becomes active and converts the unfertilized ovum into a dividing diploid cell. This zygote with two genomes is the beginning of the multicellular organism in which maternal and paternal gene expression occurs in the cells of the embryo and adult.

SEX CHROMATIN

M. L. Barr observed **chromatin bodies** in the nerve cells of female cats that were not present in cells of the male. Barr and others reported a sustained difference between the nuclear contents of human male and female cells in several kinds of tissue including epithelial cells of the buccal mucosa (lining the inside of the mouth) and neutrophils in the blood. With appropriate staining techniques, a small chromatin body could be identified in the nucleus of the cells of the female. This Barr body, which was related

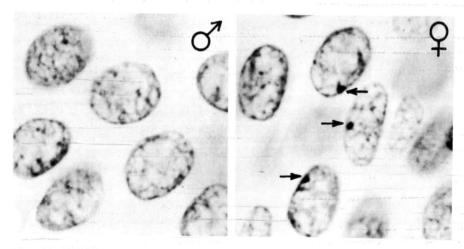

Figure 4.24 Cells (see arrows) of human epidermis showing sex chromatin bodies or Barr bodies. Left, epidermal cells of a male showing nucleoli common to cells of both sexes, right, epidermal cells of a female showing sex chromatin bodies as well as nucleoli common to both sexes. (Courtesy of M. L. Barr.)

in some way to the sex chromosomes, was not observed in the cells of the normal male. The appearance of the nuclei in the cells of the two sexes is illustrated in Fig. 4.24. With this technique, the sex of human embryos can be distinguished at early stages of development. More recent studies have shown that a sex chromatin body is present in most, if not all, somatic cells of female mice. Other female mammals also carry this body. Indeed, this cellular characteristic seems to apply to mammals generally.

Sex chromatin bodies not only distinguish normal female from male cells but they are also useful for diagnosing various kinds of sex chromosome abnormalities in human beings. In certain people, abnormal numbers of chromosomes are related to the number of sex chromatin bodies. Those who have two or more X chromosomes, have **one less chromatin body than the number of X chromosomes present.** Cells of abnormal females with only one X chromosome (Turner's syndrome) have no sex chromatin bodies and cells of abnormal males with two X and one Y chromosomes (Klinefelter's syndrome) have one sex chromatin body (Chapter 10). Abnormal females with three X chromosomes have two sex chromatin bodies in their cell nuclei.

DOSAGE COMPENSATION

For many years, geneticists have observed that, in some cases, females homozygous for genes in the X chromosomes do not express a trait more markedly than do hemizygous males. Some sort of **dosage compensation** has been suggested. This can be accomplished by any mechanism through which the effective dosage of sex-linked genes of the two sexes is made equal or nearly so. Several investigators arrived almost simultaneously at a hypothesis explaining the sex chromatin body and dosage compensation in mammals by the **inactivation of one X chromosome** in the normal female. The hypothesis was named after Mary F. Lyon (1962) who first presented it in detail, based on genetic studies of coat-color genes of the mouse and cytological observations. Hybrid female mice, heterozygous for certain coat-color genes, show a mottled effect unlike that of either homozygote, and not intermediate between the colors expressed in the parents. The fur pattern (Fig. 4.25) is a mosaic made up of randomly arranged patches of the two colors. Normal male mice and abnormal XO mice have never showed the mottled phenotype.

The mottled appearance characteristic of some other female mammals (e.g., calico or tortoise-shell cats) heterozygous for color alleles in the X chromosome are also explained on the basis of **patches of cells,** some with one, and some with the other X chromosome condensed and inactive.

The hypothesis was based on the cytological observation that the number of sex chromatin bodies in interphase cells of adult females is one less than the number of X chromosomes observed in metaphase preparations. The chromatin body is therefore a compact X chromosome. On this premise, it must be assumed that only one X chromosome is required for cellular metabolism in the female cell. An additional X becomes heteropyknotic (that is, stains more densely than other chromosomes) and is genetically inactive.

Which X chromosome becomes inactive is a matter of chance, but once an X has become inactivated, all cells arising from that cell line will have an inactive X chromosome (a sex chromatin body). In the mouse, the inactivation apparently occurs early in development, but in human embryos no sex chromatin bodies have been observed earlier than the sixteenth day of gestation. Some human traits could there-

Figure 4.25 Female mouse heterozygous for an X-linked gene for the coat color tortoise-shell. The mosaic phenotype of such females provided one of the first lines of evidence for the Lyon hypothesis. (From Margaret W. Thompson, with permission of the *Canadian Journal of Genetics and Cytology* 7:208, 1965.)

fore be influenced by both X chromosomes during the first 16 days. Later (after the 16th day), only one X would be functional in a body cell and the female would become mosaic with respect to X-linked genes. Apparently X chromosome inactivation occurs only when at least two X chromosomes are present. When several X chromosomes are in the same nucleus, all but one will be inactivated. The number of sex chromatin bodies after the period of inactivation is, therefore one less than the number of X chromosomes.

The Lyon hypothesis explains certain genetic consequences of sex-linked genes in man or other mammals: (1) dosage compen-sation for females with two X chromo-somes that express traits dependent on sex-linked genes in similar degree to males with only one X chromosome; and (2) varia-bility of expression in heterozygous females because of the random inactivation of one or the other X chromosome.

In other organisms, dosage compensation is accomplished by mechanisms other than the loss of an X chromosome. In *Drosophila melanogaster*, for example, dosage compen-sation apparently occurs through **the dif-ferential activity of some sex-linked genes.** J. Tobler found about equal activity of the enzyme tryptophan pyrrolase in the two sexes, even though only one sex-linked gene for this eye pigmentation enzyme is present in males, while females have two. It has been assumed from other experimental results on Drosophila that dosage compensation in some cases results from the action of **modifier genes** called dosage compensation genes within the X chromosome. These modifiers tend to cancel the effect of the different doses of a given gene by influencing either genetic transcription, genetic translation, or the biological activity of the protein products resulting from these processes. This influence is called **genetic regulation.**

HORMONES AND SEX DIFFERENTIATION

The hormonal system that regulates the internal or physiological environment of the organism does not directly influence the fundamental process of sex determination. It is important, however, to the development of the more conspicuous secondary sex characteristic. Sex hormones of higher animals are elaborated by the endocrine glands, particularly the ovaries and testes (gonads), although the adrenals and pituitary are also involved. The adrenals produce steroids that are chemically related to those of the gonads and that also influence the secondary sex characteristics. Ovaries and testes each have a joint function; they are responsible for the production of the **primary sex elements,** the eggs and sperm as well as production of the **secondary sex characteristics,** which are controlled by the sex hormones. Certain cells making up the basic structure of the gonad produce mature gametes. Other cells elaborate the hormones that influence the development of secondary sexual characteristics such as physiological differences in the rate of metabolism, blood pressure, heart beat, and respiration.

Early differentiation, in higher animals, including that of the sex organs, is influenced by hormones. In salamanders, for example, E. Witschi found that a peculiar type of hormonal balance is involved in the differentiation of the gonads; antagonistic substances are formed by the cortex and medulla of the embryonic gonad. The male hormone is produced by the medulla, whereas the female hormone comes from the cortex. When male and female salamanders were experimentally grafted together, the male substance suppressed the development of the ovary in the female member. When the male graft was considerably smaller than the female, however, the female hormone was dominant and genetically determined males developed ovaries. The quantity of the hormone evidently was a factor in this transition.

Partial **sex reversals** sometimes occur in adults of higher animal species (including man), indicating that at least the secondary sex characteristics of the opposite sex are potentially present throughout the life of the individual. In fowls, for example, when the ovary of the female is destroyed through injury or disease, male characteristics such as cock feathering, wattles, and crowing ability develop. Primary as well as secondary sex characteristics of chickens may be involved in the reversal. In one series of experiments, normal hens underwent complete sex reversal when their ovaries were destroyed; the male gonad developed and they fathered chicks. Primary as well as secondary sex structures and functions were thus involved in the reversal. Removal of the gonads of either sex in mammals is followed by the development of secondary sex characteristics of the other sex, although this can be counteracted artificially as is done in man.

SEX-INFLUENCED DOMINANCE

Dominance of alleles may differ in heterozygotes of the two sexes. This phenomenon is called **sex-influenced dominance.**

TABLE 4.2
**Expression of *h* Alleles in
Sex-influenced Inheritance**

Genotypes	Male	Females
h^+h^+	Horned	Horned
h^+h	Horned	Hornless
hh	Hornless	Hornless

Gene products of heterozygotes in the two sexes are apparently influenced differentially by sex hormones. For example, autosomal genes responsible for horns in some breeds of sheep behave differently in the presence of the **male and female sex hormones.** More than a single pair of genes is involved in the production of horns, but assuming all other genes to be homozygous, the example can be treated as if only a single pair were involved. Among Dorset sheep, both sexes are horned, and the gene for the horned condition is homozygous (h^+h^+). In Suffolk sheep, neither sex is horned and the genotype is hh. Among the heterozygous F_1 progeny from crosses between these two breeds, horned males and hornless females are produced. Because both sexes are genotypically alike (h^+h) the gene must behave as a dominant in males and as a recessive in females; that is, only one gene is required for an expression in the male, but the same gene must be homozygous for expression in the female.

When F_1 hybrids are mated together, a ratio of three horned to one hornless is produced among the F_2 males, whereas a ratio of three hornless to one horned is observed among the F_2 females. Genotypes and phenotypes of the two sexes are summarized as shown in Table 4.2. The only departure from the usual pattern is concerned with the heterozygous (h^+h) genotype. This genotype in the male results in the horned condition, but females with the same genotype

are hornless. Dominance of the gene is apparently influenced by the sex hormone.

Some human traits, such as a certain type of white forelock, absence of the upper lateral incisor teeth, and a particular type of enlargement of the terminal joints of the fingers, have been reported to follow this mode of inheritance.

SEX-LIMITED GENE EXPRESSION

One sex may be uniform in expression of a particular trait, yet transfer genes that produce a different phenotype in the offspring of the other sex. This is called "sex-limited gene expression." **Sex hormones** are apparently **the limiting factors in the expressions of some genes.** Premature (pattern) baldness in man, for example, has been described as a trait with sex-limited gene expression. Other types of baldness are associated with abnormalities in thyroid metabolism and infectious disease. About 26 percent of the men over 30 in the United States are bald-headed. Approximately half of these become bald prematurely, in their twenties or thirties. Baldness is known to be more common in some families than in others. Several different modes of inheritance have been associated with this trait by different investigators. For a particular type of premature baldness based on sex-limited expression, a single dominant gene is involved. This gene comes to expression only in the presence of an adequate level of

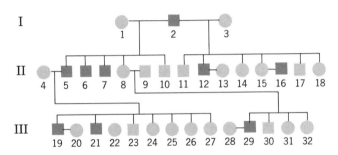

Figure 4.26 Pedigree showing incidence of premature baldness in a family group. The men represented by the darkened squares were bald before they reached the age of 35. Those symbolized by light squares were over 35 when the study was made and all had thick hair. No women in this family group expressed the trait.

androgenic hormone. The level of hormone necessary for expression of the trait is seldom, if ever, reached in women, but is attained in all normal men. Segregation of the gene determines which men will express the trait. A pedigree illustrating a hereditary pattern of baldness in one family group is presented in Figure 4.26.

Expressions of some genes are sex-limited for more basic reasons. For example, milk production among cattle and other mammals is limited to the sex that is equipped with developed mammary glands and appropriate hormones. Certain bulls are in great demand among dairy breeders and artificial insemination associations because their mothers and daughters have good milk-production records. The frequency of giving birth to fraternal twins in human families is hereditary to some extent. Mothers are immediately involved, but evidence indicates that genes from their fathers may also influence the tendency toward multiple births.

MATERNAL EFFECTS

Eggs and embryos are expected to be influenced by the **maternal environment** in which they develop. Even those removed from the body of the mother at an early stage receive the cytoplasm and nutrients in the egg from the mother, and special influences on gene action may have already taken effect. Certain potentialities of the egg are known to be determined before fertilization, and in some cases these have been influenced by the surrounding maternal tissue. Such predetermination where the traits of progeny are determined by the **genes of the mother** rather than by the genes of the individual is called the **maternal effect.** The existence of a maternal effect is commonly substantiated or disproved by **reciprocal crosses,** that is, a female from strain A mated with a male from strain B and a male from strain A mated with a female from strain B. If a maternal effect is involved, the results from reciprocal crosses will be different and the genes of the mother will be expressed.

EXAMPLES OF MATERNAL EFFECT

One of the earliest to be investigated and best-known examples of a maternal effect is that of the direction of coiling in snails, *Limnaea peregra.* Some strains of this species have dextral shells, which coil to the right, and others have sinistral shells, which coil to the left. This characteristic is determined by the **genes** of the mother (**not her phenotype**) rather than by the genes of

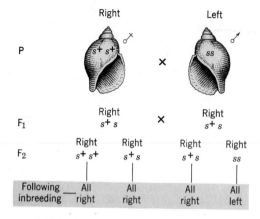

P — Right (s+ s+) ♂ × Left (ss) ♂

F₁ — Right (s+ s) × Right (s+ s)

F₂ — Right (s+ s+) | Right (s+ s) | Right (s+ s) | Right (ss)

Following inbreeding — All right | All right | All right | All left

Figure 4.27 A cross illustrating a maternal effect. The coiling pattern is controlled by the *genes* of the mother. Following inbreeding of snails with *ss* genotypes, *ss* mothers produced progeny that coiled to the left. (Data and interpretation from Boycott et al. and Sturtevant.)

the developing snail. The gene s^+ for right-handed coiling is dominant over its allele s for coiling to the left.

When crosses (Fig. 4.27) were made between females coiled to the right and males coiled left, the F₁ snails were all coiled to the right. The usual 3:1 ratio was not obtained in the F₂ because the phenotype of *ss* was not expressed. Instead, the pattern determined by the mother's genes (s^+s in the F₁) was expressed. When *ss* individuals were inbred, only progeny that coiled to the left were produced. However, when the s^+s^+ or s^+s were inbred, they produced offspring that all coiled to the right. From the reciprocal cross between left-coiling females and right-coiling males (Fig. 4.28), all the F₁ progeny were coiled to the left. The F₂ all

coiled to the right, but when each F₂ snail was inbred, those with the genotype *ss* produced progeny that coiled to the left.

Further investigation of coiling in snails has shown that the spindle formed in the metaphase of the first cleavage division influences the direction of coiling. The spindle of potential "dextral" snails is tipped to the right, but that of "sinistral" snails is tipped to the left. This difference in the arrangement of the spindle is controlled by the genes of the mother, which act on the developing eggs in the ovary. They determine the orientation of the spindle, which in turn influences further cell division, and results in the adult pattern of coiling. The actual phenotypic character, therefore, is influenced directly

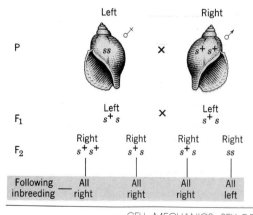

P — Left (ss) ♂ × Right (s+ s+) ♂

F₁ — Left (s+ s) × Left (s+ s)

F₂ — Right (s+ s+) | Right (s+ s) | Right (s+ s) | Right (ss)

Following inbreeding — All right | All right | All right | All left

Figure 4.28 Reciprocal cross of Fig. 4.26, illustrating a maternal effect. The *ss* P mother produced only left coiling progeny, and the s^+s F₁ mother produced only right coiling progeny. Following inbreeding the *ss* F₂ mother produced only left coiling progen. (Data and interpretation from Boycott et al. and Sturtevant.)

by the mother, with no immediate relation to the genes in the egg or the sperm. Most other snail traits, however, do not show the maternal effect pattern. The striping color pattern, for example, is also determined in the early embryo, but it is controlled directly by chromosomal genes of both parents. In this example, comparable color patterns are obtained from the results of reciprocal crosses.

A maternal effect on the intensity of eye color has been observed in the crustacean, Gammarus, sometimes called water flea or beach hopper. A pair of alleles, A and a, acting in Mendelian fashion, determine whether a representative of this species will have dark or light eyes, respectively. When AA females are crossed with aa males, the first generation young all have dark eyes and the eyes remain dark when the adult stage is reached. From the reciprocal cross (that is aa females mated with AA males), the young have light eye color like their mothers, but the adults have dark eyes. The dark pigment is produced from a substance called kynurenine in A-individuals. Since kynurenine is diffusible, it can be transferred from A-mothers to their progeny. Little or no kynurenine is present in the progeny of aa mothers, but the **pigment** can be **built** during the developmental period by progeny **carrying** A **genes.**

REFERENCES

Barr, M. L. 1960. "Sexual dimorphism in interphase nuclei." *Amer. J. Human Genet.*, 12, 118–127.

Bridges, C. B. 1925. "Sex in relation to chromosomes and genes." *Amer. Nat.*, 59, 127–137.

Dalton, A. J. and F. Haguenau. 1968. *The nucleus.* Academic Press, New York.

De Robertis, E. D. P., W. W. Nowinski, and F. A. Saez. 1970. *Cell biology*, 5th ed. W. B. Saunders, Philadelphia.

Donnelly, G. M. and A. H. Sparrow. 1965. "Mitotic and meiotic chromosomes of Amphiuma." *J. Hered.* 56, 91–98.

Dronamraju, K. R. 1965. "The function of the Y-chromosome in man, animals, and plants." *Advances in genetics*, E. W. Caspari and J. M. Thoday (eds.), 13, 227–310.

Du Praw, E. J. 1968. *Cell and molecular biology.* Academic Press, New York.

Goldschmidt, R. 1934. "Lymantria." *Bibliographia Genetica.* The Hague, Netherlands, 11, 1–186.

Harris, H. 1947. "The inheritance of premature baldness in man." *Annals of Eugenics*, 13, 172–181.

Lyon, M. F. 1962. "Sex chromatin and gene action in mammalian X, chromosomes." *Amer. J. Human Genet.*, 14, 135–148.

Mittwoch, U. 1973. *Genetics and sex differentiation.* Academic Press, New York.

Moore, K. L. 1966. *The sex chromatin.* W. B. Saunders Co., Philadelphia.

Swanson, C. P. 1969. *The cell.* Prentice-Hall, Englewood Cliffs, N.J.

Voeller, B. R. 1968. *The chromosome theory of inheritance.* Appleton-Century, Crofts, New York.

Wallace, B. 1966. *Chromosomes, giant molecules, and evolution.* W. W. Norton and Co., New York.

Warmke, H. E. 1946. "Sex determination and sex balance in Melandrium." *Amer. J. Bot.* 33, 648–660.

Westergaard, M. 1948. "The relation between chromosome constitution and sex in the offspring of triploid Melandrium." *Hereditas,* 34, 257–279.

Whiting, P. W. 1945. "The evolution of male haploidy." *Quart. Rev. Biol.,* 20, 231–260.

Witschi, E. 1957. "Sex chromatin and sex differentiation in human embryos." *Science,* 126, 1288–1290.

PROBLEMS AND QUESTIONS

4.1 Mark the true statement with a + and the false with a 0. (a) Skin cells and gametes of the same animal contain the same number of chromosomes. (b) Any chromosomes may pair with any other chromosome in the same cell in meiosis. (c) The gametes of an animal may contain more maternal chromosomes than its body cells contain. (d) Of ten chromosomes in a mature sperm cell, five are always maternal. (e) Of 22 chromosomes in a primary oöcyte, 15 may be paternal. (f) Homologous parts of two chromosomes lie opposite one another in pairing. (g) A sperm has half as many chromosomes as a spermatogonium of the same animal.

4.2 In each somatic cell of a particular animal species there are 46 chromosomes. How many should there be in a (a) mature egg, (b) first polar body, (c) sperm, (d) spermatid, (e) primary spermatocyte, (f) brain cell, (g) secondary oöcyte, (h) spermatogonium?

4.3 If spermatogenesis is normal and all cells survive, how many sperm will result from (a) 50 primary spermatocytes, (b) 50 spermatids?

4.4 In man, a type of myopia (an eye abnormality) is dependent on a dominant gene (M). Represent diagrammatically (on the chromosomes) a cross between a woman with myopia but heterozygous (*Mm*) and a normal man (*mm*). Show the kinds of gametes that each parent could produce and summarize the expected results from the cross.

4.5 Beginning with the myopic woman in Problem 4.4, diagram the oögenesis process producing the egg involved in the production of a child with myopia. Label all stages.

4.6 In what ways is cell division similar and different in animals and plants?

4.7 How does meiosis differ from mitosis? Consider differences in mechanism as well as end results.

4.8 How does gamete formation in higher plants differ from that in higher animals with reference to (a) gross mechanism, and (b) chromosome mechanism?

4.9 How is double fertilization accomplished in plants, and what is the fate of the egg and the endosperm nucleus?

4.10 In man, an abnormality of the large intestine called intestinal polyposis is dependent on a dominant gene A, and a nervous disorder called Huntington's chorea is determined by a dominant gene H. A man carrying the gene A ($Aahh$) married a woman carrying the gene H ($aaHh$). Assume that A and H are on nonhomologous chromosomes. Diagram the cross and indicate the proportions of the children that might be expected to have each abnormality, neither, or both.

4.11 Beginning with the oögonium in the woman described in Problem 4.10, diagram the steps in the process of oögenesis necessary for the formation of the egg that produced an H child. Label all stages.

4.12 Diagram completely the process of spermatogenesis involved in the production of the sperm in Problem 4.10 necessary for the production of an A child. Label all stages.

4.13 A man produces the following kinds of sperm in equal proportions: AB, Ab, aB, and ab. What is his genotype?

4.14 Would greater variability be expected among asexually reproducing organisms, self-fertilizing organisms, or bisexual organisms? Explain.

4.15 If biopsies were taken from follicle tissues of the human ovary at the following developmental periods, what stages in the process of oögenesis might be observed: (a) fifth month of intrauterine development, (b) at birth, (c) 10 years of age, (d) 17 years of age?

4.16 What difference exists between male and female determining sperm in animals with XX females and XY males?

4.17 In line with Bridges' genic balance theory for sex determination, what is the expected sex of individuals with each of the following chromosome arrangements in Drosophila: (a) 4X 4A; (b) 3X 4A; (c) 2X 3A; (d) 1X 3A; (e) 2X 2A; (f) 1X 2A?

4.18 List the expected results in terms of sex and intersex combinations from a cross between a triploid ($3n$) female fly with two X chromosomes attached and one free, and a normal diploid male. (Assume that the cross is successful and the gametes of the female will carry one or two whole sets of autosomes).

4.19 In plants of the genus Melandrium, which sex will be determined by the following chromosome arrangements (a) XY, (b) XX, (c) XY with region I removed and (d) XY with region II removed?

4.20 What sex is expected for individuals of the following genotypes in Habrobracon: (a) Xb, (b) XaXb, (c) XcXc, (d) XbXc?

4.21 How could maize plants, which are ordinarily monoecious, give rise to plants that are dioecious?

4.22 How many sex chromatin bodies are expected to occur in cell nuclei with each of the following chromosome arrangements: (a) XY, (b) XX, (c) XXY, (d) XXX, (e) XXXX (f) XYY?

4.23 (a) If newly hatched marine worms of the genus Bonellia were kept isolated from all other worms, what sex would they represent when adults? (b) If they developed in the vicinity of mature females, what would be their sex? (c) Develop an explanation for this type of sex determination.

4.24 In sheep, the gene h^+ for horned condition is dominant in males and recessive in females. If a hornless ram (male) were mated to a horned ewe (female), what is the chance that an (a) F_2 male sheep will be horned? (b) F_2 female will be horned?

4.25 The dominant autosomal gene (B) for premature baldness in man is considered to be sex limited. If a man with the genotype B^+B, married a woman with the genotype B^+B, what proportion of their (a) male and (b) female children might be expected to become bald prematurely?

4.26 In chickens, the gene h, which distinguishes hen feathering from cock feathering, is sex limited. Males may be hen-feathered or cock-feathered but females are always hen-feathered. If a cock-feathered male (hh) were mated to a homozygous (h^+h^+) hen-feathered female, what patterns of feathering might be expected among the (a) male F_2 and (b) female F_2 progeny? (hh^+ males are hen-feathered.

4.27 In a particular species of grasshoppers, two pairs of autosomes are heteromorphic; that is, they can be distinguished by microscopic observation. In one pair, one homologue is rod shaped and the other has a small hook at the end. One member of the other pair has a knob on one end. List all distinguishable combinations, with reference of these two pairs, that can be found in the sperm.

4.28 In Drosophila, the recessive gene (bb) for bobbed bristles is located in the X chromosome. The Y chromosome of Drosophila carries a homologous section in which bb or its allele bb^+ may be located. Give the genotypes and phenotypes of the offspring from the following crosses:

(a) $X^{bb}X^{bb} \times X^{bb}Y^{bb^+}$

(b) $X^{bb}X^{bb^+} \times X^{bb^+}Y^{bb}$

(c) $X^{bb^+}X^{bb} \times X^{bb^+}Y^{bb}$

(d) $X^{bb^+}X^{bb} \times X^{bb}Y^{bb^+}$

4.29 In snails of the genus Limnaea, coiling is transmitted as a maternal effect. (a) Give the phenotypes that could be associated with the following genotypes in individual snails and give the reason for each answer: s^+s^+, ss^+, and ss. (b) What might be said about the female and male parents and grandparents of snails represented by each of the three genotypes?

4.30 Diagram a cross between a female snail with dextral coiling with the genotype s^+s^+, and an inbred ss male with sinistral coiling. Carry the cross to the F_2 and represent the expected results from inbreeding each of the F_2 snails. Explain the results.

4.31 In the beach hopper of the genus Gammarus, pigment of the eyes is influenced in early stages by the genotype of the mother but later by the genes of the individual hopper. Give the expected results of the following crosses in young and adult stages: (a) dark females (AA) X light males (aa); and (b) light females (aa) X dark males (AA). (c) Give a plausible explanation for the change that sometimes occurs from light eyes in young organisms to dark eyes in later stages.

4.32 When ovaries from light-colored (aa) flour moths of the genus Ephestia are implanted into dark (AA) females, which are then mated to aa males, the aa progeny have dark eyes when first hatched but they gradually become lighter. Give a plausible explanation for such a change in eye color.

4.33 A female fruit fly known to be heterozygous for y (for yellow body color) had patches of yellow on the thorax. Is this expression more likely to be the result of (a) nonhereditary environmental modification, (b) maternal effect, or (c) nondisjunction of chromosomes resulting in a gynandromorph?

FIVE

In 1906, W. Bateson and R. C. Punnett analyzed the results of a cross between two varieties of sweet peas and observed a ratio that did not conform to the hypothesis of independent assortment. Sweet peas with purple flowers and long pollen grains had been crossed with a variety that had red flowers and round pollen grains. Purple and long phenotypes had been shown by previous experiments to be dependent on separate genes that were each dominant over the respective allele for red and round. From the F_1 dihybrid cross, Bateson and Punnett expected a 9:3:3:1 ratio in the F_2 that was characteristic of independent assortment.

Instead, progeny that had either purple flowers and long pollen, or red flowers and round pollen (parental types) occurred more frequently than expected, whereas the purple and round, and the red and long classes appeared less frequently. In another cross, sweet peas with purple flowers and round pollen were crossed with those having red flowers and long pollen. When the F_2s were classified, the parental combinations again were present in exag-

LINKAGE, CROSSING OVER, AND CHROMOSOME MAPPING

Grasshopper chromosome bivalent showing chiasmata. (Courtesy A. M. Winchester, University of Northern Colorado.)

141

gerated proportions and nonparentals (now known to be "recombinants") occurred in lesser proportion than expected on the basis of independent assortment.

Bateson and Punnett developed a theory called "coupling and repulsion" to explain their results. In their theory, the condition in which the two dominants tended to enter the gametes together in greater than random proportion was called "partial gametic coupling." The tendency for one dominant and one recessive gene to enter the gametes in the greater proportion was called "repulsion." Although the theory is now obsolete, the terms "coupling" and "repulsion" have been retained and are usefully descriptive. When both nonallelic mutants are present on one homologue and the other homologous chromosome carries the wild-type alleles (ab/++) the genes are said to be in the **coupling** configuration; when each homologue contains a mutant and a wildtype gene (a+/+b) the genes are in the **repulsion** phase. The purple-long X red-round cross represents coupling, whereas the purple-round X red-long cross illustrates repulsion.

LINKAGE AND CROSSING OVER

The chromosome theory of Boveri and Sutton, that chromosomes are the carriers of genes and that their meiotic behavior is the basis for Mendel's principles, became firmly established between 1902 and 1912. T. H. Morgan (1866–1945; Fig. 5.1) and his associates, through their investigations with Drosophila, were largely responsible for supplying experimental evidence supporting this theory. On firm cytological grounds, the Morgan group established the theory of linkage and crossing over. **Linkage** was first defined as the tendency for nonallelic genes in the same chromosome or "linkage group" to enter the gametes together (i.e., in parental combination) in higher proportion than expected from independent assortment. Nonparentals (re-

Figure 5.1 Thomas Hunt Morgan, American geneticist and embryologist. Dr. Morgan was Nobel Laureate in biology and medicine in 1933. (Courtesy of California Institute of Technology.)

combinants) were significantly fewer than the parentals and less than a frequency of 0.5. It soon became established that the number of **linkage groups** in species of animals and plants is **equivalent to the haploid number of chromosomes.**

Nonallelic genes are linked because they **reside in the same chromosome;** they, therefore, tend to remain together in the process of meiosis and enter the same gamete. The alternative, *crossing over*, is defined as the tendency for nonallelic genes to enter the gametes in proportions other than those of the parents. These are the **recombinants.** Genes cross over when chromosome parts in homologous chromosomes change places. They then enter the gametes in arrangements that differ from those in the parents. Since numerical data rather than a physical phenomenon are required to establish linkage and crossing over, they became statistical concepts.

Soon after linkage and crossing over were established in the fruit fly, other organisms were found to conform to the same pattern. The first clearly recognized linkage in a plant was reported in 1911 by G. N. Collins and J. H. Kempton in maize. They found that the gene (wx) for waxy endosperm was linked with the gene (c) for aleurone color. The principle of linkage was soon firmly established in both plants and animals and was recognized as a widely occurring alternative to independent assortment. Investigations then had to be designed to analyze the mechanism of linkage and crossing over.

It would be a simple matter for the geneticist if genes in the same member of a pair of chromosomes would always stay together and thus make linkage complete. Ordinarily, however, they do not do so. **Exchange** or crossing over between members of chromosome pairs occurs in most plants and animals. Known **exceptions** are the **male fruit fly** (Drosophila) and the **female silk moth** (Bombyx), which ordinarily have no crossing over. That is, they have complete linkage, although crossing over can be artificially induced in the male fly.

CYTOLOGICAL BASIS OF CROSSING OVER

Genetic evidence for linkage and crossing over became well established before the cytological basis was demonstrated. Linked genes were postulated to occur in linear order in their respective chromosomes. Genetic crossing over was considered a consequence of the exchange of parts between homologous chromosomes. At first this theory could not be demonstrated cytologically, however, because homologous chromosomes appeared, on microscopic examination, to be exactly alike. It was impossible to observe whether chromosome segments had changed places until visible markers could be incorporated on the chromosomes. This cytological demonstration was finally accomplished in 1931, through experiments of C. Stern (working with Drosophila), and H. B. Creighton and B. McClintock (working with maize).

Stern found flies with microscopically distinguishable X chromosomes in one of his cultures. A part of the Y chromosomes had broken off and became attached to the X chromosome. Stern recognized the value of cytological markers for demonstrating crossing over and continued his search for a second marker in the same chromosome pair. Eventually, he was provided with another distinguishing feature at the end of the other X chromosome. This marker had also arisen by fragmentation and subsequent joining of chromosome parts. A part of the X chromosome had been broken off by X-ray treatment and had become attached to the small chromosome IV. The fragmented X chromosome was con-

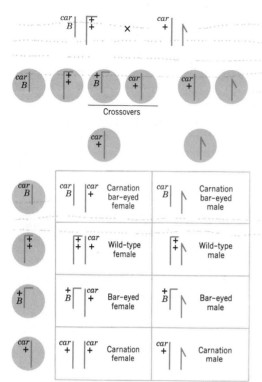

| | car B | | | | | |
| car B | | + + | + B | | car + | |

Crossovers

| car + | | | | | | |

	car B	car +	Carnation bar-eyed female	car B		Carnation bar-eyed male
	+ +	car +	Wild-type female	+ +		Wild-type male
	+ B	car +	Bar-eyed female	+ B		Bar-eyed male
	car +	car +	Carnation female	car +		Carnation male

Figure 5.2 Stern's classical experiment demonstrating the cytological basis for crossing over.

siderably shorter than the unbroken X chromosomes. Now both X chromosomes could be distinguished by microscopic examination, and a cytological demonstration of crossing over was possible.

Stern's experiment, which is one of the classics in genetics, is illustrated in Fig. 5.2. The genes, *car* for carnation-colored eye and *B* for bar-shaped eye, were made heterozygous in the cytologically distinguishable X chromosomes of the females used in the experiment. Females with red, bar-shaped eyes were mated with males having carnation-colored, normal-shaped eyes. The two X chromosomes from the female with distinct cytological markers and the one from the male with no markers could be identified in the progeny from the cross by appropriate microscopic examination. The progeny were classified and the chromosomes examined. The genetic (linkage and crossing over) and cytological (chromosome appearance) results were recorded. Material from flies that phenotypically suggested crossing over showed microscopic evidence of exchanges between homologous chromosomes. The physical or **cytological basis of crossing over** was thus established.

Creighton and McClintock, in studying maize, made use of a knob on the end of a certain chromosome and a visible irregularity on its homologue as cytological markers. Their results were generally similar to those of Stern.

MECHANISM OF CROSSING OVER
Visible crossovers or chiasmata (Fig. 4.5*d*) between homologous chromatids can be ob-

served with the microscope during the latter part of the meiotic prophase in many animals and plants. They do not remain in the same position but tend to move toward the ends of the bivalents; that is, they become terminalized. A chiasma is not synonymous with a point of crossing over, but there is a remarkable agreement between the total numbers of chiasmata and crossovers in organisms that have been investigated. A reciprocal exchange between chromatids is associated with chiasmata formation in higher animals and plants.

Historically, two different simple explanations were proposed to account for crossing over: (1) **breakage and reunion** and (2) copy choice. The best supported of the two was breakage and reunion based on what was considered to be a mechanical break-and-reunion of chromatids during the early part of the meiotic prophase. Such breaks must have occurred before chiasma formation and at the time of or following chromosome duplication. This theory required duplicated chromatids prior to synapsis. The unit threads at this stage are thin, tangled, and difficult to observe. It was postulated that they twisted around each other and broke under stress. Broken ends were further postulated to rejoin, but not necessarily with the same segment from which they were detached. When broken ends of **different chromatids became joined, an exchange in genetic material had occurred.** Single threads pulling out in the anaphase could thus include parts of different chromatids, as illustrated in Fig. 5.3. Although this theory of reciprocal breakage and rejoining will not explain all of the requirements for the mechanics of crossing over in all organisms, it was strongly supported from data obtained from higher an-

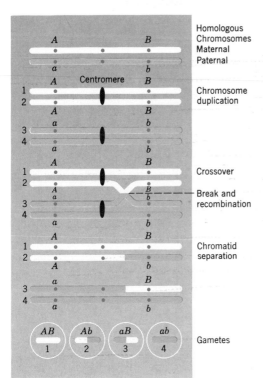

Figure 5.3 Steps in the mechanics of crossing over according to the theory of chromosome breakage and rejoining.

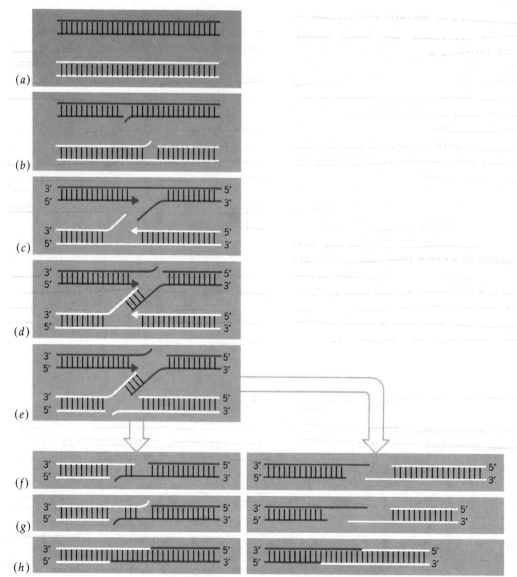

(a)

(b)

(c) 3′ ... 5′ / 5′ ... 3′ ; 3′ ... 5′ / 5′ ... 3′

(d) 3′ ... 5′ / 5′ ... 3′ ; 3′ ... 5′ / 5′ ... 3′

(e) 3′ ... 5′ / 5′ ... 3′ ; 3′ ... 5′ / 5′ ... 3′

(f) 3′ ... 5′ / 5′ ... 3′   3′ ... 5′ / 5′ ... 3′

(g) 3′ ... 5′ / 5′ ... 3′   3′ ... 5′ / 5′ ... 3′

(h) 3′ ... 5′ / 5′ ... 3′   3′ ... 5′ / 5′ ... 3′

Two double stranded recombinant DNA molecules

Figure 5.4 Hypothesis that crossing over starts with pairing between complementary single-stranded tails growing out from double-helical DNA molecules. (*a*) Two homologous DNA molecules (*b*) An endonuclease makes a nick in one strand of each molecule. (*c*) DNA polymerase synthesizes on one side of each cut strand thus exposing two single-stranded free region tails. (*d*) Two such single-stranded regions, if homologous, can base pair giving a short double-stranded bridge. (*e*) An endonuclease nicks the other strands giving one recombinant molecule and two molecular fragments with overlapping terminal sequences. (*f*) (left) DNA polymerase synthesizes the missing portions. (right) Exonuclease eats away one strand of each half

imals and plants and had impressive support from studies of bacteria and viruses. It undoubtedly represents the basic mechanism for crossing over at least in higher organisms, although other factors may also be involved. The process of breakage and reunion may, however, be **explained by enzymatic** as well as mechanical **exchange between chromatids.**

Experiments with phages T4 and lambda have shown that **specific enzymes** are involved in recombination. A current hypothesis is that a nick may occur in a single strand of each maternal and paternal double-stranded DNA region through action of an **endonuclease,** an enzyme that breaks bonds in a DNA molecule. Such cuts create free ends to which **DNA polymerase I** (Kornberg's enzyme) can add new complementary nucleotides (Fig. 5.4). These nucleotides can pair and form a short, double-stranded junction or bridge between maternal and paternal strands.

The other single strand of each molecule may then be cut by endonuclease resulting in one recombinant molecule and two molecular fragments with overlapping terminal sequences. DNA polymerase I synthesizes the missing portions of each region. Endonuclease removes nonhomologous parts of each strand and thus reveals complementary single-stranded regions that can pair. **Polynucleotide ligase,** an enzyme that catalyzes the joining of two segments of an interrupted strand in double-stranded DNA, **seals the two strands** resulting in a double-stranded recombinant molecule. Base pairing between the other two strands results in another double-stranded recombinant molecule. Gap filling is completed by such enzymes as DNA repair polymerase. Two **double-stranded recombinant DNA molecules** are thus produced, each with **exchanged regions** as compared with the original homologous molecules.

DETECTION OF LINKAGE AND CROSSING OVER

For most experimental organisms, the **test cross** is the most convenient way to determine the linkage group to which a gene belongs and the relative position of that gene within the linkage group or chromosome. A proper experimental arrangement fulfills two different requirements: (1) corresponding parts on a pair of chromosomes can exchange places, and (2) the end results can be used to identify phenotypically the crossovers that have occurred and the resulting recombinations. The chief advantage of the test cross results is the simple ratio it provides, which should be $1:1:1:1$ if independent assortment occurred. For example, the gene (vg) for vestigial wing in *D. melanogaster* is located in the second chromosome, whereas the gene (e) for ebony body color is in the third chromosome (Fig. 5.5). In a basic experiment, a homozygous normal female was crossed with a vestigial, ebony male, and F_1 females were mated with fully recessive vestigial, ebony males. The $1:1:1:1$ ratio expected when genes are

molecule revealing homologous regions. (*g*) (left) Polynucleotide ligase seals the two strands giving double stranded recombinant molecule. (right) Base pairing gives double-stranded recombinant which is completed by gap filling using DNA polymerase and ligase. The result (*h*) is two double stranded recombinant DNA molecules. (From James D. Watson, *Molecular Biology of the Gene*, Second Edition, copyright © 1970 by J. D. Watson; W. A. Benjamin, Inc., Menlo Park, Calif.)

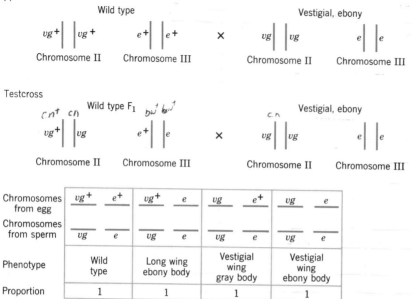

Figure 5.5 P cross between wild-type and ebony flies with genes on chromosomes. F_1 females were crossed with the full recessive vestigial ebony parental type. The resulting 1:1:1:1 ratio from the test cross indicated independent assortment for the two pairs of alleles.

not linked was obtained. Because the 1:1:1:1 ratio is characteristic of independent assortment, genes of unknown position giving this test-cross result can ordinarily be assumed to be in different chromosomes. It follows that a **significant departure from the 1:1:1:1 ratio** (determined by the χ^2 test) indicates something other than independent assortment. **Linkage** is the most likely alternative.

In an attempt to illustrate the operation of an alternative to independent assortment, a homozygous female fruit fly with straight (wild-type) wings and gray (wild-type) body was mated with a male with curled wings and ebony body. (Curled and ebony are recessive to both the wild phenotypes, straight and gray, respectively.) When F_1 females were mated with fully recessive males, the result was a ratio of about four straight, gray; one straight, ebony; one

curled, gray and four curled, ebony. This F_1 result indicated that cu and e were in the same chromosome and in the "coupling phase": $\dfrac{cu^+e^+}{cu\ e} \times \dfrac{cu\ e}{cu\ e}$. When the original cross (P) was between straight, ebony females and curled, gray males, $\dfrac{cu^+e}{cu^+e} \times \dfrac{cu\ e^+}{cu\ e^+}$ and F_1 wild-type females were mated with full recessive males, $\dfrac{cu^+e}{cu\ e^+} \times \dfrac{cu\ e}{cu\ e}$ the test-cross results yielded ratios of about one straight, gray: four straight, ebony; four curled, gray; and one curled, ebony, as expected when the gene arrangement is in the "repulsion phase." These three crosses illustrating independent combinations, coupling, and repulsion, are constructed on the chromosomes in Figs. 5.5, 5.6, 5.7, respectively.

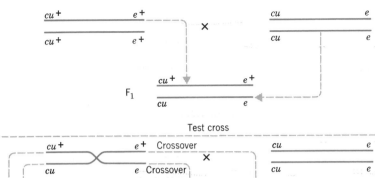

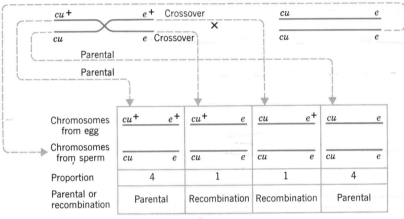

Figure 5.6 A test cross involving two genes in the same chromosome pair. The results show the parental (linkage) and recombination (cross-over) groups. This illustrates coupling.

When two genes have been identified as being in the same chromosome, their relative positions with respect to each other can be estimated by calculating the frequency of crossing over between them. In converting the proportion of crossing over to relative positions on the chromosome, **one percent of** recombination is equal to **one unit** on the **linkage map.** The number of cross-over units between two gene loci is the same as the percentage of progeny that represent an exchange between the two loci. If idealized figures (40 straight, gray; 10 straight, ebony; 10 curled, gray; and 40 curled, ebony) are supplied in the foregoing example representing the coupling phase, the parental and recombination groups would be distinguished as follows:

straight; gray 40 (parental)
curled; gray 10 (recombinant)
straight; ebony 10 (recombinant)
curled; ebony 40 (parental)

The two parental groups combined include 80 percent of the total flies and the two recombination groups, 20 percent. Therefore, the linkage strength between these two loci is 80 percent and the recombination is 20 percent. Thus the loci *cu* and *e* are about 20 units apart. Similar relations are indicated from a repulsion cross when the parental and recombination gametes are properly distinguished. Idealized figures were supplied in the above example to illustrate the concept. Data from actual experiments are not as regular as the "ideal."

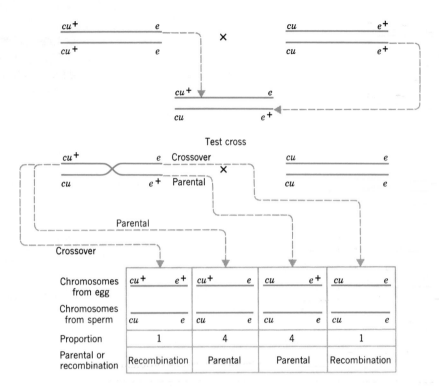

Figure 5.7 A test cross involving two genes in the same chromosome pair. The results show the recombination (crossover) and parental (linkage) groups. This illustrates repulsion.

The first observation by Bateson and Punnett that suggested a deviation from random combination was from F$_2$ dihybrid cross results rather than test-cross results. Such data are as valid as those from test crosses, but they are more cumbersome to handle because of the more complex 9:3:3:1 instead of the 1:1:1:1 ratio with which the results must be compared. However, appropriate comparisons can be made mathematically. F$_2$ data from a dihybrid cross may be more readily obtained than test-cross data in some species, especially in self-fertilizing plants such as wheat and barley. It is tedious and time consuming to make test crosses in these plants by hand emasculation and pollina-

tion, whereas it is easy merely to allow the F$_1$ plants to self-fertilize and produce F$_2$ plants. Breeders of cereal crops usually choose the simpler means of making the cross and the more complex methods of analysis and interpretation. Tables have been prepared that provide simple and effective tools for calculating linkage and cross-over values from F$_2$ data.

When cross-over values are known, **predictions can be made** as to the proportions of gametes with different gene combinations likely to result from a given cross. From the relative proportions of gametes, the offspring resulting from various crosses can be predicted. J. W. MacArthur, for example, found from appropriate test-cross re-

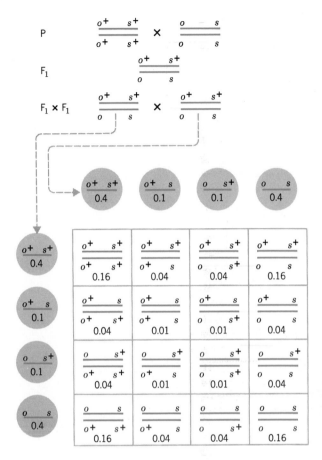

Figure 5.8 Linkage problem in the tomato illustrating the method used to predict the proportions of progeny when linkage values are known. (Data from J. S. MacArthur.)

sults with tomatoes that the locus s for compound inflorescence is about 20 units from the locus o for elongate fruit shape. With this information, it was possible to predict the F_2 results, as illustrated in Fig. 5.8. A close fit was obtained when the expected results based on the predicted 20 percent crossing over were compared with the experimental data from actual crosses.

MAXIMUM CROSSING OVER BETWEEN TWO LINKED GENES

Actually **no more than 50 percent recombination** is expected between any two loci because only two of the four chromatids in a meiotic tetrad are involved in any particular cross-over event. The recombination value between two points usually appears less than 50 percent because: (1) not every bivalent may possess a pair of cross-over chromatids and (2) double or multiple crossing over between gene pairs that are far apart along a chromosome will reduce the apparent number of recombination products. The net effect of different numbers of odd crossovers between two genes (with no gene markers between them) is no different from the effect of single

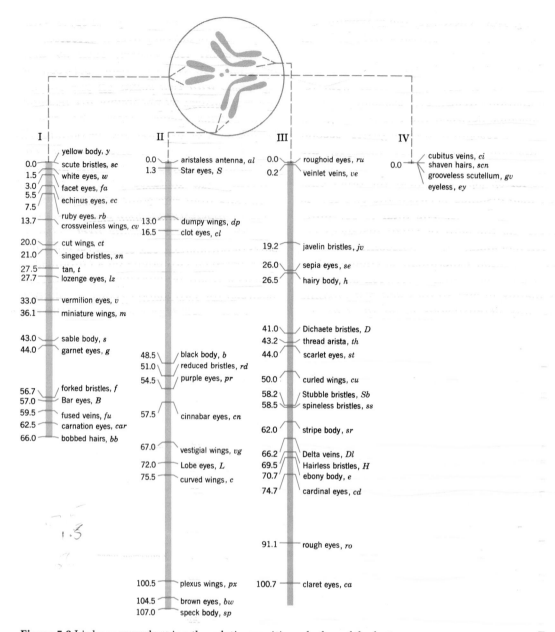

Figure 5.9 Linkage map showing the relative position of a few of the better known genes on each of the four chromosomes of *Drosophila melanogaster*. Gene symbols and descriptive phenotypes are given opposite gene locations. (Data from Bridges.)

crossing over. Therefore, the maximum detectable amount of crossing over between two genes is 50 percent.

The map distance between two loci on a chromosome is not synonomous with recombination frequency, especially when the distance between the two loci is great. On the *Drosophila melanogaster* second chromosome (Fig. 5.9) for example, the locus for *al* (aristaless) is at one end, and the locus for *sp* (speck) is near the other end. The map distance between the two loci is given as 107 units. The total of 107 units represents the *sum* of small distances between loci along the chromosomes. When a recombination experiment is conducted to determine the distance between *al* and *sp* by mating *al* and *sp* flies and test crossing the fully heterozygous F$_1$ (*al sp*$^+$ / *al*$^+$ *sp*)

with the full recessive *al sp / al sp*, less than 50 percent of crossing over is detected.

Two-point test crosses, such as those described above, are made between individuals with two points on one chromosome identified by phenotypically recognizable marker genes. From the results, frequencies of parental and recombination groups of progeny are distinguished. This information is used to determine the chromosome map distance between the two loci. Data from two-point crosses, however tend to underestimate map distances, partly because of **double crossing over.** That is, parts of chromatids can cross over and cross back between two points on a chromosome and as a result no crossing over is detected between the two marker genes. In addition, a two-point cross does not indicate the rela-

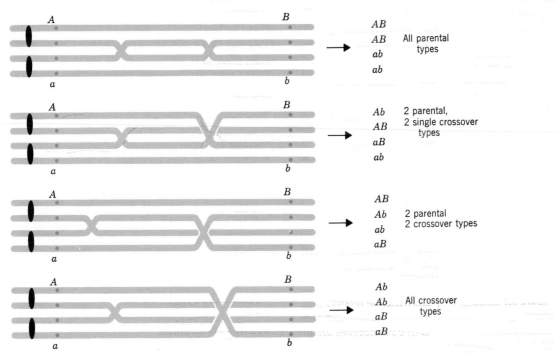

Figure 5.10 Consequences of different kinds of double crossing over between two gene loci *A* and *B*. If a third gene locus (*C*) between *A* and *B* could be studied, additional single cross-over events could be detected and evidence of double crossing over between *A* and *B* could be obtained.

tive positions of genes with respect to each other on the chromosome.

The consequences of different kinds of double crossing over between two gene loci are illustrated in Fig. 5.10. Four strands are present when crossing over occurs but only two of the four strands are involved in a particular cross-over event. If a third pair of alleles $C c$ was located between the A and B loci and available for study, more single cross-over events could be detected.

For example, if only the distance between genes a and b is known, a may be either to the right or to the left of b. To determine the order and distance between the genes on a chromosome, it is necessary either to obtain map distances between a and b and other known genes in the same linkage group, or to work with three linked genes at one time and thus determine their relative position with respect to each other. With a three-point cross, it is possible to check two adjacent chromosome areas on one chromosome simultaneously. In addition to the relative position of genes, data from crosses of this kind may identify recombination units between points on a chromosome and thus provide evidence for double crossing over. Therefore, **data from three-point crosses will correct,** at least in part, **the underestimation of map distance that is inherent with two-point crosses.**

THREE-POINT CROSS

A three-point cross may be carried out if three points or loci on a chromosome pair can be identified by marker genes, and test crosses can be conducted successfully with the experimental material under investigation. If a third marker gene, c, is in fairly close proximity to genes a and b and in the same linkage group, all three markers may be used to achieve a reasonably precise analysis of their map distances and relative positions. Matings between individuals car-

rying abc in homozygous condition and those homozygous for $a^+b^+c^+$ would be expected to result in fully heterozygous F_1 progeny. These heterozygotes, when test crossed with homozygous, full-recessive abc individuals, would be expected to produce progeny that could be classified into parental and recombination groups. With the three genes known to be linked, the results would be expected to deviate from the $1:1:1:1:1:1:1:1$ ratio expected from a trihybrid cross under the alternative hypothesis of independent assortment.

A genetics student chose a special project involving a three-point cross to check the relative positions and map distances separating three genes in the Drosophila third chromosome. The project was carried out as follows. The student first mated Drosophila females homozygous for recessive genes cu (curled), sr (striped), and e (ebony) with wild-type, cu^+ (straight), sr^+ (not striped), and e^+ (gray) males. F_1 females were subsequently test crossed with fully recessive curled, striped, ebony males. The phenotypic results of the test cross were as follows:

1. Straight, not striped, gray | $cu^+sr^+e^+$ / $cu\ sr\ e$ | 786
2. Curled, striped, ebony | $cu\ sr\ e$ / $cu\ sr\ e$ | 753
3. Straight, striped, ebony | $cu^+sr\ e$ / $cu\ sr\ e$ | 107
4. Curled, not striped, gray | $cu\ sr^+e^+$ / $cu\ sr\ e$ | 97
5. Straight, not striped, ebony | cu^+sr^+e / $cu\ sr\ e$ | 86
6. Curled, striped, gray | $cu\ sr\ e^+$ / $cu\ sr\ e$ | 94

7. Straight,
   striped, $cu^+ sr\ e^+$ 1
   gray $cu\ sr\ e$

8. Curled,
   not striped, $cu\ sr^+ e$ 2
   ebony $cu\ sr\ e$

               1926 total flies

A few basic steps may be followed in analyzing the results from a three-point cross. (1) Determine the parental genotypes by (a) reconstructing the parental cross if possible, or (b) observing the most frequent classes of test-cross progeny. Method (b) is not always an accurate way to determine parental genotypes but may be useful when data about the parents are not available. (2) When the parental cross has been established, choose the double recombinants which is the **smallest class.** Determine the order of the genes by observing the **one gene in the double recombinant class that differs from the parental genotype.** Place this as the **middle one of the three genes.** In the following illustration, for example, $cu^+\ sr^+\ e^+$ and $cu\ sr\ e$ are parental combinations. In the double recombinant, sr varies from the parental arrangement and therefore is the middle gene. The order is $cu\ sr\ e$ or $e\ sr\ cu$.

Having established the order of the genes, the distances between them may be calculated. (3) The third step for the student is to determine the number of map units between the genes from the data by calculating the percentage of crossing over within the two areas. Considering the cu and sr region first, it can be seen (Fig. 5.11) that the cross-over event in the heterozygous female between cu and sr yields classes 3, 4, 7 and 8. The total number of the recombinants of these classes is 207 ($107 + 97 + 1 + 2$) of the 1926 flies counted. Therefore, the proportion of crossovers in areas cu-sr was 207/1926 or 10.7 percent.

Between sr and e, 183 crossovers (the sum of classes 5, 6, 7, 8) occurred (183/1926 or 9.5 percent). Therefore, cu was 10.7 units from sr and sr was 9.5 units from e. These special relations are indicated along with the relative positions of the three genes in Fig. 5.11. The results obtained by the student were compared with those reflected by the Drosophila linkage map (Fig. 5.9). On the linkage map, based on extensive data and including the results of cross-checks with several different genes in the same vicinity on the chromosome, cu is placed at 50, sr at 62, and e at 70.7 map units from the end of the chromosome. By subtraction, distances between cu and sr and sr and e are, therefore, 12 and 8.7 units, respectively, compared with 10.7 and 9.5 from the results of the student. Linkage relations are based on averages and on the assumption that crossing over occurs at equal frequency in all parts of the chromosome.

When the same test cross was performed with F_1 males instead of females, the results were quite different. Only parental phenotypes (curled, striped, ebony, and wild-type) were produced. This result indicated that no crossing over had occurred in male flies.

Originally, the two genes cu and e in Drosophila, had been calculated to be 20

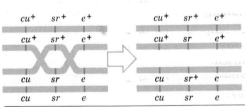

Figure 5.11 Three genes, cu, sr, and e, in their relative positions on a homologous pair of chromosomes in F_1 females. Cross-overs indicated between cu and sr and sr and e are arranged to explain the results of the three-point cross cited in the text.

MAXIMUM CROSSING OVER BETWEEN TWO LINKED GENES

units apart on the basis of single phenotypically detectable crossovers between the locus for *cu* and that for *e*. If the chromatids were involved occasionally in the two crossovers between *cu* and *e*, a section might cross over and cross back, leaving the end points *cu* and *e* in the parental arrangement. In such cases, no crossing over would be detected in test-cross results. When the third gene (*sr* between *cu* and *e*) was introduced into the investigation, it became possible to identify double crossovers between *cu* and *e*. The distance then was calculated to be $10.7 + 9.5 = 20.2$. Three-point crossover data can at least partly correct any errors in map unit calculations that are introduced by double crossing over, only when distances are great enough to permit an appreciable amount of double crossing over. In practice, a distance of 10 or less crossover (or map) units is considered by Drosophila geneticists to be a safe limit within which an appreciable amount of double crossing over will not occur.

INTERFERENCE AND COINCIDENCE

The occurrence of a crossover in any given chromosome segment is mostly a **chance phenomenon,** but the distribution is **not completely random.** The chance of two crossovers occurring simultaneously in adjacent regions should be, according to the law of probability, the product of the probabilities of separate occurrence in the two sections. H. J. Muller showed in 1916 that actual double crossovers are less frequent than would be expected on the basis of purely random distribution. He reasoned that a crossover in one place must inhibit others in the immediate vicinity. This is called **interference.** Muller developed a mathematical model to express his substantiating data, which were obtained from Drosophila and maize. The model is the ratio between the observed and expected frequency of double crossing over or

$$\frac{\text{actual double crossovers}}{\text{expected double crossovers}}$$

This mathematical measure of interference is termed the **coefficient of coincidence. Complete interference gives no double crossovers and a coincidence of 0; whereas no interference results in a coefficient of coincidence of 1.**

In view of this information, the last step in analyzing a three-point cross is to determine the amount of interference. Data cited from the double cross-over experiment performed by the genetics student can be used to calculate the **coefficient of interference** (1 − coefficient of coincidence), in the area of the Drosophila third chromosome between *cu* and *e*. On the basis of probability, $0.107 \times 0.097 = 0.0102$ or slightly more than 1 percent double crossovers should have occurred if the distribution were at random. The actual occurrence was only 3/1926 or 0.0016, and the coefficient of coincidence is 0.0016/0.0102 or 0.16. Therefore, only 16 percent of the expected double crossovers were recovered, indicating a partial interference of 84 percent. Less interference would be expected if the genes involved were further apart.

When linkage data were accumulated for mice, the model of coincidence based on Drosophila data was not adequate. A sex difference was detected in studies of linkage group X111, with males exhibiting more intense interference than females, and environmental factors (such as temperature) exerting an effect.

A different situation exists in the mold Aspergillus and bacterial phage T4. Occurrence of one crossover appears to increase the likelihood of another in the same vicinity. A coincidence greater than 1 has been reported in some investigations. This phenomenon has been called **negative interference.**

CHROMOSOME MAPPING

Three-point crosses commonly provide data for the beginning of a genetic map. These maps are graphic representations of the relative positions and distances of genes in each linkage group. Distances are expressed in percentages of recombination (or crossing over) between loci along the chromosome. When short distances (less than 10 units) are considered, simple linear addition may be used in going from one marker point to the next. Many investigators have contributed to the necessary **accumulation of recombination data.** Variations in the results of different investigators are usually explainable by appropriate checks of experimental conditions.

The momentum for chromosome mapping was established in the 1920s by Drosophila geneticists, particularly T. H. Morgan, C. B. Bridges, and A. H. Sturtevant. Some of the best-known loci in *D. melanogaster* are listed in Fig. 5.9, along with their relative positions in the four linkage groups corresponding to the four chromosome pairs. It should be noted that three (I, II, and III) of the four groups are large and one (IV) is very small. The chance that any two nonallelic genes will be linked is greater in Drosophila (with only three major linkage groups) than in organisms having more chromosomes and more chances for independent combinations. Maize (Fig. 5.12), with ten chromosome pairs corresponding to ten linkage groups, would logically be expected to have less linkage and more independent assortment than Drosophila. This expectation has been borne out by comparative studies.

After linkage relations between genes are established for a given organism, an investigator who identifies a new gene in that experimental material may determine its linkage group and position within the group. Many mutant genes have been mapped in maize, with several others being placed tentatively. Chromosomes of higher plants such as tomato, barley, wheat, rice, sorghum, morning glory, and garden pea, are quite well mapped. Progress in chromosome mapping for mammals has been comparatively slow. Mice have 20 chromosome pairs, and 20 linkage groups have been tentatively identified. Mapping of human chromosomes has been a tedious and uncertain undertaking, but new statistical and cytological techniques (Chapter 16) have resulted in great progress.

TETRAD ANALYSIS

Some organisms are more suitable than others for obtaining information about the consequences of meiosis and the positions of genes on chromosomes. In certain fungi and algae, all products of a single meiotic event can be recovered for genetic analysis. Such an analysis is called **tetrad analysis.** Much information including (1) stage in the meiotic cycle when crossing over occurs, (2) position of a gene with respect to the centromere, (3) linkage or nonlinkage between two gene loci, and (4) relative positions of linked genes can be deduced from such analyses.

ORDERED TETRADS
The mold, Neurospora, (Fig. 5.13) is especially well suited for experimental efforts to define the stage in which crossing over occurs and to investigate the meiotic division in which crossover chromatids are separated. All four of the products of meiosis in such materials are held in the ascus, or reproductive sac, in the order in

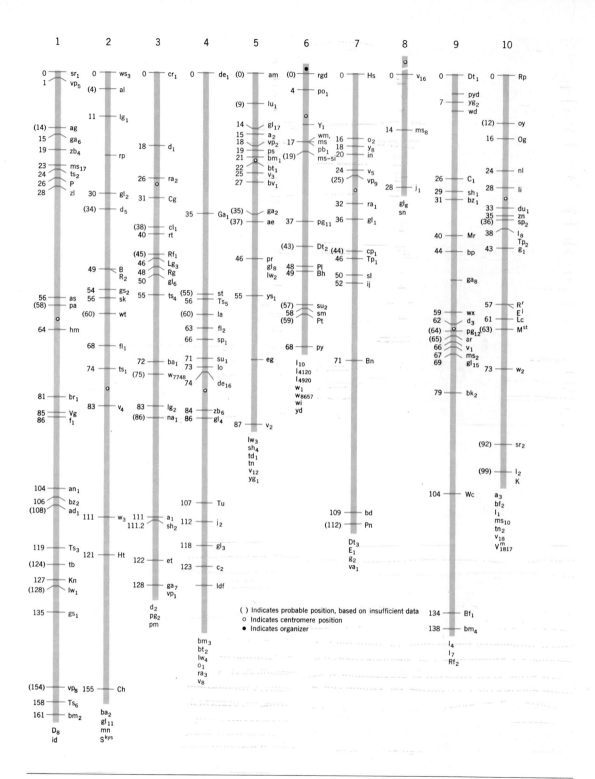

( ) Indicates probable position, based on insufficient data
o Indicates centromere position
● Indicates organizer

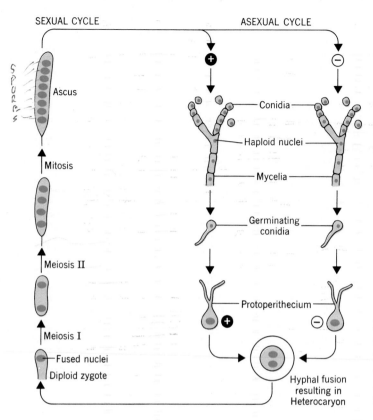

SEXUAL CYCLE   ASEXUAL CYCLE

SPORES

Ascus

Mitosis

Meiosis II

Meiosis I

Fused nuclei

Diploid zygote

Conidia

Haploid nuclei

Mycelia

Germinating conidia

Protoperithecium

Hyphal fusion resulting in Heterocaryon

Figure 5.13 Sexual and asexual cycle in Neurospora. Asexual reproduction is based on mitotic division of haploid nuclei conidia. A heterocaryon may be formed from hyphal fusion.

Figure 5.12 Linkage map of maize. Descriptive phenotypes are given below for some of the genes located in each of the 10 chromosomes. Chromosome 1: sr_1, striated-1; as, asynaptic; an_1, anther ear-1; bm_2, brown midrib-2. Chromosome 2: ws_3, white sheath-3; gl_2, glossy seedling-2; sk, silkless; Ch, chocolate pericarp. Chromosome 3: cr_1, crinkly leaf-1; d_1, dwarf-1; ts_4, tassel seed-4; ba_1, barren stalk-1. Chromosome 4: de_1, defective endosperm; Ga_1, gamete differential fertilization; Tu, tunicate; j_2, japonica-2; gl_3, glossy seedling-3. Chromosome 5: a_2, aleurone color-2; pr, red aleurone; ys_1, yellow stripe-1; v_2, virescent-2. Chromosome 6: po_1, polymitotic-1; Y_1, yellow endosperm-1; Pl, purple, sm, salmon silk, py, pigmy. Chromosome 7: o_2 opaque-2; gl_1, glossy-1; Tp_1, teopod-1; ij, iojap striping; bd, branched. Chromosome 8; v_{16}, virescent-16; ms_8, male-sterile-8. Chromosome 9: Dt_1, dotted-1, C_1, aleurone color-1; sh_1, shrunken endosperm-1, wx, waxy. Chromosome 10: Rp, rust resistant; g_1, golden-1; R, color factor. (O-Approximate position of centromere.) (Data from Rhoades)

which they were produced. The order of spores in the ascus thus indicates the pattern of events in meiosis. Spores can be picked up in order and germinated separately to demonstrate phenotypic characteristics.

This kind of analysis is not possible in Drosophila and most other organisms used routinely in genetic investigations because the gametes are produced individually from diploid germ cells and released into tubules in the testes where they become mixed. Except for special cases such as attached X-chromosomes in Drosophila, it is impossible to determine which gametes have come from which cells and the order in which they were produced. The four sperm from a primary spermatocyte, for example, are mixed with sperm from other spermatocytes. In the comparable process in the female, only one of the four products of each meiosis is functional and the other three are lost as polar bodies.

The life cycle of the mold (Fig. 5.13) includes both sexual and asexual reproduction. The asexual process is accomplished by means of spores (called conidia), which are produced on end branches of the mycelium, the interwoven threadlike mass that forms the vegetative portion of the mold. The nuclei of the mycelium are haploid and reproduce themselves by mitotic division. Sometimes a nonsexual fusion occurs between hyphae derived from different genotypes. When this occurs, cells may fuse and thus bring two nuclei into the same cell. The nuclei, however, do not come together but retain their individuality. Such a cell with two nuclei is called a **heterocaryon.** When the cell divides, each nucleus goes into a new cell. Besides its common asexual vegetative mode of reproduction, Neurospora has a **sexual phase** that involves the fusion of cells of opposite mating types (A and a). The union of two mating cells results in a **diploid zygote,** but the two nuclei do not always fuse immediately. When they do come together, a **reduction division** similar to that in higher plants and animals follows.

The first two divisions of the zygote are meiotic divisions. Four nuclei, comparable with the four spermatids of animal spermatogenesis, can be seen in the ascus at the end of the second division. A mitotic division follows the meiotic sequence, increasing the number of nuclei to eight. All divisions occur in the plane of the long axis of the ascus, and the nuclei remain in the order in which they were produced. Because the third division is mitotic, each of the four products that resulted from the meiotic divisions gives rise to two identical nuclei. The eight nuclei are lined up two by two with the identical nuclei next to each other. Each nucleus is eventually surrounded by a spore wall and becomes an ascospore. When an ascus is mature and the proper environment is provided, it ruptures and the ascospores are freed to germinate, with each giving rise to a new mycelium.

Controlled crosses are made by selecting mating types with different marker gene combinations to produce zygotes. If the ascospores are dissected out before the ascus ruptures, they may be raised in separate culture tubes. When the tubes are kept in order, that is, the first ascospore placed in the first tube, the second in the next, and so on, the pattern of segregation of genes from a single zygote can be demonstrated. The phenotypic results of the cross can then be observed or tested biochemically. Characteristics of the mycelium arising from a single ascospore reflect the gene combination received by the nucleus in meiosis.

CROSSING OVER IN FOUR STRAND STAGE

Alleles immediately next to the centromere rarely cross over. However, when separate pairs of alleles such as + and − are located some distance from the centromere on the chromosome, it is possible to detect crossing over. It is also possible to determine whether the exchange occurred in the two-strand or the four-strand stage. If crossing over occurred at the two-strand stage, either the original (parental) arrangement would be retained in the ascus or a recombination type would occur. If crossing over occurred at the four-strand stage, both parental and recombination types should occur in the same ascus. When the two pairs of alleles diagrammed in Fig. 5.14 had exchanged positions through crossing over, results were $1a+:1a-:1A+:1A-$. This indicates that **crossing over occurred in the four-strand stage,** but that only **two of the four strands were involved in a single crossover.**

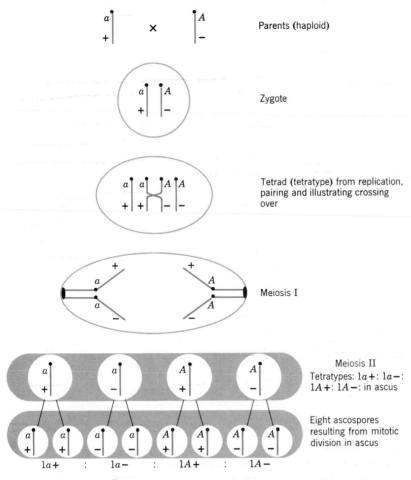

Parents (haploid)

Zygote

Tetrad (tetratype) from replication, pairing and illustrating crossing over

Meiosis I

Meiosis II
Tetratypes: $1a+: 1a-:$
$1A+: 1A-:$ in ascus

Eight ascospores resulting from mitotic division in ascus

$1a+$: $1a-$: $1A+$: $1A-$

Figure 5.14 Tetratype resulting from crossing over between genes + and − and centromeres in Neurospora.

POSITION OF GENE IN RELATION TO CENTROMERE

In addition to a life cycle that allows each of the four meiotic products to be analyzed, fungi such as Neurospora have the four meiotic products arranged linearly so that the investigator can distinguish between first and second meiotic division segregation. As shown in Fig. 5.15, the alleles, A/a and B/b, segregate at the 1st meiotic division. However, when crossing over occurs between the gene and the centromere, segregation of alleles does not occur until the second meiotic division. This is diagrammed in Fig. 5.16 for the chromosome carrying the A a alleles. Map distances between the centromere and the gene can be estimated by observing whether first or second division segregation has occurred for the particular gene.

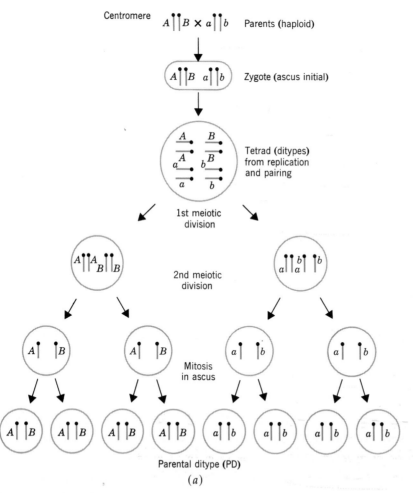

Figure 5.15 Tetrads with segregational patterns called "ditype." Each has two classes of meiotic products (a) 4AB:4ab for the parental ditype and (b) 4ab:4ab for the nonparental ditype. No crossing over has occurred.

For example, the order of meiotic products in 5.15 is 4*AB* and 4*ab*. This order is illustrative of a **parental ditype** and indicates that no crossing over has occurred. A parental ditype is defined as a tetrad that contains only two types of meiotic products, both like the parental types. In Fig. 5.16, however, the order of meiotic products is 2*AB*:2*aB*:2*ab*:2*Ab*. This is called a **tetratype** because all four meiotic products are different. It is indicative of a previous cross-over event and, in this case, is a direct product of second division segregation for gene *A*. In determining the recombination frequency for a particular gene (i.e., the map distance from the centromere), it is important to realize that since only one-half the strands in second division asci have recombined, the **recombination frequency is only one-half the frequency of 2nd division asci.**

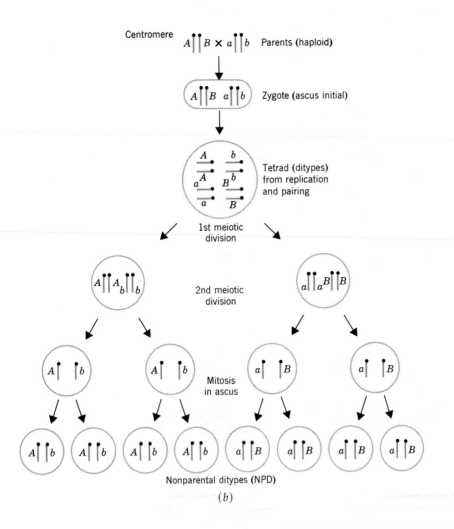

(b)

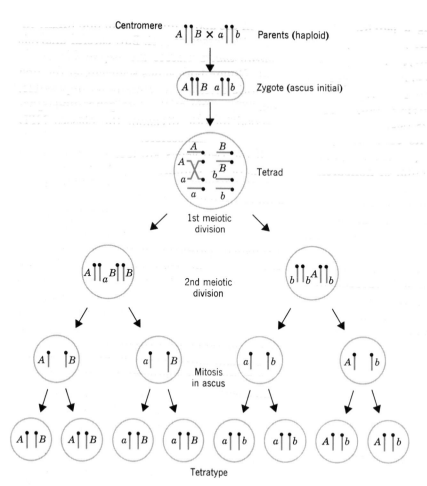

Figure 5.16 A tetrad with the segregational pattern of tetratype. The four classes of meiotic products are: $2AB:2aB:2ab:2Ab$. Crossing over has occurred in the Aa chromosomes and second division segregation is exhibited by the Aa alleles. No crossing over has occurred in the Bb chromosomes and first division segregation is exhibited by the Bb alleles.

LINKAGE DETECTION

In diploid organisms such as Drosophila, recombinant progeny were detected by observing the progeny that differed from the parental types. When two genes assorted independently (i.e., were unlinked), the recombinant progeny occurred with approximately the same frequency as did the parental types. On the other hand, when two genes were linked, the frequency of recombinants was dependent on the distance between the two genes. When the genes were less than 50 map units apart, the recombinant types occurred at a lower frequency than the parental types. The same method for detecting linkage in diploid organisms is used for detecting linkage in haploid organisms such as Neurospora. Notice from Figure 5.15 that when genes (A and B) assort independently, the probability of obtaining a **parental ditype** (PD) is equivalent to the probability of obtaining a **nonparental ditype** (NPD). Therefore, the two types will appear with approximately the same frequency, that is PD:NPD = 1:1. **Tetratypes** only appear where crossing over has occurred between one of the genes and its centromere, and hence, will usually be present at a lower frequency than the parental or nonparental ditypes (Fig. 5.16). When two genes are linked, however, different results are expected. From Fig. 5.16 observe that the only way a nonparental ditype can be obtained is by a four-strand double crossover in which all four strands are involved in two crossover events. This class will subsequently occur at a very low frequency whereas the parental ditype will occur frequently. Therefore, when genes are linked, PD ≫ NPD.

LINKAGE DISTANCE IN TWO-POINT CROSSES

Once linkage has been established, the position of the centromere relative to the two genes may be determined. This is an easy calculation if a sufficient number of ordered tetrads has been scored. To understand the process, however, it is important to understand the possible classes of asci that exist.

Asci are classified in two ways: according to tetrad types, and according to spore order. These are shown below:

tetrad types	kinds of spore order
PD	First division segregation for both genes
NPD	First division segregation for first gene and second for other
T	Second division segregation for first gene and first for other

Given the above information, one might expect 12 classes of asci. But, careful analysis shows that 5 classes are impossible to obtain. Where both genes segregate at the first meiotic division, a tetratype is impossible; and when one of the two genes undergoes second meiotic division, the classes NPD and PD are impossible.

If the two genes are located on different arms of the chromosome, the frequency of parental ditypes resulting from a two-strand double exchange should be approximately the same as the frequency of nonparental ditypes resulting from a four-strand double exchange. However, if the two genes are located on the same arm of the chromosome, the second division parental ditype frequency will be much greater than the second division nonparental ditype frequency. Triple exchange events which result in NPD are extremely rare, while a two-strand single event is not infrequent.

A method for approximating the map distance between two genes is to divide by two the frequency of the second division asci for each gene, calculate the difference between the two values, and multiply by 100. However, this method of approximating map distances between genes is inadequate because it excludes the nonparental ditypes as a recombinant class and thus underestimates the map distance. As shown in Fig. 5.17, the frequency of recombinant strands (which ultimately gives rise to recombinant asci) is equal to half the total tetratypes (T) of the gene in question, plus all of the nonparental ditypes (NPD), divided by the total number of tetrads. This can be written symbolically as

$$R_x = \frac{\frac{1}{2}(T)_x + (NPD)}{(PD) + (T) + (NPD)}$$

where (PD) is the number of parental ditypes. Although evaluating R in this manner is a satisfactory measurement of the recombination frequency, it only ap-

Type of crossover event	Example of bivalent	Tetrad type	Frequency of recomb. strands
No crossover	A ━━━━━ B a ━━━━━ b	PD	0
Two-strand single crossover	 A ━━━━━ B A ━━━━━ B a ━━━━━ b a ━━━━━ b	T	½
Two-strand double crossover	 A ━━━━━ B A ━━━━━ B a ━━━━━ b a ━━━━━ b	PD	0
Three-strand double crossover	 A ━━━━━ B A ━━━━━ B a ━━━━━ b a ━━━━━ b	T	½
Four-strand double crossover	 A ━━━━━ B A ━━━━━ B a ━━━━━ b a ━━━━━ b	NPD	1

Figure 5.17 Types of tetrads when genes such as A and B are linked. Four strands are involved in crossing over, but only two strands exchange places in a single crossover event. Positions of the crossovers are illustrated with respect to centromeres and markers. Parental ditypes have either no crossovers or two reciprocal crossovers involving only two strands. Tetratypes have either a single crossover involving only two strands or double crossovers involving three of the four strands. Nonparental ditypes involve all four strands in double crossing over.

proximates the true map distance, because interference has not been considered.

A method for estimating the position of the centromere with reference to two linked genes is based on a comparison between second division parental ditypes and second division nonparental ditypes. When, for example, a cross is made in coupling (i.e., $AB \times ab$), the second division parental and nonparental ditypes can be distinguished as shown in Fig. 5.18. Class 1 (left) is a parental ditype with genes A and B on opposite arms of the chromosomes (centromeres are between the two loci), Class II (left) is a nonparental ditype in which all four strands are involved. Since it is ex-

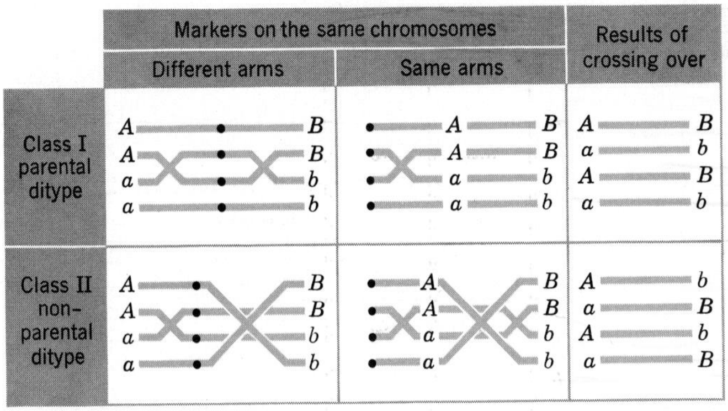

| Markers on the same chromosomes | | Results of crossing over |
Different arms	Same arms	

Figure 5.18 Comparison between second division parental and nonparental ditypes involving linked genes (A and B) from a cross made in coupling (AB × ab). Upper, parental ditypes, lower, nonparental ditypes. Left, genes A and B on different chromosome arms, right, genes on the same chromosome arm.

pected that the exchanges in Class I and Class II will occur with equal frequency, equal proportions of these types would indicate that the two genes are on opposite sides of the centromere. If, for example, the score for Class I was 15 and that for Class II was 16, the two genes would be on different arms.

Class I and Class II (right), with the two genes on the same side of the centromere, are formed by single (Class I) and triple (Class II) exchange. Since triple exchanges are rare, Class I would be expected to occur at much greater frequency than Class II. By scoring ordered tetrads, from a cross in coupling (AB × ab), comparison can be made between Class I and Class II and the centromere position can be determined. If, for example, the score for Class I was 15 and that for Class II was 0, the two genes would be on the same arm.

GENE CONVERSION

Nonreciprocal as well as reciprocal intragenic recombination has been detected in yeast, Neurospora and other fungi that lend themselves to tetrad analysis. This occurs in association with meiotic or mitotic exchange and is called "conversion." When conversion is associated with two alleles such as a and a_1 haploid progenies do not show the expected 1:1 (2:2 or 4:4) ratio but one allele is recovered more frequently than the other. If, for example, allele a is for wild type and a_1 for the mutant, the segregation may be $3a:1a_1$ or $1a:3a_1$.

A possible cause for this departure from the 1:1 ratio is a switch in templates during chromosomal replication. As a result the a gene is copied twice and the a_1 gene not at all (or vice versa). The behavior is as though one wild-type gene in the meiotic tetrad has somehow "converted" a mutant allele to adopt the wild-type sequence. Numerous examples of gene conversion have been described in fungi where four products of meiosis are recoverable. E. A. Carlson has described cases in Drosophila where two of the four products of meiosis are recoverable in attached-X (XXY) females.

SEX LINKAGE

When the parallelism was discovered between the X chromosome cycle and sex determination (Chapter 4), it was generally assumed among investigators that genes other than sex determiners were also located in the X chromosome. Indeed, several mutant genes such as the *Xg* blood alleles in man have been identified in the X chromosomes. Because of their location in the same chromosome as sex determiners, they are said to be **sex-linked.** The first extensive experimental evidence for sex linkage in a species came in 1910 with the discovery by Morgan of a white-eyed mutant Drosophila. A gene had undergone a change resulting in the alteration of an end product in the development of the fly. This change expressed itself as white eyes rather than the normal red eyes. The white-eyed male was mated with a red-eyed female. F₁ flies were all red-eyed, but the F₂ included both red and white in the proportion of about 3 red to 1 white.

This familiar ratio suggested that the gene for red eyes was dominant over the newly created allele for white. More detailed observation, however, showed that all white-eyed flies in the F₂ generation were males. About half of the F₂ males had white eyes and half had red, but all females had red eyes. The recessive gene apparently expressed itself only in the males. Morgan arrived at an explanation by **associating this gene with the X chromosome,** as illustrated in Fig. 5.19.

Because the male fly had only one X chromosome and a nearly nonhomologous Y chromosome that lacks most genes of the X, it was postulated that a single allele for white eyes was capable of expression. The word **hemizygous** is used to describe those sex-linked genes of males. Only one member of an allelic pair of genes is required for expression. Furthermore, the mutant gene present in the X chromosome

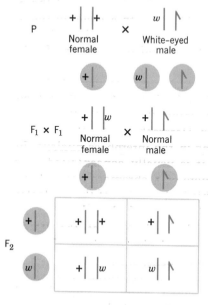

Summary: all females red, half of males white-eyed

Figure 5.19 A cross illustrating sex linkage in Drosophila. This cross is between a red-eyed female and a white-eyed male. Genes are identified with their chromosomes.

of the original white-eyed male was passed on to his daughters (he transmitted a Y chromosome to his sons). All the daughters therefore, were carriers for the allele. The F_2 hemizygous males obtained their X chromosomes from their heterozygous mothers. Half received the w^+ allele and developed red eyes and half received the w allele and developed white eyes. The equal proportion of red-eyed and white-eyed F_2 males was thus explained on the basis of the segregation of the X chromosomes from the F_1 mothers to their sons.

Could white-eyed females occur? On the basis of his hypothesis that the gene was carried in the X chromosome, Morgan predicted that a female of the genotype ww could be produced and would have white eyes. This was tested experimentally with crosses between males with white eyes and F_1 (ww^+) females with red eyes. From these crosses, half of the females as well as half of the males had white eyes, as predicted. Later studies identified many other genes of Drosophila on the X chromosome (Fig. 5.9).

The term **sex linkage** was used to describe the association or **linkage of** a hereditary trait with sex because the gene was in a sex chromosome. In most animals and plants investigated, females have two X chromosomes and are **homogametic** whereas males have an X and Y chromosome and are **heterogametic.** Some organisms, however, such as birds have a reverse pattern with XY (heterogametic) females and XX (homogametic) males. Most sex-linked genes in male heterogametic animals were found to be in the X chromosome. Some animals, however, carry genes that have visible effects in the Y chromosome. These "Y-linked" genes are transmitted directly from father to son. Because Y-chromosome linkage is comparatively rare, sex linkage usually implies X linkage or the presence of the genes in question in the X chromosome. Since all sex-linked genes are on the same chromosome, they are linked with each other as well as with the sex-determining genes.

In the absence of dominant alleles, a recessive allele, such as the one (w) responsible for the white eyes, can express itself. The crisscross pattern of inheritance, that is characteristic of sex-linked genes, means that traits appearing in males are transmitted (unexpressed) through their daughters to the males in the next generation, where they are expressed.

Cytological studies on the nature and behavior of chromosomes have supported the genetic interpretation of sex linkage. In *Drosophila melanogaster* the X and Y chromosomes can readily be identified by their appearance. The X is rod-shaped with the centromere near one end whereas the Y is hooklike, having a long and a short arm.

Although the Y chromosome is essentially devoid of genes, one small part of the short arm in *D. melanogaster* has a homologous section in the X. The bb for bobbed bristles in the X chromosome, or its allele (bb^+) may occur in the short arm of the Y chromosome. Genes with a locus in the Y chromosome as well as in a homologous part of the X are said to be **incompletely sex-linked.**

The Y chromosome is composed almost entirely of inactive **heterochromatin,** which is different from the genetically active euchromatin. Heterochromatin shows maximal staining in the interphase nucleus. In metaphase it is usually concentrated near the centromere of X chromosomes and autosomes. This is the region of the nucleolar organizer—that part of the chromosome which controls production of the RNA-containing nucleolus observed in the nondividing cells. Like the Y chromosome in Drosophila, supernumerary chromosomes in some species of animals and plants (for example, mealy bugs and maize) are composed mostly of heterochromatin.

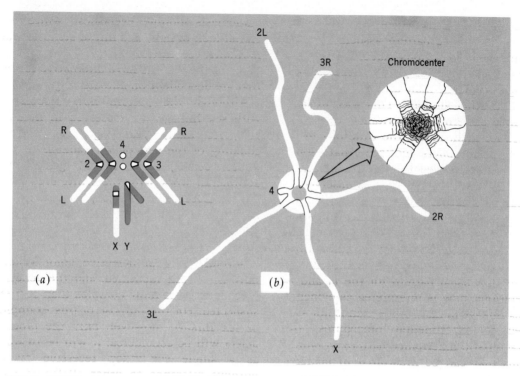

Figure 5.20 Chromosomes of *Drosophila melanogaster*. (*a*) Metaphase chromosomes in the body cells of the male; heterochromatic regions are shaded; the centromeres are indicated by clear zones set off by heavy lines; (*b*) salivary gland chromosomes. The homologous chromosomes are intimately paired along their lengths; the centromeres and centric heterochromatin are combined in the chromocenter; each chromosome arm radiates independently from the chromocenter. (From Spencer Brown, *Science*, 151:417–425, 1966 Copyright 1966 by the American Association for the Advancement of Science)

Sometimes the heterochromatin regions of different chromosomes in the same cell coalesce and form an amorphous "chromocenter." This occurs in the giant salivary gland chromosomes of *D. melanogaster* as shown in Fig. 5.20. The usual metaphase configuration observed in cells of this species other than the special giant cells is shown in Fig. 5.20*a* with heterochromatin parts shaded. Giant chromosomes from a cell of a male larva are shown in Fig. 5.20*b*. The heterochromatin near the centromere

of each autosome and the X chromosome, along with that making up all of the Y chromosome compose the chromocenter.

In summary, the **Y chromosome of many species has few if any genes.** Heterochromatin is near the centromeres in chromosomes of most species and these areas are essentially devoid of active genes. Heterochromatin may be involved however, along with other factors in regulation of gene activity (Chapter 7).

SEX LINKAGE IN MAN

The inheritance pattern associated with sex linkage is so obvious that it has become a choice example for genetic studies. Sex linkage occurs in man as well as in fruit flies and other animals. This was, in fact, the first pattern of inheritance to be recorded for man. Before the time of Christ, Greek philosophers noticed that some human traits tended to skip a generation. An inherited characteristic was observed to appear in a father but not in any of his children, either male or female, and then reappear in males of the next generation. This distinctive crisscross pattern, from father through daughter to grandson, replacing the usual pattern for the F_1 and F_2 generations, now is interpreted as evidence of sex linkage in man. Since man is not subjected to experimental procedures, the characteristic inheritance pattern in family groups that can be illustrated in pedigree charts, is the standard means of detecting sex-linked genes.

A detailed description of the inheritance pattern (now known as sex linkage) of defective color vision (commonly called color blindness) was recorded in 1777 from a study of members of a family who had difficulty in distinguishing red and green (protan defect). In 1793 the same pattern was described for the bleeder's disease, hemophilia. One type of night blindness and a form of nystagmus, an involuntary oscillation of the eyeball, were added in the late nineteenth century to the list of human sex-related traits.

Sex linkage has now been indicated for more than 120 traits in man including, in addition to those already mentioned, such important and distinctive traits as optic atrophy (degeneration of the optic nerve), juvenile glaucoma (hardening of the eyeball), myopia (nearsightedness), defective iris, juvenile muscular dystrophy (degeneration of certain muscles), epidermal cysts, distichiasis (double eyelashes), white occipital lock of hair, and mitral stenosis (abnormality of mitral valve in the heart). Some of these traits have alternative forms that are dependent on autosomal genes.

Although the pattern now associated with sex linkage was observed in human pedigrees many years ago, the understanding of the genetic mechanism was a direct consequence of Morgan's experimental work with the white-eyed mutant in Drosophila. The explanation given for sex-linked inheritance in Drosophila applies equally to traits in man that are associated with sex-linked genes.

Several kinds of defective color vision have now been identified, and the genetic mechanisms are more complex than at first suspected. For the purpose of this example of sex linkage, however, only the protan defect will be considered. It will be treated as a single sex-linked recessive allele without reference to other alleles at the same locus. A man defective in red-green color vision has a single recessive allele (rg) in his X chromosome. Since the Y chromosome carries no allele for rg, the single gene is expressed, thus resulting in the color-vision defect. If this man marries a woman homozygous for the dominant allele (rg⁺) for normal color vision (Fig. 5.21), all their daughters will receive an X chromosome from the mother carrying rg⁺ and a rg from their father and will be heterozygous carriers. Sons, with only one X chromosome, will have only one allele (rg⁺) from the mother and will be free from red-green defective color vision. The Y chromosome, carrying no allele for this trait, will be contributed by the father only to his sons. In the next generation, about half the sons of the carrier females will be normal and half will be color defective because the X chromosome carrying rg⁺ will segregate to about half of the heterozygous mother's gametes, and the other half will carry rg. If the

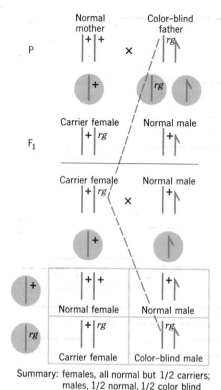

P

Normal mother
+ | +

Color–blind father
/ | *rg*

×

+

rg

F₁

Carrier female
+ | *rg*

Normal male
+

Carrier female
+ | *rg*

Normal male
+

×

+

+	+	+
Normal female		**Normal male**
+	*rg*	*rg*
Carrier female		**Color–blind male**

Summary: females, all normal but 1/2 carriers; males, 1/2 normal, 1/2 color blind

Figure 5.21 Crisscross inheritance from father through daughter to grandson (of original father) that is characteristic of a sex-linked gene. Genes are shown on the chromosomes illustrating a cross between a woman with normal vision and a color vision defective man. The symbol *rg* represents the sex-linked recessive gene for red-green color defective vision.

fathers are normal (rg^+), half of the daughters of carrier mothers will be carriers, not expressing the trait.

Segregation of X chromosomes and expression of single recessive alleles explains the higher incidence of red-green color defective males than females. About 8 percent of the men in the United States and less than 1 percent of the women are red-green color defective since they must receive two recessive alleles. Color-defective people occur in all human populations. The gene frequency, however, varies among people of different ancestral groups. Only about four percent of black men are reported to be red-green color defective.

The criteria for identifying sex-linked (X-linked) recessive genes from pedigree studies may be summarized as follows: (1) expressions occur much more frequently in males than females; (2) traits are transmitted from an affected man through his daughters to half of their sons; (3) an X-linked (sex-linked) allele is never transmitted directly from father to son; and (4) because the allele is transmitted through carrier females, affected males in a kindred may be related to one another through their mother's mutant gene.

If the X-linked allele should be dominant, such as *Xg* for a rare blood type, males expressing the trait would be expected to transmit it to all their daughters but none of their sons. Heterozygous females would transmit the trait to half of their children of either sex. If a female expressing the trait should be homozygous, all of her children would be expected to inherit the trait. Sex-linked dominant inheritance cannot be distinguished from autosomal inheritance in the progeny of females expressing the trait, but only in the progeny of affected males.

The so-called **Grandfather** method is a procedure for obtaining data on human X-chromosome linkages. Since sex-linked traits are transmitted in "crisscross" fashion from a father, through his daughters to their sons, the maternal grandfather of a male child becomes the key source in tracing a sex-linked recessive gene. The critical data include: (I) expression of a trait in a grandfather and grandson that has not appeared in the woman who was the grandfather's daughter and the grandson's mother, (b) if male offspring (grandsons) with doubly heterozygous mothers can be detected recombination frequencies may be determined from the genotype of the mother irrespective of the genotype of her mate (the father) and (c) the linkage phase (coupling or repulsion) of a doubly heterozygous female can be determined with some assurance, from her father's phenotype. A considerable amount of information about sex-linked inheritance in a family group can be determined if the maternal grandfather's phenotype is known. Additional information about phenotypes of other male and female relatives will aid greatly in analyzing sex-linked inheritance in a family group but information about the grandfather is especially significant.

INCOMPLETELY SEX-LINKED GENES IN MAN

Besides the nonhomologus part of the X chromosome that carries the usual sex-linked genes, the X chromosome of man has a section that is homologous with a part of the Y chromosome. The situation is similar to the case described in Drosophila for the section carrying the gene (*bb*) for bobbed bristles. Several genes have been postulated for this region on the basis of pedigree studies. These include the gene for total color blindness; that for xeroderma pigmentosum, a skin disease characterized by pigment patches and cancerous growths on the body; the gene for retinitis pigmentosa, a progressive degeneration of the retina, accompanied by deposition of pigment in the eye; and that for a type of nephritis, a kidney disease. These are presumably represented in the X and Y chromosomes as allelic pairs and segregate like ordinary autosomal pairs, although they do not segregate independently of sex as do autosomal genes. Even though these genes are located on the X chromosome, the usual **crisscross pattern** for sex linkage **is not expected** because of their paired (allelic) arrangement. Questions have been raised concerning the interpretation of genetic or pedigree data for incomplete sex linkage in man, but the cytological evidence is good. Chiasmata have been observed between sex chromosomes. More extensive pedigree studies will undoubtedly provide evidence for this mode of inheritance.

Y CHROMOSOME LINKAGE IN MAN

Certain published pedigrees have indicated that the Y chromosome may have a section with genes distinctive to that chromosome. Genes located in a nonhomologous part of the Y chromosome are expected to control **holandric** inheritance because they are **transmitted exclusively through the male line** (for example, "hairy pinna" Fig. 5.22). The pedigree evidence for transmission from father to son is the only criterion on which they have been predicted. Published pedigrees interpreted to show this pattern, for the most part, have not been substantiated, and there is reason to suspect that at least some of the most spectacular cases are not accurately re-

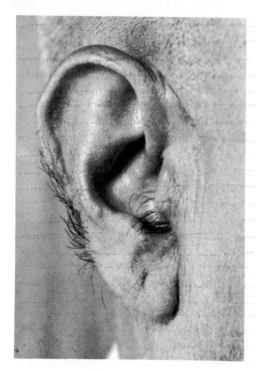

Figure 5.22 Ear of an Indian man with hairy pinna. Hairy pinna has long been considered to be an example (perhaps the only example in man) of a holandric trait. Although this example is still supported in some pedigrees, others indicate different patterns of inheritance for hairy pinna. (Courtesy of Dr. K. R. Dronamraju.)

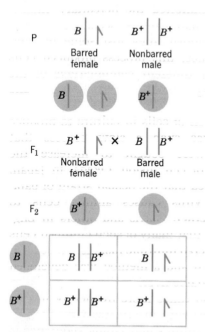

Summary: 1/2 males barred, 1/2 females barred

Figure 5.23 Cross involving a sex-linked gene in birds which have heterogametic females. The cross is between a barred female chicken and a nonbarred male.

ported. Judgment on the extent of Y-linked genes in man must be withheld until more complete evidence is available.

SEX LINKAGE IN OTHER ANIMALS AND PLANTS

Despite the great interest in sex linkage and the persistent search for cases in animals, only a comparatively few examples are on record. Rats have been studied extensively, but no well-confirmed sex linkage is known. Some 20 sex-linked genes have now been reported in the mouse. Evidence for sex linkage also has been found in some plants, including the date palm and members of the pink family of the genus Melandrium. In birds, it is the male that has two X chromosomes (homogametic) and the female only one (heterogametic). Therefore, the expression of sex-linked genes follows a crisscross pattern from mother through carrier sons to "granddaughters." This sequence can be demonstrated for such sex-linked alleles as B^+ (recessive) producing the nonbarred feather pattern in the domestic fowl (Fig. 5.23)

SOMATIC CELL HYBRIDIZATION FOR IDENTIFYING GENES WITH CHROMOSOMES

As indicated previously, linkage studies in man before the 1970s have not been productive. A few groups of genes were found from pedigree analyses to behave as if they were located in the same chromosome but very few autosomal genes were identified with particular chromosomes. As shown above, a number of genes were located on the human X chromosome and a few on the Y, because of the unique behavior of sex chromosomes. In the early 1970s all human chromosomes could be identified microscopically. Several new techniques, particularly somatic cell hybridization, have now

been successfully applied to human chromosome investigations.

Somatic cell hybridization was demonstrated by George Barski and his associates in Paris. These investigators mixed cell cultures of two different mouse cancer cells that could be distinguished by differences in cell morphology and in the shape of some of the chromosomes. After a few months, the investigators noticed that some cells in the cultures appeared to differ from those of either parental line. Each changed cell had a single large nucleus containing chromosomes of both parents. These hybrid cells had arisen by fusion of the different kinds of cells that were growing together in cell culture. It was eventually possible to develop and sustain **pure cultures of hybrid cells.** Hybrids between cell lines had two characteristics that made them especially useful for genetic analysis. First, the set of chromosomes coming from each line that was crossed remained functional in the hybrids, which therefore exhibited the hereditary characteristics of both parents. Second, as the hybrids multiplied, they lost some of their chromosomes; this process produced cells with many **different combinations** of the original **parental** chromosomes and genes.

The fusion of cells in culture to produce hybrids was such a slow process that methods were sought to increase the frequency of cell fusion. Okada, in Japan, had observed that the Sendai strain of parainfluenza virus causes animal cells to **clump together.** Harris and others in England found that many of such clumped cells would undergo multiple fusions. Procedures were refined and a method was developed by which viable hybrid cells were produced 300 to 1000 times more frequently in cultures treated with inactivated virus than in untreated cultures undergoing spontaneous hybridization. Virus-induced hybrids had

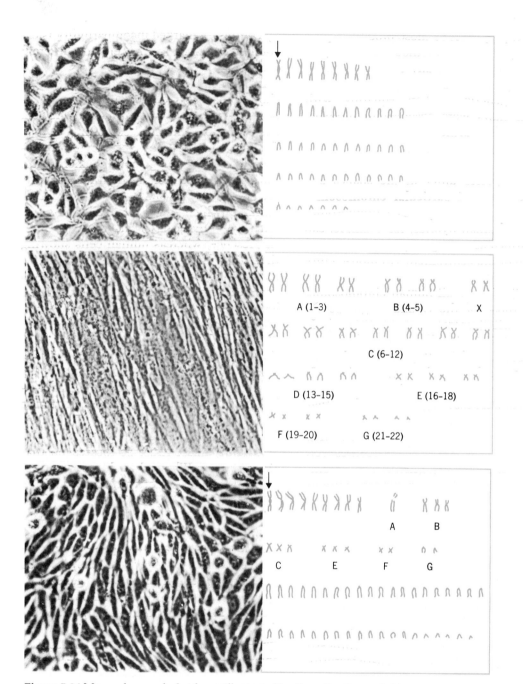

Figure 5.24 Mouse-human hybrids are illustrated by the cell cultures (left) and the karyograms (right) of the mouse parent line (top), the human parent (middle) and the hybrid (bottom). The human cells, derived from embryonic lung tissue, contain the normal number of chromosomes (46, or 23 pairs), arranged here in the usual seven groups (plus the two female sex chromosomes). Except for a tendency to align in parallel, the hybrid cells look

the same properties as those produced spontaneously. With this method it became possible to cross almost any two cells and to isolate hybrids in a short period of time.

Many genetic markers equating with parental traits were observed in interspecific hybrids. Furthermore, the number of chromosomes decreased markedly in succeeding generations. Since chromosomes of different species differ in the shapes and sizes, the **loss of particular chromosomes could be detected in hybrids.** Chromosomes in mouse-human hybrids, for example, can be identified as to origin by shape, size, and position of the centromere. All mouse chromosomes are acrocentric (centromere near one end) whereas all human chromosomes except those of the D and G groups are metacentric (centromere near center). In mouse-rat hybrid cells, more of the rat chromosomes than the mouse chromosomes were lost in later generations. In hamster-mouse hybrids, the mouse chromosomes were lost in greater numbers than the hamster chromosomes. When mouse cells were crossed with human cells (Fig. 5.24) most of the mouse chromosomes were retained in succeeding generations but only a few, 1 to 15, of the 46 human chromosomes were retained beyond the first few cell generations. After 100 cell generations diminution of chromosomes had occurred to the extent that some clones (groups of genetically identical cells descended from a single ancestral cell) had no human chromosomes.

Such investigations have shown that somatic cells of different species are compatible. Furthermore, hybrid cells synthesize hybrid enzymes that function satisfactorily.

By contrast, a marked incompatibility is encountered among germ cells of different species.

Experiments with selective media showed that **particular chromosomes could be retained or eliminated from a hybrid cell.** For example, a medium containing hypoxanthine, aminopterin, and thymine was used to select among mouse/man hybrids. Mouse cell lines lacking thymidine kinase activity were resistant to 5-bromodeoxyuridine (which kills most cells that contain thymidine kinase) and could grow in its presence, whereas human cells showing thymidine kinase activity could not grow on 5-bromodeoxyuridine. Growth on this chemical was, therefore, used to determine loss of the human chromosome that permitted growth on the selective medium. Studies of this kind have provided information on the location of genes in particular chromosomes.

By studying, for example, hybrid mouse-human cell cultures that had **lost all human chromosomes except number 17,** it was learned that the gene for the enzyme **thymidine kinase** must be located on that chromosome. When the clones were removed from the selective medium and exposed to 5-bromodeoxyuridine, none of the cells that survived contained human chromosome 17. The gene for the enzyme thymidine kinase was not present in the mouse parent. None of the enzyme was produced by the hybrid. Later generation hybrids contained only the human chromosome 17 that carried the wild-type gene for thymidine kinase and produced the enzyme. Mutants, however, for glucose-6-phosphate dehydrogenase, lactate dehy-

more like the mouse cells than the human ones. This is in keeping with the fact that the hybrid karyogram contains only 14 of the 46 human chromosomes, which are readily distinguished from mouse chromosomes. (After B. Ephrussi and M. C. Weiss. "Hybrid somatic cells." *Sci. Amer.* 220, 26–35, 1969.)

drogenase and malate dehydrogenase were also present in these hybrids. Since these mutants were not altered by wild-type alleles, the human gene loci for the wild-type enzymes were not linked in the same chromosome with the thymidine kinase locus. Studies of this kind along with other applications of new techniques have identified many genes on the human autosomes (Chapter 16).

REFERENCES

Brown, S. W. 1966. "Heterochromatin." *Science*, 151, 417–425.

Morgan, T. H., and C. B. Bridges. 1916. *Sex-linked inheritance in Drosophila*. Carnegie Inst., Wash. Publ. 237.

Creighton, H. B., and B. McClintock, 1931. "A correlation of cytological and genetical crossing over in *Zea mays*." *Proc. Nat'l Acad. Sci.*, 17, 492–497.

Lewis, K. R., and B. John, 1970, *The organization of heredity*. American Elsevier, New York.

Morgan, T. H., C. B. Bridges, and A. H. Sturtevant. 1925. "The genetics of Drosophila." *Bibliographia Genetica*, 2, 1–262. The Hague, Netherlands.

McKusick, V. A. 1971. "The mapping of human chromosomes." *Sci. Amer.* 224, 104–113.

PROBLEMS AND QUESTIONS

5.1 Suggest experiments on some organism to determine genetically (a) whether two genes are located in the same chromosome pair; (b) whether they are in the coupling or repulsion phase? (c) What are the advantages and disadvantages of the test-cross method for determining linkage relations? (d) the F_2 method?

5.2 From a cross between individuals with the genotypes *Cc Dd Ee* × *cc dd ee*, 1000 offspring were produced. The class appearing *C-D-e e* included 351 individuals. Are the genes, *c*, *d*, and *e* in the same or different chromosome pairs? Explain.

5.3 If an animal with the genotype *Rr Ss Tt* produced 1020 eggs, of which 127 are *r S t*, 121 *r S T*, and 130 *R S T*, are the three pairs of alleles in the same chromosome or independent of one another? Explain.

5.4 If the linkage strength between two loci is 70 percent, what would be the amount of crossing over between these loci?

5.5 Genes *a* and *b* are linked with 20 percent crossing over. An a^+b^+/a^+b^+ individual was mated with an *ab/ab* individual. (a) Represent the cross on the chromosomes, illustrate the gametes produced by each parent, and illustrate the F_1. (b) What gametes can the F_1 produce and in what proportion? (c) If the F_1 was crossed with the double recessive, what offspring would be expected and in what proportion? (d) Is this an example of coupling or repulsion?

5.6 If the original cross in the problem above was $a^+b/a^+b \times ab^+/ab^+$, (a) represent on the chromosomes the F_1; (b) the gametes produced by the F_1 and proportions; and (c) expected test-cross results. (d) Is this coupling or repulsion?

5.7 If the crossing over in the two problems above were 40 percent instead of 20 percent, what difference would it make in the proportions of gametes and test-cross progeny?

5.8 If Problems 5.5 and 5.6 with 20 percent crossing over were carried to the F_2 ($F_1 \times F_1$), and a^+ and b^+ were dominant over their alleles, what phenotypic classes would be produced and in what proportions?

5.9 A fully heterozygous F_1 corn plant was red with normal seed. This plant was crossed with a green plant (b) with tassel seed (ts) and the following results were obtained: red, normal 124; red, tassel 126, green, normal 125; and green tassel 123. (a) Does this indicate linkage? (b) If so, what is the percentage of crossing over? (c) Diagram the P cross on the chromosomes.

5.10 A fully heterozygous gray-bodied (b^+) normal-winged (vg^+) female F_1 fruit fly crossed with a black-bodied (b) vestigial-winged (vg) male gave the following results: gray, normal 126; gray, vestigial 24; black, normal 26; and black, vestigial 124. (a) Does this indicate linkage? (b) If so, what is the percentage of crossing over? (c) Diagram the P cross on the chromosomes.

5.11 Another fully heterozygous gray-bodied, normal-winged female F_1 fruit fly crossed with a black-bodied, vestigial-winged male gave the following results: gray, normal 23; gray, vestigial 127: black, normal 124; and black, vestigial 26. (a) Does this indicate linkage? (b) If so, what is the percentage of crossing over? (c) Diagram the P cross on the chromosomes.

5.12 In rabbits, color results from a dominant gene (c^+) and albinism from its recessive allele (c). Black is the result of a dominant gene (b^+), brown of its recessive allele (b). Fully homozygous brown rabbits were crossed with albinos carrying the gene for black in the homozygous state. F_1 rabbits were crossed to double recessive $\left(\dfrac{c\ b}{c\ b} \text{ or } \dfrac{cb}{cb}\right)$. From many such crosses the results were: black 34; brown 66; and albino 100. (a) Are these genes linked? (b) if so, what is the percentage of crossing over? (c) Diagram the P cross on the chromosomes.

5.13 In tomatoes, tall vine (d^+) is dominant over dwarf (d) and spherical fruit shape (p^+) over pear (p). Vine height and fruit shape are linked with 20 percent crossing over. A certain tall, spherical-fruit tomato plant (a) crossed with a dwarf, pear-fruited

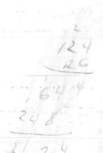

plant produced 81 tall, spherical; 79 dwarf, pear; 22 tall, pear; and 17 dwarf, spherical. Another tall, spherical plant (b) crossed with a dwarf pear produced 21 tall, pear; 18 dwarf, spherical; 5 tall, spherical; and 4 dwarf, pear. Represent on the chromosomes the arrangements of the genes in these two tall, spherical plants. (c) If these two plants were crossed with each other, what phenotypic classes would be expected and in what proportions?

5.14 Genes a and b are located in chromosome II with a crossover of 20 percent. Genes c and d are located in chromosome III with a crossover of 40 percent. An individual homozygous for $a^+b^+c^+d^+$ was crossed with a fully recessive individual. The F_1 was backcrossed to the full recessive. (a) Represent the original (P) cross on the chromosomes, (b) the F_1, and (c) the gametes that the F_1 could produce with their proportions.

5.15 A student has two dominant traits dependent on single genes, cataract (an eye abnormality), which he inherited from his mother, and polydactyly (an extra finger), which he inherited from his father. His wife has neither trait. If the genes for these two traits are closely linked, would the student's child be more apt to have: (a) either cataract or polydactyly, (b) cataract and polydactyly, or (c) neither trait? Explain.

5.16 In Drosophila, the recessive genes sr (stripe) and e (ebony body) are located at 62 and 70 map units, respectively, from the left end of the third chromosome. A striped female (homozygous for e^+) was mated with a male with ebony body (homozygous for sr^+). (a) What kinds of gametes will be produced by the F_1 female and in what proportion? (b) If F_1 females are mated with stripe, ebony males, what phenotypes would be expected and in what proportion?

5.17 In Drosophila, the gene (vg) for vestigial wing is recessive and is located at 67.0 units from the left end of the second chromosome. Another gene (cn) for cinnabar eye color is also recessive and is located at 57.0 units from the left end of the second chromosome. A fully homozygous female with vestigial wings was crossed with a fully homozygous cinnabar male. (a) How many different kinds of gametes could the F_1 female produce and in what proportion? (b) If the females are mated with cinnabar, vestigial males, what phenotypes would be expected and in what proportion?

5.18 In poultry, barring results from the dominant sex-linked gene (B), nonbarring from its recessive allele (B^+). Crested head results from a dominant autosomal gene (C), and plain head from its recessive allele (C^+). Two barred, crested birds were mated and produced two offspring: a nonbarred, plain female and a barred, crested male. (a) Give the genotypes of the parents on the chromosomes. (b) Summarize the expected result for sex, barring, and crest expressions from further matings between these two barred, crested birds.

5.19 If a sex-linked recessive and a sex-linked dominant gene with equal effect on viability were present in equal frequency in the same population in which males are heterogametic, would the recessive gene or the dominant gene express itself more frequently in (a) males? (b) females?

5.20 If a white-eyed male fruit fly should occur in a culture of red-eyed flies, how could the investigator obtain evidence to answer the following questions? (a) Is a mutant gene or an environmental change responsible for the new phenotype? (b) If a mutation has occurred, is it sex-linked? (c) Can white-eyed females occur?

5.21 The gene (w) for white eyes in *D. melanogaster* is recessive and sex-linked; males are heterogametic. (a) Symbolize on the chromosomes the genotype of a white-eyed male, red-eyed male, red-eyed female (two genotypes) and white-eyed female. (b) Diagram on the chromosomes a cross between a homozygous red-eyed female and a white-eyed male. Carry through the F_2 and summarize the expected sex and eye color phenotypes. (c) Diagram on the chromosomes and give the expected phenotypes from a cross between F_1 females and (1) a white-eyed male, and (2) a red-eyed male.

5.22 In man, red-green defective color vision results from the sex-linked recessive gene (rg) and normal vision from its allele (rg^+). A man (1) and woman (2), both of normal vision, have the following three children, all of whom are married to people with normal vision: a color-defective son (3) who has a daughter of normal vision (6); a daughter of normal vision (4) who has one color defective son (7) and two normal sons (8); and a daughter of normal vision (5) who has six normal sons (9). Give the probable genotypes of all the individuals in the family (1 to 9).

5.23 If a mother carried the sex-linked gene for protan defective color vision and the father was normal, would their sons or daughters be defective in color vision?

5.24 If a father and son are both defective in red-green color vision, is it likely that the son inherited the trait from his father?

5.25 Diagram on the chromosomes a cross between a normal woman whose father was defective in red-green color vision and a color-defective man. Summarize the expected results for sex and eye condition.

5.26 In man, the gene (h) for hemophilia is sex-linked and recessive to the gene (h^+) for normal clotting. Diagram on the chromosomes the genotypes of the parents of the following crosses and summarize the expected phenotypic ratios resulting from the crosses: (a) hemophiliac woman × normal man; (b) normal (heterozygous) woman × hemophiliac man; (c) normal (homo-zygous) woman × hemophiliac man.

5.27 A normal woman, whose father had hemophilia, married a normal man. What is the chance of hemophilia in their children?

5.28 Gene Xg is dominant and X-linked. If a woman heterozygous for this gene $(Xgxg)$ married a man carrying the allele (xg), what is the probability that (a) each daughter and (b) each son will receive the Xg gene?

5.29 In a particular family group, the gene for hairy pinna of the ear is holandric. (a) What is the chance that each daughter and each son of a man with this trait will inherit the condition? (b) If a man with hairy pinna also has the Xg blood factor and is color blind, which of these traits might be expressed in each daughter and each son?

5.30 In Drosophila, the recessive genes st (scarlet eye), ss (spineless bristles), and e (ebony body) are located in the same (third) chromosome in the following positions (map distances) from the left end of the chromosomes: st 44, ss 58, e 70. Fully heterozygous females with the genotype $st\ ss\ e^+/st^+ss^+e$ are mated with fully recessive males $st\ ss\ e/\ st\ ss\ e$. If many flies are produced and no interference occurs, what phenotypes will be expected and in what percentages?

5.31 In maize genes Pl for purple (dominant over Pl^+ for green), sm for salmon silk (recessive to sm^+ for yellow silk) and py (recessive to py^+ for normal size), are on chromosome 6 (Fig. 5.12). From the following cross:

$$\frac{Pl\ sm\ py}{Pl\ sm\ py} \times \frac{Pl^+\ sm^+\ py^+}{Pl^+\ sm^+\ py^+}$$

and the test cross between the F_1 and fully recessive, what phenotypes would be expected and in what proportions (assuming equal crossing over in all areas along the chromosome, equal viability of all gametes, and progeny, and no interference)?

5.32 In maize, the genes Tu, j_2 and gl_3 are on chromosome 4 (Fig. 5.12). If plants carrying these three genes in homozygous recessive condition are crossed with plants homozygous for the three dominant alleles and F_1 plants are test crossed to the fully recessive, what genotypes would be expected and in what proportion (assuming equal crossing over in all areas of the chromosome, equal viability of all gametes, and progeny, and no significant interference)?

5.33 A cross was made between yellow-bodied (y), echinus (ec), white-eyed (w) female $(y\ ec\ w/\ y\ ec\ w)$ flies and wild males. F_1 females were mated with $y\ ec\ w$ males. The following proportions were obtained when a sample of 1000 flies was counted:

wild (+ + +)	475
y ec w	469
y + +	8
+ ec w	7
y + w	18
+ ec +	23
+ + w	0
y ec +	0

Determine the order in which the three loci *y*, *ec*, and *w* occur in the chromosome and prepare a chromosome map.

5.34 A cross was made between yellow, bar, vermilion female flies and wild males, and the F_1 females were crossed with *y B⁺v* males. The following results were obtained when 1000 progeny were counted:

y B v and + + +	546
y + + and + B v	244
y + v and + B +	160
y B + and + + v	50

Determine the order in which the three loci occur in the chromosome and prepare a chromosome map.

5.35 Suppose that two pure breeding strains of Drosophila each of which expressed mutant phenotype(s) were crossed. The resulting female heterozygotes were then test crossed to males expressing the mutant recessive phenotypes ebony (*e*), scarlet (*s*), and spineless (*ss*). Test-cross progeny were obtained as follows:

testcross phenotypes	number
wild-type	67
ebony	8
ebony, scarlet	68
ebony, spineless	347
ebony, scarlet, spineless	78
scarlet	368
scarlet, spineless	10
spineless	54

(a) Are these genes linked? Justify your answer. (b) Write the genes given on a chromosome symbol with the genes in correct order. (c) Write the genotypes of the flies involved in the parental cross and test cross. (d) What is the map distance between the loci for ebony and scarlet? (e) What is the map distance between the loci for ebony and spineless? (f) Calculate the coefficient of coincidence.

5.36 (a) In Neurospora, gene a is 10 cross-over units from the centromere and b is 10 units beyond a in the same linkage group, a^+b/ab^+. List the different combinations of genotypes that might be expected in the haploid ascospores resulting from meiosis and the proportions of each. (b) In what proportion would loci a and b be expected to undergo segregation in the second meiotic division? (Ignore double crossing over.)

5.37 In Neurospora, mutant gene a is on one chromosome and mutant gene b is on another. When a strain carrying a was crossed with a strain carrying b only first-division segregation occurred. What kinds of ditypes would be expected and in what proportions?

5.38 In Neurospora, mutant gene a is on one chromosome and mutant gene b is on another. When a strain carrying a was crossed with a strain carrying b, a segregated in the first division and b segregated in the second division. What result is expected?

5.39 Why is tetrad analysis of molds a favorable method for demonstrating crossing over in the four strand stage?

5.40 Present evidence in support of the following propositions (if you do not agree with a proposition state your position and defend it). (a) Genes are in chromosomes. (b) Crossing over results from exchanges between parts of homologous chromosomes. (c) Test crosses are better than F_2 crosses for linkage studies. (d) The percentage of crossing over between two genes on the same chromosome is roughly proportional to the distance between them. (e) Crossing over occurs in the four-strand stage in meiosis.

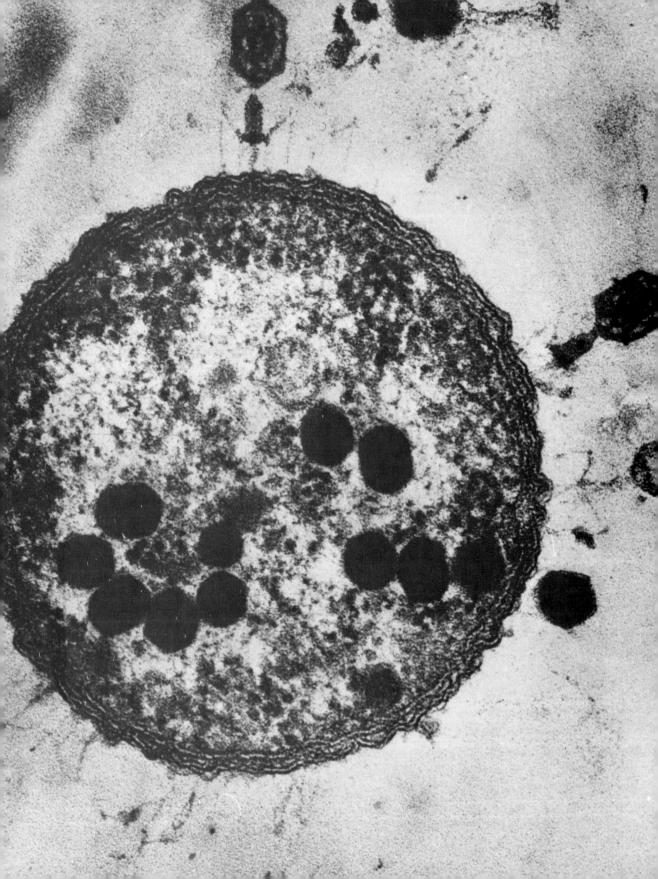

SIX

A gene is a functional genetic unit that codes for the synthesis of one polypeptide. An allele is a member of a pair or an array of mutational forms of a given gene. When several or many mutations occur in the same gene, **multiple alleles** are formed. Phenotypes representing alternate states at the same gene are recognized in populations of animals and plants. In such cases, mutations have occurred in the same gene but in different individuals or at different times. Members of a series of alleles are conventionally represented by the **same basic letter symbol** with appropriate superscripts to identify particular alleles.

Most alleles produce variations or gradations of the same trait, but some produce entirely different phenotypes. In *Drosophila melanogaster*, for example, the gene *ss* makes the bristles smaller than those produced by the wild-type gene ss^+. No effect has been observed on the legs or antennae. Another allele in the series, ss^a (aristapedia), reduces the bristles slightly, but the flies also undergo a more conspicuous phenotypic alteration;

GENETIC FINE STRUCTURE

Cross section of *E. coli* cell infected with bacteriophage T2. (Lee Simon, Institute for Cancer Research, Philadelphia.)

legs develop on the head in place of antennae. This is an exception, however, because most alleles can be recognized by comparable phenotypes.

MULTIPLE ALLELES

The individual alleles of a group that consists of more than two are called a series of "multiple alleles." A classical example of multiple alleles was discovered many years ago in rabbits. Albino (white) rabbits occur occasionally in wild (variously colored) populations. When crosses were made between homozygous colored (c^+c^+) and albino (cc) rabbits, all the F_1 progeny were colored and the F_2 were about 3 colored to 1 albino. This 3:1 ratio in F_2 indicated that only one single pair of alleles was involved; (one wild type, c^+, and one mutant allele, c) and c^+ was dominant over c.

Other rabbits, called chinchilla, appear gray because of a mixed black and white pattern. All of the F_1 progeny from crosses between fully colored and chinchilla rabbits were colored, whereas the F_2 were about three colored to one chinchilla. This ratio indicated that these genes were also alleles and the c^+ was dominant over this chinchilla gene c^{ch}.

Another fur pattern (himalayan) was characterized by a white coat and black tips on extremeties, the ears, nose and feet. When crosses were made between himalayan and fully colored rabbits, all the F_1 progeny were colored. In the F_2, the proportion was about three colored to one himalayan. Crosses between chinchilla and himalayan resulted in all chinchilla in the F_1 and about three chinchillas to one himalayan in the F_2. This result indicated that c^{ch} and c^h were also alleles and that c^{ch} was dominant over c^h. Finally, crosses between himalayan and albino produced only himalayan in the F_1 and about three himalayan to one albino in the F_2. The consistent monohybrid ratios indicated that all four genes were members of the same series of alleles. Gradation in dominance was recognized in the following order: c^+, c^{ch}, c^h, and c. Presumably, c^{ch}, c^h, and c originated somewhere in the ancestry as mutations from the wild-type gene (c^+). The phenotypes and corresponding genotypes are summarized in Table 6.1. Numbers of alleles in any series and corresponding numbers of possible diploid combinations are given in Table 6.2.

TABLE 6.1
Phenotype and Corresponding Genotypes for Multiple Alleles of c Locus in rabbits

Phenotypes	Genotypes
Full color (agouti)	$c^+c^+, c^+c^{ch}, c^+c^h, c^+c$
Chinchilla	$c^{ch}c^{ch}, c^{ch}c^h, c^{ch}c$
Himalayan	c^hc^h, c^hc
Albino	cc

TABLE 6.2
Number of Alleles in a Series and Corresponding Number of Possible Diploid Combinations. Formula: $[n(n + 1)]/2$ **Where** $n =$ **Number of Alleles**

Number of Alleles	Number of Diploid Combinations
2	3
3	6
4	10
5	15
6	21
7	28
8	36

amniotic fluid taken with a needle (am-niocentesis) can indicate whether a fetus will reach maturity in viable condition. When warranted, packed erythrocytes can be transfused into the fetus and thus insure fetal survival to a more viable maturity.

At first, the genetic mechanism of the Rh system seemed simple. A single pair of genes, R and r, was postulated to account for the difference between Rh-positive and Rh-negative individuals. New antibodies were soon discovered, however, and additional genes were postulated to explain the more complicated situation. Wiener developed a hypothesis based on a series of multiple alleles (Table 6.4). Eight alleles were included in the series and more have since been added. Much evidence has been presented in support of this hypothesis. On the other hand, R. R. Race, R. A. Fisher, and other British and American investigators explained the same data on the basis of three pairs of genes (C, D, and E) that are not alleles but are located near each other

on the same chromosome. Such genes might be expected to act like true alleles in most situations that can be tested in human beings.

Passive immunity of Rh hemolytic anemia, called erythroblastosis fetalis, has now been accomplished by use of an incomplete antibody against the Rh_0 antigen. This antibody does not agglutinate Rh-positive red blood corpuscles. Instead, the antibodies attach to antigen receptors on red cell surfaces and coat the cells. These incomplete antibodies may be injected into an Rh-negative mother immediately after she has given birth to an Rh-positive child. The coating of her cells inhibits her capacity to form Rh antibodies. Incomplete antibodies on the red cells of the mother apparently block the antigen from the red blood cells from the child which normally elicit immune response of the mother. Injected antibodies dissipate within a few months and present no danger to the mother or to her subsequent pregnancies.

TABLE 6.4
Rh Blood-Group System. Symbols Used by Fisher, Race, and Others on the Hypothesis of Three Closely Linked Gene Loci Are Given with Those of Wiener on the Hypothesis of Multiple Alleles

Fisher and Race Notations	Wiener Notations		Approximate Frequencies of Genotypes in Caucasian Populations
Genes	Type	Genes	
CDe	Rh_1	R^1	41%
cDE	Rh_2	R^2	14%
cDe	Rh_0	R^0	3%
CDE	Rh_z	R^z	Rare
cde	rh	r	39%
Cde	rh'	r'	1%
cdE	rh''	r''	1%
CdE	rh_y	r^y	Very rare

To test phenotypically for allelism in experimental plants and animals, individuals carrying a recessive mutation (*m*) suspected to be an allele of a known recessive gene (*a*), may be mated with an individual homozygous for *a*. If *m* is an allele of *a* the two genes would be in **corresponding positions on homologous chromosomes** *m/a* and the mutant phenotype would be expressed. If, on the other hand *m* is not an allele of *a*, the two genes would occupy different positions on the chromosome, each heterozygous with a dominant allele (*m*⁺ or *a*⁺). The results of the mating would be *ma*⁺/*m*⁺*a* and the phenotype of the wild type would be expressed.

If, for example, a recessive mutation (*m*) altering the shape of the wing in Drosophila was detected and traced to the same chromosome where a known recessive gene (*a*) affecting wing shape is located, the new mutant would be tested for allelism with *a*. First the new mutant would be made homozygous by inbreeding. If the F_1 progeny from subsequent matings between flies homozygous for *m* and those homozygous for *a* express the mutant phenotype or an intermediate phenotype between mutant and wild type, the mutant *m* is an allele of *a*. The cross would be:

$$P \frac{a}{a} \times \frac{a^m}{a^m} \longrightarrow F_1 \frac{a}{a^m}$$

Since alleles are conventionally identified by the **same basic symbol,** *m* would then have been symbolized a^m. In diagrams of chromosomes they would be represented in the same chromosome position with a^m expected to code for the same polypeptide and influence the same physiological process as *a*. If the F_1 progeny are wild type in appearance, the genes are not allelic and are said to complement each other. The cross is represented as follows:

$$P \frac{am^+}{am^+} \times \frac{a^+m}{a^+m} \longrightarrow F_1 \frac{a\ m^+}{a^+m}$$

This type of test called the **complementation or cis-trans test,** used to distinguish functional allelism, is discussed in the following pages. The recombination test for structural allelism is also considered later in this chapter.

SUBDIVISION OF THE CLASSICAL GENE

The Russian geneticist N. P. Dubinin (in 1939) recognized **step allelism,** that is, the graded effect on phenotypes of different alleles in Drosophila. This discovery foreshadowed the idea that "genes" could be subdivided. Investigations demonstrating subdivisions of the classical gene began (in 1940) when C. P. Oliver obtained unexpected results from a cross between two Drosophila mutants with lozenge-shaped eyes. One parent in the cross was homozygous for lz^g and the other was homozygous for lz^s. These genes were presumed to be alleles and they had been represented in exactly corresponding positions at 27.7 map units from the left end in the first (X) chromosome pair. All progeny from this cross were expected to be heterozygous $\frac{lz^g}{lz^s}$ and to have the characteristic lozenge phenotype, that is, a narrow ovoid eye. Instead, a few were wild type with large oval shaped eyes. It was then postulated that the two alleles were not in identical positions on homologous chromosomes but were located in slightly different positions and were capable of separation by crossing over. They were functional alleles (but were not structural alleles) called **pseudoalleles.** Pseudoalleles were defined as genes that behave as alleles in the allelism test but can be separated by crossing over. Each mutant pseudoallele could therefore occur in heterozygous arrangement with a wild-type allele $\left(\dfrac{lz^g\ +}{+\ lz^s}\ \textbf{trans configuration}\right)$. Crossovers had presumably occurred, placing lz^g and lz^s on the same homologue $\dfrac{lz^g\,lz^s}{+\ +}$ (**cis configuration**) and each in heterozygous condition with wild-type alleles. These genotypes gave rise to the few wild-type flies observed from the cross.

The terms *cis* (Cf. coupling) and *trans* (Cf. repulsion) distinguish the two possible configurations of two pairs of pseudoalleles. These terms came from organic chemistry where they were used to designate compounds having the same kind and number of atoms and, therefore, the same molecular formula, but which differ in configuration of atoms in the respective molecules (for instance, isomers). Cis, trans terminology is currently used in the description of subdivisions within a gene locus while the coupling, repulsion terminology is used for configurations of nonallelic genes.

M. M. and K. C. Green pursued the investigation of the *lz* locus and showed that several subloci were included in that locus. Three recombinational units, shown diagrammatically in Fig. 6.2 with their combination effects, were separable by crossing over. Markers representing these segments were symbolized lz^{BS}, lz^{46}, and lzg, and were described as pseudoalleles occupying slightly different positions on the chromosome. The recombination distance from lz^{BS} to lz^g was estimated to be .03 to .09 of a map unit. Several different states of the same sublocus, which could not be separated by crossing over, were also recognized in each of the three sites. All three of these pseudoalleles were recessive and did not express themselves in the cis arrangement when wild-type alleles were on the opposite homologue. When placed in the trans position, however, several heterozygous combinations of lozenge pseudoalleles showed mutant phenotypes. Additional sites were found in the *lz* locus that could be separated by crossing over and several behaved like the three in the original investigation. This is a type of **position effect** because the positions of the subloci with respect to each other, rather than their presence or absence, determines the phenotypic result.

The area of the first chromosome in Drosophila where the genes for white and apricot eye color are located was found, by the combined effort of several investigators, to be a complex of five sites. The gene for apricot eye color was previously considered to be an allele of the gene *w* for white eye and was originally symbolized w^a. When flies with white eyes were mated with those having apricot eyes, the eye color of the F_2 generation showed simple monohybrid ratios expected for a single sex-linked

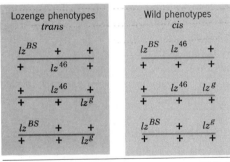

Lozenge phenotypes *trans*		
lz^{BS}	+	+
+	lz^{46}	+
+	lz^{46}	+
+	+	lz^g
lz^{BS}	+	+
+	+	lz^g

Wild phenotypes *cis*		
lz^{BS}	lz^{46}	+
+	+	+
+	lz^{46}	lz^g
+	+	+
lz^{BS}	+	lz^g
+	+	+

Figure 6.2 Combinations at the *lz* locus in Drosophila with the phenotypes of various combinations and position effects. The position effect occurs in the trans configuration. Those in the cis arrangement are wild type. (After C. P. Oliver and M. M. Green.)

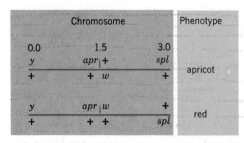

Chromosome			Phenotype
0.0	1.5	3.0	
y	$apr_|+$	spl	
—	— w —	—	apricot
$+$	$+$	$+$	
y	$apr_|w$	$+$	
—	— $+$ —	—	red
$+$	$+$	spl	

Figure 6.3 The two arrangements of the sites, w and apr, and phenotypic effects. In the trans configuration a position effect results in the mutant phenotype. (After E. B. Lewis.)

gene. The two genes were not separated by early crossover tests and there was no reason to question their allelic relationship with each other. When experiments involving thousands of flies were conducted with improved technique, however, the loci for white and apricot were separated, and the symbol for the apricot gene was changed to apr. A few individuals from crosses that ordinarily produced only apricot and white were found to have wild-type red eyes. These were explained on the basis of crossing over. Based on accumulated crossover data, the distance between the two sites was placed at about 0.01 of a map unit.

To test the hypothesis of close linkage and recombination as opposed to the possible mutation explanation, markers were placed on either side of the area in which the white (w) and apricot (apr) alleles were located and further crosses were made. The gene y for yellow body color, at 1.5 units on one side of w, and the gene spl for split bristles, located at 1.5 units on the other side, were placed in the chromosomes in heterozygous condition. These outside markers did change places in the exceptional red-eyed flies, demonstrating that w and apr were on the same homologue and that crossing over had actually occurred within the w locus, bringing the two wild-type genes together, as illustrated in Fig. 6.3. Again in this example, when the w and apr genes were together on the same chromosome and the wild ($+$) members of the pairs of alleles were on the homologous

chromosome (**cis configuration**), the eyes were wild type, **red.** When the two mutant genes were opposite each other (**trans configuration**), the eyes were **apricot.**

CIS-TRANS POSITION EFFECTS

Why did the w and apr genes behave differently in the cis and trans position? E. B. Lewis postulated that this type of position effect is a two-step process that requires two substances, and is related to the sequential nature of gene product synthesis. A linear sequence of DNA units produces a linear sequence of polypeptides that can be interrupted at a specific step by a mutation. If the mutants are recessive such as w or apr and in the cis arrangement, $\frac{w\ apr}{w^+apr^+}$ the corresponding dominant alleles w^+ and apr^+ are **free to produce wild-type polypeptides.** In the trans arrangement $\left(\frac{w^+apr}{w\ apr^+}\right)$, however, neither **chromosome segment can produce the entire wild-type product.**

Support for this hypothesis has come from other studies of Lewis on the functional alleles bx (bithorax) and bxd (bithoraxoid) located at 58.8 on the third chromosome of Drosophila. All mutations at this locus influence the small halteres (modified second pair of wings) in the direction of normal wings. Some mutations, however, change the anterior part of the halteres into winglike structures and others influence the posterior part of the halteres. A single mutant gene in the homozygous arrangement may result in partial winglike haltères. Two pairs of homozygous mutant alleles, one for the anterior and one for the posterior part of the haltères may produce a fully developed second pair of wings. Each mutation apparently results in a substance with a different but sequential effect in the change from haltères to wings. This kind of **sequential action of DNA units** in close proximity may account for the cis-trans position effects at the lz, w and other loci.

This would be accomplished if **one mRNA transcript is** produced by the region which includes w^+ and apr^+. When the two wild-type functional alleles are on the same homologue $\left(\dfrac{w^+apr^+}{w\ apr}\right)$ transcription is completed and the wild-type phenotype is expressed. When wild-type alleles are separated on different homologues $\left(\dfrac{w^+apr}{w\ apr^+}\right)$ each strand produces **only a fragment of the message** and the mutant phenotype is expressed.

COMPLEMENTATION TEST FOR FUNCTIONAL ALLELISM

Complementation is the ability of two recessive mutations in the trans position within a particular locus to restore the wild-type phenotype. Mutants in the trans position that produce the **mutant pheno**type are **noncomplementary** and those that produce the **wild type are complementary.** Among the lz mutants in Drosophila (Fig. 6.4), for example, lz^k and lz^{50} produced wild type when in the trans position. They are, therefore, complementary and represent mutation positions that affect different gene products. Mutants lz^{BS} and lz^{46}, on the other hand, in trans position produced the mutant (lozenge) phenotype. They are noncomplementary and affect the **same gene product.**

The complementation test or "cis-trans test" is performed in *E. coli* by introducing two mutant chromosomes into the same cell to see if their products will complement each other. If the mutations are noncomplementary the mutant phenotype is expressed, but if they are complementary, the wild-type phenotype is expressed because each chromosome compensates for the defect in the other. For analogy, if a student has two typewriters of the same make and model, one with a broken m key and the other with a broken p key, he may exchange the m key in the broken m typewriter (already having a good p) and thus make one functional typewriter. The m and the p keys correspond to functional alleles. One can complement the other in producing a functional unit. If on the other hand the m key in each of the two typewriters should be broken, the typewriter could

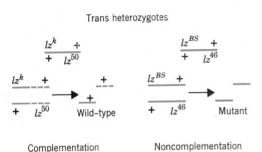

Trans heterozygotes

Complementation   Noncomplementation

Figure 6.4 Explanation for complementation and noncomplementation in the phenotypes of trans heterozygotes. Complementation occurs when mutants affect different phenotypic products (enzymes shown by solid and broken lines). Noncomplementation occurs when both mutants affect the same product (enzyme shown by solid lines). Hatched areas represent absence of function following mutation. (a) Two mutants in the same cell affect different enzymes. A wild-type gene for each enzyme is present; both enzymes are synthesized and wild type is produced. (b) Two mutants affect the same functional unit. The enzyme is not synthesized and the mutant phenotype is produced.

not be repaired by exchange of a key. The typewriter would remain nonfunctional and express a "mutant phenotype."

The complementation test does not depend on recombination but on genetic function. Functional alleles that complement each other need not be structural alleles or pseudoalleles. Furthermore, complementation is an interaction of gene products and not of genes themselves. It is based on the

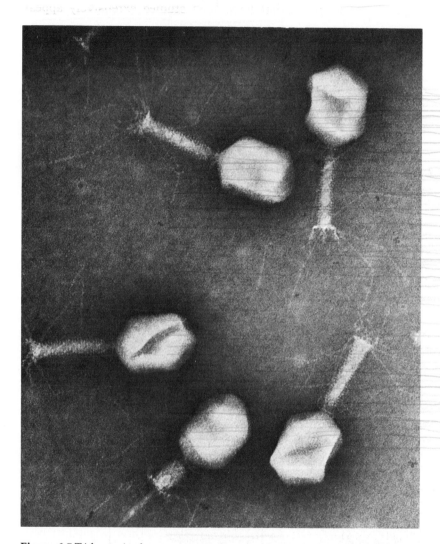

Figure 6.5 T4 bacteriophages, negatively stained. Head, tail and baseplate with tail fibres are easily distinguished. In between head and tail, the collar is visible. In one of the phages (lower left) the tail-sheath is contracted, uncovering part of the inner tail-tube. This process occurs also physiologically when phage is infecting the cell. The tube is then "pushed" through the cell envelopes, like an injection needle. Magnification ×52,000. (Courtesy E. Boy de la Tour and E. Kellenberger.)

premise that one gene (cistron) produces one polypeptide chain. The term "cistron" (coined by Benzer from cis-trans) may be equated to the gene as a functional unit. A **cistron corresponds to one functional unit on a chromosome** as determined by **complementation tests.**

GENETIC FINE STRUCTURE AND FUNCTION IN PHAGES

Phages (Fig. 6.5) are among the simplest forms that go through a life cycle, but their genetic mechanism has much in common with that of plants and animals. T phages that have been studied extensively appear

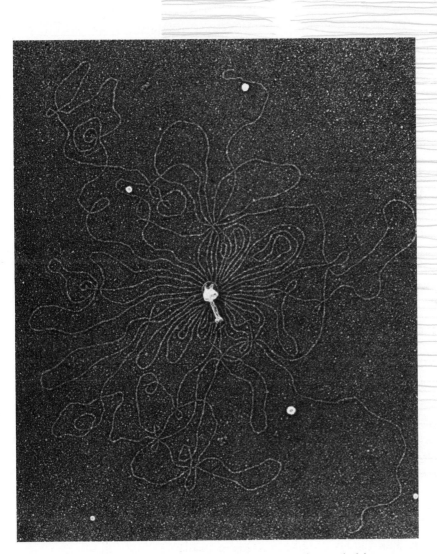

Figure 6.6 Bacteriophage T2 with coiled genetic material extruded from head. (Courtesy of Dr. A. K. Kleinschmidt. Reproduced from *Biochem. Biophys. Acta,* 61 (1962) p. 861, Fig. 1)

on electron micrographs as particles with three-dimensional enlarged heads, and short, blunt tails (but T1 and T5 have longer and more slender tails). Each is about 0.1 micron in length or approximately one-tenth the diameter of an *E. coli* cell. Digestion experiments with trypsin, a protein digesting enzyme, demonstrated that the outer part of the head is protein. The head is filled by a single molecule of coiled duplex DNA (Fig. 6.6), with about 200,000 nucleotides. About 100 cistrons are included in a single chromosome. The tail is cylindrical with an outer sheath of protein and a hollow interior through which the DNA contents of the particles are introduced into the host cell.

T phages that attack *E. coli* have been employed extensively in fine structure, complementation and recombination studies. Experiments with this material are carried out by spreading a virus suspension over a plate culture of sensitive bacterial cells. Phage particles (Fig. 6.7) become attached by their tail spikes and fibers to bacterial cells, and the phage DNA contents are extruded into the host cells (Fig. 6.8a). The invading phage DNA, on first entering the host cell, stops the metabolic activities of the host. At this stage the phage is in a noninfective, **vegetative state.** Within two to three minutes after the T4 virus DNA enters an *E. coli* cell, the chromatin region of the host cell is disrupted, and irregularly shaped masses of chromatin (Fig. 6.8b) become apparent. The bacterial DNA is degraded by an enzyme, endonuclease II that is specific for bacterial DNA. Host DNA is dispersed in the cytoplasm and moves toward the inside of the cell membrane. The

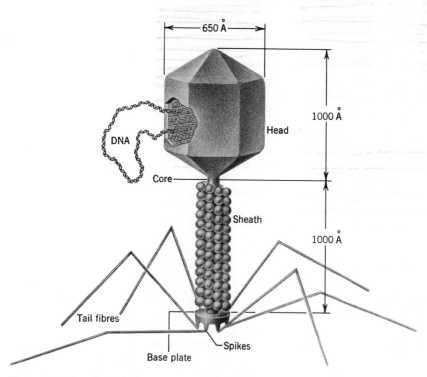

Figure 6.7 Morphological components of phage T2 and their arrangement in the intact structure. (From Nason & DeHaun "The Biological World" 1973)

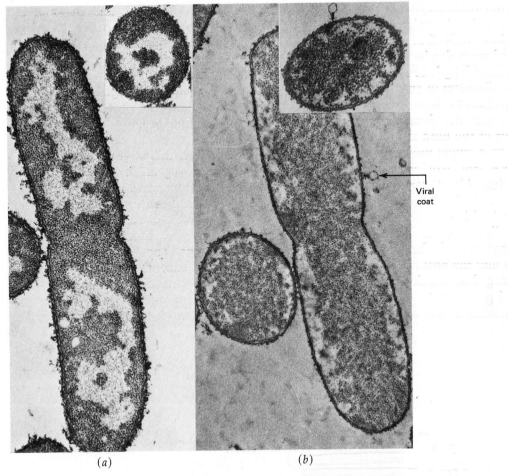

Viral
coat

(a)   (b)

Figure 6.8 Longitudinal and cross sections of *E. coli* cells illustrating degradation of the host DNA on entrance of viral DNA. (*a*) *E. coli* with (light
colored) irregular shaped chromatin masses without limiting membranes.
(*b*) Degraded DNA in *E. coli* following entrance of viral DNA. Note viral
coats without DNA at the top of cross section and around periphery of
longitudinal section. (From Peter Snustad, Journal of Virology, Vol. 10.)

phage DNA remains intact and active and takes over **full control** of the host cell.

DNA synthesis, along with other host cell activities, stops for a time but soon is reestablished under control of the virus DNA. Degraded DNA of the host is re-synthesized according to specifications from the virus DNA. The metabolic apparatus of the host cell is thus directed toward the production of phages instead of the activities needed by the bacterial cell. For phage T1 the time from introduction of the virus DNA until the bursting (lysis) of the cells and release of total, intact phage particles (virions) is about 13 minutes, and for phage T2 about 22 minutes.

When a T4 phage culture is mixed with a culture of *E. coli* and poured on a fresh agar

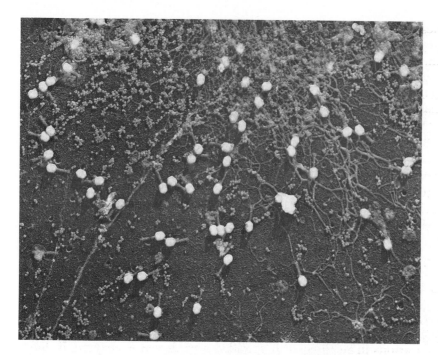

Figure 6.9 Lysis of bacterial cell freeing phage particles. Large white structures are heads of phage particles with gray tails and tail fibers. Many ribosomes (small gray structures in clusters). unused DNA fibers (long threads) and cell debris are shown. Magnification ×37,000. (Courtesy E. Kellenberger, Biozentrum der Universität, Basel.)

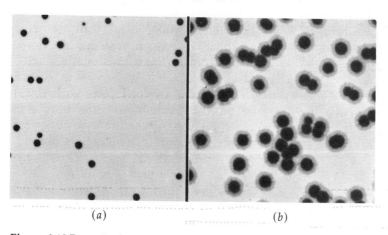

(a)   (b)

Figure 6.10 Bacteriophage growth is evidenced by clear areas or plaques on an opaque bacterial growth. (In this photograph plaques appear dark on a light background.) The size and texture of the plaque is under genetic control. (a) Plaques of T4; (b) plaques of the mutant T4r. (From *Viruses and Molecular Biology* by Dean Fraser, published by The Macmillan Company. Copyright 1967 by Dean Fraser.)

plate, the plate becomes cloudy because of the dense bacterial growth. Within a few minutes after infection, each infected bacterium releases one hundred or more mature infective **virions** (Fig. 6.9). These infect adjacent bacterial cells and a new cycle begins. As the cycles continue, areas on the culture plate where lysis is in progress become cleared of bacteria and are identified as transparent areas called plaques on an otherwise opaque culture plate (Fig. 6.10). Size and appearance of the plaques depend on the type of phage and the type of bacterial host cells.

Most T4 r mutants lyse bacterial cells rapidly and produce large plaques with sharp edges. Plaques produced by the wild-type, r^+, are smaller with more irregular edges. Three regions (I, II and III) of the phage T4 linkage map include loci in which mutations can occur. Mutations in all three regions produce large plaques when $E.$ $coli$ strain B is the host. When, however, $E.$ $coli$ strain K ($E.$ $coli$ K12 with lambda prophage integrated) is host, only T4 cells with mutations in rI and rIII form plaques (rI mutant and rIII$^+$ plaques). Mutants in the rII region can enter $E.$ $coli$ K12 cells but they are unable to reproduce and do not lyse these cells. Phage with the wild-type allele, r^+, produce their characteristic plaques with both $E.$ $coli$ strain B and strain K. Genes that control plaque size **influence the length of the T4 life cycle.**

COMPLEMENTATION WITH THE rII LOCUS

When $E.$ $coli$ cultures are infected with mixed r and r^+ phage, both kinds of virus enter the cells and reproduce. The r^+ virus acts as a "helper" and contributes a substance that the r virus can use in its own reproduction. With the two alleles in the same bacterial cells, and in the trans position, tests for functional (complementation) as well as structural (recombinational) allel-

ism can be carried out. If the mutants involve the same enzyme, **no growth** will occur, but if the mutants represent **different cistrons,** they may complement each other and thus enable both **viruses to reproduce.**

Benzer's experiments showed that while rII virus mutants could infect $E.$ $coli$ strain K, they could not destroy members of this strain. When, however, strain K was simultaneously infected with two different rII mutant viruses that could complement each other in the trans position, viral reproduction and bacterial lysis could occur. In these investigations all rII mutants fell into two complementary groups, which were identified as cistrons A and B (Fig. 6.11).

Products of both cistrons were required for bacterial lysis. When A and B mutant phage were together, mutant A apparently specified a wild-type B enzyme and mutant B specified a wild-type A enzyme. If, however, two A or two B mutants were involved, virus reproduction did not occur and the cells were not lysed. Two **different functions** were thus attributed to different **functional alleles** that produced two different **polypeptide chains.**

Complementation maps can be constructed on the basis of tests for functional alleles. Although some functional segments such as the A and B cistrons in the rII region of phage T4 correspond with linear units on the chromosome, the results of complementation tests cannot be compared with recombination data. Products of genes in structural proximity or at a greater distance may complement each other. Physical locations of genes are therefore not determined by complementation tests, but by recombination.

RECOMBINATION AMONG VIRULENT PHAGES

The rII region of the r locus in phage T4 has been studied extensively by S. Benzer from infections on $E.$ $coli$ strain B. Mixed

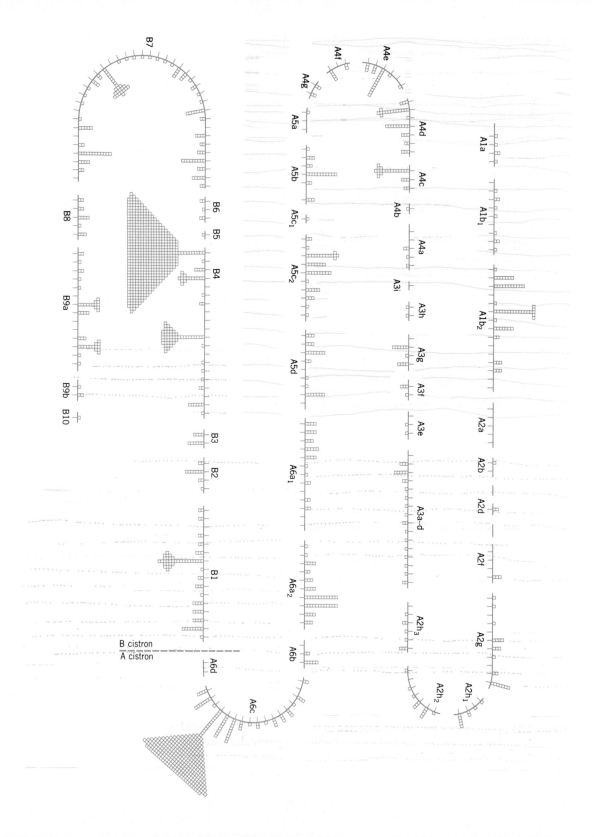

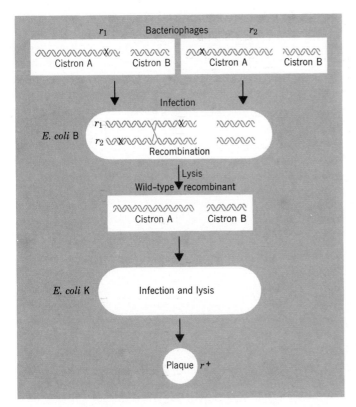

Figure 6.12 Technique used by Benzer to detect wild-type recombinants between two rII mutations affecting the same cistron.

infections of two different mutants such as r_1 and r_2 in the *rII* region would sometimes produce wild-type phage (r^+). This was attributed to reciprocal recombination. A procedure for detecting such an exchange within a cistron is illustrated in Fig. 6.12. Recombination occurred between genes in the trans position, that is, $\dfrac{r_1^+ r_2}{r_1 r_2^+} \rightarrow \dfrac{r_1^+ r_2^+}{r_1 r_2}$. Recombinants could then be isolated by transferring the doubly infected *E. coli* B to *E. coli* K, on which only r^+ is expressed. Recombination frequency was determined by the proportion of r^+ plaques on *E. coli* strain K.

Recombination frequency =

$$\frac{\text{K12 plaques}}{\text{Total plaques on B}}$$

M. Delbrück, 1969 Nobel prize winner, introduced both a T2 phage carrying the r gene and a T4 phage, which produced wild-type plaques (r^+), into a bacterial culture. The cross was thus T2r × T4r^+. The two original phage types were recovered along with T4r and T2r^+ types, which were recombinants. The results showed that genes for parental traits could recombine and produce new types. A two-factor cross, $hr × h^+r^+$ was then made between in-

Figure 6.11 Spontaneous mutations detected in the rIIA and rIIB cistrons of T4. Each square symbolizes one mutation detected at a particular locus. (After Benzer, *Proceeding of the National Academy of Science* 47:410, 1961.)

dividuals in the same strain. The *h* mutation extended the host range and made the virus T2 capable of adsorbing to and eventually lysing otherwise resistant cells of *E. coli* strain B/2. Plaque characteristics and ability to infect strain B/2 (*h*) were readily identifiable in cultures. **Recombinants** (*h⁺r* and *hr⁺*) as well as parental types (*hr* and *h⁺r⁺*) were recovered.

When the results of phage crosses were observed critically, evidence of partial heterozygosis was detected. It was partial in the sense that heterozygosis occurred only in limited regions. Mixed infections of *r* (for

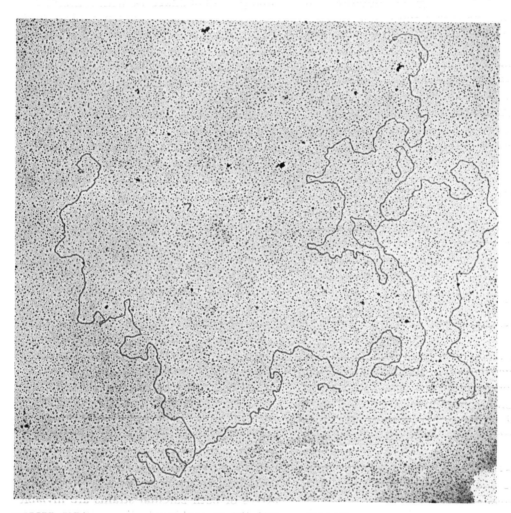

Figure 6.13 Electron micrograph of recombining bacteriophage T4 DNA molecules extracted from phage-infected *Escherichia coli*. The branches reveal the regions of synapse of the interacting chromosomes. Branch pairs of the type illustrated are one of several DNA structures characteristic of the T4 recombination pathway operable when DNA replication is prevented. Magnification ×21,400. (Courtesy of Thomas R. Broker, Stanford School of Medicine.)

rapid lysis) and r^+, for example, usually produced r or r^+ plaques, but about two percent of the resultant plaques were mottled with some r and r^+. Appropriate testing showed that both r and r^+ could be recovered from mottled plaques. The relatively high frequency with which the mottled forms occurred, ruled out regular mutations. It was shown further that the r gene was not particularly unstable, as expected if an extra high mutation rate had occurred. The best explanation was recombination based on the hypothesis that the mottled plaques were heterozygotes. Phage carry only one DNA molecule, and heterozygous recombinants occur on opposite DNA strands rather than on separate double strands (Fig. 6.13).

With no sexual process comparable to that in plants and animals, viruses must exchange genetic material in the cytoplasm of the host cell. Numerous copies of the DNA are carried in viral chromosomes within a single host cell. Genetic exchange is not restricted to two or four chromatids as in plants or animals but may occur within a large population of chromosomes. Obviously, zygote and tetrad methods are not applicable to virus studies.

Plating techniques have been devised to allow redistribution of genetic material and recovery of recombinants. Events associated with recombination must be described in statistical terms rather than in terms of single exchanges between two chromatids in the meiotic sequence. Viral chromosomes may exchange genetic material more than once and may mate with more than one chromosome at one time.

Some features of recombination and linkage known in eukaryotes have also been recognized in phages. Genetic exchanges are sometimes reciprocal events similar to those in which physical chiasmata occur between homologous chromatids. Variation in recombination frequency indicates that linkage occurs between nonallelic genes. This suggests that a phage chromosome carries genes in linear sequence as do chromosomes of eukaryotes.

Recombination data have led to the compilation of **linkage maps** showing the relative positions of many genes. A circular linkage map of phage T4 with symbols representing cistrons is shown in Fig. 6.14. The genes are arranged in sequence according to the developmental patterns in which they are functional. Some genes function only in assembling the virus; that is, the head and tail are synthesized independently and joined together by separate genes. A single molecule of DNA with a molecular weight of about 130 million daltons (units equal to the mass of the hydrogen atom) is postulated as the physical basis for the T4 chromosome.

DELETION MAPPING FOR SITES WITHIN GENES

The rII region in phage T4 studied by Benzer has been especially favorable for mapping the fine structure of genes. The main advantage is the ease with which mutants can be identified. All rII mutations result in the loss of capacity to synthesize one or more proteins necessary for development on *E. coli* K12 (lambda). The **mutants** are, therefore, **lethal on K12** host cells but they can **grow on B strain cells.** When, for example, B cells are infected with two rII mutants, the phage particles develop normally and lyse the host cells. The progeny particles are then scored on K12. Representatives of both parental strains are inviable but the recombinants thrive. This procedure is efficient for isolating recombinants. It is very sensitive with limits of sensitivity determined by rate of spontaneous reverse mutations to wild type.

To provide further control for precision mapping of genes in the rII region, Benzer

Figure 6.14 The circular linkage map of phage T4, showing the functions served by the various genes, as determined by recombination, complementation and physiological analysis of conditional lethal mutants. The inner circle shows the arrangement of the genes on the linkage map, the circumferential length of the black rectangles representing the length of the genes as estimated by recombination frequencies! Functions of the various genes are indicated: CAPSID = head structure; MD = maturation delayed; e = endolysin (lysozyme); the diagramatic drawings of structural components show the products found in lysates of the restrictive host, in cases where the actual function of the gene is unknown. On the outer circle, the white rectangles indicate genes determining early functions, and the black rectangles late functions: the criterion of early and late functions is the time of major mRNA synthesis as judged by reversal of phenotype by 5-fluorouracil (5FU). (After Hayes.)

chose particular mutations called **deletions.** These have missing segments covering more than one mutable site, and they overlap one another. The reason for choosing deletions for mapping the rII region is that they have parts missing and, therefore, they do not revert to wild type. A test based on presence or absence of wild-type phage can therefore be used to indicate recombination of chromosome parts.

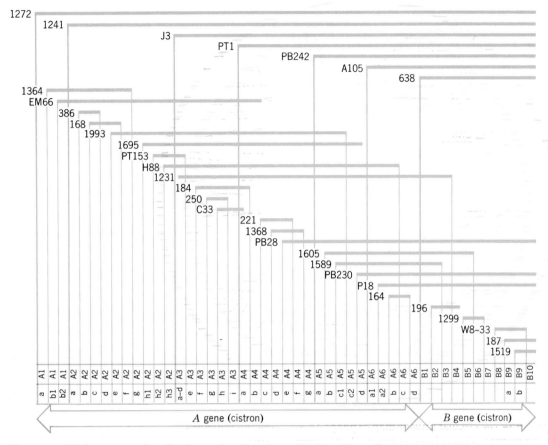

Figure 6.15 Deletions used to divide up the rII region of T4 into 47 small segments (shown as small boxes at the bottom of the figure). Some deletion ends have not been used to define a segment and are drawn fluted. The A and B genes, defined by independent complementation tests, are indicated. (From S. Benzer, *Proc. Natl. Acad. Sci.* U.S. 47:410, 1961).

When, for example three deletion mutants, (*a*, *b*, and *c*) were tested, *a* and *c* recombined and yielded wild type, but neither *a* nor *c* recombined with *b*, the results of the recombination test may be summarized as follows. (+ = recombination, − = failure of recombination):

	a	*b*	*c*
a	−	−	+
b	−	−	−
c	+	−	−

The deletions fall in the serial order, *a*, *b*, *c*, with *b* overlapping with *a* and *c*.

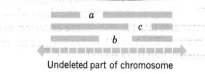

Undeleted part of chromosome

When two strains carrying deletions that overlapped one another were crossed, wild-type recombinants were never found among the progeny but if they carried deletions that did not overlap, wild-type recombinants could be formed. The relative positions of the deletions could be determined and placed on a map. Sites located within deletion *a* could be located in relative position with reference to those in deletions *b* and *c*.

Deletions actually studied by Benzer (Fig. 6.15) were identified by particular numbers (e.g., 1272, 1364). Ends of deletions delimit 47 different segments in cistrons A and B shown as AIa, AIbI, AIb2 and so on. Seven major deletions cover large sections of the rII region. When many deletion mutant strains were studied, a map was prepared. A strain carrying a mutation at an unknown site was first crossed with strains carrying the big seven deletions. If, for example, no wild-type recombinants occur in crosses with *r*1272 but they do occur with 1241 and other strains the mutant must be in the first major section. Crosses with strains carrying smaller deletions within the first section could then be used to pinpoint the mutation.

About 2000 mutations in the rII locus (Fig. 6.11) were suitable for study. These were found by deletion mapping to occupy some 308 different sites. The sites were distributed throughout the region but some "hot spots" mutated many times and some "zero spots" did not show any mutation in the course of the experiment. Some 500 sites were estimated for the rII regions. With an estimated 4000 nucleotides in the rII region and 500 sites, each site might include **8 nucleotides** on the average. Benzer's experiments led to new terms, recon, muton as well as cistron, that identify different units of DNA within the conventional gene. A **recon** is the smallest unit of DNA capable of recombination and a **muton** is the smallest unit of DNA in which a change can result in a mutation. Both a recon and a muton equate to a single nucleotide. These studies provided evidence for linearity of the fine structure of the gene. Linearity had been demonstrated for genes (Chapter 3) and these studies showed a similar linearity for small segments of DNA within the gene.

RECOMBINATION IN BACTERIA

Mechanisms for recombination of genetic material in bacteria include: (1) **direct transfer** of "naked" DNA to a member of the same or a closely related species and integration of the transferred DNA into the recipient cell's genome (transformation); (2) transfer of DNA from a donor to a recipient bacterium through the agency of a **phage vector** (transduction); (3) **mating** between a donor and recipient bacterium in which chromosomal DNA is transferred from donor to recipient (conjugation); (4) incorporation of bacterial DNA by **sex factors** (F factors) and transfer by conjugation to a recipient cell (F-mediated sexduction); (5) exchange of DNA from multiple infections of temperate phage and integration of phage DNA within a bacterial host cell (**lysogeny**) and transfer of DNA by a nonessential cell constituent (**episome**) which may be integrated with the bacterial genome or reproduce independently in the cytoplasm.

TRANSFORMATION

Initial evidence for genetic recombination in bacteria came from studies of transformation in *Diplococcus pneumoniae* (Chapter 3). DNA isolated from one bacterial strain was brought into contact with living bacteria from another strain and was absorbed by about one percent of the bacterial cells. Recipient cells were genetically transformed and capable of expressing

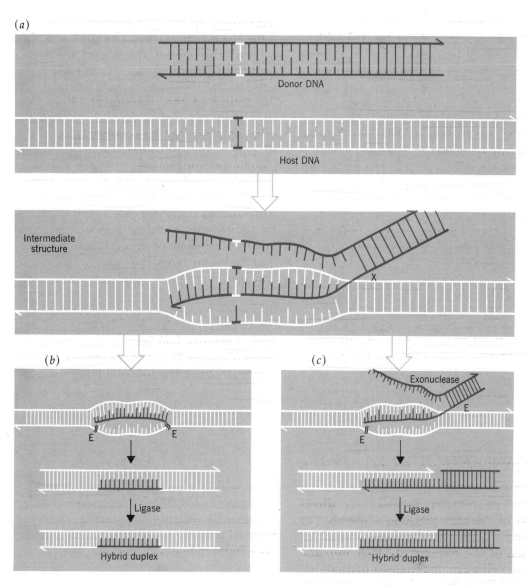

(a)

Donor DNA

Host DNA

Intermediate
structure

X

(b)

(c)

Exonuclease

E

E   E   E

Ligase

Ligase

Hybrid duplex

Hybrid duplex

Figure 6.16 Model of bacterial transformation. (*a*) Donor and host DNA form
an intermediate structure in which a strand from the donor molecule pairs
with its homologous complementary strand in the host molecule. Two
possible events (*b* or *c*) can now occur. (*b*) An endonuclease nicks the
dangling donor molecule at X to leave a small single piece of donor DNA
associated within the host chromosome. Endonucleases (E) act upon the host
strand and a piece of host DNA is lost. Finally, the nicks are sealed by a DNA
ligase. (*c*) An exonuclease digests the dangling single donor strand while
endonucleolytic nicks (E) are made in the host chromosome. Large pieces of
the host chromosome are subsequently lost as donor DNA becomes
incorporated into the helix. (From M. S. Fox, Journal of General Physiology)

phenotypes different from those on untransformed cells. The low occurrence of transformed cells was attributed to the small size of **receptor sites** and the low number of surface areas where the recipient cell could absorb DNA. Entrance of the donor DNA was facilitated by **enzymes** in the cell walls of the recipient bacterium. Entry sites therefore are areas of enzymatic activity and not simple openings in the cell wall. ·

Transformation does not occur in all bacteria but has been observed in a few genera and species. It depends on the competence of bacterial cells to absorb DNA from the extracellular environment. **Competence** is the ability of bacterial cells to take up, integrate, and express the genetic information contained in DNA of related organisms. Competent cells contain an antigen, specific for the competent stage which appears after a **competence provoking factor** (CPF) has acted on noncompetent cells. CPF is a protein that must attach itself at specific cellular sites to convert noncompetent cells to competence. Even in a transformable species, competence occurs only at specific periods of the life cycle when the cell is in a particular physiological state. At this time, the cell is in a stationary phase with a small number of nuclei, and DNA synthesis is not in progress. Only 1 to 10 percent of the cells in a "competent culture" are usually competent. Techniques based on sedimentation velocity have been developed for isolating competent cells from such a culture.

Transformation is a complex process including several integrated stages: (1) **reversible binding** of double stranded DNA at the cell surface; (2) irreversible uptake of the DNA accompained by **resistance to DNase;** (3) **conversion** of the double-stranded donor DNA into single-stranded DNA, soon after entry into the host cell (Fig. 6.16); (4) **incorporation** of the DNA into the host chromosome; and (5) **expression**, replication and segregation of the recombinant material.

Several genes are involved for the competent state of a bacterium. Competence deficient mutants for each of the five stages listed above have been isolated and classified according to the stage of transformation at which the block occurs.

Both linkage and crossing over can be detected in the transformation process. Exchanges occur only between homologous segments of donor and recipient DNA. Transforming segments of DNA are physically attached in linear sequence and, therefore, they are linked. The basic mechanisms of gene exchange (i.e., crossing over) are similar to those in sexual recombination. Two points of exchange, one on each end of the donor segment, are required before incorporation into the recipient DNA can occur. Since linkage in transformation is defined in terms of the location of different loci on the same molecule of transforming DNA, the length of the molecule is a factor in establishing linkage relations. Linkage values between markers can be altered experimentally by altering the size of the molecules with DNase and other agents. Although data in prokaryotes may be compared with those of crossing over in eukaryotes, transformation linkage distances are not as repeatable as those obtained from sexual recombination. Transformation is applicable to fine **structure mapping** but not for an overall analysis of the bacterial genome. Transformation linkage data do, however, permit estimates of the relative positions of genes in linear order in the bacterial genome.

TRANSDUCTION

Transduction is the process by which a **bacteriophage mediates** the transfer of genetic material from a bacterium (donor) to another bacterium (recipient). Isolated phage and a recipient strain of bacteria may produce transduction. This mechanism for recombination was discovered through the classical investigations of Zinder and Lederberg on mutant auxotrophic strains (mutants that cannot synthesize one or more essential nutrients) of the mouse typhoid organism *Salmonella typhimurium*. One **auxotroph** required methionine (met^-thr^+), and another required threonine (met^+thr^-). When the two strains were mixed and plated on a medium deficient for methionine and threonine, some wild types occurred and gave rise to colonies of prototrophs. Since none of the bacteria of either auxotrophic strain plated alone was able to grow into colonies on the minimal medium, mutation to the prototrophic state was ruled out. Furthermore, the occurrence was too frequent to be attributed to mutation. The **prototrophs** were judged to be the result of genetic recombination. DNA exchange between the two strains of Salmonella was mediated by a **filterable virus**, Salmonella phage P22, which one of the parent strains happened to carry.

Bacterial strains carrying several markers were then used in experiments and it was shown that each individual marker gene could be transduced independently. When, for example, phage P22 carried on a Salmonella strain that could synthesize methionine, utilize galactose and xylose, and resist streptomycin ($met^+gal^+xyl^+str-r$) was allowed to infect a strain that could not synthesize methionine, could not utilize galactose or xylose, and was streptomycin sensitive, ($met^-gal^-xyl^-str-s$), some recipients showed transduction of met^+, some of gal^+, others of xyl^+, and still others of $str-r$. Since this transduction mediated by phage P22 can involve *any* sector of the Salmonella genome, it is called generalized transduction. Further studies showed that although most genes were transferred independently, some moved in groups of two or more genes. These groups were closely linked genes. The frequency with which genes were transduced together indicated the relative distance (linkage strength) between them. This information was applicable for mapping the fine structure of the bacterial chromosome and transduction became the most useful mechanism for this purpose.

The mechanism of generalized viral transfer for bacterial DNA is an occasional error in **virus packaging.** This occurs in the host cell when the protein coats are synthesized to cover progeny phages. During the cycle of phage development within the host cell, the bacterial chromosome is degraded to small pieces of double-stranded DNA. Fragments of host DNA, the size of the phage genome are mistakenly packed into phage coats. In **generalized transduction,** any region of the bacterial chromosome apparently has an equal probability of incorporation into a transducing particle.

When lysis occurs, a few carriers of bacterial DNA (donors) are freed along with normal phage particles to inject their contents into new host (recipient) cells. When donor DNA has penetrated the new recipient, recombination may occur with the homologous regions of the recipient chromosome. Donor genes thus replace recipient genes in the recipient chromosome. When transduction brings about a change in the genotype of the recipient it is said to be "complete transduction" in contrast to "abortive transduction" (see below). The examples of generalized transduction cited above represent complete transduction. Another kind of complete transduction is called **restricted transduction.**

In restricted transduction, recombination depends on the ability of the phage genome to pair with and incorporate a particular fragment of bacterial chromosome. Limited and intermittent regions of homology exist between the genomes of the bacterium (recipient) and the phage (vector). When, for example, bacteria ordinarily sensitive to

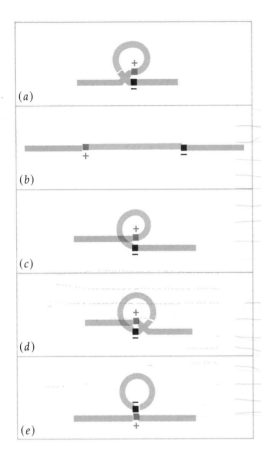

Figure 6.17 Possible mechanism for recombination by means of transduction. The bacterial chromosome is shown in black and the phage chromosome in color. Thick black segment represents the *str⁻* allelic region and the thick color segment represents the *str⁺* allelic region. The phage is inserted into the bacterium by means of recombination (*a*) to form a partially heterozygous heterogenote, (*b*). The *str* regions pair (*c*) and again recombination occurs (*d*) resulting in the bacterium containing the *str⁺* allele (*e*). (After Hayes.)

streptomycin, (Fig. 6.17, −allele) were infected with phage carrying streptomycin resistance (Fig. 6.17, +allele), the recipient cells became streptomycin resistant. The portion of the bacterial chromosome (**endogenote**) that controls resistance paired with the homologous **exogenote** from the phage and produced a partially heterozygous **heterogenote** containing the differing alleles from the endogenote and the exogenote.

In the heterogenote the +allele of the phage was included with the −allele in the chromosome of the bacterial recipient. The process of recombination by transduction was thus completed. Once altered, the recipient chromosome retained the +allele and the cell could serve as a donor for resistance, thus confirming that a genetic change had occurred in the recipient cell. Restricted transduction is "prophage mediated" by temperate phage particles whose DNA is homologous only with one or a few sites on the bacterial chromosome. Transfer of bacterial genes is restricted to those genes located adjacent to the particular site of association of the prophage DNA and the bacterial DNA of the donor cell. In restricted transduction, all transduced bacteria are **lysogenic** (i.e., they carry temperate virus in the prophage state) and the newly introduced genes are added to the genome of the recipient cell. Thus the transduced cell carries two copies of most of the phage genes and two copies of a few bacterial genes. The transduced cell is a **par-**

Figure 6.18 J. Lederberg, who has done pioneering research in the field of bacterial genetics. Among his contributions is the genetical proof that fertilization and recombination occur in bacteria. Dr. Lederberg, along with Dr. G. Beadle and Dr. E. L. Tatum, was Nobel Laureate in 1958. (Courtesy of Stanford University Medical Center.)

tial diploid. The phenotypes expressed depend on the dominant member of each pair of alleles. Restricted transduction is useful in mapping the genes by pinpointing the positions of mutations in positions near the prophage site on the chromosome.

Abortive transduction is quite different from generalized and restricted transduction in that the bacterial DNA fragment carried by the phage is injected into the recipient bacterium and may function there, but it does not become incorporated and replicate with the chromosome. The DNA segment from the donor is transferred only to one of the first two descendents of the recipient cell, thus giving rise to a "semiclone." The ratio of abortive to complete transduction is about 10 to 1.

S. typhimurium, like *E. coli,* was found to have a circular linkage map. Many recombinants have been identified in this species. Furthermore, hybrids have been obtained between *S. typhimurium* and *E. coli* in which part of the genetic material has come from each species. When phage particles were raised on different hybrids (i.e., hybrids carrying different combinations of markers), all hybrid donors gave rise to some transductants. Some hybrids produced many and others only a few transductants. After many experiments, it was possible to compare the linkage groups of the two species. Many genes were found to be similar in function, and they were located in corresponding positions in the two linkage groups.

CONJUGATION

J. Lederberg (Fig. 6.18) and E. L. Tatum, both Nobel prize winners in 1958, were the first to demonstrate sexual conjugation in bacteria. In their experiments, recombination was detected when a bacterium of one mutant strain conjugated with a bacterium of another mutant strain (Fig. 6.19). During conjugation, one cell of a pair acted as **donor** (male) and the other as **recipient** (female).

The difference between the donor and recipient mating types was associated with a fertility factor (F) present only in donors. Donor cells with F (called F⁺) may transmit the DNA factor to recipient cells (F⁻) (Fig.

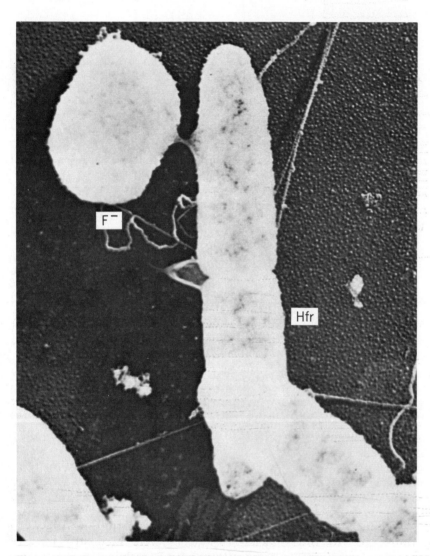

Figure 6.19 Conjugation in *E. Coli.* Magnification ×41,400. (Photograph by T. F. Anderson, E. Wollman, and F. Jacob. From *Annales Institute Pasteur*, Masson and Cie, Editors.)

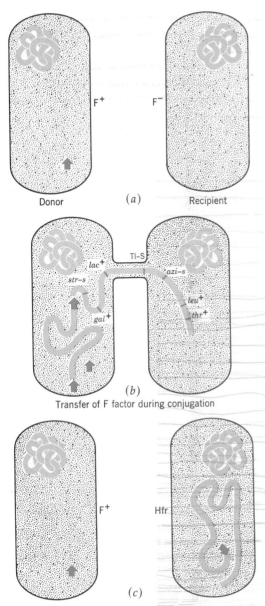

Donor   (a)   Recipient

TI-S
lac+
azi-s
str-s
leu+
gal+
thr+

(b)

Transfer of F factor during conjugation

F+   Hfr

(c)

Figure 6.20 Conjugation between F+ and F− bacterial cells with transfer of genetic material. (a) F+ donor cell with chromosome and cytoplasmic F factor, and F− recipient cell with chromosome. (b) Donor cell chromosome has replicated and is moving across the protoplasmic bridge into the recipient. F factor has replicated and one copy has become integrated into the chromosome. (c) The donor cell from which the transfer of an F factor has occurred remains an F+ donor if a copy of the F factor remains in the cytoplasm. The recipient is now Hfr with the F factor integrated into a chromosome.

6.20), thus changing recipients to donors. The donor cells probably remain donors because the **F-factor replicates** during transfer with one copy being transferred to the recipient cell and the other copy remaining in the donor cell. Recipients never change spontaneously to donors, as would be ex-

pected if a simple mutation were involved.

The F factor is cytoplasmic and is usually independent of the chromosomal genes. When F+ donors were incubated in the same culture with F− recipients, most recipients became donors, but only a small amount of recombination occurred to form

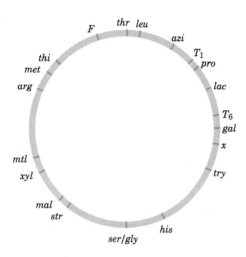

Figure 6.21 Ring-shaped linkage group in *E. coli*. The F factor is shown integrated into this chromosome.

incomplete zygotes. In a mix of the two mating types of *E. coli*, for example, recombination occurred at the rate of about one recombinant per million cells. Donors with a rate of recombination in this order of magnitude were called low-frequency recombinants (Lfr).

The F cytoplasmic factor occasionally recombines with a particular part of the bacterial chromosome. An F body incorporated into a bacterial genome (Fig. 6.21) can be transferred during conjugation. When this occurs, the frequency of recombination is greatly increased, to the order of about one per thousand cells. A cell with the F factor incorporated in a chromosome is called a high frequency recombinant (Hfr) donor or metamale. An **Hfr cell** usually remains Hfr, following transfer of DNA to a recipient cell in conjugation because copies of the integrated F factor are retained in the donor cell.

In further experiments by Jacob and others, male cells of an Hfr strain were placed in the same culture with female cells, and the two kinds of cells came together in conjugation. While the two cells of each conjugating pair were joined by a temporary protoplasmic bridge, a linear group of genes used the bridge to migrate from the donor cell into the recipient cell. This was accomplished as follows: the ring-shaped *E. coli* linkage group (Fig. 6.20) opened at the place where the F factor was located. The end opposite the F end entered the protoplasmic bridge and moved across into the recipient cell. About two hours after the beginning of the transfer, the entire linkage group was across the bridge with the F end entering last. The exchange was a **one-way** process with the genetic material always moving from the F-containing donor to the recipient cell.

A cross between genetically different Hfr and F⁻ strains demonstrated the sequence of gene exchange during conjugation. Hfr organisms used in the experiment were capable of synthesizing the amino acids threonine (thr^+) and leucine (leu^+); they were sensitive (s) to both the metabolic inhibitor sodium azide (azi-s), and the bacteriophage Tl (Tl-s); they could ferment lactose (lac^+) and galactose (gal^+); and they were sensitive to the antibiotic streptomycin (str-s). Genotypically, the Hfr organisms were thr^+ leu^+ azi-s Tl-s lac^+ gal^+ str-s, and the F⁻ organisms were thr^- leu^- azi-r (r for resistant) Tl-r lac^- gal^- str-r. The

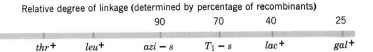

Relative degree of linkage (determined by percentage of recombinants)

		90	70	40	25
thr^+	leu^+	$azi-s$	T_1-s	lac^+	gal^+

Figure 6.22 Linkage map showing some genes of *E. coli* with the relative degree of linkage with thr⁺ and leu⁺. (From experiments of Jacob and Wollman.)

Hfr and F⁻ cells were mixed and allowed to remain together in a liquid medium for 25 minutes. The mixed cells were then plated on the first selective medium, a minimal medium containing streptomycin.

On this selective medium, the Hfr parent cells were killed by the streptomycin, and F⁻ parent cells were unable to grow because the essential amino acids threonine and leucine, which they could not synthesize, were not supplied. Since the genes (thr^+ and leu^+) for synthesizing these amino acids came from the Hfr parent and streptomycin resistance ($str-r$) came from the F⁻ parent, the surviving cells were recombinants.

Next, the thr^+ leu^+ $str-r$ recombinants were replicated (successively plated) on minimal solid media with different additives such as sodium azide and bacteriophage Tl. From experiments of this kind, Jacob and Wollman found that 90 percent of the colonies that were thr^+ leu^+ $str-r$ were also sodium azide sensitive ($azi-s$), 70 percent were $Tl-s$, 40 percent were lac^+, and 25 percent gal^+. These percentages represented the linkage strengths among the different genes and were used to develop a linkage map (Fig. 6.22). It should be noted that these percentages are not equivalent to linkage values discussed in Chapter 5 for eukaryotes, but they do indicate the relative positions of genes.

Jacob and Wollman then proceeded with a second method of mapping the gene from the mixed cultures, known as the interrupted mating experiment. Up to 60 minutes from the time Hfr cells were mixed with the F⁻ cells, samples were periodically agitated in a blender to separate the conjugants. This separation interrupted the transfer of genetic material from donor to recipient and presumably broke the linkage group. The separated cells were plated and tested to detect which genes from the donor had been integrated into the recipient. At 0 minutes from the time of mixing, 0 recombinants had occurred. The thr^+ gene became integrated in 8 minutes, leu^+ in 8½, $azi-s$ in 9, $Tl-s$ in 11, lac^+ in 18, and gal^+ in 25 minutes (Fig. 6.23).

These studies showed that transfers from Hfr donors to F⁻ recipients are not random. Rather, the genes move across the protoplasmic bridge in a regular order, as would be expected of a linkage group. In the experiments, within 50 minutes almost all

Minutes	Hfr genes transferred
0	0
8	thr^+
8½	thr^+ leu^+
9	thr^+ leu^+ $azi-s$
11	thr^+ leu^+ $azi-s$ T_1-s
18	thr^+ leu^+ $azi-s$ T_1-s lac^+
25	thr^+ leu^+ $azi-s$ T_1-s lac^+ gal^+

Figure 6.23 Time required for the transfer of different genes from Hfr to F⁻ as indicated from recombinants obtained by Jacob and Wollman.

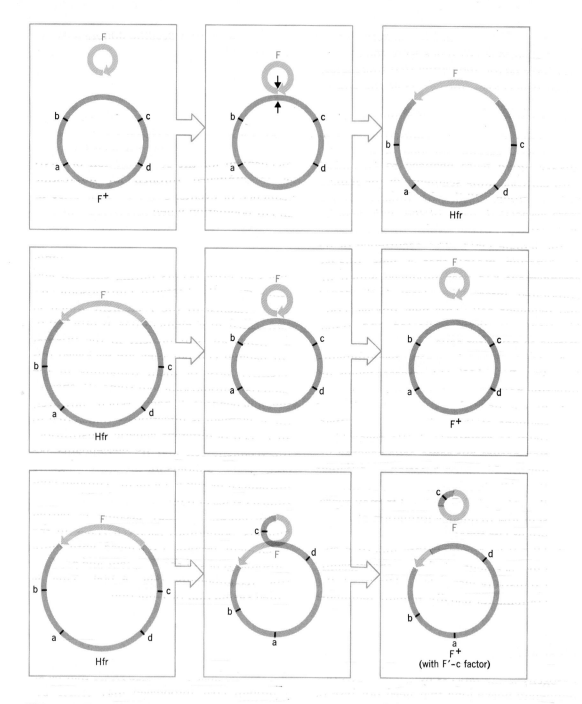

Figure 6.24 Possible mechanism for sexduction. Genetic material of a bacterium is shown in black and that of the F factor in color. (*a*) Illustration of insertion of the F factor into a F⁺ chromosome. The radial arrows identify regions of homology where pairing and recombination may occur. (*b*) Extrusion of the F factor resulting in an F⁺ cell again. (*c*) F factor being extruded along with some of the bacterial genetic material (F′) including the marker gene *c*. (After Hayes.)

the Hfr genes that could be exchanged had been transferred. No genes were transferred after the F factor, indicating that this factor, which is necessary for Hfr, is at the end of the linkage group.

Soon after migration of the linkage group was completed under noninterrupted circumstances, certain portions of the male linkage group were deduced to be paired with corresponding segments of the female linkage group. While the linkage groups were associated, parts of the male genetic material were exactly substituted for parts of the female DNA. The occurrence of recombination meant that a **new linkage arrangement** had been produced which incorporated information from both donor and recipient cell. This new linkage combination was then replicated and transmitted to daughter cells.

A comparison between recombinants from $F^+ \times F^-$ and Hfr $\times F^-$ crosses yields interesting points:

1. $F^+ \times F^-$ donor bacteria can transfer any segment of their chromosome impartially to recipients. Little recombination occurs but some cells become Lfr donors. Lfr recombination is different than Hfr because it does not involve the integrated F factor. Lfr may represent low frequency formation of a variety of Hfr types.

2. $Hfr^+ \times F^-$ recombinants inherit a specific segment of donor chromosome exclusively (depends on Hfr type). Recombinants usually remain recipients but those that become donors also become Hfr.

F-MEDIATED SEXDUCTION

Fragments of genetic material may be carried along from one bacterial cell to another in association with the sex factor F. This mechanism for the redistribution of DNA is accomplished through an interesting sequence of events. Some F^+ cells arise in an Hfr population when integrated F factors become excised from the bacterial genome and remain in the cytoplasm as autonomous F elements (Fig. 6.24a). Sometimes an error occurs during excision and genes from the bacterial chromosome remain attached to the F element which is now called an F^1 element (Fig. 6.24b). All of the genes are still present in the original Hfr haploid cell but they are packaged differently; most of them remaining in the bacterial chromosome but some attached to the independent F body. When, however, the F^1 element is infectiously transferred to an F^- cell of different genetic constitution, the recipient cell and its descendants become partial diploids for the bacterial genes introduced by the F^1 element. Through continued cycles of replication and infection an entire population may become F^1 partial diploids. Sometimes bacterial genes in the F^1 element become exchanged with chromosomal genes of the recipient and produce true recombinants (Fig. 6.24c). This is called **sexduction** or F-duction.

LYSOGENY

Lysogeny is the process by which DNA of a temperate phage (one that need not lyse or destroy host cells) and that of the bacterial host are integrated. Temperate phages establish a degree of tolerance in the bacterial cell so that the phage DNA remains for a time in the host cell. Phage DNA that lacks a protein coat (prophage) may either become attached to part of the host genome (Fig. 6.25) or be inserted into the genome of the bacterial host (Fig. 6.26).

Bacteriophage lambda is a temperate

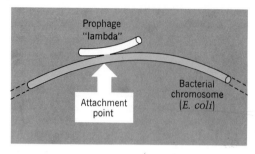

Figure 6.25 Lambda prophage attached but not inserted into the *E. coli* genome. (After Stent).

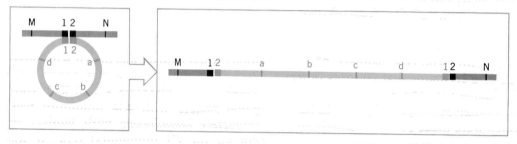

Figure 6.26 Possible mechanism of insertion of a circular lambda phage chromosome into a bacterial host chromosome. (After Campbell and after Hayes.)

phage that **lysogenizes** the bacterial host, *E. coli*. M. L. Morse and E. J. Lederberg showed that attachment of lambda DNA is specific for a position on the *E. coli* chromosome near the galactose (*gal*) locus.

At this point the viral and bacterial DNA may pair and a **single crossover** may occur. The prophage thus becomes an **addition** rather than a substitution for any given part of the bacterial chromosome. The prophage replicates with the bacterial chromosome when the bacterial cells divide.

When temperate phage infect sensitive bacteria, plaques with opaque centers are formed wherever bacteria have been lysed on a culture plate. By contrast, the plaques of virulent phage are completely clear. The opaque centers are patches of lysogenized bacteria that have not lysed but have remained intact. In general, when viral DNA prophage is associated with a bacterial genome, the cell is **immune** to infection by homologous phage. Temperate prophages may remain in host cells and divide synchronously with those cells. Several mutants have been associated with the process by which phages become prophages. Three groups of these mutants have been identified in closely linked genes in the temperate phage lambda: *C3, C1,* and *C2* (Fig. 6.27). The three genes were given numerical designations before their order in the linkage map was determined, hence the irregular sequence. These genes control the establishment and maintenance of lysogeny. Mutant phages that have undergone genetic changes in *C1* are incapable of lysogenization, and those that have mutated in either the *C3* or *C2* locus (or on either side of *C1*) have reduced lysogenic ability

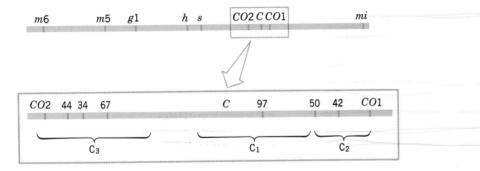

Figure 6.27 Subregions in the C part of the lambda phage chromosome. The C genes control lysogeny in host bacterial cells.

compared with wild-type lambda phage. Gene *C1* is a regulator that controls other genes including *C2* and *C3*.

When matings (multiple infections in bacterial cells) were made between lambda and some other phages, genetic recombinations were recognized. In each case, recognition of the recombination required that a number of homologous loci in the two phages be identifiable by markers. For ex- ample, when lambda was mated with phage 434, which can grow simultaneously with lambda on nonlysogenic *E. coli* K12, some hybrid phages with 434 immunity were completely homologous with lambda except for the *C1* (immunity) gene. It was concluded, therefore, that *C1* gave these two viruses their **specific immunity response to lysogenization.**

EPISOMES

DNA-containing bodies such as F agents and phage lambda in *E. coli* were named **episomes** (added bodies) by F. Jacob and his collaborators. Episomes may be virus-size or larger particles, and they may be pathogenic or nonpathogenic. They must (1) have demonstrated chromosomal (integrated) and cytoplasmic (autonomous) states and (2) be nonessential (to the host cell or organism). As cytoplasmic units, episomes replicate in the host cell independently of the bacterial chromosome. Once the episome becomes part of a bacterial chromosome, it replicates with that chromosome. Episomes cannot arise from mutations in the bacterial cell but must be introduced into the cell. They are, therefore, infections, introduced either through cell contact as in the case of the F factor, or through an infective apparatus, as in lambda phage.

Although most of the evidence for episomes has come from studies of bacteria, similar bodies have been described in Drosophila and maize. Other structures that are widespread in nature (for instance, centromeres in chromosomes and centrioles in the cytoplasm of animal cells), may be episomes or may have been derived from episomes. The central cores of these structures (for example, contrioles in centrosomes) contain DNA, and they have much in common with the episomes of bacteria.

The F factor in *E. coli* is a classical extrachromosomal episome. Not only have F factors been transferred from one bacterial strain to another, as previously discussed, but also from one species to another. *E. coli* F factors have been transferred to species of Salmonella, Shigella, and Pasteurella where they have survived and replicated. In situations where chromosomal recombination does not occur, redistribution of DNA is accomplished through episomes.

Some episomes are **colicinogenic factors** carried in some strains of bacteria. Colicinogenic bacterial cells (*col*⁺) carry proteins called colicins that have bacteriocidal effects on susceptible bacteria. Like lysogenic cells carrying lambda phage, colicinogenic cells are immune to the effects of the particles they carry. By transduction, *col*⁺ cells may transfer colicinogenic factors to *col*⁻ recipients and convert them to *col*⁺ cells. Colicins may promote conjugation between *col*⁺ bacteria and thus bring about redistribution of DNA. Another group of episomes, called resistance transfer factors, give the host cell resistance to drugs such as streptomycin and chloramphenicol.

OPERATIONAL DEFINITION OF THE GENE

The functional gene or cistron is undoubtedly the most basic, most important concept of molecular genetics. Validity of this concept has been established through experiments dealing with gene fine structure and function in both prokaryotes and eukaryotes. The cumulative evidence has led to the current concept of the gene. **The gene is a unit of function,** coding for **the synthesis of one polypeptide and operationally defined by the cis-trans test.**

REFERENCES

De Busk, A. G. 1968. *Molecular genetics.* Macmillan, New York.

Doermann, A. H., and M. B. Hill. 1953. "Genetic structure of bacteriophage T4 as described by recombination studies of factors influencing plaque morphology." *Genetics* 38, 79–90.

Dresler, S. E. 1971. *Introduction to molecular biology.* Academic Press, New York.

Hayes, W. 1968. *The genetics of bacteria and their viruses.* (John Wiley, New York) Blackwell Scientific Publications LTD, Oxford.

Fincham, J. R. S. 1966. *Genetic complementation.* W. A. Benjamin, New York.

Goodenough, U., and R. P. Levine. 1974. *Genetics.* Holt, Rinehart, Winston, New York.

Hildemann, W. H. 1970. *Immunogenetics,* Holden-Day, San Francisco.

Hotchkiss, R. D. and M. Gabor. 1970. "Bacterial transformation, with special reference to recombination processes." *Ann. Rev. of Genet.* 4, 193–224, L. Roman (ed.) Annual Reviews, Inc., Palo Alto, Calif.

Jacob, F. 1966. "Genetics of the bacterial cell." *Science* 152, 1470–1478.

Stent, G. S., 1971. *Molecular genetics.* W. H. Freeman, San Francisco.

Taylor, J. H. 1965. *Selected papers on molecular genetics.* Academic Press, New York.

Watson, J. D. 1970. *Molecular biology of the gene,* 2nd ed. W. A. Benjamin, New York.

PROBLEMS AND QUESTIONS

6.1 One inbred variety of plants has white flowers and another variety of the same species has red flowers. Outline experiments to determine whether a single pair of alleles or more than a single pair is involved in transmitting the different flower colors.

6.2 Why are mutant (alternative) genes essential for identifying wild-type alleles and locating the positions of gene loci on chromosomes?

6.3 (a) How should a series of multiple alleles be symbolized? (b) To what extent do they represent alterations of the same basic phenotype?

6.4 The following, listed in order of dominance, are four alleles in rabbits: c^+, colored: c^{ch}, chinchilla; c^h, himalayan, and c, albino. What phenotypes and ratios would be expected from the following crosses: (a) $c^+c^+ \times cc$; (b) $c^+c \times c^+c$; (c) $c^+c^{ch} \times c^+c^{ch}$; (d) $c^{ch}c \times cc$; (e) $c^+c^h \times c^+c$; and (f) $c^hc \times cc$?

6.5 In mice, a series of five alleles has been associated with fur pattern. These alleles are, in order of dominance, A^Y (homozygous lethal) for yellow fur; A^L, agouti with light belly; A^+, agouti; a^t, black and tan; and, a, black. For each of the following crosses, give the coat color of the parents and the phenotypic ratios expected among the progeny. (a) $A^YA^L \times A^YA^+$; (b) $A^Ya \times A^La^t$; (c) $a^ta \times A^Ya$; (d) $A^La^t \times A^LA^L$; (e) $A^LA^L \times A^YA^+$; (f) $A^+a^t \times a^ta$; (g) $a^ta \times aa$; (h) $A^YA^L \times Aa^t$; and (i) $A^Ya^t \times A^YA^+$.

6.6 If a series of four alleles is known to exist in a given diploid ($2n$) species, how many would be present in: (a) a chromosome, (b) a chromosome pair, (c) an individual member of the species? (d) On the same basis, how many different combinations might be expected to occur in the entire population?

6.7 Assume that in a certain animal species four alleles (c^+, c^1, c^2, and c) have their locus in chromosome I and another series of two alleles (d^+ and d) have their locus in chromosome II. How many different genotypes with respect to these two series of alleles are theoretically possible in the population?

6.8 Assume that, in a certain animal species, four alleles (c^+, c^1, c^2, and c) have their locus in chromosome I and another series of three alleles (d^+, d^1, and d) have their locus in chromosome II. How many different genotypes with respect to these two series of alleles are theoretically possible in the population? (The number of genotypes at each locus $= [n(n+1)]/2$ where $n =$ number of alleles)

6.9 A series of multiple alleles in a certain species of fish which breeds readily in the laboratory was listed by Myron Gordon as follows: P^o, one spot; P^m, moon complete; P^c, crescent; P^{cc}, crescent complete; P^{co}, comet; P^t, twin spot; and P, plain. (a) How many combinations of these alleles might be expected to occur in the population? (b) How could the allelic nature of these genes be indicated by genetic methods?

6.10 In man, a series of alleles has been associated with the blood typing groups as follows: A, A type; A^B, B type; AA^B, AB type; and aa, O type. A and A^B are dominant over a, and A and A^B together (AA^B) are codominant, each expressing a phenotype. What phenotypes and ratios might be expected from the following crosses: (a) $AA \times A^B A^B$; (b) $AA^B \times aa$; (c) $Aa \times A^B a$; and (d) $Aa \times aa$?

6.11 A case was brought before a certain judge in which a woman of blood group O presented a baby of blood group O, which she claimed as her child, and brought suit against a man of group AB whom she claimed was the father of the child. What bearing might the blood-type information have on the case?

6.12 In another case, a woman of blood group AB presented a baby of group O which she claimed as her baby. What bearing might the blood-type information have on the case?

6.13 An Rh-positive man (Rr) married an Rh-negative (rr) woman. Their first child was normal and their second child had the hemolytic disease of the newborn. (a) What genetic explanation might be offered? (b) What prediction might be made concerning future children by this couple?

6.14 An Rh-positive man (Rr) married an Rh-negative (rr) woman. Their first child was normal and their second child showed the effects of Rh incompatibility. What prediction might be made concerning future children of this couple?

6.15 If the two genes L^M and L^N for M and N blood groups, respectively, are alleles and the three genes A, A^B, and a are alleles in a different chromosome, (a) list the genotypes that are theoretically possible in the population. (b) How could these be useful in human problems of identity and paternity?

6.16 The garnet locus exhibits a cis-trans effect. What phenotype would be expected from the following genotypes?

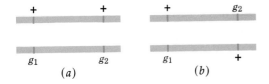

(a)   (b)

6.17 How can the different properties of plaques in *E. coli* cultures be explained? If no plaques, or only a very few, occur in a culture in which *E. coli* cells and phages are present, what implications might be drawn?

6.18 How can spontaneous mutations and recombinations be distinguished in the *r*II region of phage T4?

6.19 How was complementation between cistrons *r*II A and *r*II B detected?

6.20 Mutants a_1 and a_2 in the trans position, in a particular experiment, did not complement each other and produce the wild-type phenotype but mutants b_1 and b_2 did complement one another. (a) Are a_1 and a_2; (b) b_1 and b_2 in the same or different cistrons? Are (c) a_1 and a_2 and (d) b_1 and b_2 functional or nonfunctional alleles?

6.21 Illustrate the alleles (a) a_1, a_2 and (b) b_1, b_2 in the above problem in the cis and trans and coupling and repulsion, positions and indicate for each, normal or mutant phenotypes, and complementation or noncomplementation.

6.22 A series of revertable (single site) mutants are located in a sequence 1, 2, 3, 4, 5, 6, 7, 8, 9, 10 in a gene locus. A deletion mutation shows recombination with areas 1–4 and 4–8 but none with 6–9 or 9–10. Where is this new deletion mutation located?

6.23 How is recombination demonstrated in virulent phage?

6.24 A nutritionally defective *E. coli* strain grows only on a medium containing thymine whereas another nutritionally defective strain grows only on medium containing leucine. When these two strains were grown together a few progeny could grow on a minimal medium with neither thymine or leucine. How can this result be explained?

6.25 How can female *E. coli* cells be converted into males?

6.26 How are F factors associated with recombination of DNA other than that for sex determination?

6.27 What is lysogenization and how is it controlled?

6.28 (a) How may a virus become a prophage? (b) How does an induced prophage affect the host bacterium?

6.29 Compare, in table form similarities and differences of the mechanisms through which (a) transduction, (b) sexduction, and (c) transformation may occur.

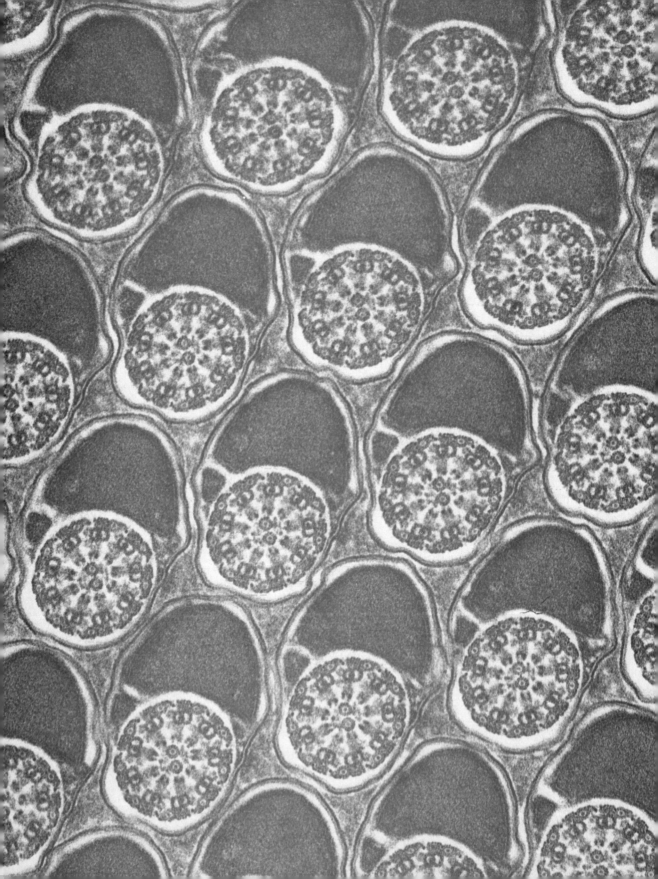

SEVEN

A living system must have control over its actions as well as the capacity to act. Biological systems gain this control through such diverse agents as enzymes, antibodies, receptors, and repressors. Despite the variations in regulatory processes, however, the basic cytoplasmic element in all cases is the **protein molecule.** By changing their shape in response to external influences, protein molecules provide "on" and "off" signals in living systems.

For example, the mechanism by which enzymes catalyze chemical reactions begins with the **binding** of the substrate to the surface of the enzyme. Next, the enzyme **catalyzes** the reaction that leads to formation of products on its surface. After these products are released, the enzyme surface regenerates and the cycle is repeated until the substrate is depleted. At no time does the enzyme act as a substrate or product, but rather it acts only as a catalyst and is never consumed. Enzyme specificity depends on the precise matching of an enzyme **active site** with a particular substrate.

Special classes of compounds bind to enzymes without forming products

GENE REGULATION AND DEVELOPMENTAL PATTERNS

Cross section of sperm tails in *Drosophila melanogaster*. Magnification ×136,000. (Richard Wilkinson.)

227

and therefore inhibit the catalytic function of the enzyme and subsequently inhibit chemical reactions. For example, an oversized molecule that binds to the surface of an enzyme distorts the protein into such a shape that the catalytic groups cannot align properly. Other compounds that may bind to the enzyme are too small, or do not have the correct chemical properties for proper alignment and they also block reactions. Some enzymes are **allosteric,** that is, regulatory enzymes with more than one binding site. Binding at one site, however, may change the shape of the enzyme so that binding either cannot occur or is facilitated at another catalytic site. Both positive and negative controls are thereby provided. Regulatory molecules not involved in a specific enzymatic chemical reaction can alter the shape of the enzyme involved and thus change its activity. Hormone molecules, for example, can turn an enzyme on by inducing it to assume a certain shape, or turn it off by inducing a different shape. Since hormones, like enzymes, are not consumed in such reactions, small quantities can be used **repeatedly** for control purposes.

INDUCIBLE ENZYMES

The best-defined systems of gene regulation are those that control enzyme production in bacteria. Bacteria are comparatively simple organisms, with neither nuclear membranes nor meiotic processes. The mechanism of regulation of bacterial protein synthesis through **control of transcription** of mRNA by regulatory proteins has been essentially resolved.

Gene activity in *E. coli* is largely regulated by inducible enzymes that are synthesized only in response to an **inducer** substance. Inducers or effectors are small organic molecules that cause the bacterial cell to produce enzymes involved in **catabolic** reactions. Usually the inducers constitute the **substrates** on which the enzymes act. These substances control the production of the bacterial enzymes in such a way that the enzymes are available when needed. Only when **regulators** are activated and the substrate (inducer) is available, are the required enzymes produced continuously. This is an efficient system because the bacterium is not burdened with proteins that would have to be stored in an inactive form when no raw material is available on which to act. When the necessary substrate is present and the proper genes for enzyme production are "turned on" the **enzymes soon appear** to carry out the particular process needed and to furnish the cell with required energy or material.

Many different proteins exist within each cell. Of these in a growing *E. coli* cell with glucose as the sole carbon source, perhaps 600 to 800 are utilized at different times and in different quantities for enzymatic activity. The quantities of enzymes associated with the first steps in glucose degradation and those involved with the syntheses of amino acids, nucleotides, and energy-rich ATP, fluctuate according to particular **activities** of the cell's cycle. Mechanisms involving interactions between the internal genetic control and the external chemical environment regulate these fluctuations.

OPERON MODEL

F. Jacob and J. Monod, both 1963 Nobel prize winners, provided experimental evidence for a specific gene regulator system for lactose fermentation in *E. coli* K12 and also devised a **model** (Fig. 7.1) for such control mechanisms. Each step in the process is explained by conventional molecular genetics. Two kinds of genes—**structural** and **regulatory**—were postulated, both of which act through **gene-controlled** proteins. Structural genes determine the actual structure of polypeptides by controlling the amino acid sequence as synthesis occurs. Several structural genes that perform sequential functions in a biochemical pathway may be located in a linear sequence or cluster. Regulatory genes control groups of structural genes in a **negative** way by preventing their activity. Each regulatory gene functions through an **operator** site, which has immediate control over a group of structural genes. The operator site is a small section of DNA which is transcribed by mRNA and is complementary to a repressor. It prevents structural gene activity by binding with this repressor. Associated with the operator is another small DNA sequence, the **promoter** to which RNA

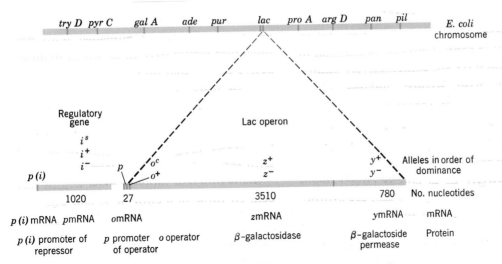

Figure 7.1 Lactose (lac) operon and its associated regulatory gene in *E. coli* chromosome. An enlarged section of chromosome including the lac region is shown in the lower part of the diagram with "i" promoter ("i" *p*), regulatory gene (i) in one region of the chromosome. The operon proper in another region includes the promoter (*p*), operator (*o*), structural gene *z* for β-galactosidase and structural gene *y* for β-galactoside permease. The regulatory gene is not connected with the operon but may be separated by a small number of nucleotides. The numbers below chromosome segments indicate the number of nucleotides in each region. Alleles for the genes in the system are listed in order of dominance. (After Jacob and Monod.)

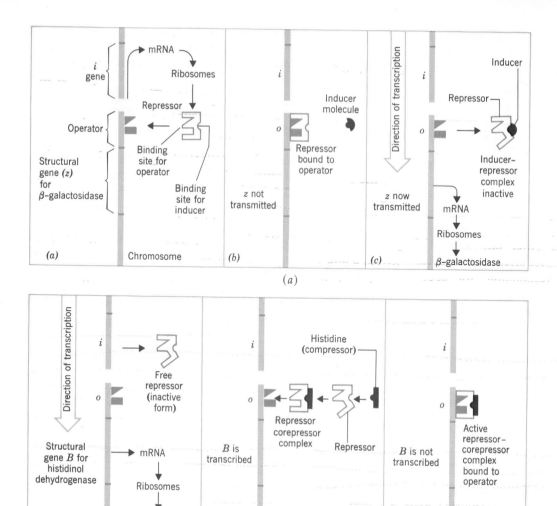

Figure 7.2 Comparison between inducible and repressible enzyme systems. (*a*) Induction of β-galactosidase activity in *E. coli*. (*b*) Repression of histidinol dehydrogenase in *S. typhimurium*. When the repressor (*b*) binds to the operator the structural gene z is not transcribed but when the inducer binds with the repressor (*c*) the operator is activated and the structural gene z is transcribed. For the repressible system, in contrast, (*d–f*) when the repressor is inactive (*d*) the structural gene is transcribed but when the repressor complex is active and bound to the operator (*f*), the structural gene is not transcribed.

polymerase binds and initiates transcription of the operator. A promoter gene also initiates transcription of the regulatory gene.

LACTOSE OPERON

In the operon for lactose fermentation (Fig. 7.2a–c), 2 functionally distinct chromosomal regions along with promoter and operator sites were identified on the *E. coli* chromosome. Two structural genes were mapped in the same chromosome area. They were found to be closely linked with each other and also with promoter and operator sites. The structural genes, z and y are responsible for specific enzymes, β-galactosidase and galactoside permease, respectively. (Another structural gene a for galactoside acetylase is a part of this operon but it has no known function in *E. coli*.) B-galactosidase splits β-galactoside (lactose) into galactose and glucose (Fig. 7.3) and galactoside permease controls the rate of entry of the β-galactosides into the bacteria.

The unique feature of this operon is that the structural genes and hence enzyme activities in the system are **controlled by an operator site** (o) which, in turn, is initiated by a **promoter** (p). The entire system is controlled by a regulatory gene (i^+) that is located in a chromosome area not necessarily attached to the operon proper. The regulatory gene (i^+) also has a promoter with a base composition that sets the rate of constant transcription of i^+ at a specific low level. It has been estimated that, on the average, each cell contains 10 copies of the lac operon repressor. This insures that the **regulatory substance** will be always present. The rate of production can be speeded up by mutation (i^Q and $i^{super\ Q}$, see below). When an inducer (effector) is present, **derepression** occurs and the system is turned on (Fig. 7.2c). The inducer is the substratum (lactose), or a closely related organic substance, on which the system acts.

The lactose operon is under **negative control**, that is, the system is usually "turned off." This is accomplished by a regulatory gene i^+ which synthesizes a small amount of repressor protein. The protein is an allosteric enzyme with two distinct binding sites, an "active" site which can attach to the operator, keeping the system repressed, and a "regulatory" site which can bind to an inducer (lactose substrate molecule). Once bound to an inducer, the repressor protein **changes shape** and cannot attach to the operator. This allows the attachment of mRNA polymerase to the promoter (Fig. 7.4a) and the system is activated as long as the inducer is present.

The operator is double-stranded DNA consisting of some 27 base pairs. It is closely linked with the promoter, but the two chromosome areas do not overlap. In

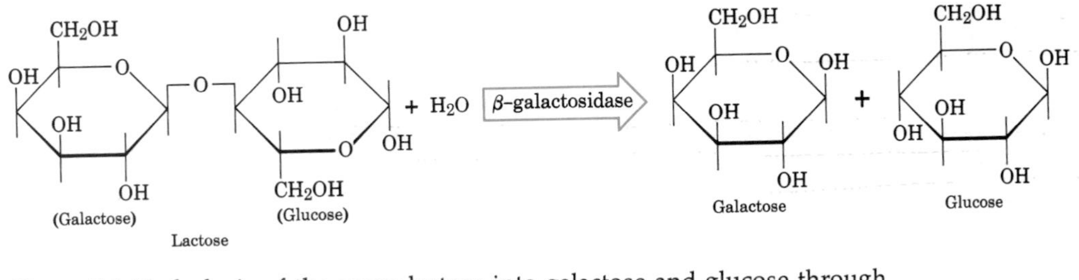

Figure 7.3 Hydrolysis of the sugar lactose into galactose and glucose through action of the enzyme β-galactosidase.

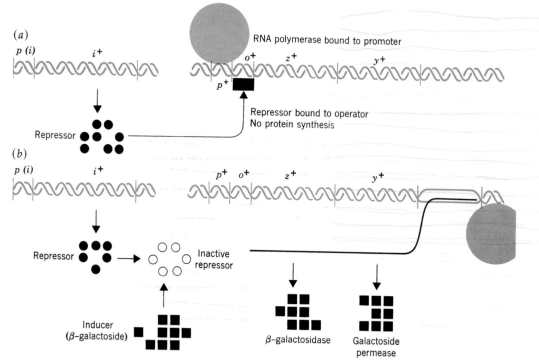

Figure 7.4 Effect of inducer in derepression of lac operon. (*a*) Operon is repressed by repressor, which binds to operator, RNA polymerase is bound to promoter and structural genes z^+ and y^+ are inactive. (*b*) inducer which is the substrate β-galactoside, inactivates the repressor, the operator is derepressed, RNA polymerase moves along the DNA strand and transcribes the message on mRNA. The structural genes z^+ and y^+ produce enzymes, β-galactosidase and galactoside permease, respectively.

the lactose operon, the operator is between the promoter and the z locus. The promoter site, however, is functionally independent of both the operator and the structural genes. The promoter region, which **binds RNA polymerase** and initiates control of the operator, activates the system through a **derepressed operator.** RNA polymerase then moves along the chromosome and transcribes β-galactosidase mRNA (Fig. 7.4*b*) according to the specifications of the z^+ gene. The enzyme is assembled on the ribosomes with appropriate amino acids. During maximal synthesis some 35–50 β-galactosidase mRNA molecules are present in each cell. The mRNA strand also includes a transcription of the structural gene y^+, which is next to z^+ in the operon. Transcribed mRNA from gene y^+ is also available for translation on the ribosomes and galactoside permease is produced. When the lactose substrate inducer is removed experimentally or is **depleted** in a natural system, β-galactosidase mRNA is no longer produced (Fig. 7.5).

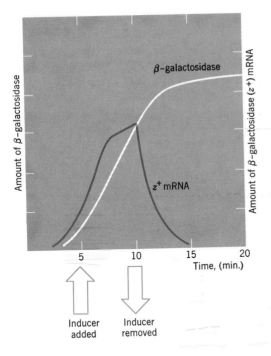

Figure 7.5 Experimental results showing relation between the inducer and z^+ mRNA and β-galactosidase in the lac operon of *E. coli*. The z^+ mRNA increased rapidly when the inducer was added and fell off abruptly when the inducer was removed. β-galactosidase production lagged slightly behind z^+ mRNA and continued to increase for a short time after the inducer was removed. (After Watson.)

MUTANTS OF THE LACTOSE OPERON

Several mutants have been recovered in the repressor (i), promoter (p), operator (o) and structural gene areas of the *E. coli* lactose operon. These provide insights into the mechanisms of the DNA units of this operon and also suggest possible genetic mechanisms of regulation in eukaryotes as well as prokaryotes. A recessive mutation (i^-) in the repressor gene alters the repressor protein and thereby **blocks repression,** permitting "constitutive" synthesis of enzymes (Fig. 7.6a) Whenever protein synthesis occurs without the negative regulation of repressors and/or operators, the products are called **constitutive proteins.** These are synthesized in response to the promoter, irrespective of environmental conditions such as the presence of an inducer. The absence of a repressor protein substance caused by a mutation is called a "genetic block."

A dominant mutant (i^s) specifies a repressor protein that is unable to bind to the inducer. The protein does, however, bind to the operator, resulting in lack of enzyme synthesis. In the presence of the recessive thermolabile repressor mutant (i^{TL}), growth at an above normal temperature results in **derepression.** Likewise, the recessive temperature-sensitive synthesis (i^{TSS}) mutant of the repressor gene decreases synthesis of the repressor at higher than normal temperatures. Dominant mutants i^Q and $i^{super\ Q}$ are mutants of the i promoter (promoter of the repressor). They increase repressor synthesis by 10 and 50 fold, respectively. The dominant mutant (i^{-d}) produces a defective repressor. The normal repressor is a tetramer composed of four identical monomers. In the presence of the i^{-d} mutation, defective monomers are made which combine with normal (i^+) monomers and affect the entire molecule so that it will not bind to the operator. Thus the dominant

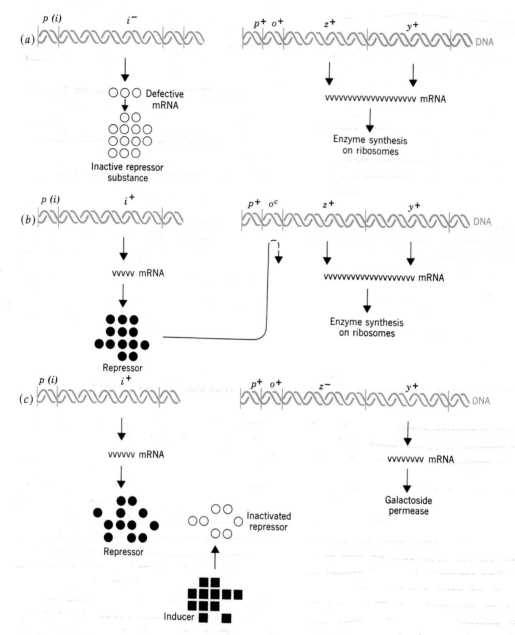

Figure 7.6 Effects of mutations in different parts of lac operon. (*a*) Mutation of regulatory gene i^+ to i^- results in defective i mRNA and inactive repressor. The operator o^+ is derepressed and z^+ and y^+ are active; their messages are transcribed on mRNA and their respective enzymes are synthesized on the ribosomes. (*b*) Mutation of o^+ to o^c prevents the operator from responding to the repressor and structural genes z^+ and y^+ act constitutively; mRNA messages are transcribed and enzymes are synthesized on the ribosomes. (*c*) Mutation of z^+ to z^- inactivates the structural gene for β-galactosidase but other parts of operon are activated. (After Watson.)

gene (present in heterozygous dosage) produces complete **enzymatic deficiency.**

A cis dominant promoter mutant (p^-) decreases the affinity of the promoter for RNA polymerase and thus inhibits synthesis of promoter enzymes. This experimental finding demonstrated that the promoter functions **independently** of both repressor and operator. Mutations at promoter sites may alter synthesis of constitutive proteins in *E. coli* systems. In the presence of a wild-type promoter, about one mRNA molecule is synthesized in every cell cycle. With a medium-level mutant promoter, about 10 mRNA molecules are synthesized and with a high-level mutant promoter about 50 mRNA molecules are synthesized under the same experimental conditions.

Operator mutant o^c, which is also cis dominant, renders the operator unresponsive to the repressor (Fig. 7.6*b*). This allows structural genes to remain active, thus permitting constitutive synthesis of enzymes. Many proteins needed in small quantities and/or in fairly constant amounts may be regulated without repressors or operators. Thus glucose degradation enzymes are apparently constitutive, and their rate of synthesis is **not controlled by inducers.** Cells carrying mutant inactive regulatory genes (i^-) or mutant operators (o^c) that prevent the inhibition of operons, produce constitutive proteins in fixed amounts **independent of need.**

The recessive mutant z^- (Fig. 7.6*c* and y^- structural genes are defective for the mRNA that controls production of β-galactosidase and galactoside permease, respectively. Recessive mutant o^o was originally postulated to be operator negative but has been shown to produce a meaningless message in structural genes near the operator.

CONTROL OF SYNTHETIC PATHWAYS: REPRESSION AND FEEDBACK INHIBITION

In some other regulator systems such as the histidine operon, it is the level of the end product of enzymatic activity rather than the substrate that controls enzyme synthesis. These are **anabolic** systems that build up organic compounds and store energy. The end product in a **repressible** system is a **corepressor** that can bind to an allosteric site on the repressor protein to activate it. The amount of manufactured material on hand thus regulates the rate at which it is accumulated. This so-called **feedback inhibition** or **end-product inhibition** slows the synthetic process when the product is in good supply but when the product is in short supply, the reaction is allowed to proceed at its normal rate. Repressible systems control gene expression through repressors and therefore exert **negative control** (Fig. 7.2*d–f*).

When, for example, the amino acid tryptophan (an end product) was present at a high level in an *E. coli* culture, it prevented synthesis of the first enzyme in a sequence within the pathway and prevented the system from functioning. When, on the other hand, the tryptophan was being utilized by the cells and its level was low, the tryptophan production system was derepressed (turned on). This metabolic system thus has an effective feedback mechanism that regulates the rate at which its end product is synthesized. Feedback inhibition is specific in that only the end product (or its analogue) can activate the process. The end product acts on a single gene, which controls an early enzyme of a pathway. The inhibitory end product has no stereochemical relation with the normal substratum of the inhibited enzyme.

Two different basic control mechanisms have thus been discovered, each regulated by the end product of a pathway. **Feedback inhibition** regulates the synthesis of small molecules (amino acids). The other process, called **repression** regulates the synthesis of large molecules (proteins). Synthesis of the enzyme tryptophan synthetase, for example, was shown by Cohn and Monod to be repressed when tryptophan was added to

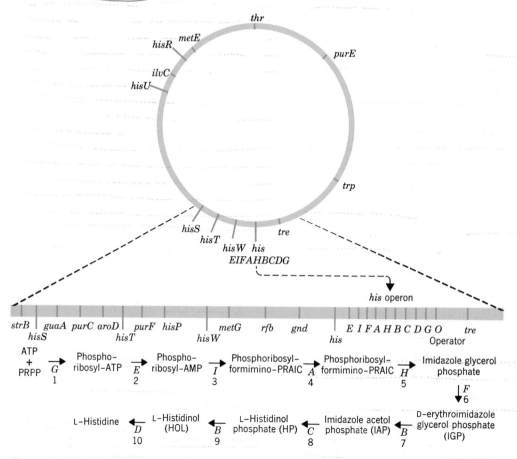

Figure 7.7 Histidine operon of *typhimurium* containing a cluster of genes involved in the synthesis of the amino acid histidine. The steps (bottom) in biosynthesis, which do not occur in the order in which the genes are located in the chromosome, are indicated by arrows (from left) and numbered from 1 to 10. The letters symbolize the enzymes in the system and the chemical names of intermediate products from phosphoribosyl-ATP pyrophosphorylase (at left) to L-histidine (at lower left). The circle in the upper part of the diagram represents the S. *typhimurium* chromosome including 5 other *his* loci which are related to histidine production but apparently do not specify enzymes that belong to the *his* operon. The *his* operon is illustrated with other regions on the linear chromosome (at the center, right) with the operator site at the far right. (After P. E. Hartman and B. N. Ames.)

the bacterial culture. Repression, like feedback inhibition, depends on the level of the **end product** in the medium, but unlike feedback inhibition, which affects only the first step in a pathway, repression affects **most,** if not all of the **steps in the pathway.**

HISTIDINE OPERON

In *Salmonella typhimurium,* an operon (Fig. 7.7) that is composed of a cluster of nine, closely linked genes, converts ATP and phosphoribosyl pyrophosphate into the amino acid histidine in a sequence of ten steps. Each gene produces a single enzyme and each enzyme (except dehydrase phosphatase) controls a single step. Dehydrase phosphatase controls steps seven and nine in the sequence. The biochemical steps do not occur in the same order in which the genes are located in the linkage map, but they do follow a sequence with each reaction producing the precursor for the step that follows. This cluster includes all the genes involved in histidine synthesis in *S. typhimurium.* The entire cluster is controlled by a single **operator site** (symbolized o at the right in Fig. 7.7). The term **coordinate repression** is used to describe such a system in which rates of production of enzymes vary together.

The amount of histidine available to the cell controls the synthetic process. When histidine can be acquired from the surrounding medium, synthesis is discontinued; when histidine is being used by the cell and the external supply is depleted, the synthetic process is activated. Experimentally, the entire operation may be **stopped by adding histidine** to the medium.

Investigations of the *S. typhimurium* histidine operon by P. E. Hartman, B. N. Ames, and others have established the linkage relations between the cluster of nine structural genes and its one operator. The histidine operon includes about 13,000 nucleotides, with the coded information for the enzymes involved with the steps in the biosynthetic pathway apparently transcribed on a single long strand of mRNA (Fig. 7.8).

More than 1000 mutants, some, of the deletion type, have been identified in the DNA area covered by the histidine operon. All cistrons in the "polycistronic" operon have been involved in the mutations. About half of the **point mutations** discovered have been shown to have dual effects: (1) activity of one enzyme is eliminated and (2) activities of all enzymes later in the sequence than the mutated gene are decreased. Such mutations are called **polarity** mutations, and probably involve single base changes. Extensive studies of polarity mutants have shown that their mRNA message begins in one location (at right, as shown in Fig. 7.7) and proceeds **stepwise** through the entire sequence. All steps later in the sequence may be affected by the mutational change, while those earlier in the sequence remain unchanged. The message thus starts from the operator end and proceeds linearly through the complete synthesis unless interrupted by a mutational alteration which blocks the enzymatic steps **above the block point.**

In wild-type organisms, the amount of each enzyme synthesized is controlled by the **operator,** and the ratio of activity of one enzyme to another remains constant. The rates of production of the intermediate substances in the series of nine coordinately repressed reactions thus stay in **balance,** and an excessive amount of any one cannot accumulate except in mutants in which the balance is upset by the blocking of one of the enzymatic steps. The steps beyond the mutation, however, are adjusted or **modulated** (relative to those of the wild-type operon) so that the enzyme ratio remains constant, although each polar mutant shows its own modulation or pattern of enzyme levels.

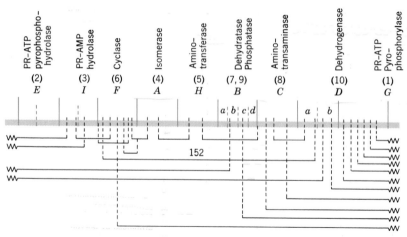

Figure 7.8 Genetic map of the histidine (his) region of the *S. typhimurium* chromosome. The thick horizontal line represents the chromosome region, divided into nine closely linked genes. The dotted lines above the chromosome indicate further subdivisions of complex loci, E, I, B, and D on the basis of interallelic complementation tests. The figures enclosed in parentheses (1)–(10) indicate the step number in the synthetic pathway of the enzyme determined by each gene.

The series of horizontal lines below the chromosomes delineate the extent of a number of multisite deletion mutations used for mapping. The wave termination of these lines at the extremities of the figure show that this end of the deletion has not been defined. (After Hayes.)

Data from mutations and from other lines of investigation involving the histidine operon have provided a basis for explaining a mechanism of genetic control. After the first gene (G) is derepressed, some 20 minutes must lapse before the last gene (D) is released to produce its enzyme. The entire sequence is required to produce **histidine.**

POSITIVE GENE REGULATION

The examples of regulation of protein synthesis in bacteria that have been cited thus far, have been basically negative mechanisms. In each case, the absence of an active repressor has permitted synthesis to proceed. A few systems have been discovered, however, that act in a **positive way.** The arabinose operon in *E. coli* (Fig. 7.9) is an example of positive control. Three structural genes, *a*, *b*, and *d*, specify mRNA for three enzymes: isomerase, kinase and epimerase. These enzymes produce L-ribulose, L-ribulose-5-P and D-xylulose 5-P, respectively, during arabinose catabolism. When arabinose is absent in the medium of *E. coli* very little if any of these enzymes is present, but when arabinose is added, all three enzymes increase in quantity. An activator or **initiator** gene (*c*) is closely linked with the three structural genes (*a*, *b*, and *d*). Although *c* occupies the position of an operator, it directly promotes rather than inhibits structural gene activity. Gene *c* apparently codes the mRNA that specifies a particular protein. The mechanism of activation is not known but does not involve derepression of the *c* region.

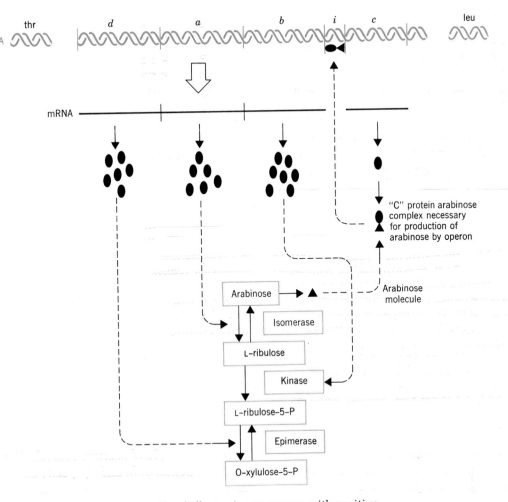

Figure 7.9 Arabinose operon in *E. coli* illustrating an operon with positive control over protein synthesis. The "C" protein, probably a polymerase, in contact with the substratum arabinose initiates the activity of the operon through the regulator gene *i*. Genes *d, a* and *b* transcribe mRNA that codes for enzymes epimerase, isomerase and kinase, respectively. (After Watson.)

The initiator of the arabinose operon is a cistron specific for DNA polymerase that separates the two chains of the DNA double helix. It is a structural gene and initiates the replication of bacterial chromosomes, F factors (Chapter 6), and other independent genetic units called **replicons**, in the cell. The initiator is influenced by signals from the cytoplasm indicating that a certain stage of cell growth has been reached. It then interacts with a particular part of the replicon DNA called the **replicator** which is associated in the bacterium with the cell membrane. This provides **coordination** between DNA replication and cell division. The replicator gives positive regulation to the structural gene to which it is attached and replication of DNA proceeds.

ABSENCE OF OPERONS IN HIGHER ORGANISMS

Operons represent groups of linked structural genes that have trigger mechanisms controlling their activity. Such systems may partially explain the mechanism of differentiation that occurs in the development of living things. Although experiments leading to the operon model were based on microorganisms, and the only operons known in detail are in bacterial systems, they suggest possible mechanisms of regulation of gene action in higher forms as well.

In higher organisms with their several chromosomes, crossing over and chromosome structural exchanges such as translocations and inversions (Chapter 9) **disrupt** clusters of linked genes. However, many

studies suggest that regulation mechanisms do exist. In one example, a gene was found to control genes responsible for anthocyanin synthesis in maize endosperm. Pollen-tube formation in the evening primrose has been found to be controlled by a structural gene which is, in turn, controlled by another gene. Further linkage studies will be required to determine whether other genes in this functional system are in **close proximity** in a chromosome.

The existence of regulatory and structural genes suggest a **more complex** level of organization than is implied by the one gene one polypeptide hypothesis. A single regulator or switch mechanism may control a whole series of chain reactions. In this kind of pleiotropism, one gene locus can control several others. The regulator is not always a special and separate entity located at a distance in the genome from the operator and structural genes which it controls. In some operons, it may be merely the beginning of

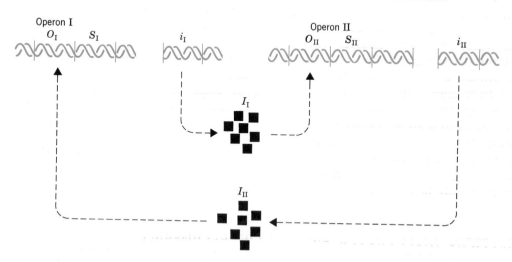

Figure 7.10 Regulation in two different but related operons. Products of regulator gene (i_{II}) are affected by inducer I_{II} but control operon I while products of regulator genes (i_I) are affected by inducer I_I but control operon II. Inducer I_{II} permits operon I to function but inhibits operon II. (After Monod and Jacob.)

a region of DNA that controls a series of closely related enzymatic events.

Repressors are not always specific for particular operators but may act on two, three, or more different operons. Monod and Jacob and others have described many combinations including a model (Fig. 7.10) for two operons sharing a mutual control of structural gene activity. Each operon includes an operator (o) and a structural gene (s). A regulatory gene (i) is functionally associated with each operon. Inducer I permits operon II to function and inhibits operon I whereas inducer II permits operon I to function but inhibits operon II. This example would seem to be a **transitional step** between regulation in prokaryotes and that in higher forms which have greater complexity. But the number of examples in higher organisms with a semblance of the operon is small. Although the operon model of gene regulation is well accepted for prokaryotes it is **not well** established for **eukaryotes.** Crossing over and other mechanisms for exchanges of chromosome parts that occur in eukaryotes tend to disrupt clusters of genes that might otherwise become operons.

In **polycistronic systems,** one **molecule of mRNA transcribes** all information carried by the structural genes of an operon. The synthesis of separate polypeptides corresponding to each gene of the operon occurs during translation on the ribosomes. In monocistronic systems, **one cistron** specifies **one polypeptide.**

Stop 3/19/72

REGULATION AND mRNA TRANSCRIPTION

In prokaryotic cells that do not store mRNA, protein synthesis is initiated directly by mRNA transcription. When the mRNA transcription is "turned on," the translation machinery becomes active and protein enzymes are soon available. When the transcription is "turned off," protein synthesis ceases. Eukaryotes, on the other hand, may have **more steps** between initiation **of transcription** and subsequent participation of mRNA in **protein synthesis.** Two major factors have been cited as influencing control of enzyme synthesis: (1) availability of mRNA in the protein-synthesizing apparatus and (2) extent to which mRNA molecules are being utilized at any given time for protein synthesis. Control of mRNA production by mRNA polymerase is called **transcriptional regulation** whereas control of the availability of mRNA molecules from storage, after transcription but before translation, is called **posttranscriptional regulation.**

Both transcriptional and posttranscriptional regulation occur during the differentiation of higher cells. Differentiation allows multicellular organisms to have specialized structures and function and thus is a fundamental process in development.

REGULATION AND ONTOGENETIC DEVELOPMENT

Ontogenetic development is the process involved in deciphering the stored genetic information and transforming the zygote into a new adult individual. It begins with the reproductive process, which has different levels: (1) **DNA replication,** (2) **cellular division,** and (3) **reproduction** of the entire organism. Replication of DNA is the most basic process since parents thus give their progeny the biological information that they possess. Most living cells can reproduce themselves, but animals and plants also have germ cells that can provide the information necessary to reproduce the entire organism. A full copy of an individual's genetic information is presumed to be provided in virtually all of its cells but only the germ cells are **totipotent;** that is, only they can reproduce other whole organisms belonging to their species.

A zygote normally develops according to the directions received from the parental DNAs, although all developmental processes are subject to environmental influences. Each viable zygote can become a complete organism. However, many, especially among lower organisms, are lost along the way to predators or accidents and do not survive. Some are **damaged** either genetically or environmentally, but survive and still carry the general characteristics of their species.

Since genes carry information necessary for the production of other members of the same species, and organisms obviously reproduce their own kind of organisms, it is logical to assume that a regular sequence of steps between genes and traits and between zygotes and mature organisms occurs in every generation. This in fact does occur. Three closely interrelated aspects of development, **differentiation, organization,** and **growth** regulation, are under the general control of DNA. Precise genetic mechanisms have, for the most part, not been defined for any of the three. The end results of similarities and differences as represented in a given species, however, indicate that genetic specifications are involved.

MODEL OF GENE REGULATION IN HIGHER CELLS

Since all the cells in almost every organism contain identical genomes, and most of the genome in an actively dividing cell is usually inactive, certain genes must be "turned off" while others are "turned on" at various times in development. Analysis of RNA from various cell types has shown that **different mRNAs** are being produced at different times by various cells. Although regulation of gene activity is essential to the differentiation that occurs in cells of higher organisms, little is known of the actual molecular mechanisms involved. A model designed to portray gene regulation of cell activity must account for the following experimental observations:

1. A simple external signal from the environment often mediates a change in the activity of a cell. Action of hormones, for example, may initiate gene actions.
2. The integrated activity of a large number of noncontiguous genes may be required for a particular state of differentiation.
3. A unique type of RNA called heterodisperse nuclear RNA (HnRNA) is produced only in the nuclei (not cytoplasm) of higher cells. It is like DNA in base composition, incorporates radioactive label rapidly (indicating a frequent metabolic turnover), and is associated with the chromosomes and not the nucleolus.
4. Genomes of higher cell types are very large as compared to bacterial genomes. A high proportion of the DNA in higher cells does not code mRNA for unique sequences of codons. Some may have regulatory and some protective functions.
5. Higher cells contain large fractions of repetitive DNA sequences (not found in bacterial genomes) that are tran-

Example (a), using redundancy in receptor genes

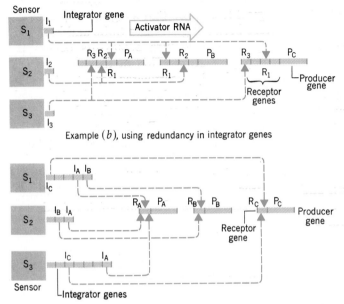

Example (b), using redundancy in integrator genes

Figure 7.11 Types of integrative systems within the model. (a) Integrative system depending on redundancy among the regulator genes. (b) Integrative system depending on redundancy among the integrator genes. These diagrams schematize the events that occur after the three sensor genes have initiated transcription of their integrator genes. Activator RNAs diffuse (symbolized by dotted line) from their sites of synthesis—the integrator genes—to receptor genes. The formation of a complex between them leads to active transcription of the producer genes P_A, P_B, and P_C. (After Roy J. Britten and Eric H. Davidson, "Gene Regulation for Higher Cells: A Theory." *Science* 105:349–357, 1969. Copyright, 1969, American Association for the Advancement of Science.)

scribed according to a cell-specific pattern in differentiated cells.

All these conditions require great **flexibility** in regulation and cannot be reconciled with the prokaryotic polycistronic operon model. A monocistronic **super-operon** such as suggested for polytene and lampbrush chromosomes (see below) would seem more appropriate to explain the great flexibility and **complexity** of higher cells. In this type of model, those portions of DNA involved in regulator functions would prevail over those carrying structural information. R. J. Britten and E. H. Davidson have developed such a theoretical model to explain the enormous genome and possible modes of regulation in higher forms of eukaryotes.

To account for the multiple changes in gene activity that can be initiated by **external signals,** two basic patterns (Fig. 7.11) were formulated. This model represents the minimum complexity that would seem to be adequate for gene regulation in higher cells. **Integrated activity** of many genes would give the needed flexibility. Most genes, however, would be involved in regulation, with relatively few acting as **producer** genes (structural genes in the operon model). Expression of each producer gene would be under the control of many receptor (promoter) genes. Transcription of a producer gene could occur only if at least one of its receptors was "activated" by forming a sequence-specific complex with **activator RNA.** This RNA would be synthesized by **sensor** (regulatory) genes that are sensitive to developmental signals.

To increase flexibility, the receptor genes as well as the integrator genes are **redun-**

dant (repeated). In example (a) (Fig. 7.11), each sensor with its integrator gene may receive stimuli from an external signal and produce an activator RNA. The activator moves to its receptor which activates a producer gene. The resulting mRNA transcript specifies protein. With only one integrator gene per sensor there would be as many copies of a given receptor gene sequence as there are producer genes in the battery. Sensor gene S, and its integrator (I_1) specify the activation of producer genes P_A and P_B and P_C; S_2 and its integrator (I_2) specify the activation of producer genes P_A and P_B; and S_3 and its integrator (I_3) specify the activa-

tion of producer genes P_A and P_C.

This system might be involved where producer genes direct the synthesis of enzymes used in tightly coordinated physiological functions, such as the ten enzymes which synthesize urea. All ten genes are needed wherever urea is produced.

In example (b) (Fig. 7.11), redundancy is present among the integrator genes of different sets. Each sensor gene has **several integrators,** as opposed to only one in example A. Upon stimulation of the sensor gene, each integrator activates a **battery of genes.** There would be as many copies of a given integrator gene as batteries that call

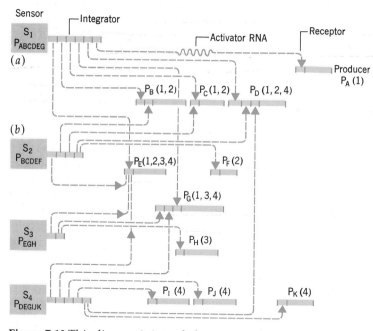

Figure 7.12 This diagram is intended to suggest the existence of overlapping batteries of genes and to show how, according to the model, control of their transcription might occur. The dotted lines symbolize the diffusion of activator RNA from its sites of synthesis, the integrator genes, to the receptor genes. The numbers in parentheses show which sensor genes control the transcription of the producer genes. At each sensor the battery of producer genes activated by that sensor is listed. In reality many batteries will be much larger than those shown and some genes will be part of hundreds of batteries. (After Roy J. Britten and Eric H. Davidson, "Gene Regulation for Higher Cells: A Theory." *Science* 165:349–357, 1969, Copyright, 1969, American Association for the Advancement of Science.)

on its producer gene. This could be a very large number for commonly required genes such as those involved in the fundamental biochemistry of each cell. The system presented in example (b) is a powerful integrative system since it can **control** a large variety of **producer** genes.

Integrated gene activity in the two control systems (Fig. 7.11) combined is illustrated in Fig. 7.12. No attempt is made to portray the actual complexity of the system in terms of the number of elements whose functions are likely to be integrated in a living system. Many batteries of genes must be coordinated to account for the massive changes that occur during differentiation. Sequential patterns of gene action, as in development, could occur as particular sensors respond to the products of producer genes. Furthermore, living systems continuously **adjust** their activities in accordance with their **internal state.** Feedback control by producer-gene products through sensors can also provide for constant adjustment of physiological functions.

In summary, any given producer gene could be turned on by a particular sensor through **redundant receptors** (a) or **redundant integrators** (b). It should be noted that, in general, genes of higher cell types are "turned off" and must be **activated,** whereas, the genes in lower organisms are apparently continuously "turned on" and must be **repressed** if activity is to be controlled. The proposed model can account for coordinated regulation of several noncontiguous genes. Upon specific stimulation such as events during development or a particular cell cycle, batteries of genes could be controlled through a single sensor gene. The Britten and Davidson model allows for considerable evolutionary flexibility in the overall number of regulatory sequences and a restricted number of sequences carrying structural information. Through interaction of specific types of nuclear RNA and DNA,

regulation could be both simplified and protected from mutational interference. The Britten and Davidson model has been a landmark for investigations of gene regulation in higher cells. It has stimulated much interest and influenced the design of many experiments. Several other models have been presented, mostly modifications or extensions of the one by Britten and Davidson. All such models are still theories with little or no experimental support.

BIOGENESIS OF mRNA IN MAMMALIAN CELLS

Eukaryotes have a different kind of RNA, **heterodisperse RNA** (HnRNA), which is similar in several respects to mRNA. A sequence of about 200 adenylic acid nucleotides (poly A residues), for example, occur in both HnRNA and mRNA, but not in other cellular RNA molecules (i.e., rRNA or tRNA). Furthermore, this poly A sequence exists only at the 3' (OH) end of both HnRNA and mRNA molecules. The HnRNA is produced in the cell nucleus where it is associated with **chromosomes** (not the nucleolus which is the source of some other RNAs). Part of the HnRNA has a rapid metabolic turnover and migrates to the cytoplasm but its cytoplasmic role has not been discovered. The mammalian cell **mRNA,** however, is apparently derived from precursor molecules of HnRNA that have **a base composition like DNA** (but with U substituted for T). Much evidence supports this mode of origin for mRNA in mammalian cells.

Models have been presented by J. E. Darnell and associates for both transcriptional and posttranscriptional regulations of the mRNA genesis. Transcriptional regulation would depend on the effectiveness of **RNA polymerase** in transcribing double-stranded DNA to HnRNA single strands. Later processing of the HnRNA would yield mRNA carrying poly A and specifications for enzymes synthesis. **Posttranscriptional** regulation implies a transcriptional overproduction of HnRNA which is **potential mRNA.** Excess HnRNA would be stored, presumably in the cytoplasm. The major regulating event would be the accumulation of mRNA from HnRNA during the part of the cell cycle when HnRNA production normally occurs. **Rate of processing** of HnRNA into mRNA could be **another regulating** device for controlling the synthesis of the relevant enzymes.

GENERAL REPRESSING EFFECT OF HISTONES

Histones are basic proteins that are associated with DNA in the chromatin of the cell. Chromatin derived from metaphase chromosomes was proved to contain 68 to 79 percent protein. One-half to three-fourths of this protein is composed predominately of histones. These proteins, which contain no cysteine or tryptophan and only small amounts of tyrosine and phenylalanine, have been known for a long time to influence gene activity. The chromatin histones are **synthesized** in the **nucleolus** of the metabolic cell and are present in both **diffuse** chromatin (in spermatids) and **condensed** chromatin (in sperm). The nonhistone chromatin proteins are present in diffuse but absent in condensed chromatin.

Although several different histones have been recognized and may be fractionated into subgroupings with particular characteristics only a few exist in nature. The same histones are present in all cells of a given organism. Likewise, nearly all orga-

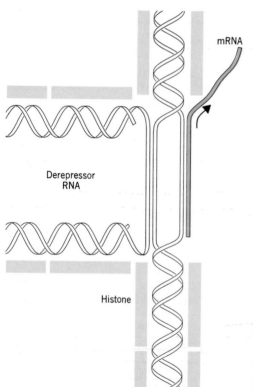

mRNA

Derepressor RNA

Histone

Figure 7.13 Model for gene repression by histone, and selective derepression. Histone (dark bars) is postulated to cover DNA double strands and to prevent enzyme activity required for transcription. The histone is displaced by action of nuclear RNA. Derepressor RNA is postulated to hybridize to the nontranscribing strand of DNA freeing the transcribing strand of DNA for mRNA synthesis.

nisms thus far studied possess similar histones. Histones act on DNA to generally repress gene activity. The extent of repression depends on the relative quantities of histone and DNA in a cell. The **quantity required for repression** is the proportion which can **complex** and thus **repress** the quantity of **DNA present.** Chicken erythrocytes, for example, have an excessive amount of histone and almost completely repress genomes, which will never be derepressed. Although these erythrocytes are nucleated, they are terminal cells and will not undergo cell division. Other cells of the chicken (for instance, liver, spleen, and heart cells) do not have histone in excessive quantities relative to their DNA, and do divide.

Chromatin in general does not contain enough histone to complex all the DNA of the particular genome with which it is associated in a cell. Usually about 80 to 90 percent of the DNA is repressed, with only 10 to 20 percent **derepressed at any one time.**

Only the part of the histone molecule that is rich in basic amino acids and proline binds to DNA. The opposite end, which is more acidic, interacts with nonhistone proteins thus allow the histone molecule to provide some **control** over **derepression** as well as **repression.** Some nonhistone proteins, perhaps in trace amounts, supply specificity for gene regulation in eukaryotes. In contrast, histone proteins do not possess sufficient heterogeneity for specific gene regulation in eukaryotes. They can be removed, however, and thus participate in the nonspecific regulation of blocks of genes in sections of the genome. Cellular agents such as phosphoproteins, acidic proteins and nuclear RNA can function as **derepressors** by effecting **histone displacement.** Certain species of nuclear RNA, for example, can select specific portions of the genome for derepression. They recognize specific base sequences on one strand of the DNA helix, and by hybridizing with a single DNA strand cause this strand to separate from the opposite DNA strand (Fig. 7.13), freeing it to serve as a template for RNA synthesis in a restricted portion of the genome.

HORMONAL CONTROL OF mRNA

Hormones have long been recognized by biologists as agents of differentiation. The control of secondary sex characteristics in animals by estrogens and androgens (Chapter 4) is the best known example. Recently, much progress has been made in explaining the mechanism through which hormones influence development of an organism. In general, they act as effectors by derepressing or releasing previously repressed genes, which can then synthesize mRNA in tissues and thus increase quantities of enzymes.

Injection of an estrogen into an immature or ovariectomized female rat, for example, causes an increase in RNA synthesis followed by increased protein synthesis. This basic effect of the RNA synthesis is reflected in alterations of the uterine wall and proliferation of the vaginal mucosa. Similarly, introduction of androgens (for example, testosterone) into immature or castrated male rats is followed by increased mRNA synthesis and enzyme production in the testes, prostate gland, and seminal vesicles.

If the antibiotic, puromycin, is introduced with the sex hormones, protein synthesis is inhibited. This antibiotic **blocks** protein synthesis by interfering with transfer RNA. Actinomycin-D also offsets the effect of sex hormones and blocks protein synthesis in bacteria and in animal cells. This antibiotic pairs with guanine in double helical DNA and prevents RNA polymerase molecules from using DNA as a template. It thus **prevents RNA synthesis** in the sex glands and prevents the induction of enzymes by sex hormones.

When rats and other mammals are treated with cortisone (a hormone from the adrenal cortex), the cortisone causes **increased production** of particular kinds of **mRNA** in liver cells, more enzymes are produced, and greater enzymatic activity is evidenced in the liver cells. This hormone is an effector that derepresses particular genes in liver cells and allows them to increase their production of mRNA. Puromycin and actinomycin-D both **inhibit** this effect of cortisone and prevent increased enzyme production in liver cells.

Likewise, evidence has been accumulated to show that thyroxine and pituitary (growth, adrenocorticotrophic, and gonadotrophic) hormones **increase mRNA** and enzyme production in their target organs. These nonsteroidal hormones probably activate membrane-bound adenylate cyclase to produce cyclic AMP from ATP. **Cyclic AMP** then may induce mRNA production. This small molecule, discovered by E. W. Sutherland, 1971 Nobel prize winner in physiology and medicine, is a **second messenger** between a hormone and its effects within the cell. ("Cyclic" refers to the fact that the atoms in a single phosphate group of the molecule are arranged in a ring, AMP stands for adenosine monophosphate.) It acts as a **promoter** molecule that binds to DNA and helps initiate protein synthesis.

Some plant hormones are effectors that increase the synthesis of mRNA. Examples are gibberellic acid, which controls dormancy in buds; and indole acetic acid, which controls growth through cell elonga-

tion. The increased rate of **RNA production** caused by each of these hormones is **suppressible by actinomycin-D.**

Carcinogenic agents such as methylcholanthrene also increase the rate of production of certain enzymes indicating that they **derepress genes for mRNA synthesis.** Such increases do not occur in the presence of actinomycin-D. This suggests that these kinds of carcinogens modify genetic repression in much the same way as do hormones. An increased enzyme protein synthesis induced by carcinogens is disadvantageous to the organism because it produces excessive growth in particular body areas in the form of tumors.

HORMONES CONTROLLING mRNA IN DROSOPHILA CHROMOSOMES

Studies of the giant salivary gland chromosomes in Diptera have shown that specific areas of these chromosomes change in appearance at successive stages of development (Fig. 7.14). Enlarged sections or **puffs** observed in some areas of the chromosomes change position as development proceeds over time, indicating that **different genes** were being **turned on** and off during development. These changes have been related to chemical activities of DNA, RNA, and protein synthesis by staining techniques, autoradiographs, and radioactive isotopes. The successive observable events in the chromosomes suggest that inducers or regulators are active at particular periods in development. When stains specific for DNA and RNA were introduced together, segments of the salivary chromosomes stained differentially. It was shown that the active areas (or puffs) carried an excess of **RNA.** RNA was being synthesized in the chromosome regions that appeared to be active. Further studies showed that **different mRNAs** were synthesized in **different puffs.** Labeled mRNA, taken from puffs and

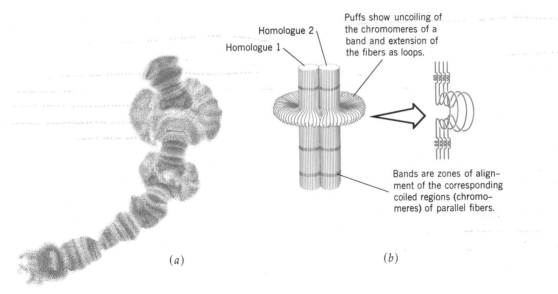

Homologue 2
Homologue 1

Puffs show uncoiling of the chromomeres of a band and extension of the fibers as loops.

Bands are zones of align-ment of the corresponding coiled regions (chromo-meres) of parallel fibers.

(a)   (b)

Figure 7.14 Development of a chromosomal puff in a larval salivary-gland cell nucleus of *Chironomus tentans*. This reversible phenomenon occurs at a specific developmental stage. Puffs carry an excess of RNA. (*a*) Part of a salivary-gland chromosome showing a large puff. (*b*) Diagrams illustrating structure of a puff. Homologues are closely paired and each chromosome is of multiple, parallel replicate fibers. (After W. Beermann and U. Clever.)

injected into other specimens, migrated to ribosomes as expected.

Hormonal control of particular puffs has been demonstrated during larval develop-ment of the flies. The insect hormone, **ec-dysone**, obtained from the silkworm (Bombyx) as well as from Drosophila and other Diptera, was found to trigger the **syn-thetic activity** that resulted in puffs early in development. In the presence of ecdysone and protein, puffing activity in one chromo-some area of young pupae increased for four to six hours and then regressed. The next chromosome area then puffed and subsided, and comparable cycles continued along the chromosome.

When ecdysone was washed out of early puffs, the puffing regressed. As protein ac-cumulates it takes over the control of later forming puffs. These facts led to the hy-pothesis that differentiation is **triggered by** hormonal stimulation of sets of genes which, in turn, produce in an **orderly sequence of mRNAs** that code for enzymes that then take over the **regulatory function** of protein synthesis.

AMPHIBIAN LAMPBRUSH CHROMOSOMES AND mRNA SYNTHESIS

Another giant chromosome that promises to be increasingly useful in efforts to corre-late cytological studies with the action of genes is the **lampbrush** type of chromo-some (Fig. 7.15) found in the oöcytes of newts (amphibians). Chromosomes of this kind are greatly enlarged, some reaching the extraordinary length of 1 mm. DNA threads form the chromosome axis and extend lat-erally as loops on either side of the main axis. Thickened portions of the lateral loops

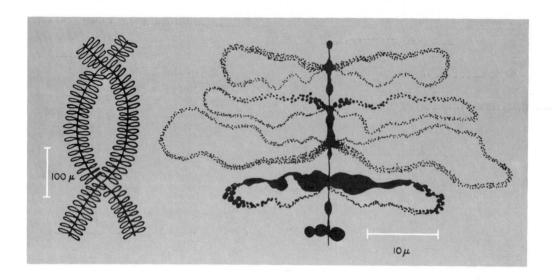

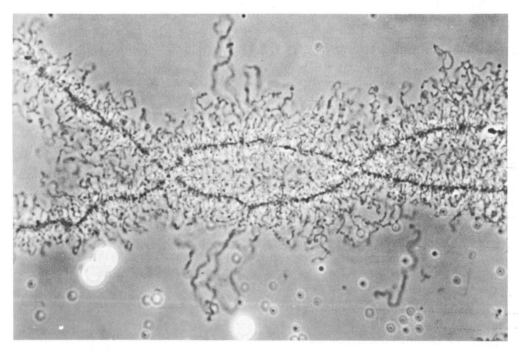

Figure 7.15 Lampbrush chromosomes from giant nuclei of Amphibian *Triturus vividescens* oöcytes. (Top left). Highly diagrammatic view of a lampbrush bivalent, showing the homologous chromosomes joined together at two chiasmata. (Top right). Semidiagrammatic view of the central chromomere axis with paired lateral loops. (From J. G. Gall, "Mutation," Brookhaven Symposium in Biology, No. 8, p. 18). (Bottom). Photograph of lampbrush bivalent. Note the numerous lateral loops of various lengths. (Courtesy of J. B. Gall.)

represent accumulations of **mRNA**. Similar structures have been observed in other organisms, but none thus far described are as conspicuous as those in newts. This observation is correlated with the fact that amphibian oöcyte chromosomes have an unusually high proportion of DNA. RNA, which may be mRNA, is produced along the loops of the lampbrush chromosomes in a regular cycle somewhat like the process involving puff formation in the salivary chromosomes of dipterous larvae.

BIOCHEMICAL MUTATIONS AND PATHWAYS

Control of biological synthesis also occurs when mutations block biochemical pathways. When a sequence of reactions is required for a particular end product and a mutation blocks the sequence at a given point (1) the **unused substratum** builds up and (2) the **products** of later steps in the pathway **are lacking**. Charles Yanofsky, for example, isolated tryptophan auxotrophic mutants from the prototrophic E. coli wild-type. These mutants were tested to see if they would grow on a minimal medium supplemented with substances other than tryptophan. The results showed that Trp⁻ auxotrophs fall into different classes with respect to their growth requirements: class TrpB⁻, grows only on tryptophan, and indole is accumulated, class TrpD⁻ grows either on tryptophan or indole and anthranilic acid is accumulated, and class TrpE⁻ grows on tryptophan or indole or anthranilic acid and no substance is accumulated (Table 7.1). Class TrpB⁻ auxotrophs must carry mutations in a gene that controls the last enzyme of the pathway, tryptophan synthetase, since that block is replaced only when tryptophan is supplied. Class TrpE⁻ auxotrophs evidently carry mutations in a gene that controls the first enzyme of a pathway, anthranilic synthetase, because these auxotrophs must possess all the enzymes necessary for converting anthranilic acid to tryptophan. Class TrpD⁻ mutants must possess a functional tryptophan B protein that enables them to convert exogenous indole into tryptophan. Results of this experiment show that different auxotrophic mutants are blocked at different stages on the tryptophan biogenesis pathway. Any auxotroph can convert into tryptophan only those intermediates whose synthesis would normally occur after the step blocked by the gene mutation.

A classical example of a biochemical pathway illustrates the control of pigment production in insects. G. W. Beadle (Fig.

TABLE 7.1
Properties of Tryptophan Auxotrophs of E. coli **When Grown on Minimal Medium Plus Supplements**

Class	No Supplement	Anthranilic Acid	Indole	Tryptophan	Substance Accumulated
Trp⁺	+	+	+	+	none
TrpE⁻	−	+	+	+	none
TrpD⁻	−	−	+	+	anthranilic acid
TrpB⁻	−	−	−	+	indole

Figure 7.16 George Beadle, distinguished researcher, teacher of genetics and university administrator. Dr. Beadle, along with E. L. Tatum and J. Lederberg, was Nobel Laureate in 1958 in medicine and physiology. (Courtesy of University of Chicago.)

7.16), B. Ephrussi, A. Kühn, and their associates provided the framework for the basic investigation. The eyes and other organs as well as the entire outer surface of the caterpillar of the wild meal moth, *Ephestia kühniella*, are dark brown. This coloring is dependent on certain ommochrome pigments. By spontaneous mutation of a single gene (v^+ mutated to v), the caterpillar loses its ability to form pigment.

If aqueous alcoholic extracts of v^+ tissue are injected into v caterpillars, these caterpillars regain the ability to form ommochromes, and normal pigmentation occurs. Tissues of wild-type organisms, therefore, are assumed to contain an **extractable substance** whose function is interposed between gene transcription and the external **phenotypic expression.**

Experimental analysis proved that the

substance that causes pigment formation under the influence of the gene v^+ was kynurenine, an amino acid that is an intermediate product of tryptophan metabolism in mammals. Numerous microorganisms have this same ability to transform tryptophan to kynurenine. Quantitative studies showed that the degree of pigment formation in the caterpillars was directly proportional to the amount of kynurenine present. Kynurenine, therefore, does not act as a catalyst in pigment formation, but as a **precursor** in the **synthesis** of the ommochrome molecule.

Next, the relation between kynurenine formation and the gene v^+ was clarified. In mammals, the transformation of tryptophan to kynurenine is dependent on a specific enzyme, tryptophan pyrrolase, found in the liver and in insects. It may be assumed that a corresponding enzyme is present in v^+ but not in v tissue. The muta-

tion of v^+ to v thus deprives the cell of the ability to convert tryptophan to kynurenine and **breaks the chain of events required for pigment formation.** With the course of tryptophan degradation blocked, this amino acid would be expected to accumulate in v caterpillars. E. Caspari has shown that the tryptophan content of v mutants is greater than that of a member of the wild race.

Results paralleling those in the meal moth were found in the fruit fly, *D. melanogaster*. The dark-red eye color of the wild-type flies also depends on a v^+ gene, and the mutation from v^+ to v (vermilion) reduces the pigmentation of the gene in Drosophila as well as in Ephestia. Another similar effect on the formation of eye pigment in Drosophila is caused by a mutation at another locus. When cn^+ mutates to cn (cinnabar), affected flies are unable to produce ommochrome. Transplantation and extraction experiments have shown that cn^+ in-

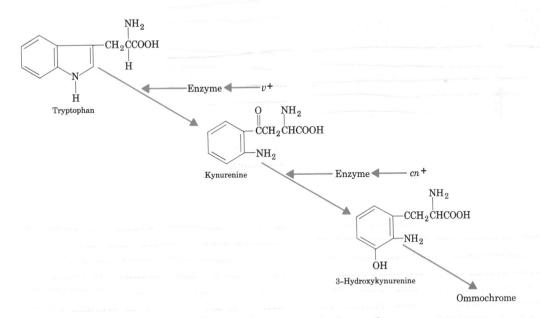

Figure 7.17 Diagram representing the steps between tryptophan and 3-hydroxykynurenine. Two genes, v^+ and cn^+, which control enzymatic action, are postulated for the steps: tryptophan to kynurenine and kynurenine to 3-hydroxykynurenine.

tervenes in the pigment-producing process after the tryptophan to kynurenine stage. Its action requires the presence of v^+ and a supply of kynurenine. The action of cn^+ can be replaced by 3-hydroxykynurenine, which evidently represents a **further link in the chain between tryptophan and ommochrome.** The 3-hydroxy form of kynurenine depends on a specific enzyme that acts only in tissue containing cn^+ genes. These two links in the chain are illustrated diagrammatically in Fig. 7.17.

A protein carrier is also required during formation of pigment. Kühn showed that a mutation of the gene wa^+ to wa, disturbed the formation of protein granules in the meal moth, and completely interrupted pigment formation. Kynurenine and 3-hydroxykynurenine were produced by the wa moths but could not be used by the wa tissues. Kühn's investigation defined another **link in pigment formation.** Three points have thus been identified at which the chain reaction leading to pigment formation may be broken. The probable actions of the three genes are illustrated in Fig. 7.18.

Pigment production in plants apparently requires **gene-controlled pathways** comparable to those in insects. Plant pigments include carotenoids (usually confined to plastids), anthocyanins, anthoxanthines, and flavocyanins. Most water-soluble red, blue, and yellow pigments found in flowers are anthocyanins or related compounds. Several of these pigments have been synthesized in a test tube, but the mechanisms by which plants make them are largely unknown. Some pigments are closely related chemically and differ only in the number of hydroxyl groups on their benzene rings. In some cases, the pH of the substratum controls the color that an indicator substance provided by a gene can produce.

ALBINISM

When pigment production is blocked, the result is albinism. A single gene (c) in homozygous condition (cc) **blocks pigment production** and gives rise to the albino phenotype. The mechanism by which this is accomplished is one of the best-known examples of the relation between gene and visible character in man. Melanin is produced through a chain of reactions (Fig. 7.19) involving the amino acids phenylalanine (one of the essential amino acids that must be supplied in the human diet) and tyrosine. The enzyme tyrosinase is involved in the oxidation of tyrosine to dihydroxyphenylalanine (dopa). Further oxidation results in the final product, **melanin.**

A lack of tyrosinase **blocks** the chain reaction in one or more places and results in **albinism.** When the recessive gene c for albinism is present in the homozygous condition, the production of tyrosinase is in-

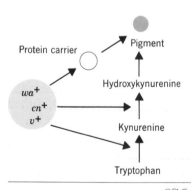

Figure 7.18 Diagram representing three steps in a chain reaction leading to pigment formation in the meal moth, *Ephestia kühniella*. (Data from Kühn.)

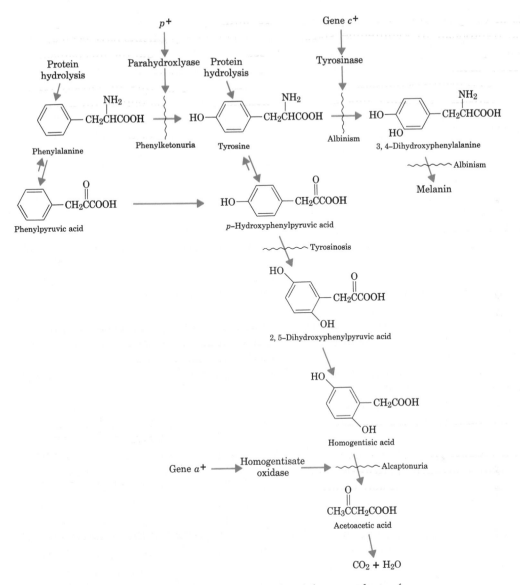

Figure 7.19 Diagrammatic representation of some breakdown products of phenylalanine and tyrosine illustrating chemical blocks which result in several different human abnormalities.

hibited. The possibility of artificially supplying an enzyme and thus overcoming albinism has been considered by many investigators. Unfortunately, no method is known by which an enzyme can be effective inside a cell unless it has been synthesized in that cell. Therefore, prospects for curing albinism by artificially supplying enzymes are not promising. Previously cited studies in Neurospora (earlier in this chapter), however, indicate that **nutritional blocks** and color variations created by mutations can be repaired or altered by the addition of appropriate chemicals. Cellular processes in higher organisms may eventually be susceptible to manipulation when more steps in the chain reactions are identified.

OTHER BLOCKS IN METABOLISM OF PHENYLALANINE IN MAN

Another hereditary block in the metabolism of phenylalanine and tyrosine results in the disease alcaptonuria, characterized chiefly by a darkening of the cartilage. In the presence of this abnormality, cartilage that is close to the body surface in the ear, wrist, and elbow shows discoloration. A type of arthritis sometimes becomes associated with the cartilage discoloration when people having alcaptonuria reach middle or old age. A more obvious but superficial symptom is the tendency of the urine of an affected person to turn black when exposed to air or when brought in contact with an alkaline substance. This phenomenon has been traced to an unusual urine component, "alcapton" or homogentisic acid. Normal people have an enzyme—homogentisate oxidase—in their livers that converts homogentisic acid to a substance which is colorless and is eventually broken down to carbon dioxide and water. Alcaptonurics lack the enzyme because of a single autosomal recessive gene in homozygous condition (aa). When alcaptonurics eat foods containing large amounts of phenylalanine, tyrosine, or p-OH-phenyl-pyruvic acid, a proportional increase of homogentisic acid is excreted in the urine. Normal people can eat similar foods in reasonable quantity with no such reaction. In alcaptonurics, an **intermediate product** has thus become an **end product** because of a hereditary block in the chain reaction.

Another abnormality resulting from a different block in the phenylalanine metabolism is phenylketonuria. This disease is characterized by a serious mental defect and is also associated with the excretion of excessive amounts (about one gram per day) of **phenylpyruvic acid** in the urine. An increase of phenylalanine in the diet is associated with an increase of the acid in the urine. Again, a single recessive gene in homozygous condition (pp) is responsible for the **metabolic block.** People with the normal allele (p^+) have parahydroxylase, which changes phenylalanine to tryosine. All phenylketonurics have light complexions because the metabolites of phenylalanine interfere with pigment production.

Although albinism, alcaptonuria, and phenylketonuria are very different phenotypically, they are related chemically. Each is the result of a **break in the chain of reactions** involved in phenylalanine and tyrosine metabolism. Mutations elsewhere in the series of reactions (Fig. 7.19) give rise to other abnormalities. Tyrosinosis, characterized by the accumulation of tyrosine in the body, represents such a disease. This rare human abnormality is probably the result of a gene mutation that leads to a block in the reaction through which p-hydroxy-phenylpyruvic acid is converted to 2, 5-dihydroxyphenylpyruvic acid. Another block in tyrosine metabolism after dopa formation, inhibits the production of adrenaline and noradrenaline. Thyroxine deficiency, associated with genetic goitrous cretinism is caused by another block in tyrosine metabolism.

PHENOCOPIES

Alterations in the environment sometimes induce **nonhereditary phenotypic changes** (phenocopies) that **resemble** those caused by **mutations.** Phenocopies and mutations can be distinguished by making appropriate crosses and determining whether a given phenotypic alteration is transmitted to offspring. Well-established phenocopies have been described in bacteria, but they have not always been clearly differentiated from mutations. Goldschmidt, who coined the term phenocopy in 1935, subjected Drosophila pupae to a higher than usual temperature (35°C) for short intervals at different periods of their development cycle. Several phenotypes indistinguishable by visible means from genetic mutants were produced. Some of these, a group of wing abnormalities paralleled the expression of the different alleles of the vestigial (*vg*) series. The imposed temperature alterations were visualized as changing the velocity of chemical reactions thus altering the end products. **The germ plasm was not affected.**

In chickens, two known mutations interfere in some fundamental way with development, and produce a phenotype called rumplessness (Fig. 7.20). The abnormality is characterized by the absence of caudal vertebrae and tail structures such as muscles, feathers, and the uropygial or preen gland from which an oily secretion is normally supplied to the feathers. Although both mutations have the same end product, their other characteristics are different. One is completely dominant whereas the other is recessive. The genes are not alleles and are not closely linked. Furthermore, birds expressing the two genes differ in structural details, chemical composition of bones, and embryonic development.

By mechanically shaking the eggs at a critical period, the rumpless phenotype may be reproduced in chickens carrying the normal alleles of the respective genes. Embryos treated at 48 hours of age with sodium methyl arsenite had rump defects and other abnormalities in greater proportion than the untreated controls. Other chemicals, including insulin, also effectively induced comparable phenocopies. Further experiments showed that an injection of nicotine amide would counteract the insulin and prevent production of the insulin-induced phenocopies. Nicotinamide apparently acts as a co-enzyme in cell respiration. The mere fact that nongenetic modifications resemble genetic mutants does not prove similar causation. When the time at which alteration can be induced (the temperature- or chemical-sensitive period) coincides with the time of gene action (which obviously occurs before visible effects can be detected), however, a similar physiological alteration is suggested. The similarity of the fundamental reaction chains that are affected must be specifically proved in each instance.

In **human** genetics, **phenocopies** add to the difficulty of distinguishing hereditary traits from those produced by **environmental factors.** Heart troubles brought on by environmentally conditioned anxiety reactions, for example, are similar to those having a hereditary cause. Similarly, eye defects following measles cannot be distinguished readily from hereditary eye abnormalities. A protozoan infection of mothers during pregnancy, known as toxoplasmosis, sometimes produces abnormalities in children resembling hereditary conditions. Several skeletal and neurological anomalies induced by irradiation parallel genetically produced phenotypes. Use of the tranquilizer, thalidomide, by pregnant women results in deformities in the fetus that resemble some hereditary ab-

(a)

(b)

Figure 7.20 Phenocopy in chickens known as rumplessness. (a) rumpless chicken: (b) normal chicken. (Courtesy of Dr. Ralph G. Somes, Jr.)

normalities. The same basic reactions may or may not be involved in phenotypic changes that differ in having either **a gene** **or some environmental agent** as their trigger mechanism.

VARIABILITY OF GENE EFFECTS

The phenotypic expression of any given gene depends on its interaction with other genes and environmental factors. When comparing the visible effects of genes, two terms, *penetrance* and *expressivity*, are useful. **Penetrance** is the **percentage** of individuals in a population **who carry the gene** in whatever combination permits its expression and who demonstrate the **phenotype.** If a dominant gene is expressed in only 70 percent of the individuals known to carry it, the penetrance of the gene would be 70 percent. The expressivity is the degree of effect or the extent to which a gene expresses itself in different individuals.

Examples given thus far have involved only genes with complete penetrance and unvarying expressivity. Whenever a gene was identified as a dominant, it was assumed to come to expression and produce a given effect in any individual in which it occurred. Recessives in homozygous condition have been assumed to express themselves in a characteristic manner. In some cases, however, genes known to be present in a proper combination are not expressed in 100 percent of the carrier individuals. Usually variable penetrance and expressivity can be explained on the basis of either modifiers that influence the action of a given gene, or environmental variations. These rationale are sometimes used to make ignorance respectable when the reasons for the failure of expression or variation in phenotype are not known.

The penetrance concept is used in human genetics to account for ratios that do not conform to strict Mendelian patterns. Dominants with reduced penetrance are dif-ficult to distinguish from recessives in family history studies. Statistical methods allow comparison of the proportion of individuals showing a trait with the proportion of those presumed to carry the gene. Data for statistical analysis of dominants can be obtained by determining the proportion of unaffected parents in whose progeny the trait appears. Further data are obtained by comparing the expected 1:1 ratio with the actual ratio of affected to normal children in all families in which one parent expressed the trait. Such human traits as blue sclerotics, stiff little fingers, and many others, have been described as depending on incompletely penetrant dominant genes.

The term penetrance has been abused in human genetics until it is in danger of falling into disrepute. Too often, variable penetrance is used as a catchall to explain presumed genetic patterns that do not follow a common Mendelian ratio. In some cases, the heritability of the trait in question may be doubtful; in others, the expression may depend on more than one gene. Most of the carefully studied human traits have been found to be too complex to be explained by single gene substitutions.

Several investigators have cited susceptibility to leukemia as an hereditary condition dependent on a dominant gene with incomplete penetrance. If a genetic basis is established for leukemia in man, and only a single gene is involved, the penetrance would have to be less than 1 percent, since only 1 of every 100 individuals carrying the presumed gene seem to contract the disease. A statistical investigation by A. G.

Steinberg, in which more than 200 families were studied, indicated that leukemia has no hereditary basis. The incidence among the relatives of leukemia patients was not significantly greater than that among the controls or in the general population.

Expressivity, or the **degree to which gene expression occurs,** may merely represent an extension of penetrance in the visible range. If genes sometimes fail to come to visible expression, variations would be expected among the phenotypes observed. Therefore, both penetrance and expressivity may be involved in a given expression. Expressivity has significance in human genetics but, like penetrance, has often been abused. In one person, a given gene may produce a slight effect, whereas in another it may result in a marked deviation from normal. It is not always possible to determine whether genetic or environmental factors are responsible for such differences.

The main difficulty in distinguishing between hereditary and environmental influences in man, is that no two human beings, except identical twins, are genetically alike. **Twins** studied thus far have furnished limited but **valuable data** on problems of expressivity. A **single dominant** gene is apparently associated with **allergy,** but the gene may be expressed in many ways in different individuals. Some people carrying the gene may develop hay fever, while others have asthma, edema, or skin rash. It is not always possible to determine whether a single gene with variable expressivity is responsible for different expressions, or whether different genes are involved in the series of phenotypes. The literature includes several examples in which a single basic gene is postulated, and interactions between modifier genes and environmental factors are suggested to account for variation. Sometimes minor and largely unknown factors and interactions must be postulated to plausibly explain variation in expressivity.

REFERENCES

Atkins, J. F., and J. C. Loper. 1970. "Transcription initiation in the histidine operon of *Salmonella typhimurium.*" *Proc. Nat'l Acad. Sci.* U. S. 65, 925–932.

Beermann, W. (ed) 1972. *Developmental studies on giant chromosomes.* Springer-Verlag, New York.

Bonner, J. 1965. *The molecular biology of development.* Oxford University Press, New York.

Britten, R. J., and E. H. Davidson. 1969. "Gene regulation for higher cells: A theory." *Science* 165, 349–357.

Brown, D. D. 1973. "The isolation of genes." *Sci. Amer.* 229:20–29.

Darnell, J. E., W. R. Jelinek, and G. R. Molloy. 1973. "Biogenesis of mRNA: genetic regulation in mammalian cells." *Science* 181:1215–1221.

Davidson, J. N. 1972. *The biochemistry of nucleic acids.* Academic Press, New York.

Ephrussi, B. 1972. *Hybridization of somatic cells.* Princeton University Press, Princeton, N.J.

Jacob, F. and J. Monod. 1961. "Genetic regulatory mechanisms in synthesis of proteins." *J. Mol. Biol.* 3, 318–356.

Koshland, Jr. D. E. 1973. "Protein shape and biological control." *Sci. Amer.* 229:52–64.

Loomis, W. F. (ed.) 1970. *Papers on regulation of gene activity during development.* Harper and Row, New York.

Phillips, D. M. 1970. "Insect sperm: their structure and morphogenesis." *J. Cell Biol.* 44, 243–277.

Smith, H. H. (Com. Chairman). 1965. *Genetic control of differentiation.* Brookhaven Nat'l Lab., U.S., Brookhaven, N.Y.

Romrell, L. J., H. P. Stanley, and J. T. Bowman. 1972. "Genetic control of spermiogenesis in *Drosophila melanogaster:* an autosomal mutant (ms (2)10R) demonstrating disruption of the axonemal complex." *J. Ultrastructure Research* 38, 578–590.

Sirlin, J. L. 1972. *Biology of RNA.* Academic Press, New York.

PROBLEMS AND QUESTIONS

7.1 How can inducible and repressible enzymes be distinguished?

7.2 From the standpoint of economy of the organism, which method of genetic control would seem to be most efficient: feedback inhibition from the finished product or presence of the substrate in the medium?

7.3 In the lactose operon in *E. coli* (Fig. 7.1), what is the function of each of the following genes or sites and how does each perform its function: (1) regulator, (2) operator, (3) structural gene z, and (4) structural gene y?

7.4 What would be the result of inactivation by mutation of the following genes or sites in the *E. coli* lactose operon: (a) regulator, (b) operator, (c) structural gene z, and (d) structural gene y?

7.5 Groups of alleles associated with the lactose operon (Fig. 7.1) are as follows (in order of dominance for each allelic series): repressor, i^s (superrepressor), i^+ and i^- (constitutive); operator, o^c (constitutive, cis dominant) and o^+; structural, z^+ and z^-, y^+ and y^-. (a) Which of the following genotypes will produce β-galactosidase and permease if the inducer (lactose) is present: (1) $i^+o^+z^+y^+$, (2) $i^-o^cz^+y^+$, (3) $i^so^cz^+y^+$, (4) $i^so^+z^+y^+$, (5) $i^-o^+z^+y^+$? (b) Which of the above genotypes will produce β-galactosidase and permease if the inducer is absent?

7.6 For each of the following partial diploids indicate whether enzyme formation is constitutive or inductive

(a) $\dfrac{i^+\,o^+\,z^+\,y^+}{i^+o^+z^+y^+}$, (b) $\dfrac{i^+\,o^+\,z^+\,y^+}{i^+o^cz^+y^+}$, (c) $\dfrac{i^+\,o^c\,z^+\,y^+}{i^+o^cz^+y^+}$, (d) $\dfrac{i^+\,o^+\,z^+\,y^+}{i^-o^+z^+y^+}$, (e) $\dfrac{i^-\,o^+\,z^+\,y^+}{i^-o^+z^+y^+}$

7.7 Write the partial diploid genotype for a strain that will (a) produce β-galactosidase constitutively and permease by induction and (b) produce β-galactosidase on induction but not permease either constitutively or on induction even though a y^- gene is known to be present.

7.8 Constitutive mutations produce elevated enzyme levels at all times; they may be of two types, o^c or i^-. Assume that all other DNA present is wild type. Outline the way by which the two constitutive mutants can be distinguished with respect to: (a) map postion; (b) regulation of enzyme levels in o^c/o^+ versus i^-/i^+ partial diploids; (c) the position of the genes affected by an o^c mutant versus the genes affected by an i^- mutant in a partial diploid.

7.9 How could the histidine operon in *S. typhimurium* have developed and been maintained in evolution?

7.10 How could (a) polarity and (b) modulation be demonstrated? (c) If a polarity mutation should inactivate gene A in the histidine operon, how would the enzymatic steps be affected?

7.11 Are operons expected to be more common in bacteria or in higher organisms? Why?

7.12 If the operon theory had been established at the time the two classical hypotheses, (1) segregation of nuclear elements and (2) intervention of cytoplasm, for the mechanics of differentiation were being considered, which hypothesis would have been most acceptable?

7.13 Distinguish, with examples, between regulation based on negative mechanisms and those based on positive regulatory mechanisms.

7.14 How do prokaryotes and eukaryotes differ with respect to regulations of mRNA transcription?

7.15 Why is the Britten and Davidson model more acceptable than the operon model for explaining regulation in cells of higher animals?

7.16 (a) How do histones regulate gene activity? (b) Where are they synthesized? and (c) how are they related to chromosomes in the cell cycle?

7.17 How can estrogens, androgens, cortisone, and gibberellic acid regulate gene activity?

7.18 Why is inhibition by actinomycin-D significant in identifying hormones with the regulation of gene activity?

7.19 How may carcinogens be involved in tumor formation?

7.20 Why are salivary chromosomes in the larvae of Diptera useful in studying hormonal regulation of gene activity?

7.21 (a) Why are such organisms as molds, bacteria, and viruses favorable materials for biochemical genetic study? (b) What type of experimental material would be most suitable for a study of

(1) operons, (2) repressing effect of histones, (3) hormonal control of mRNA synthesis and (4) chromosome puffs?

7.22 At birth, rabbits of the Himalayan breed are all white, but as they grow older the extremities (paws, nose, ears, and tail) become black. When the white fur is shaved from a spot on the body of the adult and the rabbit is kept in a cool place, the new hair that grows in the shaved spot is black. The temperature of the body proper is about 33°C, but in the extremities it is about 27°C. (a) How may genetic and environmental factors be involved? (b) Formulate an explanation for the difference in pigmentation in different areas of body surface.

7.23 Ordinarily, Drosophila eye-disc transplants placed in hosts with different genotypes develop according to the genotype of the transplant. For example, transplants from larvae with the genotype for white eye develop while in wild-type hosts, and transplants from wild-type larvae develop wild-type red in hosts with genotype for white. Beadle and Ephrussi performed transplantation experiments on larvae with vermilion (*v*) and cinnabar (*cn*) eyes. The phenotypes for these two mutants are similar. They have bright red color because they lack the brown pigment that is a part of the wild-type red. (a) When discs from wild-type larvae were transplanted into vermilion or cinnabar hosts they developed wild type, but when *v* or *cn* transplants were placed in wild-type hosts they also developed wild type. (b) When discs from *cn* larvae were placed in *v* hosts no brown pigment was formed, and the eyes were bright red. From the reciprocal transplant, that is, *v* in *cn* hosts, brown pigment was produced and the eyes were wild type. Formulate an explanation for these results.

7.24 The father of two albino children has made widespread inquiries among geneticists and physicians concerning a possible cure for albinism. The steps in pigment production have been elucidated and it seems feasible to him that something might be added to the diet or given by injection which would supply the missing step or steps in melanin production in his children. Evaluate the possibility of such a development.

7.25 How is (a) differentiation (b) organization and (c) growth involved in the development of an animal?

7.26 If all cells in a given organism carry the same genes, how can gene expressions that are localized in time and space be explained?

7.27 (a) How can phenocopies be used to study gene action? (b) What values and limitations do they have for investigations of this kind? (c) How can an investigator determine whether an altered phenotype is a mutation or a phenocopy?

7.28 Develop a plausible explanation for the occurrence of rumplessness in chickens in association with each of the following: (a) without treatment; (b) mechanical shaking; and (c) injection with insulin.

7.29 What significance may phenocopies have in medical genetics?

7.30 Why are studies involving changes in environmental factors such as temperature more useful in studying the action and products of some genes, such as those controlling flower color in the primrose, than others, such as those controlling blood type in man?

7.31 A series of crosses in a particular controlled environment from which all progeny were known to carry a dominant autosomal gene A (causing an eye abnormality), resulted in 1400 abnormal and 600 normal flies. What is the penetrance of gene A in this experiment?

7.32 Ten nutritional mutation groups in Neurospora, $arg - 1$ to $arg -10$, will grow when arginine is added to the minimal medium. Those mutants identified below with a + sign will grow when the following substances are added to the medium:

mutational groups	minimal medium	glutamic semi-aldehyde	ornithine	citrulline	arginine
$arg -8, -9$	−	+	+	+	+
$arg -4, -5, -6, -7$	−	−	+	+	+
$arg -2, -3$	−	−	−	+	+
$arg -1, -10$	−	−	−	−	+

(a) Suggest a metabolic pathway for the synthesis of arginine?

(b) What products are expected to accumulate following mutation of (1) $arg -1$, and (2) $arg -2$?

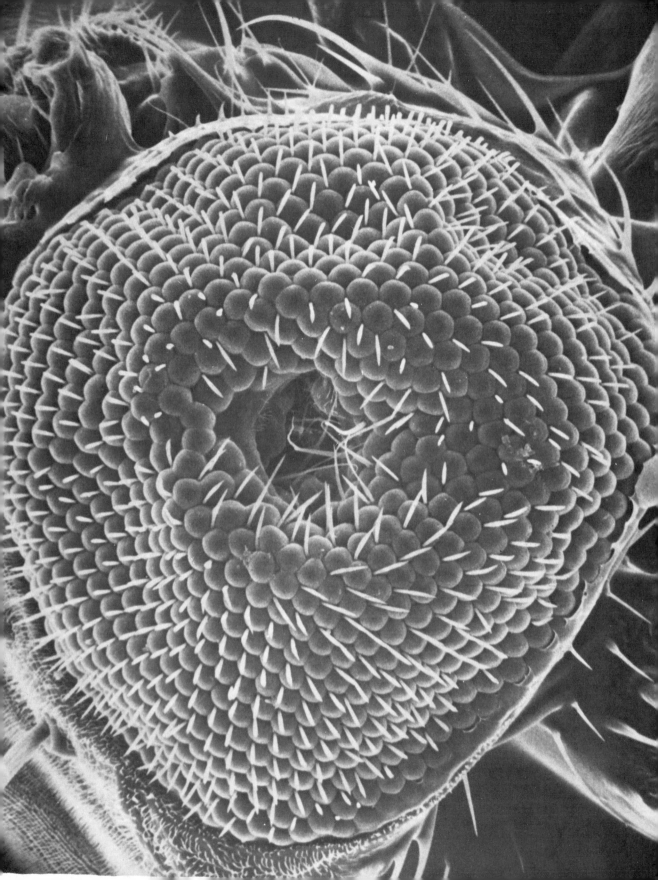

EIGHT

Mutation is a process by which a gene undergoes a structural change. The same term describes the modified gene resulting from the mutation process. Most commonly, however, the word mutation indicates a phenotypic change resulting from a changed gene. The word mutant is used to identify an individual expressing the phenotypic change resulting from mutation. Mutational changes usually occur in nucleotides (*mutons*) or larger units of DNA or RNA (in RNA viruses). Historically the word "mutation" has been used to describe a phenotypic change regardless of any genetic implications. For this discussion a mutation will be defined as a structural **change in a gene.** Mutants produced by DNA base substitutions are the most common type at the molecular level.

A change involving only a nucleotide or a single base will be designated a **point mutation** in contrast to one involving a larger segment of a gene or chromosome. An exchange of one purine (adenine or guanine) for the other, or of one pyrimidine (cytosine or thymine) for the other is called a **tran-**

MUTAGENESIS

Abnormal compound eye, an expression of the tumorous head trait in *Drosophila melanogaster.* (550×). (Courtesy Nabil N. Youssef.)

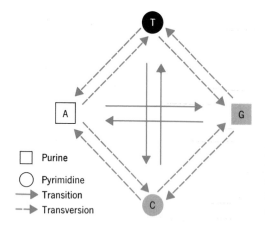

- □ Purine
- ○ Pyrimidine
- → Transition
- ⇢ Transversion

Figure 8.1 Four possible transitions and eight possible transversions resulting in point mutations; T, thymine, A, adenine, C, cytosine, G, guanine.

sition whereas a substitution of a purine for a pyrimidine or vice versa is called a **transversion** as illustrated diagrammatically in Fig. 8.1. Four different transitions and eight different transversions are possible. Reading **frame shifts** resulting from addition or subtraction of a nucleotide in a DNA strand are also included among the point mutations but are additions or deletions.

Since each gene (*cistron*) usually consists of hundreds of nucleotide sites, the possible number of allelic states is very great. The model of two allelic states (e.g., *A* and *a*) with comparable rates is inadequate to account for mutational change at the molecular level. The probability that a mutant site will revert to a previous allelic state by mutation is very much smaller than the probability of its mutating further to a new allelic state. Reversion to a particular state requires a change to a particular base (A, T, G, or C) whereas a new allelic state may occur by a change to any one of the four bases.

SPONTANEOUS MUTATIONS

Mutations can either occur spontaneously, without a **known** cause, or be induced by mutagenic agents in the environment. So called "spontaneous" mutations are probably caused by physical or chemical agents. They are usually infrequent occurrences for any particular gene (Table 8.1), but over a period of time they accumulate in populations. A map of spontaneous mutations that had occurred in two adjacent cistrons, *rIIA* and *rIIB* of bacteriophage T4 was illustrated in Fig. 6.11. The *rII* region includes about 2800 nucleotides; mutations were detected at 308 different sites. The spontaneous mutation rate in the *rII* region was found to be on the order of one mutant per 10^8 phage particles per generation. Each mutational change produced a gene differing from the original gene and consistently self-replicating its changed form. A lethal expression or a changed phenotype might be expected to occur in response to a gene mutation. In organisms with paired chromosomes, however, homozygous or other appropriate arrangements of chromosomes would be required for recessive genes to come to expression.

During the normal metabolic processes of

TABLE 8.1
Spontaneous Mutation Rate at Specified Loci in Different Species

Species	Trait	Mutations per 100,000 Gametes
Bacteriophage T4	gene 42 *ts* reversion	0.3
Escherichia coli (K12)	to leucine independence	0.00007
E. coli (K12)	to streptomycin resistance	0.00004
Salmonella typhimurium	to threonine resistance	0.41
Diplococcus pneumoniae	to penicillin resistance	0.01
Saccharomyces cerevisiae	lysine reversion	0.4
Neurospora crassa	to adenine independence	0.0008
Zea mays (corn)	sh^+ to *sh* (shrunken)	0.12
Z. mays	wy^+ to *wy* (waxy)	0.00
Drosophila melanogaster	ey^+ to *ey* (eyeless)	2.0
D. melanogaster	e^+ to *e* (ebony body color)	2.0
Homo sapiens (man)	retinoblastoma	2.0
H. sapiens	Huntington's Chorea	0.5

an organism, each biochemical reaction is catalyzed by an enzyme, and each enzyme is controlled by one or more genes. If a gene mutation produces an altered protein, the ultimate result might be a **metabolic block** or inborn error of metabolism (Chapter 7). Mutations also include framing errors such as those induced by proflavin (later in this chapter) and deletions or duplications involving various lengths of DNA, as well as larger structural (Chapter 9) and numerical (Chapter 10) changes in chromosomes.

MUTATIONS AND PHENOTYPIC CHANGES

Because units of DNA ordinarily cannot be observed and compared directly, some detectable phenotypic alteration must be associated with a mutation or it will go unrecognized. Genetic analyses and comparisons among phenotypic results of mutation are greatly expedited if the mutant genes (traits) of two carrier parents can be combined in one individual. Such recombinations permit comparisons of phenotypes as well as of the influence of one expression or another, for example, dominance and recessiveness. Minor changes presumably occur in genes as a matter of course without producing any visible alterations. Some mutations are associated with slight effects on the **fertility** or **viability** of the organism. These alterations ordinarily go undetected unless critical comparisons can be made. They may, however, influence natural selection and thus represent a factor in evolution.

The normal development of any organism is influenced by numerous genes. But, because the entire organism develops as a unit, the action of individual wild-type genes often is not apparent. The existence of mutant alleles substantiates the existence of wild-type genes when the mutants' influence on developmental reactions produces a **visible** difference from the unmutated allele. Mutations thus may provide the basis for **postulating wild-type alleles.**

Mutated as well as unmutated genes tend to be self-perpetuating since the process of gene duplication precedes each cell division. Once established, some mutated genes (e.g., white eye, w in Drosophila) are as stable as the original genes which mutated, but others (e.g., a forked allele, f^{3n} in Drosophila) are more subject to further change.

Not all mutations are immediately detectable because many, perhaps the great majority, are recessive and must become homozygous before they can be expressed. Because a phenotypic effect is the only readily observable evidence of mutations, a sudden **phenotypic change** that subsequently proves to be heritable is an accepted indication that a mutation has occurred in an organism.

SOMATIC AND GERMINAL MUTATIONS

Mutations may occur in any cell and at any stage in the cell cycle. The immediate effect of the mutation and its ability to produce a phenotypic change are determined by its dominance, the type of cell in which it occurs, and when it happens relative to the life cycle of the organism.

If the mutation occurs in a somatic (body) cell that can produce cells like itself but not the whole organism, the mutant change would be perpetuated only in **somatic cells** that descended from the original cell in which the mutation occurred. The "Deli-

cious" apple and the navel orange, for example, originally were **mosaics** in somatic tissues. Changes that give these two fruits their desirable qualities apparently followed spontaneous mutation in single cells, which constituted only a very small part of the body of the individual apple and orange trees that were involved. In each case, the cell carrying the mutant gene gave rise to more of its own kind, eventually producing an entire branch on its respective tree which had the characteristics of the mutant type. Fortunately, **vegetative propagation** was feasible for both the Delicious apple and the navel orange, and today numerous progeny from grafts and buds have perpetuated the original mutation. Descendants of the mutant types are now widespread in apple orchards and orange groves.

If dominant mutant genes occur in **germ cells,** their effects may be expressed immediately in progeny. If the mutants are **recessive or hypostatic,** their effects may be obscured. Germinal mutations, like somatic mutations, may occur at any stage in the cycle of the organism but they are more common during some stages than others. If the mutation arises in a **gamete,** only a single member of the progeny is likely to have the mutant gene. If, on the other hand, a mutation occurs early in **gametogenesis** (a premeiotic event), several gametes may receive the mutant gene and thus enhance its potential for perpetuation. In any case, the dominance of the mutant gene and the stage in the cycle when the mutation occurs are major factors in determining the extent of any expression that follows.

The earliest recorded dominant germinal mutation in domestic animals was that observed by Seth Wright in 1791 on his farm by the Charles River at Dover, Massachusetts. Wright noticed a peculiar male lamb with unusually short legs in his flock of sheep. It occurred to him that it would be an advantage to have a whole flock of these

(a)   (b)

Figure 8.2 Short-legged sheep of the Ancon breed. (a) Short-legged sheep; (b) sheep with normal length legs. (Courtesy of Australian News and Information Bureau.)

short-legged sheep, Fig. 8.2, which could not get over the low stone fences in his New England neighborhood. Wright used the new short-legged ram for breeding his 15 ewes in the next season. Two of the 15 lambs produced were short-legged. Short-legged sheep were then bred together and a line was developed in which the new trait was expressed in all individuals. The mutation that gave rise to the short-legged sheep was obviously of the germinal type because the cell carrying the mutation had the capacity to reproduce the entire organism. Germinal mutations have since been described in a variety of animals and plants.

FREQUENCY OF SPONTANEOUS MUTATIONS

Each gene has its own characteristic mutational behavior. Some genes undergo mutations more frequently than others in the same organism. Those with unusually rapid mutation rates are called unstable or **mutable** but a wide range of mutation rates exists among genes that are considered stable. Spontaneous mutation rates for particular loci in a few different species are given in Table 8.1. The average mutation rate for bacteria is in the order of 1 in 10 million (10^{-7}) per cell generation. For fruit flies, the average for mutation in a particular locus is in the order of 1 in 100,000 with a range from 1 in 20,000 to 1 in 200,000 flies. Since most of the data on mutation rate in fruit flies have been obtained from experiments with males, questions have arisen concerning a possible sex difference in overall mutation rates. Bruce Wallace has shown through extensive experiments that mutation rates are not significantly different in the two sexes of *D. melanogaster*. Mutation rates for **different strains,** however, **do differ.**

Estimates for man indicate a somewhat greater frequency than those cited for stable genes in most other organisms. Samples collected thus far have been small, and the methods used were indirect and subject to large errors. Genes associated with such human traits as intestinal polyposis and muscular dystrophy have been estimated to mutate once in 10^4 to 10^5 people. A human generation is equal to about 50 to 100 cell generations. By expressing any mutation rate as a probability of mutation per cell per generation, a mutation rate is defined independently of exact physiological conditions and stage in life cycle. This definition is based on a time unit proportional to a cell's division time. When expressed in terms of **cell generations,** rates for fruit flies and man are generally **comparable** with those for bacteria.

MUTABLE GENES

Most genes are relatively stable and **mutate infrequently,** but in some organisms a few genes mutate spontaneously so often that individuals carrying them are **mosaics** of mutated and unmutated genes. The *R* gene in maize, for example, was found by R. A. Emerson to undergo alteration much more frequently than others. The change from R^r to r^r for example, was found to occur at the rate of 50 per 100,000 gametes. These "mutable" genes are either more unstable than others or they are influenced by other factors in the genetic environment. McClintock's conclusion from extensive studies on maize was that a mutable gene is not an autonomous entity, but is derived from an agent that is integrated at the site of the mutable gene to **cause instability.**

Examples of highly mutable genes have been found in both plants and animals, but they are expressed more commonly in plants. They occur frequently in somatic tissue and occasionally in germ cells. Somatic mutations may show their effects as color variegations (mosaics) in such plant parts as endosperm, leaves, and petals. Many common plants including the larkspur, snapdragon, sweet pea, four o'clock, and morning glory have color variegations suggesting unstable or mutable genes.

The classical investigations of M. De-

merec (1941) on the mutable gene, miniature-alpha, in *Drosophila virilis* provided the first substantial data on the genetic properties of mutable genes in animals. These properties are: (1) mutation occurs primarily before meiosis, (2) mutation occurs both in females and males, implying that meiotic crossing over is not involved, and (3) mutation is strongly influenced by neighboring genes. More recently, M. M. Green and others have described several mutable gene systems involving the white eye cistron in *D. melanogaster*. White-crimson (w^c), for example, mutates to wild-type and phenotypes other than white-crimson at a frequency of $1/10^3$. This mutable system and several others at the *w* locus are associated with chromosome alterations, particularly deficiencies. This has led to the hypothesis that a **controlling element from the cytoplasm** is integrated into the chromosome at the site of mutability and that this agent is responsible for chromosome aberrations.

MUTATOR AND ANTIMUTATOR GENES

Genes in maize were shown by McClintock to influence the stability of other genes. These have been called **mutator** genes. A striking example of the action of a mutator gene with a specific effect on a basic color gene in maize was described by M. M. Rhoades. The color of maize leaves and other plant parts is dependent on a complex of three complementary genes, symbolized *A, C,* and *R*. All three dominant members of the unit, that is, *ACR*, must be present when a color other than green is produced. The color may be purple if gene *P* is also present, or red (*pp*), or some other color depending on what other genes are included along with *A–C–R*. Plants with *aa, cc,* or *rr* and the other genes in the normal complement are green. Plants with the genotype *aaC–R–* would be expected to be

Figure 8.3 Kernels of corn on a cob. Light-colored kernels with purple dots carry the mutator gene *Dt*. This gene increases the mutation rate of *a* resulting in *A* which controls the purple color. The deep purple-colored kernels carry *A* as well as other genes necessary for purple color. (From *Crop Science*, 1968, *The Mutants of Maize*, by M. G. Neuffer, Loring Jones, and M. S. Zuber. Courtesy of M. G. Neuffer and the Crop Science Society of America.)

green, but in the presence of the mutator gene, *Dt*, they are variegated. Light-colored corn kernels have purple spots indicating places where somatic mutations have occurred (Fig. 8.3). The *Dt* gene produces its effect by influencing the *a* gene of the *aa* genotype to mutate to *A*. Patches of cells scattered throughout the plant carry *A* genes resulting from these mutations. On the leaves and kernels, these patches give a speckled appearance. The size of the spots depends on the stage of development at which the mutations occurred. Green cells contain unmutated genes, that is *aa*. A factor in the genetic composition thus influences the mutation rate of a specific gene.

J. F. Speyer and others have found broad-spectrum mutator genes in bacteriophage T4. Temperature sensitive (*ts*) mutants of gene 43, for example, control mutation rates of all types of point mutations (transitions, transversions, and frameshifts). Mu-

tator genes function in DNA replication and transcription by **altering polymerase activity**, producing mutagenic base analogs and modifying DNA bases, thus influencing the mutation rate of other genes. One mutator at T4 locus 43 (*ts* L88), for example, produces a DNA polymerase that utilizes **incorrect** nucleotides at a higher frequency than does the wild-type enzyme. Both transitions and transversions are promoted by the *ts* L88 mutation.

Several mutations in gene 43 exhibit powerful negative or **antimutator** activities, particularly along the pathway $A:T \rightarrow C:G$ transversions. Experiments have shown that DNA polymerases (gene 43 products), isolated from *E. coli* that were infected with mutator, antimutator and wild-type strains of T4 bacteriophage, discriminate between adenine and 2-aminopurine during DNA synthesis *in vitro*. Significantly larger amounts of 2-aminopurine are incorporated into DNA by wild-type and mutator than by antimutator **enzymes. This suggests a mechanism for reducing mutation rates in some organisms.**

Decreases as well as increases in mutation rates, as compared with wild type, could thus be accomplished by these mutants. With both mutators and antimutators operating in the same system, particular spontaneous mutation rates may be optimized through natural selection. Mutators and antimutators thus become relative terms because no standards are available for optimum rates.

R. C. von Borstell and others measured spontaneous mutation rates in yeast and isolated many mutator genes with different kinds of mutator and antimutator activity. Both transitions and transversions were included among the mutations that controlled mutation rates. From one investigation it was concluded that one mutator functioned by inducing **missense** mutations that directed the incorporation of a different amino acid into a polypeptide, thus making an enzyme nonfunctional. The mutator "phenotype" (the mutation rate of a particular gene) might be the result of a selective process. Results of studies on yeast have borne out the conclusions of those on viruses and bacteria: *that spontaneous rates are under genetic control of the cell itself*, occurring through misrepair, nonrepair, misreplication, and/or misalignment of bases in DNA (discussed Chapter 3 and later in this chapter).

PLEIOTROPY

The term **pleiotropy** refers to the situation in which a gene influences more than one single trait. Many such instances have been discovered. In fact, all genes (whether mutant or nonmutant) may be pleiotropic, with their various effects simply not yet recognized. Even though a gene may have many end effects, it has only one primary function, that of producing **one polypeptide.** This polypeptide may give rise to **different expressions** at the **phenotypic level.**

Cystic fibrosis, for example, is a hereditary, metabolic disorder in children that is controlled by a single, autosomal, recessive gene (Chapter 16). The gene apparently specifies an enzyme that produces a unique glycoprotein. This glycoprotein results in the production of mucous with abnormally high viscosity. Overly viscous mucous interferes with the normal functioning of several exocrine glands including those in the skin (sweat), lungs (mucous), liver, and pancreas. The syndrome or group of symptoms that characterizes the disease is related directly or indirectly to the abnormal mucous. Abnormally high levels of sodium chloride occur in the sweat. Mucous stagnates in tubules of the lungs, frequently becoming infected and giving rise to bronchitis. Secreting cells in the liver and the pancreas are impaired, curtailing production of fat-emulsifying agents and digestive enzymes and thus interfering with digestion and absorption of food. Several different phenotypic effects thus result from the action of a single pleiotropic gene.

ACCUMULATED MUTATIONS OVER PERIODS OF TIME

Variation at the molecular level can be evaluated by (1) studying the product of a single gene in different individuals, (2) measuring the amount and kind of variation within a species, and between closely related and more distantly related species,

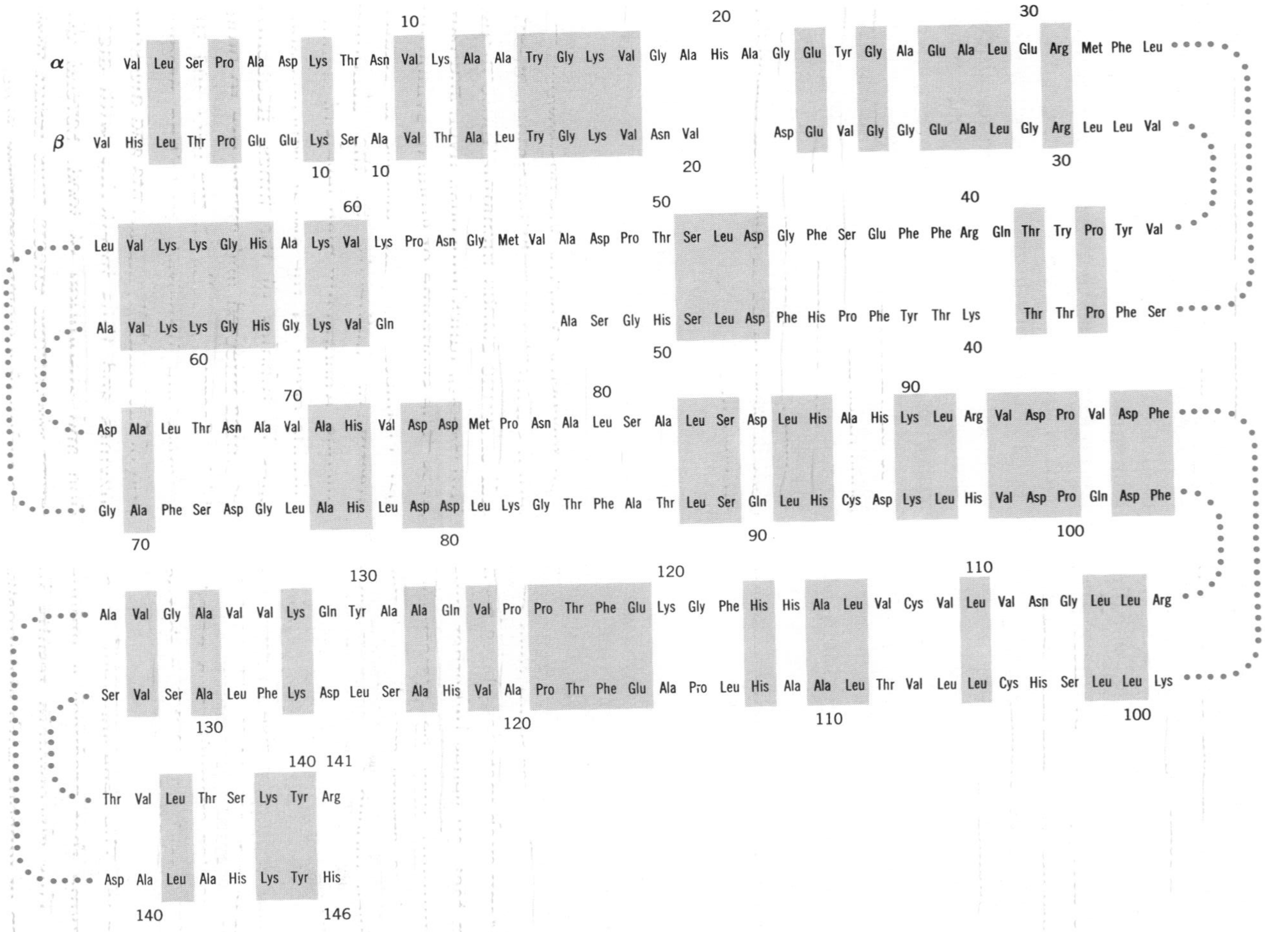

Figure 8.4 The amino acid sequences of the α and β peptide chains of hemoglobin A. The amino acids enclosed in boxes are identical and occupy corresponding positions along the peptide chains. The amino acids are numbered sequentially from the N-terminus. (After Ingram.)

and (3) by estimating gene changes over time. Variations resulting from mutation are mostly dependent upon chemical distinctions between mutant and wild-type proteins. They are detected by so-called **fingerprinting,** which is an enzymatic fragmentation of polypeptides, and identification of amino acids in the peptides by electrophoresis and chromatography. Fingerprints can be compared to determine differences in the amino acid content of different peptides, and along with other techniques, the precise sequence of amino acids can be discovered. Hemoglobin and cytochrome C have been studied extensively by these methods and will be used to illustrate fingerprinting.

THE HEMOGLOBIN MOLECULE

Hemoglobin is the oxygen-carrying substance in the red corpuscle of chordate animals. It is a conjugated protein consisting of the protein globin combined with the iron pigment heme. Several different forms of hemoglobin occur in the human population. Most of them were detected initially by their behavior in an electric field, while some were distinguished chemically by fingerprinting and by other laboratory procedures.

Each form of **hemoglobin** is controlled by a **particular gene.** The human adult hemoglobin A molecule is composed of four polypeptide chains: two identical *alpha* chains and two identical *beta* chains. Each alpha chain has 141 amino acids and each beta chain has 146, making a total of 574 amino acids (Fig. 8.4). The 574 include, however, only 19 different amino acids. Numbers 20 (his) and 21 (ala) on the alpha chain have no counterparts on the beta chain, and numbers 1 (val), 39 (gln), 54 (val), 55 (met), 56 (gly), 57 (asn), and 58 (pro) on the beta chain have no counterparts on the alpha chain. Among the 137 paired amino

acids (146 − 9 = 137), 61 are alike and 76 are different. Comparisons of the variation in the **nucleotide** (gene) **sequence** with that in the corresponding **amino acid** (protein) **sequence** (Fig. 8.4) shows a closer relation among the genes than among the proteins. This is to be expected because the genes are coded in a **four-letter** language (ACGT) whereas the proteins are in a **20-letter** language (20 different amino acids) which magnifies the potential difference.

Hemoglobin F, with two alpha and two gamma chains, is in the blood of the developing human fetus but is normally replaced with hemoglobin A (alpha and beta chains) during the first six months after birth. Seventy-one percent of the amino acids on the gamma chain are identical with those in corresponding positions on the beta chain. A correspondence of 39 percent exists between alpha and gamma chains, whereas it is 42 percent between alpha and beta chains. **The lack of correspondence** in both cases is attributed to **spontaneous mutations** that have occurred over long periods of time.

ABNORMAL MUTANT HEMOGLOBINS

When the beta chain of normal human hemoglobin was degraded by trypsin (Fig. 8.5), the amino acids in part four were found to be arranged in the following order: val, his, leu, thr, pro, **glu,** glu, lys, (Fig. 8.6). Hemoglobin S of sickle-cell patients was found to have all amino acids in the same order except for number six, **glutamic acid,** which was replaced by **valine.** All of the other 145 residues of the beta chain were normal in S hemoglobin. When gene *Si* is heterozygous, the normal hemoglobin A predominates but a small quantity of abnormal hemoglobin S is also present. The red cells may become sickle-shaped when oxygen tension is low even while the pa-

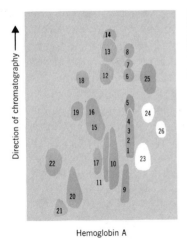

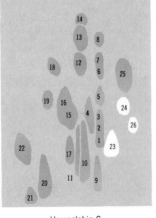

Hemoglobin A   Hemoglobin S

Direction of chromatography →

Figure 8.5 Peptide fingerprints of hemoglobins A and S. The hemoglobins were first digested with trypsin; then the peptides were subjected to electrophoresis, followed by paper chromatography at right angles. White spaces indicate peptides that are visible only after heating. Only peptide No. 4 was found to differ in the two hemoglobins. (After Ingram.)

tient is generally in good health (Fig. 8.7). In *Si/Si*, hemoglobin A is replaced completely by hemoglobin S, and the patient has the severe form of the disease characterized by **hemolytic anemia.**

Hemoglobin S in the cells precipitates when it is deoxygenated and produces **crystalloid aggregates** that distort blood corpuscles. Fragments of hemoglobin S can also be distinguished from fragments of hemoglobin A by electrophoretic mobility. When placed in an electric field under appropriate conditions, hemoglobin fragments with glutamic acid in position six have a negative charge and migrate toward the positive pole. By contrast, hemoglobin S fragments with valine replacing glutamic acid

have no net charge and do not migrate (Fig. 8.8). Fragments of another type of hemoglobin, hemoglobin C, with lysine in the number six position on the beta chain have a **positive charge** and migrate to the negative pole. Fragments of different forms of hemoglobin with single amino acid substitutions can thus be identified by their difference in **electrophoretic mobility** as well as by identifying a particular amino acid in position six.

The known deleterious effects of abnormal (S) hemoglobin on the erythrocytes (and the patient) seem to be out of proportion to the causative single amino acid substitution in a total of 574. The amino acid sequence, however, influences hemoglo-

Kind of Hemoglobin	Amino Acids Numbered in Order							
	1	2	3	4	5	6	7	8
+	val	his	leu	thr	pro	glu	glu	lys
S	val	his	leu	thr	pro	val	glu	lys
C	val	his	leu	thr	pro	lys	glu	lys

Figure 8.6 Results of fingerprinting one part (peptide 4) of the beta chain for three kinds of hemoglobin (+, S, and C). The only alteration is in position 6 where glutamic acid of hemoglobin-+ is replaced by valine in hemoglobin-S and lysine in hemoglobin-C.

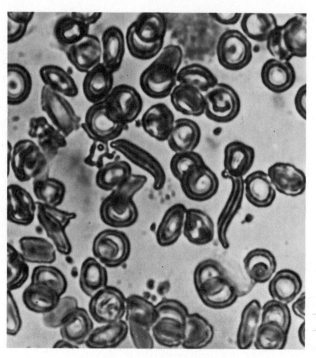

Figure 8.7 Photomicrograph of normal and sickled erythrocytes. Erythrocytes with abnormal hemoglobin become elongate, filamentous and some become shaped like a sickle. Such red corpuscles have a short life span. They clump together, interfere with capillary circulation and are rapidly destroyed. (Walter Dawn)

bin's formation, its tertiary and quaternary structure, and subsequent effects. One hypothesis for the cause and effect mechanism states that the genetic substitution of valine for glutamic acid at the sixth position in the two beta chains allows a bond to form within the molecule that changes its conformation so that molecular **stacking** results. Electron micrographs show hemoglobin S molecular threads to be hollow cablelike structures, which supports the stacking theory.

Alleles associated with normal hemoglobin A (Si^+), sickle-cell S (Si), and hemoglobin C (Si^c) are apparently members of a large series of alleles. Evidence now indicates that several other kinds of hemoglobin in the same series have originated through **gene-controlled substitutions** of amino acids on both the alpha and beta chains.

SUBSTITUTIONS IN BASE TRIPLETS

In abnormal hemoglobin, a mutation has substituted one amino acid for another. A model (Fig. 8.9) may be presented for such changes. In the DNA code, glutamic acid is specified by the triplets CTT and CTC. These DNA triplets code for GAA and GAG respectively in the mRNA transcript. A **single base substitution** of an A for a T in the second position of the DNA triplet (CTC) for glutamic acid could result in a triplet CAC message, which codes for valine and produces the abnormal hemoglobin S.

The mutational change from glutamic acid to lysine could be explained by other single nucleotide substitutions. If the cytosine in the first position of a DNA triplet (CTT) for glutamic acid were changed to a thymine (TTT), the resulting mRNA codon

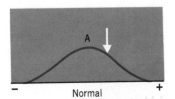

- Charge   No charge   + Charge
Glutamic acid   Valine   Lysine

(a)

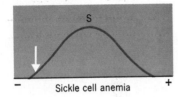

A

−   Normal   +

S

−   Sickle cell anemia   +

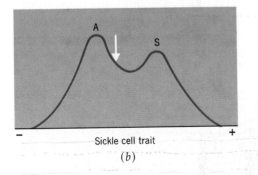

A

S

−   Sickle cell trait   +

(b)

Figure 8.8 Electrophoretic pattern for different hemoglobin molecules. (a) Structural formulas of carbon chains of three amino acids, glutamic acid, valine, and lysine. These amino acids are replaced in the hemoglobin molecule by the substitution of three alleles: Si^+, Si, and Si^c, respectively. The three amino acids behave differently in an electric field. (b) Curves representing electrophoretic pattern for homozygous (Si^+/Si^+) normal individual (A); a person (Si/Si) with sickle cell anemia—(S) and one (Si^+/Si) with the sickle cell trait (AS).

(AAA) would produce lysine. Likewise, if the other DNA triplet (CTC) for glutamic acid should undergo a substitution in the first position, from C to T (TTC) the resulting mRNA codon (AAG) would also code for lysine, and abnormal hemoglobin C (Si^c) would be produced. The substitution of thymine, a pyrimidine, for cytosine, an-

other pyrimidine, would be a transition type of a mutation.

A single base change in the DNA sequence specifying a substitution of one amino acid for another at a specific point in a particular protein molecule could account for the disease **sickle-cell anemia.** An actual case of gene-controlled altered hemoglobin

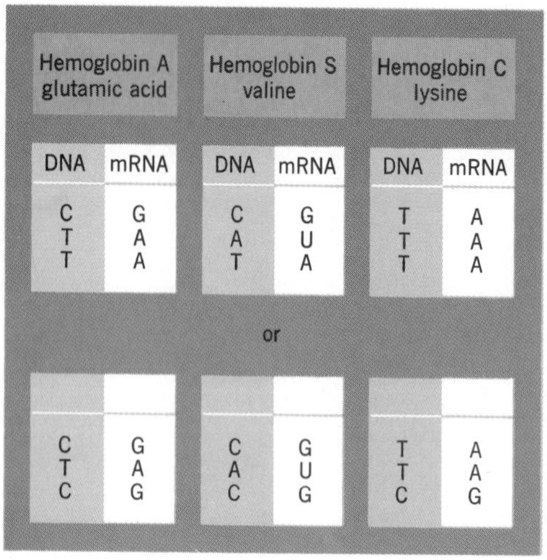

Hemoglobin A — glutamic acid | Hemoglobin S — valine | Hemoglobin C — lysine

DNA	mRNA	DNA	mRNA	DNA	mRNA
C T T	G A A	C A T	G U A	T T T	A A A

or

DNA	mRNA	DNA	mRNA	DNA	mRNA
C T C	G A G	C A C	G U G	T T C	A A G

Figure 8.9 Diagram illustrating possible base substitutions to change hemoglobin A to hemoglobin S or C.

can thus be explained by a single DNA base substitution. Linus Pauling has described sickle-cell anemia appropriately as a "molecular disease."

Nucleotide (base) alterations of this kind in proteins undoubtedly account ultimately for much of the natural variation that occurs in microorganisms, plants, animals, and man. The phenotypic expressions of these alterations along with the recombinations provide the basis on which natural selection operates in the process of evolution (Chapter 13).

Base substitutions have been accumulated in proteins over long periods of time and thus facilitate efforts to survey the **past history of proteins** as well as that of the organisms of which they are a part. Toward this end, the proteins hemoglobin and cytochrome have been investigated extensively in organisms that diverged at different levels of the phylogenetic scale.

CHANGE OF HEMOGLOBINS IN TIME

The hemoglobin molecule has undergone mutational changes over time. **Myoglobin,** a protein found in muscle tissue, for example, is presumed to be the primitive protein that gave rise to the globin part of all hemoglobins. A section of DNA made up of about 468 base pairs was apparently the ancestral information-carrying unit from which all the globin components of the various hemoglobins were derived. The following striking similarities among hemoglobins support the view of a common origin. (1) All hemoglobins that have been examined have two histidine residues that are 29 residues apart. These bind the heme group to the polypeptide chain. (2) All the DNA units specifying globin have about 9 loci in common and in the same position. (3) Closely related species (based on other criteria) have very similar hemoglobins while those that are far apart on the evolutionary scale have greater numbers of amino acid differences. (4) The separate chains have different lineages. Greater differences exist between the alpha and beta chains of man than between the alpha

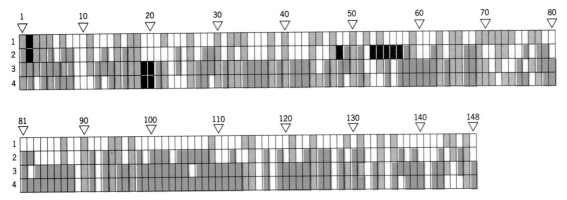

Figure 8.10 Amino acid replacements in human globin series: (1) myoglobin. (2) alpha hemoglobin. (3) beta hemoglobin, (4) gamma hemoglobin chains. Light gray areas indicate similarities; white areas indicate differences; red areas indicate identity of alpha, beta and gamma hemoglobin sites with each other; black areas indicate gaps. (From Thomas H. Jukes: *Molecules and Evolution,* New York: Columbia University Press, 1966, reprinted by permission of the publisher.)

chains of man and those of the horse. (5) Examples of single base changes, which control single amino acid substitutions in either the alpha or beta chain of contemporary human hemoglobins, probably illustrate the mechanism through which similar substitutions occurred in the past. (6) The absence of the terminal six amino acids of primitive myoglobin (seen in the lamprey) as compared with modern myoglobin, and the alpha, beta, and gamma chains of hemoglobin, can be explained by single changes that produced code-terminating triplets.

Human myoglobin and human alpha, beta, and gamma hemoglobin (Fig. 8.10) have 21 widely scattered homologous sites. Many **substitutions** have occurred in **different species,** but the number of changes generally coincides with the phylogenetic position of the particular species. This indicates that the hemoglobins and myoglobins have evolved from a single precursor molecule. The four hemoglobin chains connected with the heme prosthetic group proved such an efficient oxygen-carrying vehicle that they were maintained during evolution.

CHANGE IN THE CYTOCHROME C MOLECULE

Cytochrome is widely distributed in animal and plant tissues, where it is involved in oxidative-phosphorylation. Cytochromes A, B, and C are distinguished by their absorption spectra. Cytochrome C is a protein electron carrier that has been identified in many different animals. Some 35 different

amino acid sequences of mammalian type cytochrome C have been compared and are remarkably similar. **Similarities and differences** in such animals as lamprey, tuna, chicken, rabbit, cow, horse, rhesus monkey, chimpanzee, and man have been detected. Cytochrome Cs from man, horse, chicken,

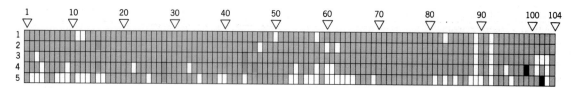

Figure 8.11 Amino acid replacements in possible coding sequences for cytochrome C: (1) human, (2) horse, (3) chicken, (4) tuna, and (5) yeast. Gray areas indicate similarities: white areas indicate amino acid differences: red areas indicate differences from (1) but identity of other chains with each other: and black areas indicate gaps. (From Thomas H. Jukes: *Molecules and Evolution,* New York: Columbia University Press, 1966, reprinted by permission of the publisher.)

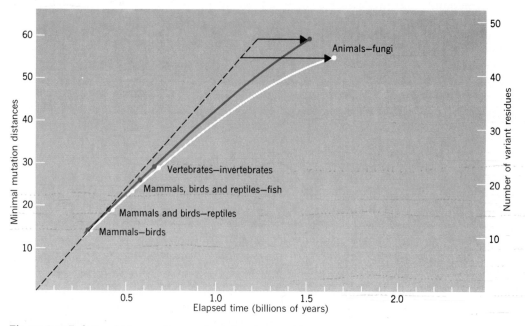

Figure 8.12 Relation between the number of variant residues among cytochromes C of different classes and phyla of organisms and the time elapsed since the divergence of the corresponding lines of evolutionary descent (dashed line and curve indicated by white circles, ordinate to the right); similar relation of the number of variant nucleotide positions in the corresponding structural genes (dashed line and curve indicated by red circles, •; ordinate to the left). The straight dashed line applies to both residue and nucleotide variations and is calculated on the basis of a value of 280 million years for the time elapsed since the divergence of the avian and mammalian lines of descent. The solid curves show the relationship corrected for probable multiple changes in single residue (white circle) or nucleotide (red circle) positions, assuming a total of 76 variable residues. Reproduced with permission from "Comparative Aspects of Primary Structures of Proteins," Annual Review of Biochemistry, Volume 37, page 740. Copyright ©1968 by Annual Reviews, Inc. All rights reserved.

tuna, and yeast are compared in Fig. 8.11.

Most cytochromes have 104 amino acids in their polypeptide chains. Some amino acids are in the same positions in all species studied. Glycine, for example, occurs at eight sites (1, 6, 29, 34, 41, 45, 77, and 84), lysine at six sites, and from one to three representatives of all other amino acids (represented by residues 70 to 80) have been found in all samples. These series of amino acids in present-day animals must have descended from ancestral proteins. The probability of their occurring by chance is extremely remote. Differences as well as similarities, however, are noted. In the horse, cow, and rabbit, for example, alanine is in position 15, but man has serine in this position. Such changes have arisen through mutation.

The finding of certain residues in the same place in all samples suggests that they must have important and specific functions.

For example, arginine is located in positions 39 and 91 in all samples. The amino acid sequences in cytochrome C are identical in man and the chimpanzee; man and the rhesus monkey differ in one residue. There are 12 to 13 amino acid differences between tuna and man. Cytochrome C from man differs from that of yeast in 46 to 48 residues. Assuming that each difference is the result of a single mutational change somewhere in the ancestral history, time estimates are possible for divergences. A **phylogenetic tree** (Fig. 8.12) based on the cytochrome C gene alone resembles those based on studies from other areas of biology. Spontaneous mutations arising from single base substitutions have apparently brought about profound changes in hemoglobin, cytochrome, and other proteins over long periods of time. Some other studies do not yield nice phylogenetic trees.

INDUCTION AND DETECTION OF MUTATIONS

The most serious limitations to early studies of mutations were (1) the low rate at which spontaneous mutations occur and (2) the high proportion of lethals among mutations of genes that control vital functions. The first difficulty was resolved by **inducing** mutations with ionizing irradiation and other mutagenic agents. The second problem has been met for certain experimental materials by ingenious methods devised for detecting and **preserving** mutations that would ordinarily be lethal.

INDUCED MUTATIONS

In 1927, H. J. Muller (Fig. 8.13) demonstrated that the mutation rate of *D. melanogaster* could be markedly increased by X-ray treatment. Expressions of induced mutations seemed to be the same as those

of comparable mutations that occurred spontaneously, but the frequency was increased as much as 150-fold. For the first time in history, a particular agent from the external environment had been shown to increase the frequency of mutations. It was suggested that the effect of irradiation was destructive and that the induced type of mutation might result from the loss of a minute part of the chromosome. Many X-ray induced mutations are **deletions** in DNA and, therefore, are not expected to undergo **reversion**. The evidence that a mutation is not a deletion in Drosophila is based on reverse mutations.

The gene (w^+) for red eye pigment, for example, mutated in such a way that white or an intermediate color between red and white was produced. Later the mutant gene was reported to have reversed itself in such

Figure 8.13 H. J. Muller, distinguished research investigator and teacher in genetics. Professor Muller has made many contributions to basic mechanisms in genetics through his numerous investigations. He won the Nobel prize in 1949 for experimental demonstration of induced mutations in Drosophila. (Wide World Photos.)

a way that red or an intermediate color was obtained in the offspring. For a mutation to be reversible, the chemical structure of the mutant gene must remain intact with no essential part permanently altered or destroyed. Results of appropriate crosses demonstrated that the mutant allele had changed back to the wild-type allele (w^+).

When the f (forked) locus in Drosophila was investigated, a few reverse mutations were detected. No significant difference was observed in the frequency of reversals on allele f^1 when spontaneous mutations were compared with those induced by X-ray treatment. One study comparing the frequency of reversals in spontaneous and X-ray induced mutations showed a reverse mutation frequency of one in 65,000 flies for spontaneous mutations and one in 68,000 for X-ray induced mutations. The difference was not significant. On the other hand, when allele f^{3n} was studied, the reverse mutation frequency for spontaneous mutations was one in 12,000 compared

with that of one in 3000 for X-ray induced mutations. Drosophila alleles at some loci have been classified as high, medium, and low in their tendency to undergo reversals.

In the pink bread mold (Neurospora), **reversible processes** are associated with certain "nutritional mutations" (discussed later in this chapter) from one allele to another. One mutant, for example, cannot grow on a medium that lacks adenine. However, when numerous samples of adenine-requiring organisms were introduced on an adenine-free medium, a few organisms grew. These had apparently changed back to adenine-independent forms. The problem then was to determine whether the mutant gene had actually mutated back to wild-type or whether some other change had circumvented the requirement. Critical analyses have shown that some apparent reversals involve entirely different loci. Another locus can suppress the mutant gene or otherwise take care of the change created by the mutation. Although the phenotypic result is the same as a reverse mutation the change involves another locus that may be some distance from the site of the primary mutation in the genome. A **suppressor** is thus a secondary mutation that restores a function lost due to a primary mutation which is located at a mutational site different from that of the primary mutation.

Effective methods for detecting mutations became more necessary when more mutations were induced. Muller led the way in devising methods for detecting mutations in Drosophila. He made use of the high proportion of lethal mutations in his procedure for detection. After ten years of experimentation, Muller developed the now classical *ClB* method for detecting **sex-linked lethal** mutations in Drosophila (Fig. 8.14).

The *ClB* stock carries a chromosome inversion ("C") (Chapter 9) in heterozygous condition, that is, a chromosome rearrangement that acts as a crossover suppressor; a recessive lethal (*l*); and a dominant gene (*B*) which is a marker expressing bar eye. The lethal is associated with the inversion. The test is made by crossing *ClB* females with males suspected to carry lethal mutations in the X chromosome. Half of the eggs produced by *ClB* females carry the inversion and half receive the X chromosome without the inversion. Half of the sperm receive the Y and half carry the X chromosome suspected to carry the lethal mutation. At fertilization, about half of the zygotes receive a Y chromosome while about half receive the X in question. All male-producing zygotes from *ClB* females that receive the *ClB* chromosome express the lethal *l* gene and die, leaving only half as many male as female flies. The F_1 females expressing the bar eye and carrying the X chromosome being tested are then mated to wild-type males. Half of the males from this cross receive the *ClB* chromosome and die. If the other half of the males also die a lethal mu-

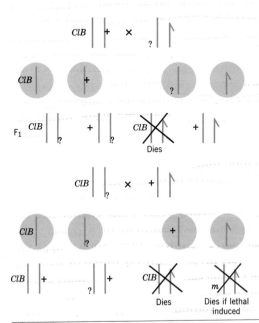

Figure 8.14 *ClB* method for detecting sex-linked lethal mutations in Drosophila. A *ClB*/+ female is crossed with a male that may have a lethal mutation in the X chromosome (?). *ClB* F_1 females are then pair mated with wild-type males. If an X chromosome in question (?) carries a lethal mutation (*m*) the males from the cross do not survive and only females appear.

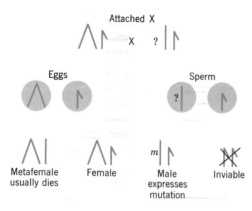

Attached X

Eggs

Metafemale
usually dies

Female

Sperm

Male
expresses
mutation

Inviable

Figure 8.15 Attached-X method for detecting sex-linked visible mutations. The cross is between an attached-X female carrying a Y chromosome and a male that may carry a mutation in the X chromosome (?). The F₁ males that survive carry the X chromosome in question (?). If a mutation (*m*) with a visible phenotype is present it should be expressed in all surviving F₁ males.

tation has occurred and the ? on the suspected X chromosome (Fig. 8.14) may be changed to *m* for mutant gene.

In addition to the detection of sex-linked lethal mutations, these studies have resulted in methods for maintenance and even preservation of "lethal" mutations in the laboratory. Since most lethals are recessive, they can be maintained in heterozygous condition. Balanced lethal systems (Chapter 9) have been developed to maintain recessive lethal genes in an enforced heterozygous arrangment.

Attached-X stocks of Drosophila, in which compulsory nondisjunction occurs, have proved valuable for detecting sex-linked **visible** mutations. Advantages of the attached-X method can be observed from the accompanying diagram (Fig. 8.15). Females with attached-X chromosomes are crossed with males in which mutations are suspected to have occurred. First-generation males express the phenotypic change produced by a recessive sex-linked mutant gene which expresses a visible phenotype.

DETECTION AND PRESERVATION OF NUTRITIONAL MUTANTS

Wild-type Neurospora can synthesize all of its nutritional requirements from a **minimal medium** with only certain in-

organic materials, sugar, and the vitamin biotin. G. W. Beadle and E. L. Tatum devised a method for detecting nutritional mutants on **selective media** (Fig. 8.16). The technique was based on the premise that an orderly sequence of steps is followed by the organism in the synthesis of a particular nutrient. It was postulated that a mutation creating a block in a biosynthetic pathway would result in a new growth factor requirement by a mutant strain. If, for example, a mutant strain cannot synthesize pantothenic acid, it will be **lethal** in a minimal medium lacking pantothenic acid. When, however, pantothenic acid is added as a supplement to the minimal medium the mutant can not only be **detected**, but can be **maintained.**

Beadle and Tatum irradiated wild-type Neurospora. Isolates from the irradiated cultures were grown on **complete** media containing all the substances that might be essential metabolites for Neurospora with various nutritional defects. From these cultures, asexual spores (conidia) were transferred to minimal medium. Many isolates from irradiated wild-type Neurospora cultures were unable to grow on minimal medium. It was suspected that a **nutritional mutation** had occurred in the cell from which each mutant strain was derived. Mutants were crossed individually with wild-

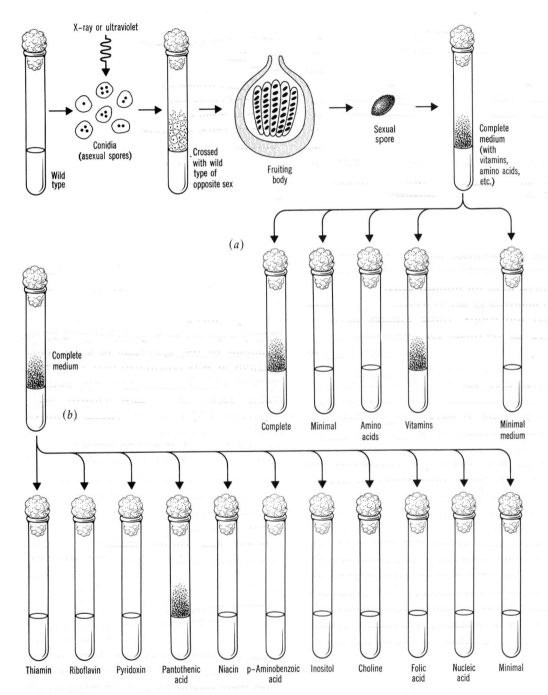

X-ray or ultraviolet

Wild type

Conidia (asexual spores)

Crossed with wild type of opposite sex

Fruiting body

Sexual spore

Complete medium (with vitamins, amino acids, etc.)

(a)

Complete medium

(b)

Complete   Minimal   Amino acids   Vitamins   Minimal medium

Thiamin   Riboflavin   Pyridoxin   Pantothenic acid   Niacin   p-Aminobenzoic acid   Inositol   Choline   Folic acid   Nucleic acid   Minimal

Figure 8.16 Method of detection of biochemical mutants in Neurospora. The mutant in this case fails to grow on minimal medium or on minimal medium enriched by a mixture of amino acids, but it does grow when vitamins are added (a). The screening test (b) with all vitamins added individually to minimal medium indicated that the mutant lacked the ability to synthesize pantothenic acid. (From Beadle, *Science in Progress,* by permission of Yale University Press).

type. Evidence of gene mutation was obtained from the results of these crosses. In some cases a mutant allele was identified and assigned to a particular **gene locus.**

Transfers were then made to minimal media containing (1) all vitamins known to be utilized by Neurospora, (2) all amino acids and (3) other compounds utilized by Neurospora. If growth occurred in the vitamin medium, for example, but not in the other two test media, the mutant was presumed to be deficient in the synthesis of some vitamin. Further screening tests were then provided on minimal media to which single vitamins were added. By this procedure, it was possible to identify particular nutritional mutants.

These studies on Neurospora resulted not only in the detection and preservation of nutritional mutants but they also demonstrated that genes control fundamental reactions through which nutrients are utilized by the organism. The investigators moved on to the next question: how do genes

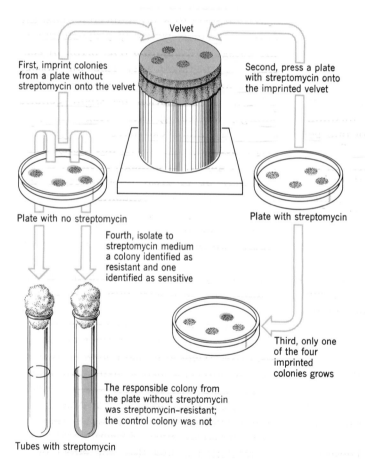

First, imprint colonies from a plate without streptomycin onto the velvet

Velvet

Second, press a plate with streptomycin onto the imprinted velvet

Plate with no streptomycin

Plate with streptomycin

Fourth, isolate to streptomycin medium a colony identified as resistant and one identified as sensitive

Third, only one of the four imprinted colonies grows

The responsible colony from the plate without streptomycin was streptomycin-resistant; the control colony was not

Tubes with streptomycin

Figure 8.17 The replicate-plating technique which demonstrates the preadaptive nature of mutation. For diagrammatic reasons, too few colonies are indicated on the plates. (From Sager and Ryan, *Cell Heredity,* John Wiley and Sons, Inc., 1961.)

control biochemical reactions? A particular mutant gene could block a certain enzymatic reaction in a biochemical pathway. This led to the conclusion that single genes act through specific enzymes. Furthermore, gene-controlled enzymes are specific for single biochemical reactions in the metabolism of the organism. The **one gene-one enzyme hypothesis** emerged from these investigations. This hypothesis has since been refined to: **one gene-one polypeptide**, which forms the basis for biochemical genetics.

DETECTION OF MUTANTS IN BACTERIA

Ingenious methods have been devised to facilitate the detection of mutations that make bacteria resistant to certain chemicals that can be added to the medium. Mutants occurring in the order of one per ten million cells were difficult to find by older methods. J. and E. M. Lederberg developed a method for direct transfer of all colonies on a plate to one or more plates, which is now called the **replicate plate method** (Fig. 8.17).

The same methods that greatly facilitated the detection of mutations in bacteria simultaneously provided tools for demonstrating that mutations are heritable and not acquired characteristics. In one of the Lederberg experiments, for example, bacteria initially were grown on the usual standard solid medium. When the organisms had established themselves and covered the plate, direct transfers were made to several plates to which streptomycin had been added. Only a very few of the vast numbers of organisms on the original culture **survived** on the selective medium. These few multiplied and **produced colonies**. The same pattern was found in all of the replicate plates, showing that the organisms in one colony were **resistant** to the strep-

tomycin. When the colony that had given rise to the resistant colonies on the replicated plates was identified on the original plate, which had not been treated with streptomycin, the members of this colony were found to be resistant. The mutation for resistance had occurred before the organisms were introduced to the poison. Mutants resistant to streptomycin were merely identified by the medium on the plates to which the same culture was transferred.

MUTATION RATES

With respect to X-ray induced lethals, the mutation rate is generally **proportional** to the **dosage of irradiation** (Fig. 8.18). In general, the proportionality rule for dosage holds for **quantity** of damage. Small amounts of irradiation do slight genetic damage, whereas greater dosages do proportionally more extensive damage. The proportionality rule does not hold, however, for **quality** of damage. A single mutation may be of great importance to the individual and his descendants.

When Drosophila spermatozoa were used as experimental material in irradiation studies, **no intensity effect** relative to genetic damage could be noted. Regardless of whether a given amount of irradiation was administered in a large dose or in several small doses, the end results were the same. However, W. L. Russell showed that the effects in mice are related to the stage in the life cycle. When spermatogonial cells were treated, intensity had a significant effect, suggesting a different mechanism in these cells as compared with that in sperm. Recent studies on mice have shown that the total mutagenic effect of fractionated doses of irradiation is less than the effect of an equally large single dose. Low-level, long continuing (chronic) irradiation on bacteria, for example, produced less damage than do large single doses of the same total inten-

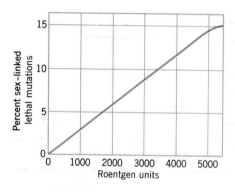

Figure 8.18 Relation between dosage of irradiation and frequency of sex-linked lethal mutations. The frequency of mutations was shown to be proportional to the number of roentgen units (units of quantity of X-rays) to which the living tissues were subjected.

sity. A **repair system** is apparently involved in the differential response between chronic and acute radiation.

As pointed out earlier, there are different kinds of mutations and undoubtedly many different kinds of gene changes. An early attempt to explain mutations involved the "hit" theory, which postulated that a specific gene was hit and changed by a mutagenic agent. The change could be destructive and therefore irreversible, or it could be the substitution of a base. A single alpha particle (a helium nucleus emitted by radioactive elements) is capable of creating a mutation. In the wasp, Habrobracon, it has been shown that one hit in the nucleus by an alpha particle will inactivate a gene. Although the hit theory is attractively simple, it does not explain all of the facts.

The **single-hit** theory implies that **one ionization** produces one **mutation,** that is, a linear relationship exists between mutations produced and total dosage. It does not imply anything about the relative efficiency with which an ionization produces a mutation and does not imply that the ionizations have no other detrimental effect on the cell or the organism. The **receptivity** of a cell at

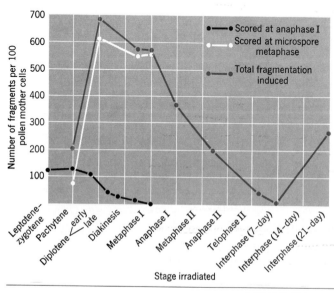

Figure 8.19 Relation between the number of chromosomal fragments induced in meiotic cells of *Trillium erectum* at a specific dosage of radiation (50r) and the stage at which the irradiation occurred. (From A. H. Sparrow, *New York Academy of Sciences Annals*)

different stages of its metabolic cycle is an important factor in determining rates for induced mutations. A. H. Sparrow has shown marked variations in numbers of chromosome fragments, assumed to be directly related to mutational changes at different stages in meiosis and cleavage (Fig. 8.19) in the plant Trillium. Chromosome aberrations were induced about 60 times more frequently at metaphase than at interphase. Nondividing cells in the Trillium showed little radiation damage whereas rapidly dividing cells were very sensitive.

Oxygen tension and **temperature change**, when associated with irradiation, also may significantly **alter** the frequency of mutations. Low oxygen tension decreases mutations. Oxygen can magnify the effect of radiation, but only if it is present during the irradiation. Oxygen has less effect with intense than with moderate conditions of ionization. Environmental agents that protect germ cells from radiation damage often do so by lowering the oxygen concentration of tissue, and those that enhance the effectiveness of radiation add oxygen.

MUTATIONS INDUCED BY ULTRAVIOLET RAYS

Ultraviolet (UV) rays are a significant but less potent factor than X-rays in inducing mutations. Because of long wavelength and therefore less energy, UV rays do not **penetrate** as well as X-rays, but they are readily **absorbed** in some compounds, particularly those containing **purines** and **pyrimidines.** The pyrimidines, thymine and cytosine, are especially receptive and are changed by absorption of UV rays (Fig. 8.20). These changes in bases may cause the primary mutagenic effect of UV not by altering the DNA itself but by forming thymine dimers which distort the DNA helix and block future replication. The relation between dosage of UV rays and mutation rate is generally not linear like that of X-rays. Evidence from many studies on different ma-

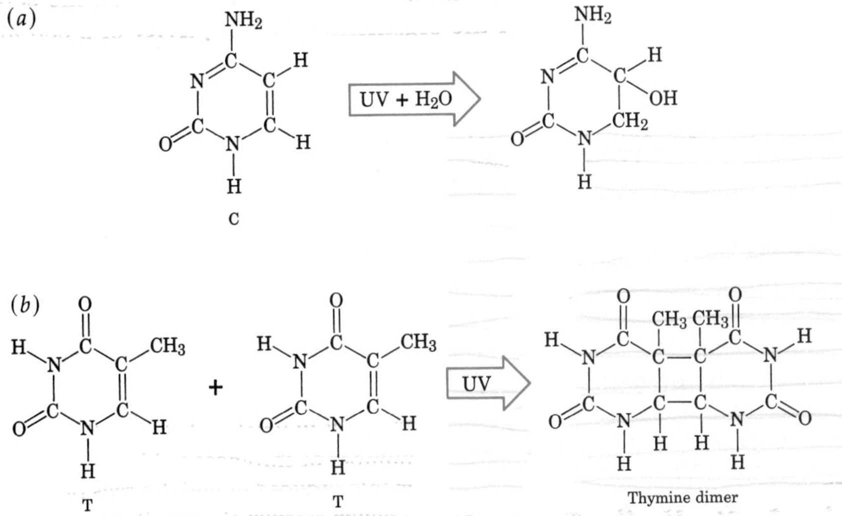

Figure 8.20 Direct effects of UV on pyrimidines; (a) hydrolysis of cytosine to a product which may cause mispairing and thus result in mutational change. (b) joining of thymine molecules in the formation of a thymine dimer which blocks replication and thus creates a mutation.

terials indicates that the mutational effect of UV rays is indirect and that it involves precursors and enzymes producing DNA rather than DNA itself.

UV ray induced alterations of DNA are most evident when replication is in progress, indicating that errors occur during the time of base incorporation. At this time, extraneous bonds between neighboring pyrimidines disrupt the regular duplication process. In one series of UV ray induced mutations in bacteriophage, most changes were found to be $C \rightarrow T$ transitions. The pyrimidine-altered mutants were then further treated with ultraviolet and other mutagenic agents, and induced to revert to the original wild type. The mechanism for reversion was shown to be a change in the $A:T$ base pair at the mutant site. Base T was altered to hydroxymethylcytosine, which paired with G and replicated as C. A reversible mutational change from a $C:G$ base pair to a $T:A$ pair and back to $C:G$ had thus been demonstrated.

REPAIR MECHANISMS FOR UV DAMAGE

Damage to the DNA caused by ultraviolet irradiation usually consists of a **covalent bond** forming between adjacent pyrimidines, particularly adjacent **thymine residues.** Covalently bonded dimers behave like jammed zippers and block replication. This defect can be repaired enzymatically by a

post replication recombinational process. The enzyme **endonuclease nicks** one strand carrying the defect and excludes the defective thymine-thymine dinucleotides along with the nucleotides on either side (Fig. 8.21). Newly synthesized nucleotides, complementary to the **undamaged strand,** are then inserted to fill the gap.

Polynucleotide ligase seals the ends of the excision strand and completes the chromosome repair. Defects of this kind apparently occur quite frequently but can be repaired if only one strand of the DNA duplex is damaged. If the repairs are made immediately, mutations do not occur. Sometimes, however, errors of the transition or transversion type occur in the repair process and result in permanent damage or mutation.

UV radiation has been associated with the incidence of skin cancer in man. It is a normal component of sunlight in the atmosphere. Purines and pyrimidines strongly absorb UV light which may produce excited states and increase the level of spontaneous mutations. Eukaryotic cells possess a photoreactivating system that repairs thymine dimers induced by UV as well as an excision process (Fig. 8.21). An endonuclease recognizes and acts on thymine dimers and is involved in DNA repair. People with an inherited malignant disease, xeroderma pigmentosum, for example, have greatly reduced endonuclease activity. These people are apparently unable to excise UV-induced

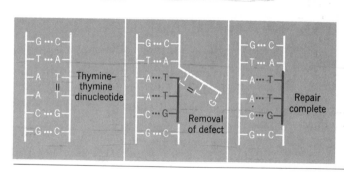

Figure 8.21 Repair mechanism for UV damage to one strand of a DNA duplex. The damaged segment of one strand is peeled off and new nucleotides complementary to those of the undamaged strand are inserted in the gap.

thymine dimers from the chromosomes of their skin cells. As these cells are exposed to UV from sunlight, their DNA becomes damaged and the unrepaired damage leads to cancerous conditions.

Many people, particularly those with a light complexion and unprotected skin, receive more UV damage in their skin DNA than can be repaired and suffer from skin cancer.

CHEMICAL MUTAGENS

Certain **chemical agents** such as mustard gas and other members of the chemical family of nitrogen sulfur mustards, ethyl methane sulphonate (EMS), ethyl urethane, phenol, methylcholanthrene, and dibenzanthracene, can **increase mutation rate.** Formaldehyde, which is a common preservative in zoology laboratories, has a slight but significant mutagenic effect. When chemical mutagens were first discovered by Charlotte Auerbach and her associates during World War II, their data were classified. Since then, numerous chemicals have been studied experimentally and a long list of mutagenic agents has been compiled. A few of these, such as nitrous acid, base analogs (purine or pyrimidine bases that differ slightly in structure from the normal base), and acridines (organic molecules that bind to DNA), have specific effects. Most chemical mutagens identified thus far, however, have general effects comparable to those of irradiating agents. They apparently induce a general **instability** of some kind that results in chemical changes within the DNA. Chemical mutagenic agents can (1) **penetrate** cells and (2) **alter** the chemical **structure** of the DNA within the cells.

The impressive list of physical and chemical mutagens include few commonly encountered environmental agents. Whether environmental agents of some kind are responsible for spontaneous germinal mutations is still unknown. Spontaneous somatic mutations, however, are often associated with factors in the environment. Examples are lung cancer and cigarette smoke; skin cancer and tar products; and testicular cancer, which has been reported to be more prevalent among chimney sweeps than in the general population.

INDUCED TRANSITIONS

Some **environmental** agents tend to create unstable conditions in particular bases and thus induce **specific** kinds of alterations. Nitrous acid (HNO_2), acts as a mutagen of extracellular bacteriophages by altering adenine and cytosine. By oxidative deamination, for example, nitrous acid may cause the removal of an amino (NH_2) group in adenine, forming hypoxanthine as illustrated in Fig. 8.22. Such replacement of a **single base** constitutes a mutation. Following this alteration, a new arrangement of bases would occur in the replication process.

Adenine ordinarily pairs with thymine in DNA, but when changed to hypoxanthine it has properties like guanine and specifies cytosine. Cytosine in turn specifies guanine. These transitions from A to G and T to C equate with a base pair change of **A:T to G:C.** A substitution of this kind results in a changed codon on each DNA strand, and may result in a **mutant** protein.

BASE ANALOGS

Base analogs are purine or pyrimidine bases that differ slightly from the structures usually found in DNA and RNA. They may be incorporated into DNA or RNA in place of normal bases and cause mutations. 5-bromouracil (BU), for example is a mutagenically active pyrimidine analog that

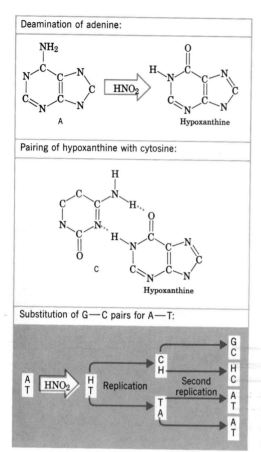

Deamination of adenine:

A → HNO₂ → Hypoxanthine

Pairing of hypoxanthine with cytosine:

C — Hypoxanthine

Substitution of G—C pairs for A—T:

A/T → HNO₂ → H/T → Replication →
C/H → Second replication → G/C
H/C
→ T/A → A/T
A/T

Figure 8.22 Action of nitrous acid on adenine resulting in the substitution of G-C pair for A-T.

may be carried into a bacterial cell by a bacteriophage or may enter a cell without a phage.

5-bromouracil has no effect on extracellular bacteriophages, but it **induces mutations in bacteriophage-infected** *E. coli* in which the bacteriophage has taken over the replicating machinery of the bacteria. This mutagen is inserted into the **DNA** chain of the **replicating virus** by enzymatic activity in the bacterial cell. Mutational change apparently occurs as the new strands of DNA are synthesized. (By contrast, the changes induced by nitrous acid are in bases that are incorporated in the DNA chain, and the errors are induced by

subsequently altered bases). 5-bromouracil is a T analog but mispairs with G at a greater frequency than does T. Once incorporated BU behaves as T and pairs with A. The new result is a transition from guanine to adenine and a base pair change from G-C to A:T, then AT → GC, (Fig. 8.23). Because replication errors occasionally occur in which 5-bromouracil pairs with G, this **mutation is reversible.**

A base alteration similar to that induced by 5-bromouracil has been obtained in the messenger RNA of bacteriophage-infected *E. coli* cells grown in a medium with 5-fluorouracil. In this example, the fluorouracil was incorporated into mRNA in

Pairing of 5–bromouracil with adenine:

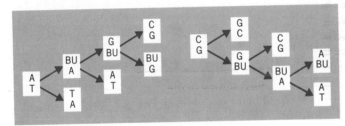

5–Bromouracil   A

Pairing of 5–bromouracil with guanine:

5–Bromouracil   G

Base pair changes:

Figure 8.23 Pairing relationships of 5-bromouracil which result in a C-G pair substituted for A-T, and A-T substituted for C-G.

place of uracil, and then behaved like cytosine in pairing with guanine. When the mutants were maintained on a medium containing 5-fluorouracil, the alteration in some mutants was reversed. The altered base, which had behaved like cytosine, changed back to uracil and paired with adenine. The explanation was that the 5-fluorouracil, which had been incorporated at the U sites in mRNA, occasionally **corrected the coding error,** thus rendering the change in mRNA reversible. The mutagen, hydroxylamine, has an effect on mRNA opposite to that of 5-fluorouracil. Whereas 5-fluorouracil produced a change from U to C, hydroxylamine produces a C to U effect.

READING FRAMESHIFT MUTATIONS

In a series of experiments, Crick et al. made use of *r* mutants in phage T4 that exhibit two distinct phenotypes: (1) they **cannot grow on E. coli** strain *K12* when it is lysogenic for lambda phage, and (2) they cause **rapid,** atypical **lysis of E. coli strain B.** The wild-type phenotype of the *r* gene is presumed to be protein-mediated normal bacterial lysis. Some *r* mutants induced by base analogs produce partial wild-type function. They are capable of producing plaques on strain B that closely resemble wild-type plaques. These are called leaky mutants. Nonleaky *r* mutants are induced by acridine dyes. They do not grow on strain K12 and they produce large, clear plaques on strain B. The *r* mutant strains induced by acridine dyes which produce nonleaky mutants are known to produce deletions or insertions of one or a few nucleotides in a chromosome.

When the investigators grew a nonleaky acridine-induced *r* mutant (*FC* O) strain on *E. coli* B they observed a few wild-type plaques. Phages isolated from these plaques could infect *E. coli* K cells. The first indication was that the mutant phages had reverted to wild-type. In further studies they were found to carry **two mutations,** the original *FC* O mutation and a second mutation. By mixed infection, the new phage strain was crossed with wild-type and rare recombinants with either *FC* O or a new mutation could be isolated. This did not indicate a reversion to wild-type. On the other hand, it suggested a double mutant strain called pseudo wild-type that could generate a lytic response like the wild-type response. It carried the *FC* O mutation and also a second mutation that suppressed the effect of *FC* O. Recombinant phages that carried only the suppressor mutations were nonleaky for the *r* phenotype. They had the properties of acridine-induced *r* mutant strains. **Insertions or deletions** of a nucleotide were apparently responsible for these mutants.

The *FC* O strain was postulated to carry an insertion (+) and the *FC* O suppressor, a deletion (−) in the *r* gene. If the reading of the message begins from a fixed point and continues sequentially from that point, one codon at a time, a + mutation would shift the reading frame **forward** and a − mutation would shift the frame **backward.** If for example, the correct reading frame is UAU ACC UAU UUG . . . , the insertion (+) of an A in the second codon would change the reading to UAU AAC CUA UUU . . . On the other hand a deletion (−) of an A in the

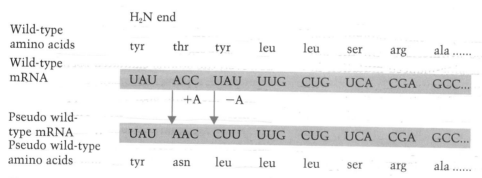

Figure 8.24 Effect of frameshift mutations on amino acid sequence. An insertion of the base adenine (+A) in the second codon from left has changed the wild-type codon that specifies amino acid thr to the pseudo wild type that specifies asn. In the third codon, the wild-type tyr has been changed by a deletion of the base adenine (−A) to leu in the pseudo wild type. Deletion (−A) has restored the wild type in the fourth frame and on to the end of the sequence. Based on experiments with trytophan synthetase A protein in *E. coli*.

Alteration of Nucleotides	Reading Frame Sequence	r Phenotype
— or +	out of phase beyond alteration	nonleaky mutant
— — or + +	out of phase beyond alteration	nonleaky mutant
— +	part back in phase	pseudo wild type
— — — or + + +	part back in phase	pseudo wild type

Figure 8 25 Changes in the reading frame of mRNA transcript because of acridine-induced insertions (+) and deletions (−) of nucleotides. (After Crick et al.)

third codon would change the reading to UAU ACC UUU UGC. . . . If the + and − should occur in the same sequence, the inserted nucleotide (+) in codon 2 would be corrected by the deleted nucleotide (−) in codon 3. In this case codon 4 and those beyond, would read correctly as shown in Fig. 8.24. In this example, only two codons in the message would be in error and a pseudo wild-type protein might be expected.

A single deletion or insertion near the beginning of a gene would disrupt the reading of most of the genetic message and a **nonfunctional** (nonleaky) protein would most likely be synthesized. On the other hand, a base analog-induced mutation in a gene would change only a **single nucleotide** in the mRNA and result in a single incorrect amino acid in a sequence of correct amino acids. The resulting polypeptide might be capable of partial functioning which would sustain a **leaky** phenotype.

In further experiments, the investigators infected *E. coli* B cells with phage (−) sup-pressor strains that had been collected. These produced r phenotypes but some yielded pseudo wild-type plaques formed by suppressed strains with double mutations. The second mutations in these examples were **suppressors** of − mutations and were of the + class. Eventually about 80 r strains were accumulated, all designated (+) or (−). Crosses were then made between different + and − strains. Most − mutations were suppressors of + and most + mutations were suppressors of −. Two suppressors of the same sign produced nonleaky r phenotypes and thus could **not suppress each other.** When three mutations of the same sign, +++ or −−−, were present, however, the strains exhibited pseudo **wild-type** phenotypes (Fig. 8.25). **Three** mutations of the same sign could suppress one another but two could not. This demonstrated further that the **coding ratio** is three. Each codon consists of three nucleotides; three insertions or three deletions will restore the correct reading frame.

PRACTICAL APPLICATION OF MUTATIONS

Even though most mutations make the organism less efficient and are thus disadvantageous, the possibility of developing new desirable traits through induced mutations has intrigued many plant breeders. The German plant breeder, H. Stubbe, has

Figure 8.26 Heads of barley demonstrating the effects of resistance to loose smut. (Left) A head from a strain that is smut resistant. (Center) heads from a strain that carries some resistance to loose smut. (Right) head destroyed by loose smut from a susceptible strain. (Courtesy of Wade Dewey.)

reported induced mutants in barley, wheat, oats, soybeans, tomatoes, and fruit trees, that may improve presently cultivated strains. Barley mutants, for example, have been obtained that provide increased yield, **resistance to smut** (Fig. 8.26), stiff straw, increased protein content, and hull-less seeds.

Some success has come from inducing mutations for disease-resistance in otherwise useful varieties. In studies to develop disease-resistant mutations in grain, seeds from irradiated plants were treated with toxin from a disease organism. The seeds were then germinated on moist filter paper. Disease-susceptible seeds were **killed by the toxin** and did not germinate. A mutation for

resistance had occurred in a few seeds, however, and these survived and perpetuated that resistance in their progeny (Fig. 8.27).

One application of induced mutations comes from concentrated efforts to improve the yield of penicillin by the mold Penicillium. When penicillin was first discovered, the yield was low and production was seriously limited. Then millions of spores were irradiated and a few of the surviving colonies produced considerably more penicillin than the average. Such improvement was possible because large numbers of spores could be treated and the subsequent selection could be made efficient.

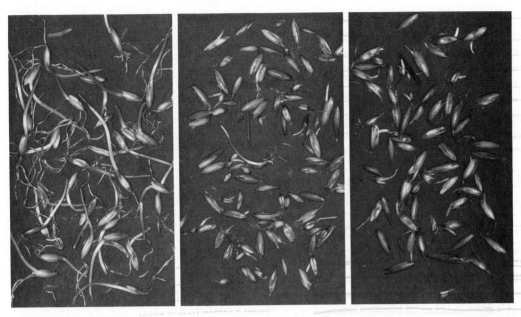

Figure 8.27 Results of an experiment to induce a mutation for disease resistance in grain. (Left) Seeds of susceptible variety treated with a disease toxin and killed. (Center) Seeds irradiated with 30,000 r before treatment with disease toxin. One (sprouted in center) survived and perpetuated disease resistance in its progeny. (Right) Untreated control, all seeds germinated. (Courtesy of Brookhaven National Laboratory.)

MUTATIONS AND MAN

Purposeful artificial selection is not practiced in man and, therefore, the possible advantages cited for domestic animals and plants do not apply to mankind. Variations do exist in populations, however, and presumably they originated through past mutations. Since most mutations are **detrimental** in the usual environment, it would seem advantageous, from the standpoint of short-term effects, for man to **avoid** excessive **exposure** to mutagenic agents. The danger level from irradiation has been estimated differently by those concerned with different aspects of the problem. Two types of danger should be considered: (1) the immediate damage to the exposed person, and (2) the more insidious damage to the DNA in his or her **reproductive cells** which would affect future generations. The immediate damage is indicated by burns and other direct or secondary effects on body tissues. When doses are on the order of 50 mr (milliroentgens) or lower, no immediate damage can be detected, although some unseen harmful effects such as induction of leukemia and general shortening of the life span may occur. Doses that exceed safety standards may be prescribed by physicians as therapeutic measures, such as for cancer treatment. In these cases, possible benefits must be carefully weighed against possible damage.

Effects of the second type of damage will only be observed in future generations. There is reason to believe, however, that exposure to high-energy irradiations of any kind, **at any dosage level,** is potentially harmful. Mutations have generally been proportional to the dosage and the effects have been cumulative in the few clear experimental results to date.

The relation between dosage and effects cannot be measured in man at present because of the complexity of the subject and the special difficulties of dealing with the genetics of man. Problems have been recognized in investigations concerning the survivors of the Hiroshima and Nagasaki bombings. The Atomic Bomb Casualty Commission is investigating the bomb's effect on the people exposed to atomic bomb irradiation and their descendents. Preliminary reports, including data on children born to parents who survived the bombing, have revealed a significant increase in the incidence of **leukemia.** The normal sex ratio has been **altered,** and the changes have been interpreted as resulting from induced **sex-linked lethals.** Considering the nature and complexity of man and the mutation data from other organisms, effects on descendants of survivors are not expected to be easily detectable. Most of the data available on mutation rates and the nature of mutations have come from other organisms and only by inference are they applied to man. The general effects of irradiation, however, seem comparable for all organisms, since the genetic mechanism is very similar. In the absence of specific data, the facts learned from other organisms should be considered relevant to people.

The origin of so-called "spontaneous" mutations, which proceed from unknown causes, is of great significance. J. V. Neel and W. J. Schull have outlined methods for **estimating mutations rates** in human populations and the sources of error with which the methods are associated. The average spontaneous mutation rate for man at any particular locus has been estimated to be about **1 per 100,000 eggs or sperm.**

Cosmic rays from outer space strike man as well as other subjects. Ultraviolet light and radioactive substance are in man's environment. Muller and others, however, calculated from fruit fly experiments that only 1/2000 of the spontaneous mutations could be explained by these factors. The greater proportion of spontaneous mutations (as in microorganisms) must be dependent on *errors* in pairing, replication and repair of bases.

Since man is larger than the fly and his life span is considerably longer, some 30 percent of man's spontaneous mutations might be expected to be produced by the natural background. This figure is based on the assumption that he has the same sensitivity to mutations as the fly. If he is more sensitive, as suggested by extrapolation from mouse studies, a higher proportion of his spontaneous mutations could be accounted for in this way. The effects of fallout and other present or potential sources of irradiation on human populations depend on man's sensitivity, which at present is estimated to be at least as great as that of experimental mammals investigated. Increased exposure to irradiation will undoubtedly be detrimental to future generations.

Spontaneous mutations may occur because: (1) **chemical reactions in cells** occasionally behave atypically and produce new configurations in DNA, (2) chance **irradiation** may alter DNA, or (3) various chemicals in the **environment** may change DNA. No satisfactory explanation for spontaneous mutations exists at present, but it is not unlikely that **environmental factors** cause most or all of those mutations.

REFERENCES

Auerbach, C. 1962. *Mutation, an introduction to research on mutagenesis.* Oliver and Boyd, Edinburgh.

Drake, J. W. 1970. *The molecular basis of mutations.* Holden-Day, San Francisco.

_____ 1973. (ed.) "The genetic control of mutation." *Genetics* 73, supplement, 1–205.

Fishbein, L., W. G. Flamm, and H. L. Falk. 1970. *Chemical mutagens.* Academic Press, New York.

Hollaender, A. 1971. *Chemical mutagens.* Plenum Press, New York.

Jukes, T. H. 1966. *Molecules and evolution.* Columbia University Press, New York.

Vogel, F. 1970. *Chemical mutagenesis in mammals and man.* Springer-Verlag, New York.

PROBLEMS AND QUESTIONS

8.1 Identify the following point mutations represented in DNA and in RNA as (1) transitions, (2) transversions, or (3) reading frame shifts. (a) A to G, (b) C to T, (c) C to G, (d) T to A, (e) UAU ACC UAU to UAU AAC CUA, (f) UUG CUA AUA to UUG CUG AUA.

8.2 Both lethal and visible mutations are expected to occur in fruit flies that are subjected to irradiation. Outline a method for detecting (a) sex-linked lethal and (b) visible mutations in irradiated Drosophila.

8.3 How can nutritional mutations in Neurospora be detected?

8.4 How can mutations in bacteria causing resistance to a particular drug be detected? How can it be determined whether a particular drug causes mutations or merely identifies mutations already present in the organisms under investigation?

8.5 Published spontaneous mutation rates for man are generally higher than those for bacteria. Does this indicate that individual genes of man mutate more frequently than those of bacteria? Explain.

8.6 A precancerous condition in man (intestinal polyposis) in a particular family group is determined by a single dominant gene. Among the descendants of one woman who died with cancer of the colon, ten people have died with the same type of cancer and six now have intestinal polyposis. All other branches of the large kindred have been carefully examined and no cases have been found. Suggest an explanation for the origin of the defective gene.

8.7 Juvenile muscular dystrophy in man is dependent on a sex-linked recessive gene. In an intensive study, 33 cases were found in a population of some 800,000 people. The investigators were confident that they had found all cases that were well enough advanced to be detected at the time the study was made. The symptoms of the disease were expressed only in males. Most of those who had it died at an early age and none lived beyond 21 years of age. Usually only one case was detected in a family, but sometimes two or three cases occurred in the same family. Suggest an explanation for the sporadic occurrence of the disease and the tendency for the gene to persist in the population.

8.8 Products of somatic mutation such as the navel orange and the Delicious apple have becomes widespread in citrus groves and apple orchards but they are uncommon in animals. Why?

8.9 If a single short-legged sheep should occur in a flock, suggest experiments to determine whether it is the result of a mutation or an environmental modification, and if it is a mutation, is it dominant or recessive?

8.10 How might enzymes such as DNA polymerase be involved in the mechanism of both mutator and antimutator genes?

8.11 How could spontaneous mutation rates be optimized by natural selection?

8.12 A mutator gene Dt in maize increases the rate at which the gene for colorless aleurone (a) mutates to the dominant allele (A) for colored aleurone. When reciprocal crosses were made (i.e., seed parent $dt/dt,a/a \times Dt/Dt,a/a$ and seed parent $Dt/Dt, a/a \times dt/dt,a/a$) the cross with the Dt/Dt seed parent produced three times as many dots per kernel as the reciprocal cross. Explain these results.

8.13 A single gene change blocks the normal conversion of phenylalanine to tyrosine. (a) Is the mutant gene expected to be pleiotropic? (b) Explain.

8.14 How can normal hemoglobin, hemoglobin S, and hemoglobin C be distinguished?

8.15 If CTT is a DNA base for glutamic acid, what DNA and mRNA base triplet alterations could account for valine and lysine in position six of the beta hemoglobin chain?

8.16 Why is sickle-cell anemia called a molecular disease?

8.17 Assuming that the beta hemoglobin chain originated in evolution from the alpha chain, what mechanisms might explain the differences that now exist in these two chains? What changes in DNA and mRNA codons would account for the differences that have resulted in unlike amino acids in corresponding positions?

8.18 Outline the evidence that all human hemoglobin chains are related and that they originated from myoglobin.

8.19 How does the evidence for evolution of proteins obtained from cytochrome C compare with that from hemoglobins?

8.20 In a strain of bacteria, all organisms are usually killed when a given amount of streptomycin is introduced into the medium. Mutations sometimes occur that make the bacteria resistant to streptomycin. Resistant mutants are of two types: some can live with or without streptomycin, others cannot live unless this drug is present in the medium. Given a nonresistant strain, outline an experimental procedure by which resistant strains of the two types might be established.

8.21 One sample of fruit flies was treated with X-rays at 1000 roentgens (r). The mutation rate of a particular gene was found to be increased by two percent. What percentage increase would be expected at 1500 r, 2000 r and 3000 r?

8.22 Why does the frequency of chromosome breaks induced by X-rays vary with the total dosage and not with the rate at which it is delivered?

8.23 One person was in an accident and received 50 roentgens (r) of X-rays at one time. Another person received 5 r in each of 20 treatments. Assuming no intensity effect, what proportionate numbers of mutations would be expected in each person?

8.24 How does ultraviolet light produce reversible mutations?

8.25 How does nitrous acid induce mutations? What specific end results might be expected on DNA and mRNA from treatment of viruses with nitrous acid?

8.26 Are mutational changes induced by nitrous acid more likely to be transitions or transversions?

8.27 How does the action and mutagenic effect of 5-bromouracil differ from that of nitrous acid?

8.28 How has induced mutation been put to practical use in improving crops of economic value?

8.29 Evaluate the effects, immediate and potential, that might come from intense, mass irradiation of people.

8.30 How do acridine-induced changes in DNA result in inactive protein?

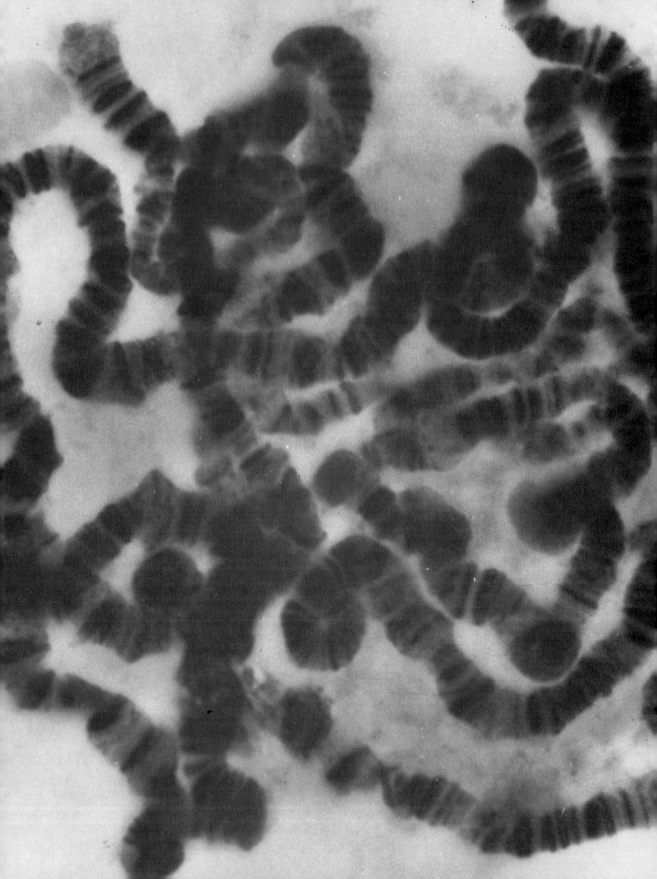

NINE

A review of chromosome anatomy and the chromosome cycle can help explain the nature and effect of structural **rearrangements** and **position effects.** During metaphase, when chromosomes are most easily observed, they are discrete rods or ovals, but in early prophase and late telophase they are long thin threads. A coiling process occurs in the transition from prophase to metaphase. The threads uncoil during telophase. The remarkable process of **DNA replication** presumably occurs either in interphase or in the early meiotic prophase. Most structural genes are single copies of DNA.

In general, the amount of cellular DNA is related to the phylogenetic position of a species but some organisms have a **disproportionate amount.** *E. coli* has 80 times as much DNA as phage lambda. The slime mold (Dictyostelium) has 7 times as much DNA as *E. coli* while some mammals have 700 to 800 times that of *E. coli*. Man has some 800 times more DNA than *E. coli* but the unicellular Euglena has nearly as much as man and the congo eel *Amphiuma means* has 26 times more than man. The role of this **excess DNA** ob-

CHROMOSOME STRUCTURE AND MODIFICATION

Salivary gland chromosomes of *Drosophila melanogaster*. (R. A. Boolootian, Science Software Systems, Inc.)

served in **eukaryotes** is the subject of much interest. It is repetitive in sequences of moderate length and from 10^2 to 10^5 copies, or in satellite fractions, in sequences of only a few nucleotides and more than 10^6 repetitions.

REPETITIVE DNA

Some animals have a considerable quantity of DNA that has **no mRNA** to match and does not transcribe a message but does have functional significance. It is DNA chemically but not functionally in terms of possessing structural genes that code for enzymes. Simple sequences of 6–8 nucleotides have been **repeated many times** in the cells of some higher animals.

Eukaryotes can usually be distinguished from prokaryotes by the presence of some DNA with a buoyant density different from the usual and constant nuclear DNA. This is called **satellite DNA**. When, for example, the DNA of bacteria is isolated and centrifuged to equilibrium in a cesium (CsCl) gradient, a single band usually appears in the tube. This band corresponds to a GC content of about 50 percent. When large pieces of prokaryote DNA are randomly fragmented, the proportion of **GC** and **AT pairs** is about **equal**. By contrast, when DNA of mammalian cells is subjected to the same treatment **a main band** and one or more smaller bands is usually observed. About 40 percent of the DNA represented by the main band is composed of GC pairs. The smaller **satellite bands** frequently observed, often have either a higher or lower GC content. These bands represent the satellite DNA that is located in the cell nucleus. Considerable variation

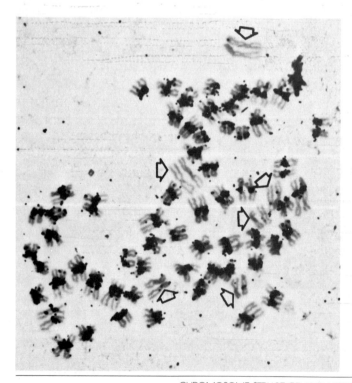

Figure 9.1 Autoradiograph showing the hybridization of ^{3}H RNA transcripts of the DNA satellite HS-β of the kangaroo rat (*D. ordii*) to metaphase chromosomes. Most of the chromosomes show hybridization localized in or around the centromere. Chromosomes (arrows) not labeled are endoreduplicated chromosomes. Magnification $\times 44,300$. (Courtesy of D. M. Prescott and *Chromosoma* 42: 205–213, 1973.)

exists in the proportion of GC and AT pairs in the satellite DNA of different eukaryotes. Satellites found in the crab, for example, are composed almost entirely of **AT pairs** (3 percent GC) whereas satellites of the barnacle have 45 percent AT (55 percent GC).

Satellite DNA can be observed microscopically in metaphase chromosomes with appropriate hybridization and centrifugation techniques. Prescott and associates, for example, demonstrated the location of satellite DNA in the chromosomes of the kangaroo rat, *Dipodomys ordii* (Fig. 9.1). Most, if not all of the satellite DNA was found to occupy the **centromere** regions and sometimes the entire short arm of some metacentric chromosomes.

Small fractions of repetitive DNA have been found in Drosophila, guinea pigs, mice, and men. Repetitive DNA is usually located in major **heterochromatin** regions, near centromeres and telomeres but is also distributed widely in chromosomes. An immense variety has been detected with predominantly A and T bases. In *Drosophila virilis*, for example, Joseph Gall has described repetitive sequences of seven bases: $5'ATAAACT3'$. Mutations of the transversion type were detected in positions 2 (T to C) and 6 (C to T), indicating that repetitive DNA can mutate like other DNA. Other investigators found sequences of 6 bases $5'TTTTTC3'$ in the mouse. Repetitive DNA in some quantity is apparently a characteristic feature of **all higher animals.**

How could repetitive DNA have any selective advantage? It is in heterochromatin regions of higher organisms where it may influence some gene activities such as position effects. This material may also initiate **receptor genes** such as those postulated in the Britten and Davidson model (Chapter 7). Some repetitive DNA apparently has a **structural** or **organizational** role in the chromosome. Repeats of a few nucleotides are found in large numbers (10 million copies) in the **nucleolus** of some organisms. They may protect the nucleolar organizer which is a vital organelle in the secondary constriction region of the chromosome. Indeed, repetitive DNA may also **protect** eukaryotic genes. It may also play a role in chromosome **pairing** and **recombination.**

CHROMOSOME INACTIVATION

Some chromosomes and parts of chromosomes become **inactivated** and fail to respond to the stimuli for **gene transcription.** They share the properties of the noncoding, heterochromatic DNA described above. If the unresponsiveness becomes permanent, hereditary changes are expected to follow. Such inactivation does not include lampbrush chromosomes and chromosome puffs (Chapter 7) which have regular active and inactive stages but does include the loss of one or an entire set of chromosomes in the developing organism. In female mammals, for example, one **X chromosome** is permanently inactivated early in development. The **paternal** X chromosome in marsupials is the one to be inactivated but in Eutheria it may be either the paternal or maternal X (Lyon hypothesis, Chapter 4).

In the male mealy bug (*Planococcus citri*) embryo, the entire set of paternal chromosomes becomes inactivated and remains inactive during subsequent development. The chromosomes of all such embryos are euchromatic in the cleavage divisions immediately following fertilization. Shortly thereafter, at blastula stage, the paternal chromosomes become heterochromatic in those embryos destined to become male. The **paternal grandparent,** therefore, makes no genetic contribution through the father to his progeny. The male (functional haploid) **mealy bug** embryo contains four times as much heterochromatin, 1½ times as many cells, and goes through **one more cell division** than the female embryo. In spermatogenesis first division, the euchromatin and heterochromatin sets of chromosomes both divide equationally. They segregate during the second division in such a way that only the euchromatin products form sperm. The heterochromatin products degenerate. Only two of the four cells resulting from the meiotic divisions become viable sperm. All chromosomes in female embryos remain in the euchromatic state. Further development, including oögenesis, is regular.

In aphids and fungus gnats (Sciara) different events occur in the eggs of the two sexes. No **heterochromatization** occurs in the female aphids that produce by parthenogenesis only female progeny during the summer months. In the fall, however, a chromosome is **inactivated** and both **males** and **females** are produced by the usual sexual process. Fungus gnats produce two different kinds of X chromosomes while the autosomes in the two sexes are all alike. The **XXAA** zygotes all become **males** while the **XX'AA** zygotes all become **females.**

BREAKAGE AND JOINING OF CHROMOSOMES

Structural changes presuppose breaks in the chromosomes. More than one break can occur in a single chromosome or set of chromosomes, and the broken parts may then reunite in new arrangements. Any **broken end** presumably **may unite** with any other **broken end,** thus potentially resulting in **new linkage arrangements.** The loss or addition of a chromosome segment may also occur in the process. More than one type of deviation or **aberration** may occur at the same time. For example, a section may be broken off and lost during the formation of an inversion or translocation, and thus simultaneously produce a deficiency.

Identifying chromosomal aberrations by observation is a major problem for the **cytogeneticist** because members of a pair usually lack any visible means of identifying different areas along their length. Maize chromosomes are exceptions. They have become one of the most favored materials for microscopic study because meiotic prophase chromosomes have deepstaining bodies, called **heteropyknotic knobs** (Fig. 9.2), distributed along their length. With the aid of these visible markers, many chromosome changes have been detected in maize. The acetocarmine **smear method,** first applied to other plant materials by J. Belling in 1931, greatly facilitated studies of the maize chromosomes. This technique permits whole chromosomes to be fixed, stained, and spread on a microscope slide in one operation. It provides a way to compare **individual chromosomes** within a chromosome set or genome, as well as **whole sets** from different organisms. The ten maize chromosome pairs were first described in their meiotic prophase stages by McClintock on the basis of a series of studies that utilized the acetocarmine method.

In the 1920s, **attached-X** chromosomes and attached-X and -Y chromosomes were discovered in Drosophila. H. J. Muller and Edgar Altenburg induced structural changes in Drosophila chromosomes with X-rays and then detected translocations. The first cytological demonstration of plant chromosome rearrangements was made in maize by Barbara McClintock (in 1930). Working

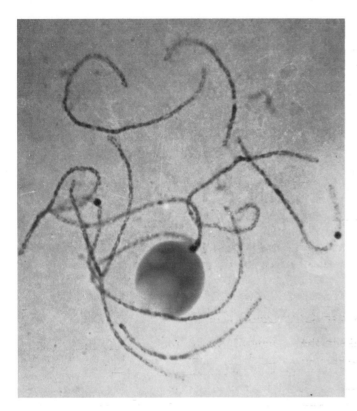

Figure 9.2 The ten chromosomes of maize showing large nucleolus, centromeres and some of the more conspicuous knobs which are used to identify particular chromosomes. (From M. M. Rhoades, reprinted with permission from the *Journal of Heredity*.)

with **pachytene** and other meiotic prophase stages that present large chromosomes for microscopic observation, she eventually demonstrated that irregular configurations made by chromosome rearrangements in the pairing process led to four different kinds of structural changes. The four types were: (1) **deficiencies** (parts of chromosomes lost or deleted), (2) **duplications** (parts added or duplicated), (3) **inversions** (sections detached and reunited in reverse order), and (4) **translocations** (parts of chromosomes detached and joined to nonhomologous chromosomes). Comparable demonstrations were later made with the giant polytene chromosomes of Drosophila.

Structural modifications of chromosomes are common in nature and have apparently played a significant role in evolution. They occur spontaneously, that is, without any obvious cause. The frequency of structural changes is increased by **ionizing radiations** and **chemical mutagens.**

GIANT POLYTENE CHROMOSOMES IN DIPTERA

In cells of higher organisms, chromosomes occur in pairs, one member having come originally from each parent. Chromosomes normally do not appear in paired arrangement, however, except during the brief period of synapsis, which is a part of the meiotic process. In the other stages, the maternal and paternal member of each set are present but unpaired in the same nucleus. In a few exceptional cases such as **giant chromosomes** of larval diptera, chromosome mates are continuously attracted to each other and **stay together in pairs.**

Large coiled bodies about 150 to 200 times as large as gonad cell chromosomes were observed in the nuclei of glandular tissues of dipterous larvae as early as 1881 by E. G. Balbiani. He described banded structures in the nuclei of cells of larval midges in the genus Chironomus but did not attach any genetic significance to the observation. Three years later, J. B. Carnoy made further morphological observations, and in 1912, F. Alverdes traced the development of these structures from the early embryo to a late larval stage. In 1930, D. Kos-

toff suggested a relation between the bands of these structures and the linear sequence known to occur among genes. The anatomical significance of the nuclear bodies was further studied by E. Heitz and H. Bauer in 1933 in the genus Bibio, a group of March flies whose larvae feed on the roots of grasses. These authors identified the bodies as giant chromosomes occurring in pairs. They described the morphology in detail and discovered the relation between the salivary gland chromosomes and other somatic and germ cell chromosomes. They also demonstrated that comparable elements occurred in the giant chromosomes and in the chromosomes of other cells of the same organism.

It was largely because of the work of T. S.

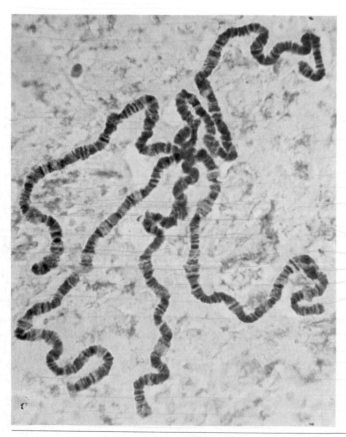

Figure 9.3 Salivary gland chromosomes of *Drosophila melanogaster.* (Courtesy of Berwind P. Kaufmann, reprinted with permission from The Journal of Heredity.)

Drosophila phenotypes:
a. wild-type male, lateral view
b. wild-type female, lateral view (note differences in foreleg bristles (sex comb) and caudal end of abdomen)
c. vestigial wings
d. ebony body color (left) and wild type, dorsal view
e. black body and curved wing
f. dicaete wings

Courtesy Carolina Biological Supply Company.

a.

b.

c.

d.

e.

f.

PLATE
2

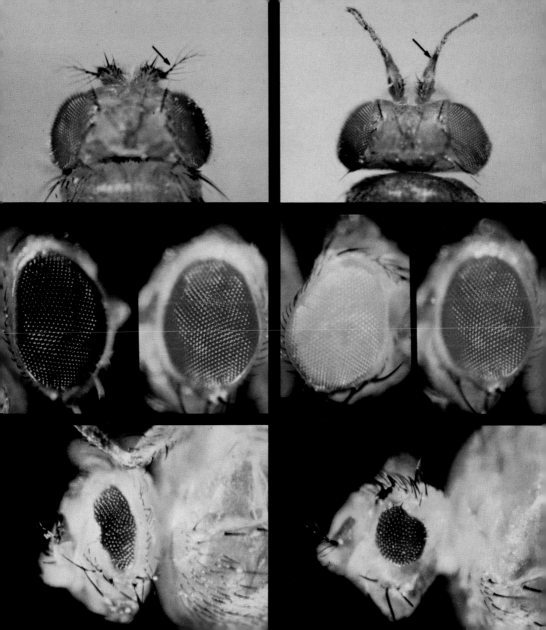

Albinism—that is, absence of normal pigmentation produced by melanin in animals and chlorophyll in plants; usually genetically or cytoplasmically controlled:

a. albino molly
b. normally pigmented male molly displaying dorsal fin to female (note front part of fin imperfectly regenerated after injury)
c. albino grey squirrel
d. normally pigmented grey squirrel
e. albino and normal corn seedlings in 3:1 genetic ratio

(a *and* b *Jane Burton/Bruce Coleman,* c *and* d *J. Markham/Bruce Coleman,* e *Grant Heilman.*)

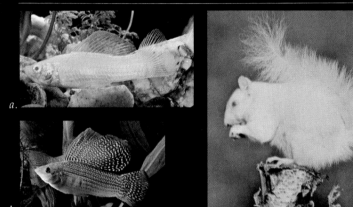

a.

c.

b.

d.

e.

PLATE

5

Animal behavior patterns with genetic and evolutionary implications in insects:
a. Viceroy butterfly which beautifully mimics
b. Monarch butterfly
c. walkingstick on leaves camouflaged as twig
d. butterfly camouflaged as a dead leaf
e. stick insect camouflaged as branches of a plant

a *A. J. Dignan*, b *H. N. Darrow*, c *Alan Blank*, d *Thase Daniel*, e *M. F. Soper, all from Bruce Coleman.*

PLATE
6

Painter that Drosophila salivary gland chromosomes (Fig. 9.3) were first used for cytological verification of genetic data. Painter related the **bands** on the giant chromosomes to genes, but he was more interested in the morphology of the chromosomes and implications concerning speciation than in associating chromosome sections with particular genes. Bridges, beginning in 1934, made extensive and detailed studies of the salivary gland chromosomes and, in the course of his investigations, developed a tool of practical usefulness in relating **genes to chromosomes.** In applying this method to Drosophila melanogaster, he prepared a series of cytological (chromosome) maps (Fig. 9.4) to correspond with the linkage maps already available. He was constructing maps of all four chromosomes of D. melanogaster at the time of his death in 1938.

FEATURES OF POLYTENE CHROMOSOMES

The unusual size of the salivary gland chromosomes is explained at least superficially by the type of growth that occurs in larval glandular tissues in dipterous insects. Salivaries and other glands grow by enlargement rather than by duplication of individual cells. This can be demonstrated by cell counts and measurements taken at different stages in development of a larva. As a larva develops, the chromosome threads (chromonemata) in the cells **duplicate** themselves repeatedly, producing bundles. A great many chromonemata, having originated by duplication of chromosomes in the original cells involved in the formation of the gland, are present in a single giant chromosome. The chromosomes are thus many-stranded (polytene), **cablelike structures.** Bands running crosswise on the giant chromosomes represent an accumulation (through continuous duplication) of baso-

philic stainable regions (chromomeres). Giant polytene chromosomes thus correspond in linear structure with other chromosomes of the same species. The difference is that the duplicate strands are held together in bundles through a special process and they do not separate out to new cells through cell division. Giant chromosomes are basically like other chromosomes, with DNA and protein in each strand.

The other feature of the giant salivary chromosomes that makes them valuable for study is their continuous state of **somatic synapsis.** If one member of a homologous pair is altered by deficiency, duplication, inversion, or translocation, an irregularity occurs in pairing. Characteristic and observable irregularities make it possible to recognize different kinds of chromosome modifications and to identify their location on the chromosome.

APPLICATIONS OF POLYTENE CHROMOSOMES

Polytene chromosomes allow comparisons between **cytological** chromosome maps and **linkage** data maps. Cytological maps are obtained by placing the genes in their visually observed positions on the chromosome. By contrast, linkage maps are deduced from cross-over data. When the two types of maps were prepared for D. melanogaster, they were remarkably parallel (Fig. 9.4) except for one major difference—linkage maps cannot include or allow proper spacing in parts of the chromosome where few genes are located and crossing over occurs at less than the usual frequency. Chromosome sections (heterochromatin) near the **centromere regions** and elsewhere were found to cross over less frequently than other parts near the free ends of the chromosomes.

In salivary gland preparations of D. melanogaster, the major portion of het-

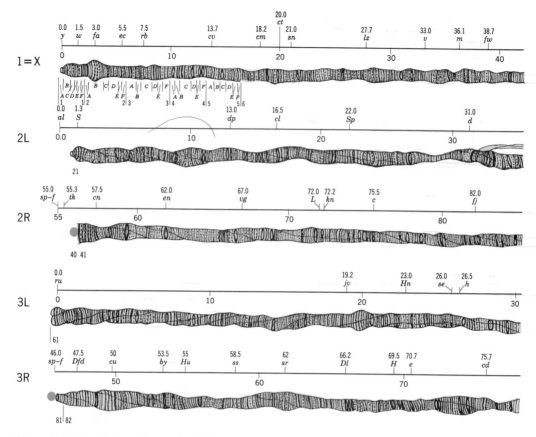

Figure 9.4 Map of the salivary gland chromosomes of Drosophila melanogaster. The linkage map is superimposed on the cytological map. On the left end of the first (X) chromosome the numbering system used to identify particular bands is illustrated. (From Bridges, reprinted with permission from the Journal of Heredity.)

erochromatin of all chromosomes coalesces into one central body or **chromocenter.** The chromocenter of a female cell includes the heterochromatin sections of the four paired chromosomes. In the male cell, the entire Y chromosome is also included in the chromocenter. When the heterochromatin arrangement is taken into account, salivary gland chromosomes are remarkably similar to chromosomes found elsewhere in larvae or in adult flies. About 5000 single cross bands have been noted on the four pairs of salivary gland

chromosomes in *D. melanogaster.* This number was considered by H. J. Muller to be a minimum approximation of the number of genes in that insect. B. Judd has now demonstrated by studies of deficiencies that each band corresponds with a **unit of genetic function** (complementation unit, Chapter 6).

Some genes have been associated with individual bands. Bridges' system (Fig. 9.4) of designating parts of chromosomes with numbers, subdivisions with letters, and bands within subdivisions with numbers

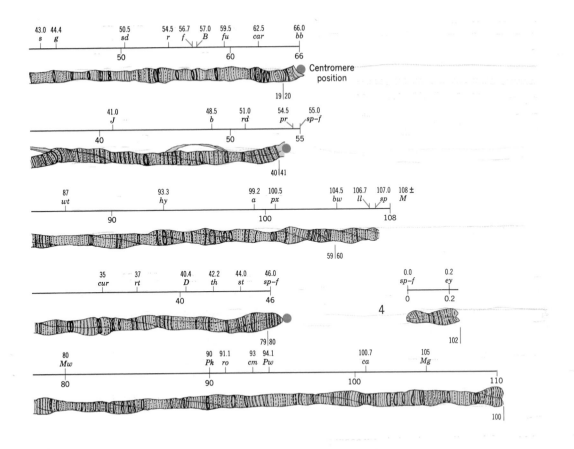

made it possible for investigators to discuss precise locations. In this system, fairly uniform divisions are numbered in order throughout each entire chromosome set from 0 at the beginning of the X chromosome to 102 at the end of chromosome 4. Subdivisions within the areas are identified with letters from A to F, and bands within subdivisions are numbered from left to right. For example, the gene (w) for white eyes is in bands 3C2. In linkage units this gene is located at 1.5 in the X chromosome. Linkage data do not correspond exactly with cytological locations, as shown in Fig. 9.4, but the linear sequence of genes can be verified from salivary preparations.

The main use of polytene chromosomes is in **locating genes** and identifying **structural changes** in the chromosomes. They are also useful for studying effects of environmental agents on chromosomes. M. Diaz and C. Pavan, for example, have demonstrated effects of protozoan and viral infections on particular areas of polytene chromosomes. Particular sections disintegrated and nucleolarlike bodies were formed. Polytene chromosomes are also used for physiological studies of gene action (mRNA synthesis, Chapter 7).

Chromosome modifications, particularly missing sections or **deficiencies**, are useful tools for locating genes on chromosomes. In general, chromosome locations are detected by: (1) identifying the deleted section in the salivary chromosome by microscopic observation, and (2) making appropriate matings and analyzing the results. The test matings must be designed in such a way that genes along the chromosome are **heterozygous.** When recessive genes in such an arrangement express themselves, a plausible explanation is that the region in the homologous chromosome carrying the dominant allele has been deleted. The term **pseudodominance** describes the expression of a **recessive gene** that occurs because the dominant allele, which would ordinarily suppress it, is missing.

DEFICIENCIES

In 1917, C. B. Bridges observed that a sex-linked recessive gene in Drosophila came to expression when it was presumed to be in a heterozygous condition. He postulated that a section of the homologous chromosome containing the dominant allele was missing; that is, a deficiency had occurred in a chromosome.

A single break near the end of a chromosome would be expected to result in a terminal deficiency. If two breaks occur, a section may be deleted and an intercalary deficiency created. Terminal deficiencies might seem less complicated and more likely to occur than those involving two breaks. Instead the great majority of deficiencies detected thus far are of the **intercalary** type within the chromosome. No truly terminal deficiencies have been found in Drosophila, although they have been described in maize. In either type of deficiency, the chromosome set is left without the genes carried in the deleted portion unless the deleted part becomes fused to a chromosome that has a centromere. Without a centromere, a chromosome section cannot move to the pole of the spindle during cell division but lags in the dividing cell and is excluded from the chromosome group when the nuclear membrane forms around the chromosomes of a daughter cell.

When an intercalary part of a chromosome is missing, a buckling effect may be observed microscopically in the paired salivary gland chromosomes (Fig. 9.5). The mechanism is illustrated diagrammatically in Fig. 9.6. Large deficiencies are more readily detected than small ones, but with good optical equipment and patience, an investigator may be able to see single bands of the salivary preparations and thus to identify minute heterozygous deficiencies. By identifying the part of the polytene chromosome in which the buckle occurs and then studying the phenotype of flies carrying a recessive gene in the homologous chromosome opposite the deficiency, the gene can be spatially positioned on the chromosome. Chromosome deficiencies have greatly facilitated the **checking of linkage maps.** Many genes are now precisely located on actual chromosomes in *D. melanogaster* and other species of Diptera because of the effective use of this technique.

A somatic cell that has lost a small chromosome segment may live and produce other cells like itself, each with the deleted section of a chromosome. Phenotypic effects sometimes indicate which cells or portions of the body have descended from the originally deficient cell. If, on the other hand, the deficient cell is a gamete that is

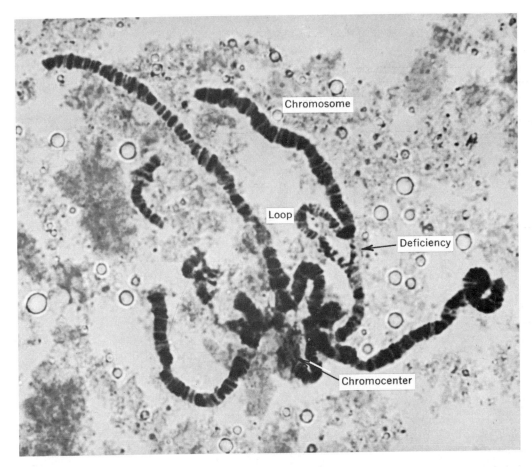

Figure 9 5 Salivary gland chromosomes of Drosophila illustrating a heterozygous deficiency in the X-chromosome. (Courtesy of Jack Tobler.)

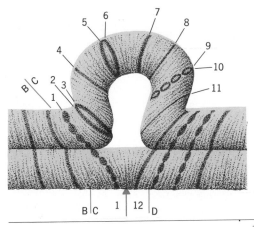

Figure 9.6 A deficiency loop in the paired X-chromosomes from a salivary gland cell of a larva heterozygous for Notch. Only a short section of the chromosome pair is shown. In this figure, subsection C of section 3 is presented. Note that bands 3C2 through 3C11 are missing from the lower chromosome. (Redrawn from *Principles of Human Genetics*, 3rd edition, by Curt Stern. W. H. Freeman and Company. Copyright © 1973.)

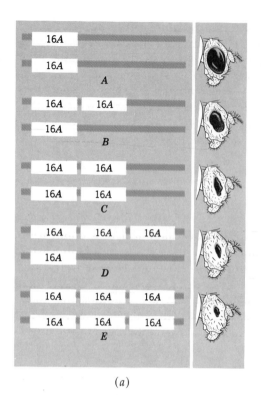

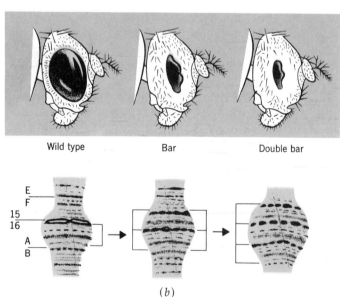

Figure 9.7 Effects on eye size of different arrangements of duplicated sections 16A in *D. melanogaster* X chromosome. (*a*) Positions of 16A segment on chromosome related to size of eye. (*A*) Wild-type female; (*B*) bar female, heterozygous; (*C*) bar female, homozygous; (*D*) double-bar female,

subsequently fertilized by a gamete carrying a nondeficient homologue, all cells of the resulting organism will carry the deficiency in heterozygous condition. Recessive genes on the nondeficient chromosome in the region of the deficiency may express themselves (pseudodominance). Heterozygous deficiencies usually decrease to some extent the general viability of the flies that carry them. Flies carrying deficiencies in homozygous condition usually die. Some very small homozygous deficiencies, however, have proved **viable** in Drosophila. Such deficiencies occur in the region in which the *w* (white eye) and *fa* (facet eye) genes are located near the end of the X chromosome.

DUPLICATIONS

Duplications represent **additions** of chromosome parts arranged in such a way that sections are longitudinally repeated. They provide a means for studying the effects of different arrangements of **multiple** chromosome **segments** or genes that normally occur only singly or in allelic pairs. Some chromosome segments behave as dominants and some as recessives with respect to certain phenotypes. Others show intermediate inheritance and still others have cumulative effects. Duplications provide a means for determining effects of chromosome segments when three, four, or more similar sections are present in individual animals or plants.

The first duplication to be critically examined involved the *B* (bar) locus in the X chromosome of Drosophila. In the presence of a single *B* (heterozygous) in a female, the eye is somewhat smaller than the normal eye, and the sides are straighter, giving an oblong or bar appearance. In the homozygous condition or in the male, the eye is considerably smaller. Bridges and Muller discovered independently that the bar phenotype was the result of a duplication involving a part of the X chromosome already present in the wild-type flies. Both men observed not only the effect of a duplication producing bar eye, but also a duplication resulting in an extreme decrease in the size of the eye, which was called **double bar.** By using the salivary chromosome technique, they identified the segments of the chromosome actually involved in the duplication. The different phenotypes and the corresponding segments of the salivary gland chromosome pairs are illustrated in Fig. 9.7. Section 16A is present in flies with the wild-type eye. When this section is duplicated, it produces the bar phenotype, but when it is represented three times in a single chromosome, the double-bar phenotype results.

Each additional duplicate segment of Section 16A made the **eye smaller.** Other duplications that have since been found in Drosophila, work in the opposite direction. These **suppress** the effects of **mutant genes** and make the fly appear more normal with respect to certain traits. Further studies demonstrated that duplications need not occur in the immediate vicinity of the section duplicated to exert an influence. Chromosome fragments may become attached to

heterozygous showing position effect; (*E*) double-bar homozygous. (After Morgan, Sturtevant, and Bridges.) (*b*) Appearance of wild type, bar and double bar eye (above) and corresponding chromosome segments as seen in salivary gland chromosomes. (After Bridges.)

entirely different chromosomes. Through the assortment of such chromosomes in gametes, duplications may be carried to succeeding generations.

INVERSIONS

Inversions occur when parts of chromosomes become detached, turn through 180°, and are reinserted in such a way that the genes are in **reversed order.** Some inversions presumably result from entanglements of the threads during the meiotic prophase and chromosome breaks that occur at that time. For example, a certain segment may be broken in two places and the two breaks may be in close proximity because of a loop in the chromosome. When they rejoin, the wrong ends may become connected. The part on one side of the loop connects with a broken end different from the one with which it was formerly connected. This leaves the other two broken ends to become attached, as illustrated in Fig. 9.8. The part within the loop thus becomes turned around or inverted. It is not known whether all inversions occur in this way, but this is a plausible explanation for many chromosome inversions.

Inversions may be perpetuated in the pairing process at meiosis and segregated into viable gametes. As indicated in earlier chapters, chromosome **pairing is essential** to the production of fertile gametes. The mechanism by which homologous chromosomes heterozygous for inversions accomplish such pairing in the meiotic sequence is remarkable. The part of the uninverted chromosome corresponding to the inversion forms a **loop.** A similar loop is formed by the inverted section of the homologous chromosome but in **reverse direction.** If, for example, the loop of the uninverted section is formed with the gene sequence in a clockwise direction, the inverted part will form in a counterclockwise direction. In this way, corresponding parts come together even though one of the sections is inverted, as illustrated diagrammatically in Fig. 9.9.

Inversions have been associated with the **supression of crossing over.** Before Drosophila chromosomes were studied extensively, investigators had already identified genetic crossover suppressors in this organism. These were first considered to be genes that somehow interfered with crossing over. It was later shown that the locations of inversions and crossover suppressors coincided and that the apparent suppression of crossing over was associated directly with inversions. It was shown further that the main process was not a

Figure 9.8 Mechanism by which chromosome inversions might be produced. If a chromosome in prophase of a cell division is folded in such a way that a circle is formed and two broken ends occur, the two broken ends may join, placing parts within the circle in reverse sequence as compared with the original chromosome.

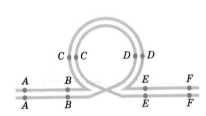

Figure 9.9 Pairing of an uninverted chromosome with a chromosome in which an inversion has occurred. C and D parts are in reverse position. Only two of the four strands are shown. A loop may be observed in salivary gland preparations of Drosophila which carry heterozygous inversions.

suppression of crossing over, although instances of physical crossing over may be reduced. The principal effect was that crossover gametes that did occur were not recovered, that is, the **zygotes died** before they could be detected.

The mechanism through which an inversion can remove the crossover gamete has been described in chromosomes with the centromere outside the inverted area (paracentric; Fig. 9.10) and in those with the centromere inside the loop (pericentric; Fig. 9.11). As shown in the diagram, chromosomes that carry **paracentric** inversions cannot cross over within the loop without producing fragments of chromosomes lacking centromeres (**acentric**) and chromosome complexes with two centromeres (**dicentric**). These result either in fragments that lag in the center of the spindle, or chromatid bridges that tie together the two homologues involved and interfere with the division process. In either case, the chromosomes do not separate properly to their respective poles. Crossovers within the loops of **pericentric** inversions result in **duplications and deficiencies.** The fate of these cells varies in animals and plants. Following a single crossover within the inversion loop of a maturing plant cell, the gametes receiving crossover chromatids are inviable, thus crossing over is effectively suppressed. In animals, unbalanced zygotes produced by abnormal gametes die, thus eliminating crossover chromatids from the population. The apparent suppressing effect of inversions on crossing over can thus be explained mostly on the basis of secondary results that follow crossing over in inverted segments.

Polytene chromosomes of Drosophila have been especially useful for detecting heterozygous inversions in the flies. A characteristic loop in these chromosomes is illustrated in Fig. 9.12. Loops in giant chromosomes of salivary gland tissues presum-

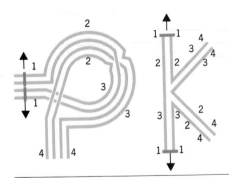

Figure 9.10 Meiotic metaphase and anaphase illustrating mechanism through which a paracentric inversion acts as a "crossover suppressor." Dicentric (1 2 3 1) and acentric (4 3 2 4) chromosomes, which are also unbalanced, are formed by crossover chromatids.

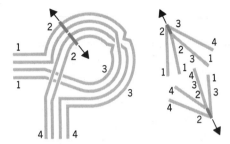

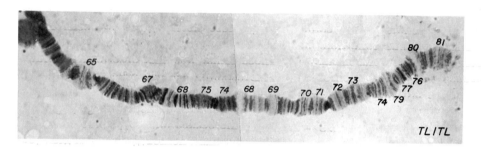

Figure 9.11 Diagram illustrating the "crossover suppressor" action of a pericentric inversion. The resulting chromatids are out of balance and therefore zygotic lethals result.

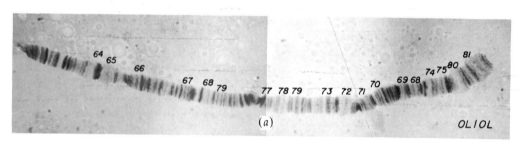

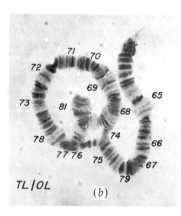

(*a*)

(*b*)

Figure 9.12 Salivary gland chromosomes illustrating an inversion. (*a*) homozygous chromosomes TL/TL and OL/OL. (*b*) heterozygote, TL/OL with inversion loop. (C. D. Kastritsis and D. W. Crumpaker. Reprinted with permission from the *Journal of Heredity*.)

ably resemble those in meiotic prophase chromosomes where observation is much more difficult.

Chromosomes with inversions have practical applications in maintaining Drosophila stocks. They are used as "balancers," that is, chromosomes that can be **placed opposite** homologous chromosomes carrying certain genes that are **homozygous lethal.** Crossing over is suppressed in such chromosomes and it is possible to maintain a gene in heterozygous condition that could not be kept in homozygous condition. In laboratory stocks carrying several mutants, it is advantageous to keep the chromosomes intact without crossing over. However, because recessive genes are not expressed when they are in heterozygous condition, frequent checks must be made to insure that the gene is not lost from the stock. Appropriate outcrosses are conducted occasionally to check for the presence of the gene or genes in the stock. The mechanism through which some laboratory stocks are kept balanced and heterozygous involves **balanced lethal mutations.**

INVERSIONS AND BALANCED LETHAL MUTATIONS

Sometimes through the process of mutation, **two recessive lethals** occur at different loci on each member of the same pair of homologous chromosomes. If the loci involved are near each other, or if chromosomal aberrations "suppress" crossing over between them, an **enforced heterozygous condition** may be established. Since individuals homozygous for each lethal die, only those heterozygous for both lethals survive. The first documented case of a balanced lethal was that described by Muller in Drosophila. He had been maintaining in this laboratory the *Bd* (beaded) stock producing flies with scalloped wings (Fig. 9.13). Since the beaded phenotype was controlled by a dominant homozygous lethal gene (*Bd*) the progeny from the crosses between beaded flies were two-thirds beaded and one-third normal, as illustrated in Fig. 9.14. All those homozygous for beaded were dead, and thus a **2:1 ratio** replaced the 3:1 ratio, which would otherwise have been expected. This pattern is similar to that described in Chapter 2 for creeper chickens.

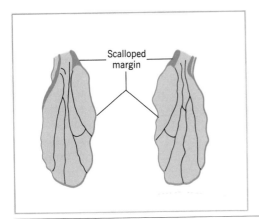

Figure 9.13 Pair of scalloped wings from Drosophila beaded (*Bd*) stock.

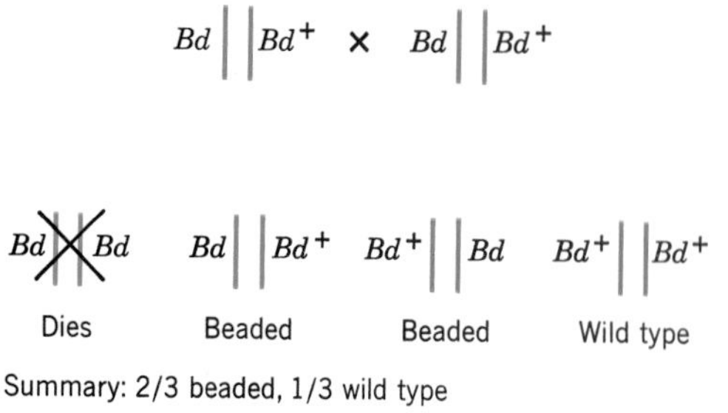

$$Bd \,\|\, \|Bd^+ \ \times \ Bd \,\|\, \|Bd^+$$

$$Bd \bowtie Bd \qquad Bd\,\|\,\|Bd^+ \qquad Bd^+\|\,\|Bd \qquad Bd^+\|\,\|Bd^+$$

Dies Beaded Beaded Wild type

Summary: 2/3 beaded, 1/3 wild type

Figure 9.14 A cross between two *Bd* flies. Since the gene *Bd* is homozygous lethal, all surviving beaded flies are heterozygous and one-fourth of the progeny from beaded parents die before hatching. One-fourth of the total and one-third of the flies which survive are wild type (not beaded).

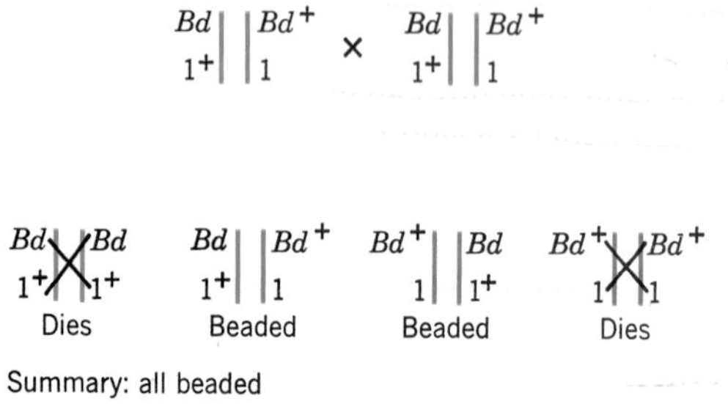

$$\begin{array}{cc} Bd\|\,\|Bd^+ \\ 1^+\|\,\|1 \end{array} \ \times \ \begin{array}{cc} Bd\|\,\|Bd^+ \\ 1^+\|\,\|1 \end{array}$$

$$\begin{array}{c} Bd\bowtie Bd \\ 1^+\ \ 1^+ \end{array} \quad \begin{array}{c} Bd\|\,\|Bd^+ \\ 1^+\|\,\|1 \end{array} \quad \begin{array}{c} Bd^+\|\,\|Bd \\ 1\|\,\|1^+ \end{array} \quad \begin{array}{c} Bd^+\bowtie Bd^+ \\ 1\ \ 1 \end{array}$$

Dies Beaded Beaded Dies

Summary: all beaded

Figure 9.15 A "balanced lethal" cross. A cross between two beaded flies after a new lethal mutation (*l*) had occurred on the same chromosome with *Bd*. All progeny homozygous for *Bd* and those homozygous for *l* died. All of the survivors expressed the beaded phenotype.

Abruptly, with no visible change in the beaded phenotype, all of the flies in the cultures began to appear beaded. In an attempt to explain the change that had occurred, Muller postulated that a new lethal (*l*) had been created by mutation in the homologous chromosome opposite *Bd*. An inversion associated with the gene for beaded "suppressed" crossing over, and only two kinds of gametes were produced by the beaded flies, one carrying the gene (*Bd*) for beaded, the other carrying the new lethal (*l*). The cross representing the new arrangement is illustrated in Fig. 9.15. When these chromosomes segregated and fertilization occurred, some zygotes became homozygous for *Bd* and were therefore lethal; others became homozygous for *l* and they also died. Only those heterozygous for both *Bd* and *l* survived. Therefore, the heterozygous condition was enforced. This explanation was substantiated and the term "balanced lethal" was used to designate such as arrangement involving two lethals in the same chromosome pair in which recovered crossovers are infrequent or entirely absent.

Permanent heterozygotes described in *Oenothera lamarckiana* are also maintained in a balanced lethal system. These plants are characterized by gene complexes made up of several chromosomes joined through translocations (Fig. 9.16). The ordinary segregation of independent chromosomes does not occur. Different complexes, however, have different gene combinations, and when gametes are formed the complexes separate as units. Two kinds of hybrids, that is, twin hybrids, are produced when plants carrying such complexes are outcrossed. Appropriate names have been associated with the different gene complexes. *O. lamarckiana* is composed of the complex called *gaudens* and *velans*. The *gaudens* complex has genes for green bud, nonpunctate stem, broad leaf,

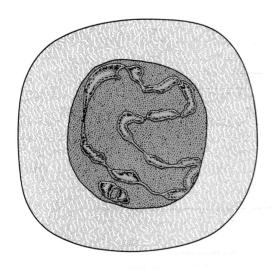

Figure 9.16 Chromosome complex in *Oenothera lamarchiana* in late meiotic prophase showing a ring of 12 chromosomes and one pair of chromosomes. (After Cleland.)

and red flecks on the leaves. The *velans* complex carried genes for red-striped bud, punctate stem, narrow leaf, and no red flecks on the leaves. Half the gametes carry the *gaudens* gene combination and half the *velans*. If the complexes were capable of independent segregation with respect to each other, crosses between individual plants would be expected to produce three combinations in the following proportions: 1/4 *gaudens gaudens*, 1/4 *velans velans*, and 1/2 *gaudens velans*. Actually, only *gaudens velans* survive. This leads to the conclusion that each of the complexes in homozygous condition is lethal. **Only the heterozygotes survive** and maintain themselves as **balanced lethals.**

INVERSIONS IN SPECIES FORMATION

Differences in chromosome complements between different races and species of Drosophila depend on the number, extent, and location of inversions. Inversions that evidently occurred spontaneously in nature, have become established in populations. Furthermore, the degree of separation between taxonomic groups is correlated with the number of inversions present. Whether this is true in organisms other than Drosophila is not known. No other group has been studied as completely as Drosophila, and further observations are necessary before comparisons can be drawn.

TRANSLOCATIONS

Sometimes parts of chromosomes become detached and reunited with **nonhomologous chromosomes** thus producing **translocations.** The definition for translocations includes exchanges between nonhomologous parts of the same chromosome pair. Exchanges between the X and Y chromosomes, for example, are translocations.

Reciprocal translocations occur when parts of chromosomes belonging to members of two different pairs become exchanged. Part of chromosome I, for example, may be detached from its linkage group and attached to chromosome II, while a section of chromosome II becomes attached to chromosome I. Reciprocal translocations have been described in a number of plants and are important factors in the evolution of certain plant groups such as Datura and Oenothera. Translocations do not ordinarily involve a loss or an addition of chromosome material but only **rearrangements** of chromosome parts that were already present.

Translocations can be detected from genetic data by noting altered linkage arrangements brought about by exchanges of parts between different chromosomes. If, for example, gene *a* is ordinarily linked with *b* and *c*, following translocation it may be linked with *s* and *t*, which are ordinarily in a different linkage group. It should be noted here that evidence such as this based on translocations strengthens the theory that genes are in chromosomes. Historically, evidence from this source, along with the studies of nondisjunction have supported the chromosome theory of inheritance.

A consequence of **heterozygous translocations** is sterile pollen or ovules, because some gametes from a plant carrying a het-

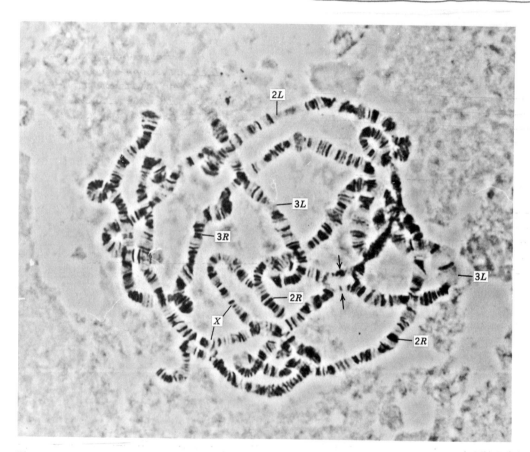

Figure 9.17 Complete complement of chromosomes from salivary gland preparation of *Drosophila melanogaster* showing a translocation between the right arm of chromosome 2 and the left arm of chromosome 3. Break points are marked with arrows. Chromosomes are identified as follows: left arm of 2nd, 2L, right arm of 2nd, 2R, left arm of 3rd, 3L, right arm of 3rd, 3R. X chromosome, X. (Courtesy of Burke H. Judd.)

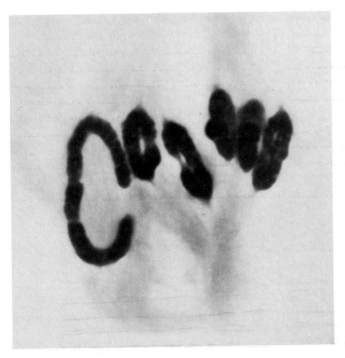

Figure 9.18 Meiotic chromosomes of barley translocations. Five bivalents and a reciprocal quadrivalent are evident. The quadrivalent chromosome is the result of a reciprocal translocation. (Courtesy of W. S. Boyle.)

erozygous translocation will be unbalanced and inviable. Such plants are called **semi-sterile.** The imbalance causing the sterility occurs at the time when chromosomes separate to the poles in meiosis. Some gametes are deficient in chromosome parts and some carry duplications.

Cytological evidence for translocations can be obtained from microscopic studies of polytene chromosomes in Drosophila and meiotic prophase stages in plant materials. The cross configuration marked by arrows in Fig. 9.17 is a characteristic cytological pattern for translocations.

Pollen mother cells of plants carrying heterozygous translocations show rings of four or more chromosomes instead of regular pairs. New structural arrangements created by the translocations result in particular configurations. Homologous parts pair together, and if homologous parts happen to be located on entirely different chromosomes, these chromosomes are held together during synapsis.

Characteristics of translocations and a method of detecting them are illustrated further from an actual study on barley. The alleles (r^+ and r) for rough and smooth awns and those (s^+ and s) for long and short rachilla hairs are in linkage group 7. Alleles (n^+ and n) for hulled and hull-less and those (l^+ and l) for lax and club head are in group 1. When the progeny of a certain plant were classified, the results indicated that s was linked with n. This genetic evidence suggested that a translocation joining chromosomes 1 and 7 had occurred. When the plants were observed in the field and studied in the laboratory, at least 30 percent of the pollen was found to be sterile. Further investigations showed that many ovules also were sterile. Cytological studies of pollen and mother cells (Fig. 9.18), showed one quadripartite ring and five bivalents instead of the seven bivalents usually observed in barley preparations. This cytological evidence demonstrated that a **reciprocal translocation** had occurred. The

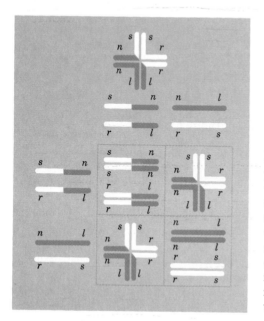

Figure 9.19 Explanation to account for the results of barley crosses. A reciprocal translocation had occurred between chromosomes 1 and 7, bringing the genes *n* and *s* into the same linkage group. The crosslike configuration at the top of the diagram represents the pairing of the translocated chromosomes (only two of the four strands are shown). Two viable male and two viable female gametes of the selfed plant are represented at the top and left, respectively, of the Punnett square. In the four squares of the Punnett, some zygotes from the selfed plant are illustrated with the chromosomes as they would appear at synapsis. The two zygotes which have crosslike configurations are heterozygous for the translocation and produce plants which are partially sterile, i.e., semisteriles, because of the deficiencies and duplications of chromosome parts occurring in the gametes. (Data from R. W. Woodward and W. S. Boyle.)

explanation for the **altered linkage** grouping and semisterility, and the mechanism of gamete and zygote formation involving the chromosomes carrying the translocation are illustrated in Fig. 9.19.

It has been demonstrated in some plants that complex groupings of chromosomes result from this type of rearrangement. In the pairing that occurs in meiosis, chromosomes and chromosome parts must find their mates and pair properly. If the mate of a certain chromosome segment happens to be translocated to a nonhomologous chromosome, these two chromosomes will be held together during pairing. Rings and sometimes chains of chromosomes are thus produced. This is the probable explanation for the chromosome complex in the

evening primrose (Oenothera), which was studied by Hugo de Vries and which permitted him to elaborate the mutation theory (Chapter 8).

Many of the examples used here to illustrate structural modifications of chromosomes have come from plants such as maize, in which chromosomes are large, comparatively few in number, and lend themselves well to cytological investigation. The giant polytene chromosomes of diptera have made these insects especially useful for cytogenetic studies. Examples could be cited, however, in many other higher plants and animals, and a large body of cytological data is accumulating for molds, yeasts, and protozoa.

Translocations have been detected in

studies of **human chromosomes.** At the Yale-New Haven Hospital, for example cytological studies were conducted on 4500 infants born consecutively during one year. Leucocytes from umbilical cord blood of each infant were grown *in vitro* and prepared for microscopic observation of chromosomes. Six translocations were detected. None of these were associated with phenotypic anomaly. Other translocations in human chromosomes have resulted in duplications and deficiencies and thus have been associated with phenotypic change. One such example is a translocation between chromosome 15 and chromosome 21 resulting in a duplication of a part of 21 on 15. When this chromosome is present in the zygote along with the usual two of number 21, Down's syndrome is produced (Chapter 10).

POSITION EFFECTS

When a chromosome rearrangement involves no change in the amount of genetic material, but only in the order of DNA units the term **position effect** is used to describe any associated phenotypic alteration. Along with gene mutations, position effects represent a source of genetic variation. The extent to which chromosome rearrangements such as **inversions** and **translocations** are associated with new phenotypic variation, however, is open to question. In this regard, it must be remembered that chromosome modifications, particularly inversions, curtail recombination and thus would be expected generally to restrict genetic variation.

Nevertheless, several well-established position effects are on record. The first example, from the studies of Sturtevant and Bridges on the bar-eye duplication in Drosophila (Fig. 9.7) also demonstrated a **dosage** effect. These investigators found a relation between the number of chromosome sections (16A) present and the number of facets in the eye. Further critical experiments, however, showed that it is not a strictly proportional relation. The arrangement of the chromosome segments with respect to each other, as well as their presence or absence, influences the size of the eye. When section 16A was duplicated in such a way that two extra similar sections were present side by side in the same chromosome, the effect was different from that produced by one extra segment in each member of the pair (homozygous). The effect of different arrangements was demonstrated by manipulating the chromosomes through appropriate matings and counting the facets in the eyes of the female offspring.

When section 16A was duplicated and the extra segment occurred in homozygous condition (Fig. 9.7c), the number of facets in the eyes averaged 68. When the two 16A sections were side by side, however, the eyes averaged 45 facets. When the sections were homozygous, small eyes with an average of 25 facets occurred and were called **ultra-bar** eyes. Since the same number of 16A units is present in the eyes represented by Fig. 9.7c and 9.7d, the difference depends on the arrangement or **position** of the genes with respect to each other. This phenomenon was interpreted as a position effect.

Several well-established and many possible position effects have now been described in Drosophila. E. B. Lewis has shown that all known position effects fall into two classes: (1) **stable** and (2) **variegated.** Stable position effects are uniform phenotypic effects resulting from changes of specific segments of chromosomes. The bar-eye position effect is an example. Variegated position effects result in the diversification of a trait usually evidenced in a particular structure or area of the body. Specks of different colors, for example, may occur in the eyes of Drosophila following rearrangements of the w (white eye) locus. Inversions or translocations that place w^+ in heterochromatin may cause white variegation or mosaicism for eye color.

Several eye-color combinations depending on genotype and environmental conditions have resulted from experiments with Drosophila. In some cases, red color patches occur on a light background and, in others, light patches occur on a red background. Variegated position effect in Drosophila is associated with chromosome structural change. The action of a gene in euchromatin is depressed when the gene is transferred to **heterochromatin.** A position effect usually involves two chromosome breaks, one in the heterochromatin region and one near the gene partially or entirely suppressed. Regulation of this gene activity occurs at the chromosome rather than the cell level and results from suppression of gene transcription. **Variegated transcription** of linear DNA results in variegated expression or position effect.

This explanation of position effects was supported by experiments of H. J. Becker on Drosophila eye development. With X-rays he induced mitotic exchanges of eye-color genes at time intervals during development. Because each affected sector was made up of the cell progeny of a single cell carrying an exchange product, it was possible to trace the developmental sequence of the tissue composing the eye. Becker then induced structural changes in chromosomes and found that alterations in the position of genes with reference to heterochromatin, created effects that appeared the same phenotypically as those of X-ray-induced disturbances.

J. Tobler has obtained similar results at the enzymatic level. He relocated the gene v^+, which specifies the enzyme tryptophan pyrrolase, by translocation from a euchromatin to a heterochromatin region. When two-day-old adult flies were assayed for tryptophan pyrrolase, the results clearly showed that the enzyme activity of the translocated gene (v^+) was intermediate between that of wild type and a null allele v^{36f}. **The heterochromatin** environment apparently **effects gene transcription.** Cytochemical studies are in progress to determine how a break in continuity disrupts transcription of DNA.

Most of the evidence thus far obtained on variegated-type position effects has come from Drosophila. D. G. Catcheside, however, has described a case in Oenothera in which a gene (P^5) that produces broad red and narrow green stripes on the sepals was translocated to a different chromosome. Variegated patches of red occurred in place of the broad red stripe. This situation may be similar to those described in Drosophila. Other possible examples of variegated-type position effects have been suggested in Datura and maize.

CHROMOSOME CHANGES IN MAN

Until the middle 1950s, the chromosome number in man was recorded as 48. Suitable material for human chromosome studies was difficult to obtain, and existing techniques were not satisfactory for critical chromosome counts. Human chromosomes are small; they tend to overlap on the microscope slide, and can be extremely difficult to distinguish.

In 1956, two cytologists, J. H. Tjio and A.

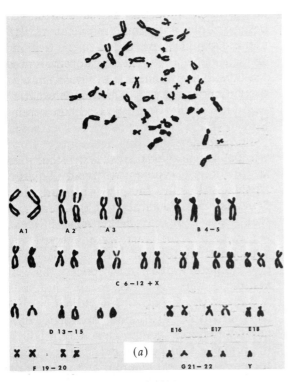

(a)

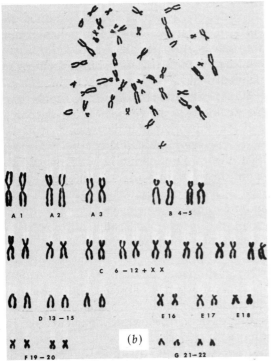

(b)

Figure 9.20 Male and female human metaphase chromosomes as they appear in microscope preparations and as they are classified (karyotyped) in seven major groups according to criteria established at the Denver (1960), London (1963), and Chicago (1966) conferences. (a) Metaphase and karyotype of a normal male cell, with both X and Y chromosomes present. (b) Metaphase and karyotype of a normal female cell, with the 2 X chromosomes placed among the C group chromosomes (Courtesy of Arthur D. Bloom, Division of Genetics, College of Physicians and Surgeons of Columbia University.)

Levan, working in Sweden, published a paper that gave 46 as the $2n$ number (Fig. 9.20). Their first counts were made from cell culture preparations from the lungs of four different human embryos. A pretreatment with certain solutions had spread the chromosomes of dividing cells and made it possible to observe each chromosome separately. The investigators examined 265 dividing cells and concluded that the chromosome number in human embryo lung tissue was 46. Similar results were obtained later with two more embryos.

In the same year, two English investigators, C. E. Ford and J. L. Hammerton, reported 23 pairs of chromosomes in testicular preparations of each of three adult Englishmen. In 1961, S. Makino and M. S. Sasaki studied 1422 cells from 54 different human embryos, ages 2 to 7 months, and found virtually all to have 46 chromosomes. Many specimens have now been obtained from widely separated parts of the world and the number 46 has become well established as the $2n$ number. No persistent difference in chromosome number has been detected among races or ethnic groups of mankind.

A group of cytogeneticists meeting at **Denver,** Colorado in 1960, adopted a system for **classifying** and identifying human chromosomes. Chromosome **length** and **centromere position** were the criteria for classification. The Denver classification, with refinements made at the **London** Conference (1963) and the **Chicago** Conference (1966), has become a standard for human chromosome studies. In spite of standardized preparations and numerous specimens for comparison, it was still impossible in 1966 to identify consistently every chromosome pair. The 22 pairs of autosomes were divided successfully into **7 groups** identified with letters A to G (Fig. 9.20). All autosomes could be placed satisfactorily with a group, but the numbering within the

groups was more or less tentative. Experienced observers could distinguish the X chromosome from members of the C group and although the Y chromosome showed considerable variability in different preparations it could usually be distinguished from members of the G group, which it resembled in size.

Chromosome **banding techniques** along with methods of identification established at previous conferences finally distinguished all 46 human chromosomes. Bands are defined as parts of chromosomes that appear lighter or darker than adjacent regions with particular staining methods. The centric region and other heterochromatin areas are stained by C-staining methods. These are methods which demonstrate "constitutive heterochromatin." Q-staining methods employ quinacrine compounds and produce flourescent Q-bands along the chromosomes. G (Giemsa) staining methods result in G-bands, and with some Giemsa techniques the R (reverse) staining method results in R-bands. Refinements in staining techniques, now permit not only all human chromosomes but parts of chromosomes to be identified.

With current techniques Lejeune and his colleagues discovered a chromosome **deficiency** in man that has been associated with the **cri-du-chat** (cat's cry) syndrome (Fig. 9.21). The name of this syndrome came from a mewing cat-like cry from babies with the disorder. Other characteristics are microcephaly (a small, broad head), broad face and saddle nose. Mental, motor and growth retardation are included in the syndrome. I.Q.'s of children studied were in the range of 20 to 40. Death occurs in infancy or early childhood, so the chromosome deletion is not transmitted to offspring. The chromosome deficiency is in the short arm of No. 5 chromosome. Abnormalities similar to those of the cri-du-

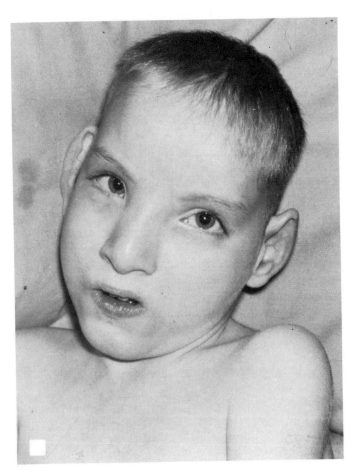

Figure 9.21 Patient with *cri-du-chat* syndrome which is associated with a deletion of part of the short arm of chromosome No. 5. The syndrome includes microcephaly and catlike cry. (Courtesy of Irene A. Uchida, Department of Pediatrics, McMaster University.)

chat syndrome but without the cat cry, have been associated with No. 4, the other member of the B group.

Another abnormality in man, which fragments chromosome No. 22 (Fig. 9.22), was named the **Philadelphia** (Ph¹) chromosome after the city where it was discovered by I. M. Tough and colleagues. This chromosome **deficiency** is associated with a specific disease, chronic granulocytic (myelogenous) leukemia and is identified by the lack of a part of the long arm of chromosome No. 22 as detected in bone marrow cells.

CAUSES OF CHROMOSOME CHANGES

Cause and effect relations between breakage in human chromosomes and specific phenotypes is mostly unexplored territory. Several human traits that have in the past seemed totally unrelated to chromosome structure have now been associated with **translocations, deletions** and other chromosome **aberrations.** Two human phenotypes named after G. E. Bloom and G. Fanconi are controlled by recessive genes but are associated generally with a **high incidence of chromosome breaks** and rearrangements.

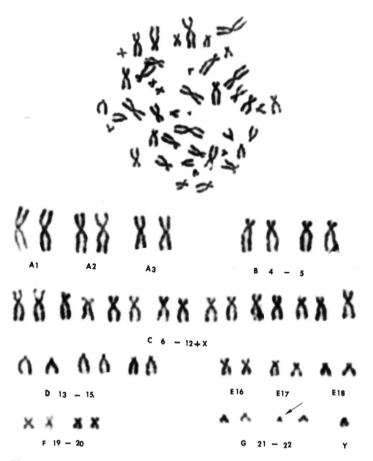

Figure 9.22 Metaphase and karyotype of a bone marrow cell from a patient with chronic myelogenous leukemia. The arrow indicates the Philadelphia chromosome (Ph¹) which has a partial deletion of the long arm of a No. 22 chromosome. This is the only instance in which a malignant disease in man is associated with a specific chromosomal aberration. (Courtesy of Arthur D. Bloom, Division of Genetics, College of Physicians and Surgeons of Columbia University.)

The Bloom syndrome is characterized by excessive sensitivity to sunlight, coupled with congenital telangiectasia of the face (dilation of small blood vessels with tumors involving blood vessels), stunted growth, and increased incidence of leukemia. Leukemia itself is associated with chromosomal breakage, particularly that resulting from irradiation. The Fanconi syndrome is characterized by anemia and other blood changes, irregular pigmentation, and malformations of extremities and several internal organs.

Chromosome structural changes are caused by X-rays and other kinds of irradiation, viruses such as rubella (German measles) and a number of drugs and other chemicals. Most of these environmental agents produce breaks that can lead to positional rearrangments but some (e.g., chlorpromazine) cause only breaks without rearrangements. Nonrearrangement breaks tend to become reconstituted without permanent aberrations. It is, therefore, impossible to interpret the significance of chromosome irregularities that might appear in a particular, single karyotype without information about environmental conditions affecting the person represented by the **karyotype.** Additional samples from the same person at different times are helpful. A valid interpretation of the significance of chromosome irregularities that cannot be attributed to a particular cause or associated directly with a phenotype, however, is difficult to obtain.

Chromosomal aberrations are related to the **age of the parents,** particularly the **mother,** at the time of birth of the person being studied. Many factors may be involved. Increased chromosome irregularities are mainly caused by breakdown in cell division, either mitotic or meiotic. Both structural and numerical changes in chromosomes result from aging effects on cell physiology. Indeed, chromosomal change must be a major factor in the **aging process.**

STANDARDIZATION OF HUMAN CYTOGENETICS

In 1971 a conference on the standardization of human cytogenetics was held in **Paris.** By this time technical developments, particularly chromosome banding techniques, had made possible the **identification of each of the human chromosomes.** A standardized nomenclature was needed to describe the chromosomes and chromosome regions revealed by the new techniques. Conferees agreed that a + or − sign should be placed before the appropriate symbol to signify added or missing whole chromosomes and these signs should be placed after a symbol to signify increase or decrease in length. For chromosomes with a long and a short arm, the letter q symbolizes the long arm and p the short arm. For example, 46, XY, 1q+ symbolizes an increase in the length of the long arm of chromosome No. 1 in the metaphase chromosomes arranged in a classified sequence (karyotype) of a male with 46 chromosomes, while 47, XY, +14p+ symbolizes a male with 47 chromosomes, including an additional chromosome No. 14 with an increase in the length of its short arm.

It was agreed that the number of the rearranged chromosome or chromosomes always should be placed in parentheses. For example, the descriptive symbols 46, XX, r(18) represent a karyotype of a female with 46 chromosomes, including a ring chromosome No. 18; 46, X, i(Xq), is a karyotype of a female with 46 chromosomes, including one normal X chromosome and an isochromosome (chromosome with arms that are mutually homologous) for the long arm of the X.

Names of chromosome aberrations were abbreviated for convenience in presenting chromosome formulas: def (deficiency), dup (duplication), inv (inversion) and tra (translocation). The bands were numbered in order along the short arm (p) and the long arm (q) of the chromosome. A chromosome change extending from a given band to the end of the arm is identified by an arrow to ter (terminal). Recombination chromosomes (rec) are identified such as 46 rec (3) for a recombined third chromosome in a karyotype of 46 chromosomes.

P. W. Alderdice, for example, described a syndrome associated with duplication and deficiency in chromosome 3. These chromosome aberrations were identified through chromosome banding studies of a

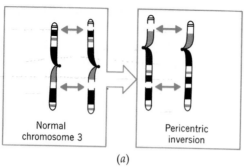

Normal
chromosome 3

Pericentric
inversion

(a)

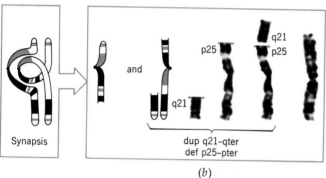

Synapsis

p25

q21

q21
p25

and

dup q21–qter
def p25–pter

(b)

Figure 9.23 Pericentric inversion in the human chromosome 3 associated with duplication and deficiency. (a) Normal chromosome 3 compared with inverted chromosome. (b) Synapsis configuration of normal and inverted chromosome and deficiency and duplication illustrated by banding of chromosome 3. (Courtesy of P. W. Alderdice.)

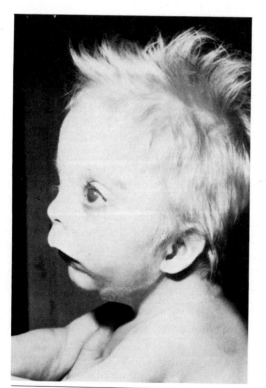

Figure 9.24 Child with duplication and deficiency in chromosome 3. The duplication-deficiency syndrome includes cone-shaped skull (coming to a point on top), distended veins on scalp, facial hair, prominent eyebrows, glaucoma (hardening of eyeball), short nose, anteverted nostrils, low set abnormal ears, high arched cleft palate, short neck and limbs, foot abnormalities. (Courtesy of P. W. Alderdice.)

pericentric inversion (p 25 q 21) (Fig. 9.23). Among 53 phenotypically normal carriers of the inversion, 13 male carriers fathered 94 pregnancies; 71 normal, 14 stillborn or neonatal deaths, 9 spontaneous abortions. The 8 females reported 46 pregnancies: 29 normal children, 8 stillborn or neonatal deaths, 7 spontaneous abortions and 2 living retarded children. These were karyotyped as 46 rec (3) dup (q 21 → q ter) def (p 25 → p ter). The children cannot sit up, turn over, nor eat solid food. Syndrome symptoms associated with the duplication and deficiency in chromosome 3 are illustrated in Fig. 9.24.

REFERENCES

Bergsma, D. (ed.) 1972. "Advances in human genetics and their impact of society." *Birth Defects*, 8(4), 1–118.

Bridges, C. B. 1935. "Salivary chromosome maps with a key to the banding of the chromosomes of *Drosophila melanogaster*." *J. Hered.*, 26, 60–64.

Burnham, C. R. 1962. *Discussions in cytogenetics*. Burgess Publishing Co., Minneapolis.

Darlington, C. D. 1965. *Cytology*, J. and A. Churchill, London.

Eggen, R. R. 1965. *Chromosome diagnosis in clinical medicine*. Charles C. Thomas, Springfield, Ill.

Judd, B. H., M. W. Shen, and T. C. Kaufman. 1972. "The anatomy and function of a segment of the X chromosome of *Drosophila melanogaster*." *Genetics* 71: 139–156.

Lewis, E. B. 1950. "The phenomenon of position effect." *Advances in Genetics*. M. Demerec (ed.). Vol. 3., Academic Press, New York.

Patterson, J. T., and W. S. Stone. 1952. *Evolution in the genus Drosophila*. Macmillan, New York.

Swanson, C. P., T. Merz, and W. J. Young. 1967. *Cytogenetics*. Prentice-Hall, Englewood Cliffs, N.J.

Turpin, R. and J. Lejeune. 1969. *Human afflictions and chromosomal aberrations*. Pergamon Press, Oxford.

Wright, S. W., B. F. Crandall, and L. Boyer (eds.) 1972. *Perspectives in cytogenetics*. Charles C. Thomas, Springfield, Ill.

PROBLEMS AND QUESTIONS

9.1 DNA of one cell type has 50 percent GC base pairs and DNA from another cell type has 5 percent GC base pairs. From these data what might be inferred about the source of the cell types?

9.2 How are heterochromatin and euchromatin involved in the development of males and females in the mealy bug?

9.3 (a) What genetic evidence first suggested chromosome structural changes? (b) Why did it take so many years to obtain cytological verification for the genetic evidence? (c) How was cytological verification obtained?

9.4 Compare the methods now available for cytogenetic studies of the fruit fly with those available for maize.

9.5 What characteristics of Drosophila salivary gland chromosomes make them especially suitable for cytogenetic studies?

9.6 What are the advantages of the acetocarmine smear technique as compared with fixing, sectioning, and staining methods for chromosome studies?

9.7 Formulate a plausible explanation for the origin of the giant salivary gland chromosomes in the developing larva of the fly. What do the cross bands represent?

9.8 What is the difference between a linkage and a cytological chromosome map?

9.9 Describe or illustrate with appropriate sketches how a recessive gene may be expressed through pseudodominance.

9.10 How can the extent of a chromosome deficiency be determined (a) genetically, (b) cytologically?

9.11 If a trait such as "waltzing" in mice known to depend on a single recessive gene (v) should appear in an animal considered to be heterozygous for the gene, how could it be determined (a) genetically and (b) cytologically whether a mutation had occurred or a deficiency was present in the chromosome opposite v?

9.12 How can (a) paracentric and (b) pericentric inversions act as "crossover suppressors?" Describe or illustrate. (c) Is crossing over really suppressed?

9.13 In *D. melanogaster,* the gene *Bd* is dominant with respect to a wing abnormality but is homozygous lethal. Another homozygous lethal gene (l) is located on the homologous chromosome and a cross-over suppressor prevents crossing over between *Bd* and *l*. What results would be expected from a cross between two flies with the genotype Bdl^+/Bd^+l?

9.14 In barley, a_n for white seedlings and x_c for yellow seedlings are on the same chromosomes with a crossover value of 10 percent. Plants with the genotype $a_n^+x_c^+$ are green. Homozygous a_n

and x_c plants die in the seedling stage. A plant with the genotype $a_n x_c^+/a_n^+ x_c$ was selfed. Give the expected results.

9.15 Reciprocal crosses were made between flies with striped sr bodies and wild-type flies sr^+, no differences were found between results of crosses in either the F_1 or the F_2 progeny. Homozygous sr females were then mated to males bearing the balanced curly and plum dominant genes on chromosome II, and stubble and dichaete dominant genes on chromosome III. F_1 progeny showing both curly and stubble were mated. Some progeny had striped bodies and the curly wings. (a) Is sr autosomal or sex-linked? (b) Which chromosome is it on?

9.16 Describe or illustrate with sketches the appearance of the following heterozygous chromosome modifications in salivary preparations: (a) deficiency, (b) duplication, (c) inversion, and (d) reciprocal translocation.

9.17 What (a) genetic and (b) cytological evidence would indicate that a translocation was present in a plant material such as barley?

9.18 How is pollen sterility associated with translocations? Illustrate.

9.19 In a Drosophila salivary chromosome section, the bands have a sequence 1 2 3 4 5 6 7 8. The homologue with which this chromosome must pair has a sequence 1 2 3 6 5 4 7 8. (a) What kind of a chromosome change has occurred? (b) Describe or draw a diagram to illustrate the possible pairing arrangement.

9.20 Other chromosomes have sequences as follows: (a) 1 2 5 6 7 8; (b) 1 2 3 4 4 5 6 7 8; (c) 1 2 3 4 5 8 7 6. What kind of chromosome modification is present in each? Illustrate with diagrams the pairing of these chromosomes with their normal homologues in salivary preparations.

9.21 Chromosome I in maize has the sequence A B C D E F, whereas chromosome II has the sequence M N O P Q R. A reciprocal translocation resulted in the following arrangements A B C P Q R and M N O D E F. Diagram the expected pachytene configuration and describe the causes of the pollen sterility that might be expected.

9.22 How could a phenotypic effect, such as the number of facets in expressions of bar eye in Drosophila, be demonstrated as a position effect?

9.23 (a) How do variegated position effects originate and (b) how can they be explained?

9.24 What is the (a) theoretical and (b) practical significance of the Philadelphia chromosome?

9.25 Describe with standard nomenclature the following human chromosome complements: (a) male karyotype with all normal chromosomes except one missing, No. 21, (b) female karyotype with a translocation between long arms of a D- and a G-group chromosome and short arms of D and G missing, (c) male karyotype with two translocations involving interchange of both whole arms of chromosomes Nos. 5 and 12.

9.26 What is the significance of chromosome inversions and translocations in evolution?

9.27 Why were critical chromosome studies in man not carried out before 1956?

9.28 (a) Why was it difficult to identify individual human chromosomes? (b) What criteria were used by the conference groups at Denver, London, Chicago and Paris to distinguish chromosomes?

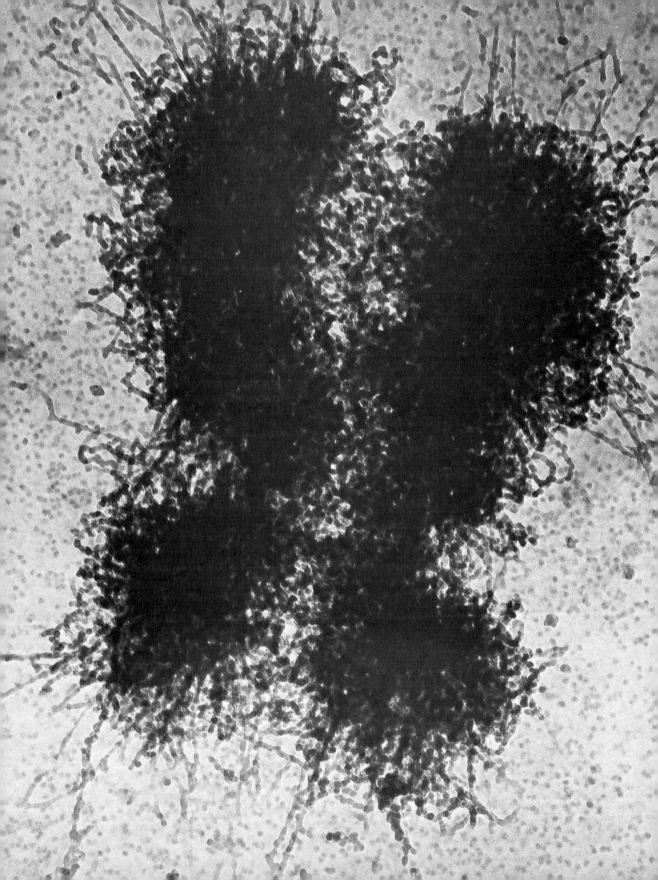

TEN

Somatic cells of higher plants and animals usually have chromosomes in pairs $(2n)$; that is, two of each kind of chromosome are present in each cell. Mature germ cells, having undergone reduction division, normally have one member of each pair (n). Many individual plants and animals, however, have local areas of somatic tissue characterized by a multiple of the basic chromosome number. A **doubling process** in cell division is the usual explanation for these deviations.

With the exception of sex differences, somatic doubling, and minor variations that occur in natural and experimental populations, all members of a species of plants or animals have the same basic chromosome number. The range of reported chromosome numbers in animals extends from 2 pair in a rhabdocoel *Gyratrix hermaphroditus* and some mites, midges, and scale insects, to more than 100 in some butterflies and Crustacea. The Crustacean, *Paralithodes camtschatica*, for example, has 208 chromosomes or 104 pairs. The reported range in plants is from 2 pairs in the small composite plant

VARIATIONS IN CHROMOSOME NUMBER

Two human metaphase chromosomes. (38,250×). (Courtesy of Gunther F. Bahr, Armed Forces Institute of Pathology.)

Haplopappus gracilis to several hundred in some ferns. A species of fernlike plants of the genus Ophioglossum is reported to have 768 chromosomes. Chromosome numbers of a few well-known plants and animals are listed in Table 10.1.

Whereas all individuals within a species, with the exceptions noted above, have the same chromosome number, different species within a genus often have different numbers. Cytological investigations of chromosomes help to unravel problems of species formation. The evolutionary path of a certain species can sometimes be followed by comparing the numerical and structural relations of its chromosomes with those of other species within the genus. The essential genetic material is a major factor in determining evolutionary patterns, with the chromosome number itself representing the number of packages into which the **DNA is divided.** Chromosome number is probably more constant, however, than any other single morphological characteristic that is available for species identification.

Changes in the number of chromosomes may be reflected in phenotypic variations, which constitute a useful tool for iden-

TABLE 10.1
Chromosome Numbers of Some Common Animal and Plants

Species	Number of Chromosome Pairs
Plants	
Garden pea, *Pisum sativum*	7
Sorghum, *Sorghum vulgare*	10
Maize, *Zea mays*	10
Johnson grass, *Sorghum halepense*	20
Alfalfa, *Medicago sativa*	16
Barley, *Hordeum vulgare*	7
Oats, *Avena sativa*	21
Tomato, *Lycopersicon esculentum*	12
Tobacco, *Nicotiana tabacum*	24
Trillium, *Trillium erectum*	5
Animals	
Gypsy moth, *Lymantria dispar*	31
Mouse, *Mus musculus*	20
Rabbit, *Oryctolagus cuniculus*	22
Cow, *Bos tarus*	30
Horse, *Equus caballus*	32
Donkey (ass), *Equus asinus*	31
Dog, *Canis familiaris*	39
Monkey, *Macaca rhesus*	21
Gorilla, *Gorilla gorilla*	24
Chimpanzee, *Pan troglodytes*	24

tifying the influence of individual chromosomes. If, for example, phenotypically distinguishable individuals with different chromosome numbers can be identified in natural populations or produced experimentally, it is sometimes possible to determine the effect of adding or removing certain chromosomes. Some plants with increased chromosome numbers have phenotypic changes in morphological or physiological characteristics that are of practical importance to man. Tomato plants, for example, with chromosome numbers above $2n$ are larger and produce more desirable fruit than do corresponding varieties with the usual $2n$ number.

Chromosome changes are classified in terms of obvious additions or eliminations of parts of chromosomes, whole chromosomes, or whole sets of chromosome (genomes). Two main classes are **euploidy** and **aneuploidy** ("ploid," Greek for unit, "eu," true or even; and "aneu," uneven). Euploids have chromosome complements consisting of whole sets of genomes. The basic chromosome number of euploid organisms is represented by the monoploid (n). Euploids with chromosome numbers above the monoploid level may be diploid ($2n$), triploid ($3n$), tetraploid ($4n$), or have some other "polyploid" number. The symbol x designates the basic number of chromosones, that is, the smallest number in a chromosome set. In garden peas, for example, $x = 7$ (haploid) and $2x = 14$ (diploid). An aneuploid is an organism with a chromosome number that is not an exact multiple of the monoploid (n) number. Among the aneuploids are monosomics ($2n - 1$) and trisomics ($2n + 1$).

NONDISJUNCTION AS A CAUSE OF ANEUPLOIDY IN MAN

With techniques now available, it is possible to observe and count human chromosomes with considerable precision. Total numbers above and below the usual 46 can be detected. A particular chromosome added or deleted can be identified. This can be accomplished from samples of cells sloughed off from the fetus and taken from the amnion (amniocentesis) during the 8th to 16th week of pregnancy. The cells are grown in cell cultures. Chromosome counts are made from dividing, cultured cells representing the fetus. As a basis for human chromosome studies and genetic counseling it is vital to know the frequency and significance of particular chromosomal abnormalities (Table 10.2).

Early cytogenetic studies were carried out on small samples of mostly institutionalized patients. More recent investigations have involved all babies born in a particular hospital over a period of time. Such work has provided data on the frequency and significance of chromosomal variations from more reliable samples in populations. For example, the Yale-New Haven Hospital study was based on a relatively unbiased sample of 4500 infants born consecutively during one year from a population area of about 400,000 people. Leucocytes from the cord blood of each infant were grown *in vitro* and then prepared for microscopic examination. Studies of this kind have shown that at least one in every 200 newborns on the average has a numerical chromosome irregularity. About half of the aneuploids detected thus far involve the sex chromosomes and half the autosomes.

TABLE 10.2
Aneuploidy Resulting from Nondisjunction in the Human Population

Chromosome Nomenclature	Chromosome Formula	Clinical Syndrome	Estimated Frequency at Birth	Main Phenotypic Characteristics
47, +21	2n + 1	Down	1/700	Short broad hands with Simian type palmar crease, short stature, hyperflexibility of joints, mental retardation, broad head with round face, open mouth with large tongue, slanting eyes.
47, +13	2n + 1	Patau	1/5000	Mental deficiency and deafness, minor muscle seizures, cleft lip and/or palate, polydactyly, cardiac anomalies, posterior heel prominence.
47, +18	2n + 1	Edwards	1/4000 to 1/18000	Multiple congenital malformation of many organs, low set, malformed ears; receding mandible, small mouth and nose with general elfin appearance; mental deficiency; horseshoe or double kidney; short sternum. 90% die in the first 6 months.
45, X	2n − 1	Turner	1/5000	Female with retarded sexual development, usually sterile, short stature, webbing of skin in neck region, cardiovascular abnormalities, hearing impairment.
47, XXY _1 Barr_ 48, XXXY _2 Barr_ 48, XXYY _1 Barr_ 49, XXXXY 50, XXXXXY	2n + 1 2n + 2 2n + 2 2n + 3 2n + 4	Klinefelter	1/500	Male, subfertile with small testes, developed breasts, feminine pitched voice, mental deficiency, long limbs, knock knees, rambling talkativeness, frequent early death.
47, XXX	2n + 1	Triple X	1/700	Female with underdeveloped genitalia and limited fertility. Frequent mental retardation.

THE DOWN SYNDROME, (47, +21)

The best-known and most important chromosome-related syndrome was formerly known as "mongolism" but is now designated the **Down syndrome.** It was named after Langdon Down who first described the symptoms in 1866. Individuals he studied, and those with the same syndrome who have since been observed (Fig. 10.1) were short in stature (about four feet tall), had slanting eyes (thus the earlier name "mongolian"), broad short skulls, stubby hands (particularly the fifth digit) with the simian crease on the palm and a single crease on the fifth digit (Fig. 10.2), a dermatoglyphic pattern in apes and monkeys, large tongues with a distinctive furrowing, and general loose jointedness, observed particularly in the ankles. They were characterized as low in mentality. Through the investigation of J. Lejeune in 1959, this became the first chromosomal disorder to be explained in man.

A small autosome in the G group, now known to be number **21** (Fig. 10.3), is added to the normal complement in cases where the Down syndrome is recognized. This is a **trisomic** for No. 21; all other chromosomes are "disomes." Chromosome No. 21 has been difficult to distinguish from No. 22. Number 21 was considered to be slightly larger than No. 22 when they were first arranged in a karyotype on the basis of size. Critical study has now shown that No. 22 is larger than No. 21 but the Paris conference agreed to leave them in their present positions in the standard karyotype.

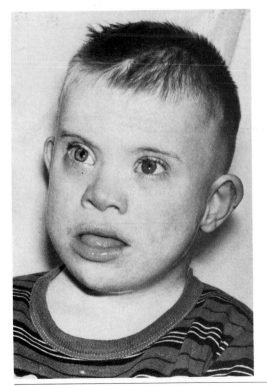

Figure 10.1 Facial features of a child with the Down syndrome. Note slanting eyes, open mouth with large tongue, protruding ears, and broad head. (Courtesy of Irene A. Uchida.)

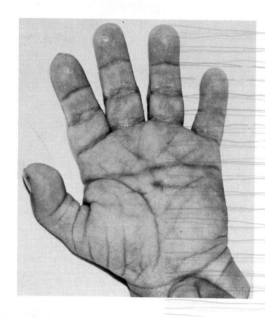

Figure 10.2 Hand of child with the Down syndrome showing simian crease on palm and single crease on little finger. (Courtesy of Irene A. Uchida.)

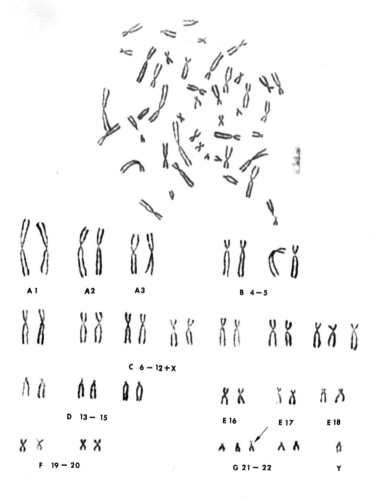

A 1   A2   A3   B 4 – 5

C 6 – 12 + X

D 13 – 15   E 16   E 17   E 18

F 19 – 20   G 21 – 22   Y

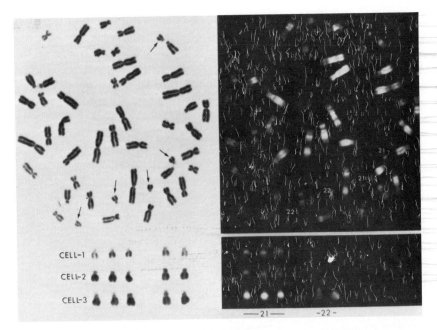

Figure 10.4 Photograph showing the difference between chromosome No. 21 and No. 22 in a trisomic with the Down syndrome. The spread on the left is stained with orcein. The one on the right has been stained with quinacrine dihydrochloride and photographed through a fluorescent microscope. This technique shows up the distinctive configuration of guanine in chromosomes. The group G chromosomes at the bottom came from three different cells from the same child with the Down syndrome. (Courtesy of Irene A. Uchida and C. C. Lin.)

Both 21 and 22 have small satellites that can be observed only in the best of preparations for microscopic study. Differential staining and photography through a fluorescent microscope have now distinguished between 21 and 22 (Fig. 10.4), and have allowed No. 21 to be identified with the Down syndrome. Trisomy of No. 21 is apparently the result of primary **nondisjunction**, which occurs at the reduction division in the meiosis of the mother. Paired chromosomes do not separate properly to the poles at anaphase and, as a result, one egg receives two No. 21 chromosomes and the first polar body receives no No. 21 chromosomes.

The Down syndrome occurs once in about 700 live births among European people. Incidence at conception is estimated to be considerably higher (7.3 per 1000), the difference being reflected in fetal loss due to spontaneous abortion.

Figure 10.3 Metaphase chromosomes and karyotype of a cell from a patient with trisomy G, No. 21 (see arrow) known clinically as the Down syndrome. (Courtesy of Arthur D. Bloom, Division of Genetics. College of Physicians and Surgeons of Columbia University.)

About 40 percent of the Down syndrome victims are born to women over 40 years of age, whereas this group of mothers produces only about four percent of all births in the general population. The mean maternal age for all babies born in the population according to Penrose is 28.17 years compared with a mean age of 34.43 years for mothers of babies with the Down syndrome (Table 10.3). In the New Haven Newborn Study, which utilized a large sample, the trisomy for chromosome No. 21 (of 0.69 per 1000) was less than previously reported. This was presumably because of the lower maternal age (25.7 years) in the New Haven sample as compared with samples of populations in several countries.

Some patients with the syndrome have had a total of only 46 instead of 47 chromosomes, but in such cases a translocation has joined chromosome No. 21 with another chromosome in the same complement (Fig. 10.5). When a translocation such as the D/G example cited in Chapter 9 can be associated with the Down syndrome in an infant, the risk figures of a second occurrence in the same family can be calculated. With assurance that the translocated chromosome will remain intact and assuming that one parent has a normal chromosome complement, the risk is 1/3 that each additional child will have the Down syndrome. Genotypes of parents who have already produced a child with the translocation Down syndrome are presumed to be: $DD/GGG \times DD/GG$. Further gametes from the first parent are expected to be: (1) (D/GG), (2) (D), (3) (G), (4) (DD/G), (5) (D/G) and (6) (D/G). If each is fertilized by the normal gamete (DG) from the second parent, the expected zygotes are: (1) DD/GGG, (2) DDG, (3) DGG, (4) DDD/GG, (5) DD/GG and (6) DD/GG If the monosomics (2) for G and (3) for D, and the trisomic (4) for D, are lethal, (5) is normal and

TABLE 10.3

Average Age of Mothers Bearing Children with the Down Syndrome from Samples Taken in Various Countries (From Penrose and Smith, 1966)

Country	Number of Cases	Mean Maternal Age Down Syndrome	Controls
Australia	1119	33.7	28.0
Canada	312	34.9	27.8
Denmark	518	34.6	28.3
England	2605	35.1	28.4
Finland	946	34.9	28.2
Formosa	20	36.6	28.3
Germany	225	34.2	27.1
Japan	321	33.2	28.2
Sweden	1242	35.4	28.7
United States	1788	33.3	27.7
U.S.S.R.	345	33.7	27.9
All	9441*	34.43	28.17

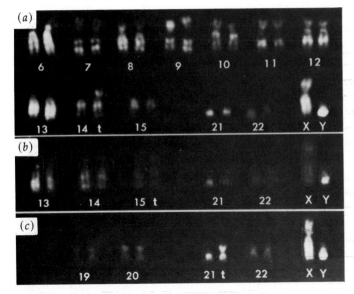

(a)

6 7 8 9 10 11 12

13 14 t 15 21 22 X Y

(b)

13 14 15 t 21 22 X Y

(c)

19 20 21 t 22 X Y

Figure 10.5 Partial karyotypes from three different translocation patients with the Down syndrome. The top band (*a*) is from a child with a translocation of No. 21 to a No. 14 chromosome; (*b*) involves 21 and 15, and (*c*) shows centric fusion of two No. 21 chromosomes. (Courtesy of Irene A. Uchida and C. C. Lin.)

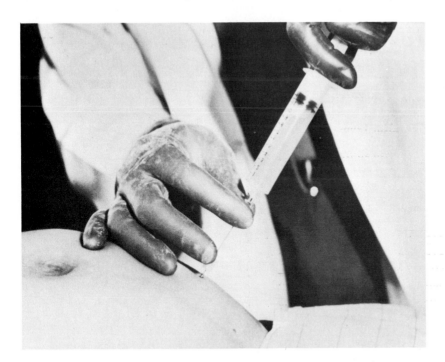

Figure 10.6 A physician taking a sample of fluid from the amnion of a pregnant woman for prenatal diagnosis of the Down syndrome. The procedure is called amniocentesis. (From *Human Genetics* by Richard A. Boolootian. Copyright © 1971 John Wiley & Sons.)

(6) is a balanced translocation carrier, **zygote (1)** is the only kind of zygote expected to carry the translocation and to express the syndrome. Thus one-third of the viable zygotes from the cross, representing possible future children in the family, are expected to have the Down syndrome.

AMNIOCENTESIS FOR DETECTING ANEUPLOIDY

The Down syndrome is sufficiently well understood to permit genetic counseling. A fetus may be checked in early stages of development by a process called **amniocentesis** (Fig. 10.6) to see if the chromosome abnormality exists. This is accomplished by taking a sample of amniotic fluid with a needle. The fetal cells are cultured and after a period of two to three weeks, chromosomes in dividing cells can be observed. If the chromosome arrangement for the Down syndrome is established, consideration may be given to termination of the pregnancy. Questions arise immediately concerning the desirability of widespread **screening** of pregnant women for trisomy of No. 21. Age of the mothers is a major factor for consideration (Table 10.3). The risk for the mothers less than 25 years of age is about one in 2500 births, at 40 years of age 1/100, at 45, 1/50. If all pregnant women of age 45 were checked, about one in 50 would be expected to be positive for trisomy of chromosome 21. Pregnancies in women over 45 are

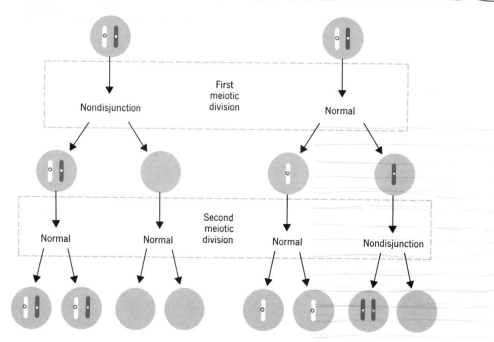

Figure 10.7 Nondisjunction at the first and second meiotic divisions. Nondisjunction at meiosis I produces gametes containing both or neither of the members of homologous pairs of chromosomes. Nondisjunction at meiosis II produces gametes containing (or lacking) two identical chromosomes both derived from the same member of the homologous pair. It is assumed that other chromosome pairs not shown in this diagram behave normally in the first and second meiotic divisions.

a special group of **high-risk** pregnancies.

In one study by Nadler and Gerbie, 155 "high-risk" pregnancies (chromosomal translocation carriers, maternal age more than 40 years, previous history of the trisomic Down syndrome) were diagnosed by amniocentesis between their 13th and 18th weeks of fetal gestation. No fetal or maternal complications were encountered and the risk to the infant and mother were considered to be low. Ten cases of trisomy for the Down syndrome were detected. The parents involved in all 10 cases chose therapeutic abortion.

Most, and perhaps all, cases of aneuploidy in man can be explained basically by nondisjunction at meiosis, as illustrated in Fig. 10.7. Older women are more likely than younger women to evidence chromosome damage and meiotic irregularities. An ovary may form about 400,000 eggs, but only about 400 will mature. All the eggs remain in prophase I from the time they are formed in the female fetus until the time of life when they become mature and capable of fertilization and development. The longer the time before fertilization, the greater the chance for **damage** and **irregularity**. Many drugs and other environmental factors can cause chromosome anomalies.

AUTOSOMAL TRISOMY D AND E

Syndromes reflecting autosomal trisomy have been established in the D and E chromosome groups. The Patau syndrome (47, +13) was described by K. Patau in 1960. This syndrome occurs in about one in 5000 newborns. It is rare in children and nonexis-

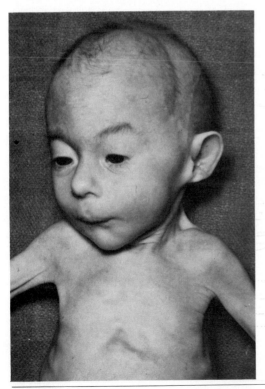

Figure 10.8 Facial features associated with trisomy 18. (Courtesy of Irene A. Uchida.)

tent in adults because the severe symptoms result in early death. Most of the deaths occur within the first three months but a few victims have lived for as long as five years. Symptoms include small brain, apparent mental deficiency, deafness, and numerous other external and internal abnormalities.

J. H. Edwards and his colleagues (in 1960) described a syndrome (Fig. 10.8) associated with chromosome trisomy (47, +18). The Edwards syndrome includes apparent mental deficiency and multiple congenital malformations involving virtually every organ system. Most infants with the Edwards syndrome die at an early age; some 90 percent within their first six months. Nearly all are deceased before they reach one year of age, but a few have been reported to be alive in their teen years. The sex ratio of three affected females to one affected male is attributed to the greater survival of genetically more balanced XX as compared with XY individuals.

The incidence of trisomy 18 has been variously estimated in different studies. The New Haven Newborn Study reported 0.23 per 1000 births. Another study showed the incidence to be one in 18,000. In these studies only a few cases have been observed and it is not known whether racial or other population groups differ in incidence. In general, the incidence of this deformity is greater among infants of older than those of younger women as expected if the cause of this trisomy is primary nondisjunction in meiosis.

SEX CHROMOSOME ANEUPLOIDY

THE TURNER SYNDROME (45,X)

This monosomic has a chromosome complement of 44 autosomes and one X chromosome. The chromosome anomaly is associated with an abnormal female phenotype described in 1938 by H. H. Turner and associates, and known as the **Turner** syndrome (Fig. 10.9). It occurs in about 0.23 per 1000 live births. A rough estimate for adults in the general population is one in 5000. Adults have virtually no ovaries, limited secondary sexual characteristics and they are sterile. Microscopic sections of ovaries of XO adult females show fibrous streaks of tissue representing remnants of the ovaries. XO females have short stature, low-set ears, hypoplasia, abnormal jaw formation, webbed neck, and a shieldlike chest. These symptoms can be recognized in affected infants. Mental deficiency is not usually associated with this syndrome. Epithelial cells of XO patients are chromatin negative (Chapter 4), as expected when only one X chromosome is present.

XO monosomics probably originate from exceptional eggs or sperm with no X chromosomes or from the loss of a sex chromosome in mitosis during early cleavage stages after an XX or XY zygote has been formed. This latter probability is supported by the high frequency of mosaics resulting from **postzygotic** events in patients with the Turner syndrome. **Mosaics** with XO/XX chromosomes show symptoms of the Turner syndrome but are usually taller than XO females with fewer anomalies. They show more **feminization**, more normal menstruation and have greater fertility. Many cases of the somatic Turner phenotype without the typical XO chromosome constitution are now known. Most of these

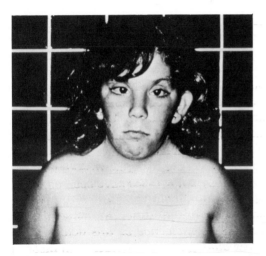

Figure 10.9 Physical characteristics of the Turner syndrome include short stature, "webbing" (folds) of the neck; and a broad, shield-like chest with widely-spaced nipples and underdeveloped breasts. There are often abnormalities of the eyes and the ears are low-set on the head. The uterus is small and the ovaries are represented only by fibrous streaks. Such females are sterile. (From *Human Genetics* by Richard A. Boolootian. Copyright © 1971 John Wiley & Sons.)

have one X chromosome and a fragment of a second X chromosome. Both arms of the second X chromosome are apparently necessary for normal ovarian differentiation. Individuals with only the long arm of the second X are short in stature and show other somatic symptoms of the Turner syndrome, whereas those with only the short arm of the second X have normal stature and do not show the Turner syndrome. When the short arm is missing the patient shows symptoms of the Turner syndrome. This indicates that the Turner somatic phenotype is controlled by genes on the **short arm** of the **X**.

Fragments of the Y chromosome also occur in some individuals with the somatic Turner phenotype. People with one X and a Y fragment, not including the Y short arm, have only streak ovaries but are normal in somatic phenotype. This suggests that male determining genes are in the short arm of the Y chromosome and those for the somatic Turner phenotype are in the long arm. Major somatic features of the Turner phenotype occur in some males as well as females. The male Turner syndrome is characterized by defective development of the testes, sterility and limited male sec-

ondary sexual characteristics along with somatic features of the Turner phenotype. These people have 46 chromosomes, one X and one or more parts of the Y incorporated into the chromosome complement.

THE KLINEFELTER SYNDROME, (47,XXY)

An extra Y chromosome in addition to the usual female (XX) complement (trisomic) has been associated with the abnormal male syndrome described (in 1942) by H. F. Klinefelter and known as the **Klinefelter** syndrome. It is estimated to occur in 1 per 500 live births and is characterized by small testes. Individuals with this syndrome are phenotypically males but with some tendency toward femaleness, particularly in secondary sex characteristics. Such features as enlarged breasts (Fig. 10.10), underdeveloped body hair, and small prostate glands are a part of the syndrome. Usually mental retardation is also recognized. Presumably the XXY constitution originates either by fertilization of an exceptional XX egg by a Y sperm or of an X egg by an exceptional XY sperm. Studies of the Klinefelter syndrome and of other abnormal sex chromosome conditions indicate that the Y chromosome in human beings, unlike that in Drosophila, is male determining.

The most common karyotype (about three-fourths of the cases) for the Klinefelter syndrome, is 47, XXY, but the symptoms of the syndrome will usually occur whenever more than one X chromosome is present along with a Y chromosome. More complex karyotypes associated with the Klinefelter syndrome include: XXYY, XXXY, XXXYY, XXXXY, XXXXYY and XXXXXY. All patients with the Klinefelter syndrome have one or more sex chromatin bodies in their cells. Higher numbers of sex chromosomes are associated with greater mental abnormality. Mosaicism, particularly XY/XXY, also detected in many patients with the Klinefelter syndrome is associated with less severe **physical, reproductive** and **mental** anomalies.

ANEUPLOIDY OF X CHROMOSOMES AND MENTAL DEFICIENCY

Other irregular combinations of sex chromosomes have also been recognized. About one percent of all mentally defective women in institutions have been shown to be **trisomic** for the X chromosome (47,XXX). This chromosome abnormality occurs in about 0.69 per 1000 live births in the general population. Individuals that have been studied with tetrasomic X chromosomes (48,XXXX) are all mentally defective females. The degree of mental deficiency apparently increases with the

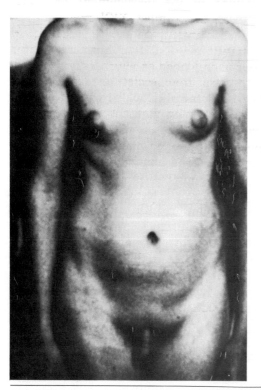

Figure 10.10 Males with the Klinefelter syndrome have small testicles, very little body hair, and tend to be long-legged and thin. They often develop breasts (gynecomastia) like females. They are sterile and often mentally retarded. (From *Human Genetics* by Richard A. Boolootian. Copyright © 1971 by John Wiley & Sons.)

number of extra X chromosomes present.

In the general population, multiple X chromosomes occur in the proportion of more than two per 1000. Individuals with the "triple X syndrome" are comparable in some ways to Drosophila metafemales (XXX). In Drosophila, however, such individuals are strikingly abnormal and sterile, whereas human XXX individuals are visibly indistinguishable from individuals with the normal XX arrangement. They may, however, be mentally abnormal.

The best-known symptoms in this syndrome are abnormalities associated with functional processes such as menstruation. One patient cited by P. A. Jacobs was a 37-year-old female who reported that the first suggestion of an abnormality was highly irregular menstruation. When the abdominal wall was opened the ovaries appeared as if they were postmenopausal. Microscopically, they showed deficient ovarian follicle formation. Of 63 cells observed, 51 had 47 chromosomes; the extra chromosome was an X. Nondisjunction in the production of the egg from which this woman developed was postulated as the mechanisms for the occurrence of the extra chromosome. When the somatic cells of this woman were studied microscopically, further evidence for X chromosome trisomy was obtained. Buccal smears showed two sex chromatin bodies in the epithelial cells as expected if three X chromosomes were present.

TRISOMIC (47, XYY) AND BEHAVIOR

P. A. Jacobs and her associates reported in 1965 that seven XYY males were detected in a population of 197 male, mentally subnormal inmates of a penal institution in Scotland. The XYY men were **unusually tall,** with an average height of 73.1 inches, compared with 67 inches for XY men in the same prison. Numerous other studies, mostly in institutionalized populations, have since confirmed that a high proportion of XYY individuals are tall, subnormal in intelligence (with I.Q.'s ranging from 80 to 95) and **antisocial.** The aggressive behavior that brought them into conflict with the law was usually against property rather than people. Although several investigators and authors have associated antisocial behavior of XYY men with the extra Y chromosome, further consideration must be given to the problem of causation.

XYY trisomy is one of the most common forms of aneuploidy in man, occurring about once in 500 live births in the general European population. These cannot all be accounted for in the criminal population. Furthermore, some XYY men have been described as perfectly normal in behavior. **Environmental factors** are presumed to be involved in the development of aggressiveness. Since most XYY men are subnormal in intelligence and excessively tall in stature, particular environmental situations in childhood or adulthood may lead to withdrawal from society or aggressive behavior. Unfavorable social conditions such as frustration in personal accomplishment and taunting from associates may encourage physical aggression as a means of adaptation. The XYY abnormality is not transmitted in inheritance. Males with this trisomy have not been found to transmit the extra Y chromosome to their sons. This extra chromosome seems to be weeded out in gametogenesis. A wide range of physical and mental abnormalities has been detected in XYY men but most of these are irregular in occurrence and do not form the syndrome pattern. Tallness of stature and **mental dullness,** however, are fairly constant characteristics among XYY men.

CHROMOSOMAL MOSAICS

Individuals who have at least two cell lines with different karyotypes derived from the same zygote, originate from a chromosome irregularity after fertilization. A chromosome may lag in a mitotic division and become incorporated in the same nucleus with the chromosome from which it divided. One daughter cell would thus receive one too many and the other would be one deficient, as illustrated in Fig. 10.11. Each cell would give rise to a cell line with its irregular chromosome number. Proportions of cells representing the different cell lines would vary in different tissues, making the extent of the mosaicism and the effect on the organism difficult to predict.

Many sex chromosome mosaics have been detected in human beings. The main phenotypic characteristic is extreme **variability.** Some sex chromosome mosaics that have been reported are: XO/XX, XO/XY, XX/XY, XXY/XX, XX/XXX, XXX/XO, XXXX/XXXXY, and several other combinations reflecting two or three cell lines. Mild to severe phenotypic symptoms have been associated with these sex chromosome mosaics.

ANEUPLOIDY FROM NONDISJUNCTION IN DROSOPHILA

Bridges' example of trisomic X chromosome (Chapter 4) was explained on the basis of primary nondisjunction of the X chromosomes (Fig. 4.19). When fertilized with normal sperm, eggs with two X chromosomes produced two kinds of trisomics, **XXX** and **XXY.** The metafemales (XXX) were always **sterile** and usually **inviable,** but the XXY combinations produced females phenotypically indistinguishable

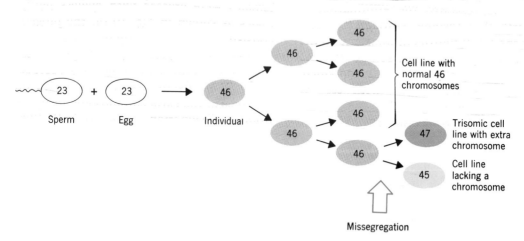

Missegregation

Figure 10.11 Mechanism through which chromosomal mosaicism may occur. A normal individual with 46 chromosomes is formed when sperm and egg unite. During the third cell division missegregation gives rise to a cell with 47 chromosomes and a cell with 45 chromosomes. Each of these cells initiates a cell line, thus forming a mosaic individual with cells carrying different numbers of chromosomes.

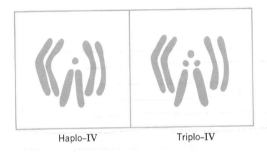

Figure 10.12 Trisomic called "triplo-IV," produced by the addition of a fourth chromosome, and monosomic called "haplo-IV," resulting from elimination of one fourth chromosome in Drosophila.

Haplo-IV   Triplo-IV

from wild-type females. Zygotes with normal pairs of autosomes but single X or Y chromosomes were monosomics whereas those with a single X produced the exceptional red-eyed males, which were normal in appearance but were sterile. Zygotes with a single Y chromosome were inviable. Later investigations by Bridges showed that chromosome IV, which is very small, could be added or eliminated without seriously affecting the viability of the flies. When, however, the large II and III chromosomes were lost or added, the resulting cells were always inviable.

A monosomic $(2n - 1)$ called **haplo-IV** was obtained through the elimination of one fourth chromosome. Through the addition of the small fourth chromosome to the wild-type complement (Fig. 10.12), trisomic $(2n + 1)$ called **triplo-IV** was obtained. Haplo-IV flies were small with slender bristles and deviating in several minor respects from the wild type. Triplo-IV flies

were slightly larger than the wild type with bristles that were appreciably more coarse than those of wild-type flies.

Chromosome IV genes, such as *ey* for eyeless (a phenotype characterized by small or missing eyes), behave differently with different chromosome combinations, as expected. When $2n$ eyeless flies (*ey/ey*) were outcrossed with haplo-IV (Fig. 10.13), the F_1 generation consisted of normal $2n$ and eyeless haplo-IV flies. When eyeless flies were crossed with triplo-IV, none of the first-generation progeny was eyeless. About half were triplo-IV, and half were normal (diplo-IV) as expected. When F_1 triplo-IV females were crossed with eyeless (*ey/ey*) males as shown in Fig. 10.14, the progeny consisted of about half triplo-IV and half $2n$. Normal and eyeless phenotypes were present in a ratio of about 5/6 to 1/6, respectively. This **trisomic ratio** is explained on the basis of the extra chromosome resulting from nondisjunction in the triplo-IV flies.

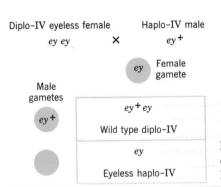

Diplo-IV eyeless female   Haplo-IV male

ey ey   ✕   *ey* +

Male gametes

ey +

	ey Female gamete
ey + *ey*	
Wild type diplo-IV	
ey	
Eyeless haplo-IV	

Figure 10.13 Cross between a diplo-IV eyeless female and a haplo-IV male.

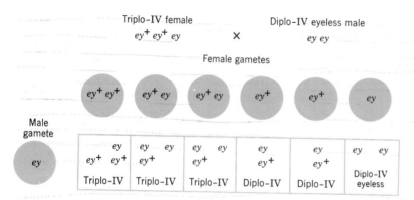

Triplo-IV female   Diplo-IV eyeless male

$ey^+ \; ey^+ \; ey$ × $ey \; ey$

Figure 10.14 Cross between a triplo-IV female and a diplo-IV eyeless male.

ANEUPLOID SEGREGATION IN PLANTS

The first critical study of aneuploid plants was initiated (in 1924) by Blakeslee and Belling when they discovered a "mutant type" with 25 chromosomes in the common Jimson weed *Datura stramonium*, which normally has 24 chromosomes in the somatic cells. At the meiotic metaphase, one of the 12 pairs was found to have an extra member; that is, one trisome was present along with 11 disomes (i.e., $2n + 1$). This trisomic plant differed from wild-type plants in several specific ways, particularly in shape and spine characteristics of seed capsules. Because the complement was composed of 12 chromosome pairs differing in the genes they carried, 12 distinguishable trisomics were possible in Jimson weeds. Through experimental breeding, Blakeslee and his associates succeeded in producing all 12 possible trisomics. These were grown in Blakeslee's garden and each was found to have a distinguishable phenotype that was attributed to the extra set of genes contained in one of the 12 chromosomes.

One of the 12 trisomic types, known as Poinsettia, had several distinguishing traits, including morphological characteristics of seed capsules, that were attributed to the basic trisomic arrangement. It was also possible to identify some traits that were deter-

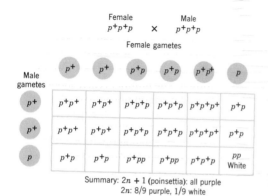

Figure 10.15 Cross between two purple trisomic plants of the genus Datura. Illustrating a trisomic ratio of 17:1.

mined by genes on particular chromosomes by trisomic ratios, that is 5:1, 17:1 (Fig. 10.15) and 35:1 in contrast to 3:1 and 1:1 expected from monohybrid crosses in regular $2n$ plants. The extra chromosome in Poinsettia, for example, was found to carry the locus for alleles p^+ and p for purple or white flowers, respectively. Any one of three chromosome arrangements with the dominant gene p^+ ($p^+p^+p^+$, p^+p^+p, or p^+pp) produced Poinsettia plants with purple flowers whereas only one, the fully recessive trisomic (ppp), gave rise to Poinsettia plants with white flowers. The $2n$ plants had two chromosome arrangements (p^+p^+ and p^+p) for purple and one (pp) for white.

Plant trisomics experience interesting complications when undergoing meiosis. **Two** chromosomes ordinarily go to **one** pole and **one** goes to the **other** in the reduction division of megasporogenesis, thus giving rise to different kinds of gametes, some with two and some with one member of the trisome. Trisomic ratios reflect increased proportions of progeny carrying wild-type dominant alleles. In the Jimson weed, developing megaspores tolerate extra chromosomes, and form gametes with relatively little loss of viability. When additional chromosomes above the n number enter developing microspores, however, the resulting spores cannot successfully compete against those with the normal haploid complement. A cross between two purple Poinsettia plants $p^+p^+p \times p^+p^+p$ (seed parent (female) written first) may be constructed as illustrated in Fig. 10.15. Female gametes are of four kinds, two haploid and two carrying an extra chromosome, in the proportion: $2p^+$, $2p^+p$, $1p^+p^+$, $1p$. Because male gametes

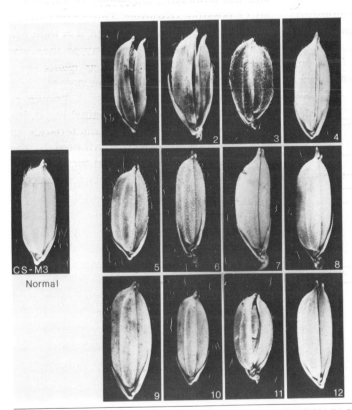

CS-M3
Normal

Figure 10.16 A set of 12 primary trisomics of rice, variety CS-M3, showing the grain characteristics of each type along with the diploid normal. The numbers correspond to the extra chromosome in each trisomic in order of its length in the karyotype. (Courtesy of Sharafot H. Khan and J. N. Rutger, University of California, Davis, and ARS, USDA, Davis, California.)

receiving extra chromosomes are nonfunctional, only two kinds, p^+ and p, occurring in the proportion $2p^+$ and $1p$, were involved in fertilization. All Poinsettia $(2n + 1)$ resulting from the cross carried at least one p^+ gene and had purple flowers. The $2n$ plants occurred in the proportion of 8 purple to 1 white. Complete sets of trisomics have since been discovered for other plants including rice (Fig. 10.16). The different trisomics may be distinguished by characteristics of the grain as well as other features of the plants.

TETRASOMICS AND NULLISOMICS

Aneuploids more complex than trisomics have been produced in some species, but ordinarily these are highly inviable. Tetrasomics $(2n + 2)$ have been identified in wheat, but no phenotypic characteristics were observed that could distinguish them from trisomics. Another variation that could be expected to occur in an aneuploid series is the complete absence of a certain kind of chromosome $(2n - 2)$. Plants in which a chromosome pair is completely missing are called **nullisomics.** These may be produced occasionally in nature but seldom survive long enough to be recognized or to perpetuate the chromosome type. E. R. Sears experimentally produced all of the 21 possible nullisomics in wheat, *Triticum aestivum.* By associating certain phenotypes with corresponding chromosome arrangements, nullisomics have been used effectively in locating several different genes in wheat.

EUPLOIDY

Although aneuploids differ from standard $2n$ chromosome complements in single chromosomes, euploids differ in multiples of n or x if $n = x$. The symbol n represents the haploid chromosome whereas x represents the smallest possible number in a chromosome set. Bread wheat, for example, has 42 chromosomes; $n = 21$ but $x = 7$, the lowest haploid chromosome number. **Monoploids** (n) carry one genome, that is, one each of the normally present chromosomes. The n or x chromosome number is usual for gametes of diploid animals, but unusual for somatic cells. Monoploidy is seldom observed in animals except in the male honey bee and other insects in which male haploids occur.

By contrast, plants have a **gametophyte** stage in their cycle which is characterized by the **reduced** (n) chromosome number. In higher plants, this stage is brief and inconspicuous, but in some lower plant groups it is the major part of the cycle. Occasionally, plants in natural populations or experimental plots can be recognized as monoploids by observation and verified by cytological procedures. These plants are usually frail in structure with small leaves, low viability, and a high degree of sterility. Sterility is attributed to **irregularities** at **meiosis.** Obviously, no pairing is possible because only one set of chromosomes is present. Therefore, if the meiotic process succeeds at all, the dispersal of chromosomes to the poles is irregular and the resulting gametes are highly inviable. Because monoploids undergo no segregation and carry a single set of genes, they are valuable experimental tools when they can be produced successfully. Microörganisms that are propagating monoploids are especially useful in genetics (Chapter 6). Diploid plants with two genomes (2n) are most common among euploids. Normal chromosome behavior in animals and plants is based on diploids, which are used in the following examples as standards for comparison.

POLYPLOIDY IN ANIMALS

Organisms with three or more genomes are polyploids. Fully one-half of all known plant genera contain polyploids, and about two-thirds of all the grasses are polyploids, but polyploids are rarely seen in animals. One reason is that the **sex balance** in animals is much more delicate than that in plants. As we noted Chapter 6, the addition of chromosomes above the diploid number gives rise to intersexes that do not reproduce. Sterility in animals is virtually always associated with a departure from the diploid number. The few animals (such as the brine shrimp, *Artemia salina*) that show evidence of polyploidy utilize **parthenogenesis** to escape the hazard of anomalous gametes.

An exceptional case of triploidy in salamanders related to *Ambystoma jeffersonianum* has been reported. Female salamanders of this particular group having large erythrocytes and erythrocyte nuclei produced some triploid larvae with 42 chromosomes, whereas those with small erythrocytes and erythrocyte nuclei produced diploid larvae with 28 chromosomes. Field observations and laboratory studies indicated that distinct, persisting populations of triploid females had become established in parts of the range occupied by this species complex.

Although animals composed entirely of polyploid cells are rare, many diploid animals have polyploid cells within certain tissues of their bodies. In teleostean embryos, for example, giant nuclei, presumably polyploid, have been observed in many species. Giant polyploid nuclei occur in particular tissues of a wide range of diploid animals (e.g., liver and kidney in man).

SOMATIC AND GERM CELL DOUBLING AS A CAUSE OF POLYPLOIDY IN PLANTS

Two basic irregular processes have been discovered by which polyploids may evolve from diploid plants and become established in nature. (1) **Somatic doubling:** cells sometimes undergo irregularities at mitosis and give rise to meristematic cells that perpetuate these irregularities in new generations of plants. (2) Reproductive cells may have an irregular reduction or equation division in which the sets of chromosomes fail to separate completely to the poles at anaphase. Both sets thus become incorporated in the same **restitution** nucleus, which doubles the chromosome number in the gamete. Both of these irregularities occur in nature. Once polyploidy is established, intercrossing among plants with different chromosome numbers may give rise to numerous chromosome combinations that are then under the influence of natural selection. All degrees of viability are encountered, from lethal combinations to those that compete favorably with diploids in particular environmental situations.

Two main kinds of polyploids, **autopolyploids** and **allopolyploids**, may be distinguished on the basis of their source of chromosomes. Autopolyploids occur when the same genome is duplicated. Apparently, this occurs rather frequently in single cells of many plants, but these cells usually do not survive. Allopolyploids result when different genomes come together through hybridization. Usually it is impossible to determine whether the genomes are alike and therefore whether the polyploids are autopolyploids or allopolyploids unless information from ancestral history is available and detailed chromosome studies are performed.

The presence of varying numbers of quadrivalents (i.e., four homologous chromosomes instead of the usual two pairing with one another in synapsis) suggests autopolyploidy. Unequal segregation of chromosomes in quadrivalents is one reason why **autopolyploids are sterile** to varying degrees. In the meiotic prophase, chromosomes must pair with one another throughout their entire length. When four similar chromosomes are present, they usually pair with different chromosomes at different places along their length, thus complicating the meiotic process and frequently resulting in unequal disjunction and other meiotic irregularities, thus giving rise to nonviable cells. Unequal chromosome pairing is not, however, the whole basis for sterility in autopolyploids. Unequal disjunction occurs and there are other, perhaps more important phenomena, but none of these is well enough established to permit discussion here. Chromosomes in some autopolyploids appear to pair properly and form bivalents rather than quadrivalents.

Some plant groups have a series of chromosome numbers based on a multiple of a basic number. In the genus Chrysanthemum, for example, the basic number is 9, and species are known that have 18, 36, 54, 72, and 90 chromosomes. In Solanum, the genus of nightshades including the potato, *S. tuberosum*, the basic number is 12. Members of this genus include species with 24, 36, 48, 60, 72, 96, 108, 120, and 144 chromosomes. In spite of such conspicuous examples, however, where autopolyploidy would appear superficially to be involved in the origin of plants with different chromosome numbers, it is doubtful that autopolyploidy alone has played a major role in the evolution of plant groups. Inviability and sterility would seem to preclude the perpetuation of true autopolyploids in nature. Autopolyploidy combined with allopolyploidy, however, produces **autoallopolyploids** and has apparently been an important process in the evolution of some plants.

Triploids (3n) with three genomes can occur when unreduced (2n) and normal (n) gametes unite. Reduction is commonly missed in many diploid plants, resulting in gametes with more than a single genome. Triploids do not ordinarily become established in nature because of irregularity during meiosis, which results in sterility and low survival. Plants that can be **vegetatively reproduced**, however, may be preserved as triploids. Gravenstein and Baldwin apples are perpetuated by grafting and budding and thus maintain their triploid characteristics. Tulips with three sets of chromosomes (3n) are also propagated by vegetative means. Triploids occurring in grasses, vegetables, and flower garden varieties are less stable and less fertile than corresponding diploids.

AMPHIDIPLOIDS RESULTING FROM HYBRIDIZATION

Tetraploids (4n) with four genomes frequently originate from a doubling of diploids, and may arise through intercrossing among various polyploids. They also may result from the duplication of somatic chromosomes following irregularities at mitosis. If the spindle does not develop properly in the mitotic sequence of a diploid, and cell division fails to follow chromosome duplication, a single nuclear membrane may develop around the two sets of chromosomes that ordinarily would produce daughter nuclei to form a single restitution nucleus. If this tetraploid cell perpetuates itself through normal mitosis, the increased chromosome number may become established in a group of cells or tissues within the organism. When such plants are capable of vegetative reproduction, they may be manipulated to produce whole tetraploid plants. Failure at reduction division in oöcytes of some plants also results in polyploids. Chromosome irregularity in mature pollen is rare because developing male gametes with irregular chromosome numbers do not compete favorably with normal gametes.

The low fertility and marked phenotypic variation associated with chromosome irregularity are illustrated in Fig. 10.17. A cross was made between Polish wheat *T. polonicum*, a tetraploid with large, amber kernels, and Marquis, a hexaploid with hard, red kernels. The entire F_2 from the cross is shown in the illustration. Only a few plants were produced from an extensive experiment and those observed varied widely.

Allopolyploidy has occurred in various

Figure 10.17 Heads of wheat resulting from a cross between Polish with 28 chromosomes and Marquis with 42 chromosomes. These are the only heads produced from an extensive experiment, indicating the low fertility encountered in crosses between plants with different chromosome numbers.

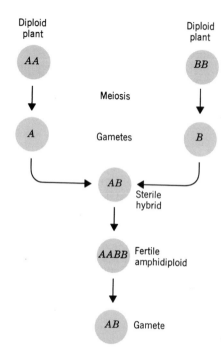

Figure 10.18 The production of a fertile amphidiploid (*AABB*) by doubling of the chromosomes of a sterile hybrid (*AB*) produced by crossing two normal diploid plants (*AA* and *BB*). *A* represents a haploid set of chromosomes from diploid plant *AA* and *B* represents a haploid set of chromosomes from diploid plant *BB*. Since *A* chromosomes are not homologous with *B* chromosomes, meiosis is highly irregular in hybrid *AB*, resulting in a high degree of sterility. The fertile amphidiploid (*AABB*) may be produced by the rare union of diploid (*AB*) gametes produced by the hybrid *AB*. If the amphidiploid meiosis is normal— *A* chromosomes pair with *A* chromosomes and *B* chromosomes pair with *B* chromosomes. (Reprinted with permission from *BioScience*, Vol. 21, No. 9, 1971.)

plant groups, and some present-day plants have resulted from this kind of hybridization. Some polyploid species that are established in nature have genomes that correspond more or less completely to the combined chromosome complements of two different but related diploid plants (Fig. 10.18) These **amphidiploids** have undergone hybridization somewhere in their ancestral history. Allopolyploidy thus represents a method by which new species may be formed almost immediately, whereas autopolyploidy alone results in cell-division anomalies and "dead ends" with reference to evolution.

POLYPLOIDY AND PLANT EVOLUTION

Early in the nineteenth century, some seeds of the American saltmarsh grass (*Spartina alterniflora*) were accidentally transported by ship to Bayonne, France, and Southampton, England. The American species became established in the same localities where a European saltmarsh grass (*S. maritima*, formerly *S. stricta*) was growing. A new saltmarsh grass, *S. townsendii*, (currently known as *S. anglica*) commonly called **Townsend's grass,** was later identified in these localities. By 1907 it had become common along the coast of southern England and northern France. Townsend's grass was more vigorous and aggressive than either the American or the European species and crowded out the native grasses in many places. It therefore was

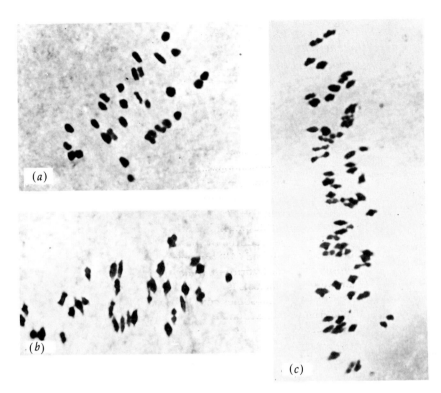

Figure 10.19 Metaphase I (paired) chromosomes of two parent species and the amphidiploid hybrid, Townsend's grass. (*a*) European marshgrass *Spartina maritima*, $2n = 60$; (*b*) American saltmarshgrass *S. alterniflora*, $2n = 62$; (*c*) Townsend's grass *S. townsendii* (*S. anglica*), $2n = 122$ (some plants had $2n = 120$ and some $2n = 124$). (Courtesy of C. J. Marchant.)

intentionally introduced into Holland to support the dikes and it was also imported into other localities for similar purposes. Townsend's grass was considered to be a hybrid between the American and European species but, unlike most hybrids, it was fertile and true breeding. The European grass had $2n = 60$ chromosomes, the American species $2n = 62$, and Townsend's grass $2n = 122$ (some plants had $2n = 120$ and others 124). (See Fig. 10.19.) These facts suggested that a cross had occurred in which allopolyploidy was involved. Townsend's grass was an amphidiploid with the sum of the diploid chromosomes carried by the two species. This evidence, along with the high fertility and intermediate appearance, indicated that the new plant arose from natural **hybridization and doubling** of chromosomes. The chromosome doubling had presumably given the hybrid its **fertility** and ability to survive.

The primrose, *Primula kewensis*, is an allotetraploid with 36 ($2n$) chromosomes. It was derived from a cross between two diploids *P. floribunda* ($x = 9$) and *P. verticillata* ($x = 9$). Plants from these two species crossed readily, producing hybrids with 18 chromosomes in their vegetative cells, 9 from *P. floribunda* and 9 from *P. verticillata*, but the hybrids were sterile. Eventually, however, a branch on a hybrid

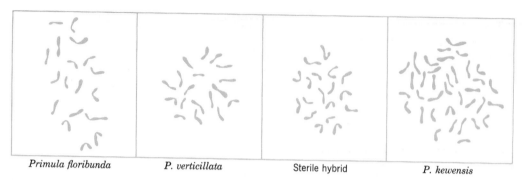

Figure 10.20 Chromosomes of *Primula floribunda*, *P. verticillata*, the sterile hybrid, and the allotetraploid, *P. kewensis*. The two parents and the hybrid each have 18 chromosomes, whereas the fertile tetraploid *P. kewensis* has 36 chromosomes.

plant developed from a cell in which the chromosome number was doubled (36) so that each chromosome had a homologous partner. This branch was propagated and gave rise to a fertile primrose plant with cells containing 36 chromosomes. In Fig. 10.20, metaphase chromosomes of the two diploid parents, the sterile diploid hybrid, and the fertile allotetraploid are shown.

INDUCED POLYPLOIDY

Polyploids have been **induced experimentally** by several methods in various plants. Anything that interferes with spindle formation during mitosis might result in a doubling of the chromosomes. Induced polyploidy was first demonstrated by subjecting growing plants to a higher than usual temperature. Maize and some other plants responded to such treatment with an increase in chromosome number of certain cells. Some of these cells gave rise to germinal tissue and whole plants were propagated. Other such cells were cultured artificially and polyploid plants were produced. When, for example, the buds of tomato plants were removed, cells of some shoots developing from the scar tissue were tetraploid. These were propagated and whole plants with 4*n* chromosomes were produced.

The method of inducing polyploidy in plants that has become most widely used was developed by A. F. Blakeslee, A. G. Avery, and B. R. Nebel. These investigators found that an alkaloid, **colchicine**, extracted from the autumn crocus, *Colchicum autumnale*, could disturb spindle formation during cell division. When root tips or other growing plant parts were placed in appropriate concentrations of colchicine, chromosomes of the treated cells duplicated properly, but spindle formation was inhibited and the cytoplasmic phase of cell division did not occur. Instead, restitution nuclei with different numbers of chromosomes were produced in the treated tissue. Some cells had a completely doubled chromosome number. When these cells were propagated, tetraploid plants were produced and tetraploid seed was obtained.

In some plants, growing areas at the stem tips and lateral buds have three distinct cell

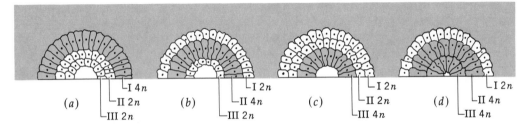

Figure 10.21 Three cell layers found in the stem tips of a plant. These diagrams illustrate a mechanism by which chromosome doubling may be reflected in various parts of the plant, including the reproductive cells.

layers, as illustrated in Fig. 10.21. Cells from each layer are much alike in early stages of development, but later they give rise to separate tissues in stems, leaves, and other organs. The outer layer (I) becomes the epidermis, the middle layer (II) gives rise to the reproductive cells (eggs and pollen), and the inner layer (III) produces the internal parts of stems and leaves. Colchicine placed in the medium of the growing tips may interfere with division and result in transformation of $2n$ cells into $4n$ cells in one or more layers. Cells usually divide vertically in such a way that the number within a given layer is increased. In examples a, b, and c, the $4n$ cells are restricted to the first, second, and third layer, respectively. Sometimes cells divide horizontally and the daughter cells enter a new layer. Thus, a doubled ($4n$) cell in layer III may give rise to a $4n$ cell in layer II, as shown in d. This pattern of irregularity may extend $4n$ cells to the reproductive tissue and thus perpetuate $4n$ cells in a new plant. By propagating tetraploid tissue, it is possible to produce tetraploid plants.

EXPERIMENTAL PRODUCTION OF POLYPLOIDS

In an early experiment, the Russian cytologist G. D. Karpechenko synthesized a polyploid from crosses between two common vegetables belonging to different genera, the radish, *Raphanus sativus*, and the cabbage, *Brassica oleracea*. Although these plants were only distantly related, they were enough alike to be crossed successfully with the intervention of the experimenter. Both had 9 pairs of chromosomes. The diploid hybrid had 18 chromosomes, 9 from each parent, but it was sterile and could not perpetuate itself, largely because of the failure in pairing between the unlike chromosomes in meiosis. Some unreduced gametes were formed, however, and in the F_2 population Karpechenko recovered some tetraploids. When the chromosomes of the F_1 hybrid were doubled in this way, a fertile polyploid named Raphanobrassica was produced with 18 radish and 18 cabbage chromosomes. Because two sets of chromosomes were now present from each parental variety, pairing was quite regular. Normal gametes were produced and a high degree of fertility was obtained. This experiment had theoretical significance because it demonstrated a method by which fertile interspecific hybrids could be produced. It also suggested the possibility of incorporating desirable genotypes from two different species into a new polyploid species. Seed capsules of the

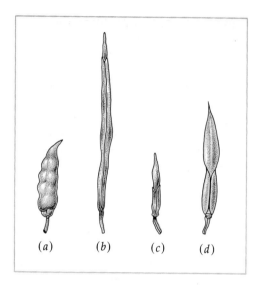

Figure 10.22 Seed pods of (*a*) radish (Raphanus), with 18R chromosomes; (*b*) cabbage (Brassica); with 18B chromosomes; (*c*) sterile diploid hybrid with 9R + 9B chromosomes; and (*d*) tetraploid resulting from chromosome doubling, with 18R + 18B chromosomes. (After Karpechenko.)

(*a*)   (*b*)   (*c*)   (*d*)

parents, the sterile hybrid, and the tetraploid plants of Karpechenko's experiment are shown in Fig. 10.22. Unfortunately from the practical standpoint, Raphanobrassica had the foliage of a radish and the root of a cabbage.

INTERSPECIFIC HYBRIDIZATION IN THE ORIGIN OF BREAD WHEAT

Among the cultivated varieties of wheat, three different chromosome numbers are represented: 14, 28, and 42 ($x = 7$). For example, the primitive small-grained einkorn type of Europe and Asia, *Triticum monococcum*, has 14 chromosomes in its vegetative cells. Its yield is low and it is of comparatively little value. An emmer wheat (durum) *T. dicoccum*, grown chiefly in northern Europe but also in the United States, has 28 chromosomes. It has thick heads with large hard kernels and is used mainly for macaroni, spaghetti, and stock feed. The bread wheats, *T. aestivum* with 42 chromosomes, were postulated by J. Percival in England to have come from a cross between emmer wheat and goat grass (Aegilops), both of which are native to the Babylonian region where bread wheat originated.

When techniques for artificial chromosome doubling became established, investigations of the origin of bread wheat confirmed Percival's theory. Experimental evidence obtained by E. S. McFadden, E. R. Sears, and H. Kihara established the origin of one type of bread wheat, *T. spelta*. These investigators doubled the chromosome numbers of an emmer wheat and of a wild goat grass, *Aegilops squarrosa*, with colchicine, and made a cross. The hybrid had 42 chromosomes, 28 from the wheat parent and 14 from the goat grass parent. It was phenotypically similar to primitive forms of bread wheat. The decisive genetic test was made by crossing the synthesized wheat with *T. spelta*. Hybrids were fertile and evidenced normal chromosome behavior. These experiments indicated that a moderately useful wheat and a useless weed had, at some time in the past, hybridized in nature and produced forerunners of man's valuable crop, wheat.

ORIGIN OF NEW WORLD COTTON

Crosses can be made between distinct species of cotton, Gossypium. The hybrids show a wide range of vigor and fertility, making the material favorable for studies of origins. Three cytological groups have been found to correspond with the major world distributional areas. Old World cotton has 13 pairs of large chromosomes. American cotton, which originated in Central or South America, has 13 pairs of small chromosomes. New World cotton (the cultivated long-staple type) has 26 pairs, 13 large and 13 small. Evidently, **hybridization** and **chromosome duplication** occurred somewhere in the ancestry of the New World cotton.

J. O. Beasley used the colchicine tech-nique and succeeded in doubling the chromosomes of a hybrid between the Old World and American cotton. The resulting hybrids, with four sets of chromosomes (amphidiploids), crossed readily among themselves and produced fertile plants resembling New World cotton. The process by which the valuable polyploid cotton may have originated in nature was thus duplicated in the laboratory.

STERILE HYBRID GRASS MADE FERTILE BY CHROMOSOME DOUBLING

Interesting experiments that made use of induced polyploidy and cytological analysis in certain grasses have been conducted by W. S. Boyle, A. H. Holmgren, and their as-

Figure 10.23 Parent plants representing different genera and a hybrid produced by these plants (a) *Hordeum jubatum;* (b) F$_1$ hybrid; (c) *Agropyron trachycaulum.* (Courtesy of W. S. Boyle.)

sociates at Utah State University. A completely sterile perennial grass was observed in Cache Valley, Utah, and in several other locations. On the basis of its sterility and morphological characteristics, the grass was tentatively identified as a natural, sterile hybrid between two genera in the tribe Hordeae. Examination of the Hordeae species in the vicinity of the hybrids indicated that the two parents were probably *Agropyron trachycaulum* and *Hordeum jubatum* (Fig. 10.23). Cytological studies on the two presumed parents and the hybrid supported this view. Both *A. Trachycaulum* and *H. jubatum* were found to be allotetraploids ($2n = 28$). During meiosis of both species, normal pairing of chromosomes occurred to form 14 paired chromosomes (bivalents). The sterile hybrid was also a tetraploid with the same chromosome number ($2n = 28$), but its chromosome behavior during the meiotic process was highly irregular (Fig. 10.24). Fourteen unpaired chromosomes (univalents) and seven bivalents were frequently observed at metaphase in pollen mother cells. Many lagging chromosomes remained in the center of the spindle during anaphase, and numerous small micronuclei, reflecting chromosome irregularity, were observed following division. No viable pollen was produced. These observations indicated that **meiotic irregularity** was a major factor in the sterility.

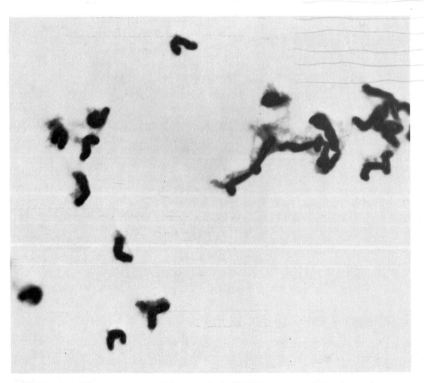

Figure 10.24 Chromosomes of the sterile hybrid between *A. trachycaulum* and *H. jubatum* during meiosis. Seven bivalents and 14 single chromosomes were frequently found as in this photograph. (Courtesy of W. S. Boyle.)

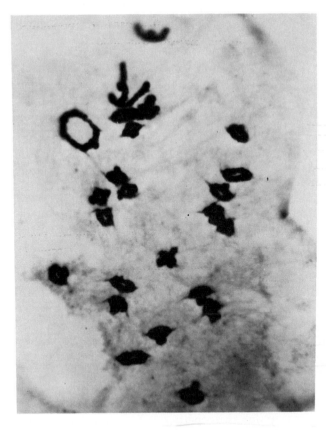

Figure 10.25 Diakinesis stage of colchicine-induced octoploid. In this photograph the following chromosome associations are present: 1 ring of 4; 22 rings of 2; 3 rods of 2; and 2 single chromosomes, making a total of 56. (Courtesy of W. S. Boyle.)

Sterile hybrids were produced through controlled reciprocal crosses between *A. trachycaulum* and *H. jubatum,* thus confirming the predicted parentage. The hybrids were then treated with colchicine. Some stalks or clums with double chromosome numbers (octoploids) were produced and set seed. All this seed was viable and produced **fertile** plants considered to be autoalloploids. This name was applied because both autopolyploidy and allopolyploidy had entered into the formation of the fertile octoploid. Chromosome studies indicated that the two parents carried a genome in common. Therefore, the genomic formulas of the parents were represented as **AABB** and **AACC.** The sterile hybrid was **AABC** and the colchicine-induced octoploid was **AAAABBCC** (Fig. 10.25).

PRACTICAL APPLICATIONS OF POLYPLOIDY

Induced polyploidy has not been exploited to a great extent. Practical applications may become more common as additional data are accumulated. By artificially induced polyploidy, **disease resistance** and other desirable qualities have been incorporated into some commercial crop plants. Tobacco, *Nicotiana tabacum,* for example, is susceptible to the tobacco mosiac virus (TMV), whereas *N. glutinosa* appeared at first observation to be resistant. Further investigation, however, showed that in *N. glutinosa* the virus killed the cells that were invaded and the virus particles became isolated in the dead cells. The apparent resistance thus was attributable to hypersensitivity. When the two tobacco species were crossed, the hybrid was found to be "resistant" to the virus but totally sterile. When the chromosomes were doubled, it was possible to secure a fertile polyploid, "resistant" to the virus.

Some varieties of plants that serve man's purposes more effectively than others have now been identified as polyploids. Many polyploids were selected and cultivated because of their large size, vigor, and ornamental values, before their chromosome numbers were known. Giant **sports** from twigs of McIntosh apple trees that were found to be tetraploid (4*n*) were propagated into whole trees, which produce extra large fruit. The texture of the giant apples is as fine as that of diploids, but the yield is inferior. Mass selection of seedlings may overcome this difficulty. Bartlett pears, several varieties of grapes, and cranberries have also produced sports with giant fruits. Some of these show promise of practical usefulness. With colchicine treatment, a number of polyploids have been developed artificially. This technique has provided a way to explore the mechanism involved in polyploid formation and to make use of the good qualities of polyploids. Tetraploid (4*n*) maize is more vigorous than the ordinary diploid and produces some 20 percent more vitamin A. Its fertility is somewhat reduced, but this drawback responds to selection. Polyploid watermelons have been developed by colchicine treatment by Kihara and others. The tetraploid with 44 chromosomes is large and has practical value. Triploid watermelons with 33 chromosomes are especially desirable because they are sterile and have no seeds. Among the flower garden varieties, 4*n* marigolds and snapdragons are widely cultivated.

Polyploid plants respond to artificial selection and hybridization, as do diploid species. The recent history of plant breeding has been characterized by a marked improvement in many polyploid plant crops. The yield of wheat, for example, has increased appreciably. This has been accomplished by developing disease-resistant strains and breeding for increased hardiness and greater efficiency under various environmental conditions available in wheat growing areas. A constant threat to the wheat crop is rust, a fungus that attacks the stems and leaves of the growing plants and destroys the ripening grain. Spores are borne by wind, and when conditions are right they spread like fire through wheat fields. The disease can be combated by developing rust-resistant strains and by eradicating barberry bushes, which are hosts to the spores during the spring months. But the rust keeps evolving new varieties that destroy previously resistant grain, thus perpetuating the job of plant breeders.

The advantage of incorporating rust resistance is illustrated in Fig. 10.26. The larger kernels at the left are from a new strain of rust-resistant spring wheat. At the right are shown kernels of wheat, similar in other respects, but not resistant, which are dwarfed from the infection with stem rust. The number of kernels of grain per plant as well as the size of the kernels is decreased by rust infection. Investigators in agricultural experiment stations are constantly alert for **new rusts.** When a new one is found, the standard wheat varieties are tested against it. If they are not resistant, breeding programs are initiated immediately to develop new strains resistant to that particular rust.

In 1953, for example, race 15B of wheat stem rust rose to epidemic proportions in

Figure 10.26 Wheat kernels illustrating the advantage of rust resistance. Left, rust-resistant wheat; right, rust-susceptible wheat infected in nature, similar in other respects to the strain represented at left. (Courtesy of Utah State Experiment Station.)

the United States. Some 65 percent of the durum wheat crop was destroyed and some 75 percent in 1954. At the same time, some 25 percent of the bread wheat in the United States was destroyed because of blight. Surveys conducted over a period of 11 years before 1953 showed that race 15B was present at a very low level. During the years preceding 1953 this strain evidently accumulated the genetic traits that enabled it to multiply and to become a major component of the rust population. Strains resistant to 15B were developed in a short time and since 1954 this rust has been con-trolled. Breeding for rust resistance in wheat is a continuing program.

This example illustrates the **uneasy equilibrium** existing between plant crops and strains of pathogens in nature. When a new crop variety is introduced, diseases to which it is resistant are suppressed. Those to which it is susceptible will thrive and multiply. Likewise, as variants of the pathogen arise by mutation or other genetic means, those that encounter a susceptible host increase whereas those that are not virulent on the prevalent host plants do not survive, or persist only in limited numbers.

REFERENCES

Bergsma, D. 1972. "Paris conference: Standardization in human cytogenetics." *Birth Defects* 8(7), 1–46.

Blakeslee, A. F. 1941. "Effect of induced polyploidy in plants." *Amer. Natur.* 75, 117–135.

Brown, W. V. and E. M. Bertke. 1969. *Textbook of cytology.* C. V. Mosby, St. Louis, Mo.

Darlington, C. D. and K. R. Lewis. 1966. *Chromosomes today.* Plenum Press, New York.

Dawson, G. W. P. 1962. *An introduction to the cytogenetics of polyploidy.* Blackwell Sci, Publ., Oxford.

Jacobs, P. A., W. H. Price, and P. Law (eds.) 1970. *Human population cytogenetics.* University Press, Edinburgh.

Lilienfeld, A. M. and C. H. Benesch. 1969. *Epidemiology of mongolism.* Johns Hopkins Press, Baltimore.

McFadden, E. S. and E. R. Sears. 1949. "The origin of *Triticum spelta* and its free-threshing hexaploid relatives." *J. Hered.* 37, 107–116.

Müntzing, A. 1961. *Genetic research.* Lts. Forlag, Stockholm.

Raven, P. H. and H. Curtis. 1970. *Biology of plants.* Worth Publishers, New York.

Sears, E. R. 1948. "The cytology and genetics of wheats and their relatives." *Adv. in Genet.,* 2, 240–270. M. Demerec (ed.) Academic Press, New York.

Swanson, C. P., T. Merz, and W. J. Young. 1967. *Cytogenetics.* Prentice-Hall, Englwood Cliffs, N.J.

Uchida, I. A. and C. C. Lin. 1973. "Identification of partial 12 trisomy by quinacrine fluorescence." *J. Pediat.* 82(2), 269–272.

PROBLEMS AND QUESTIONS

10.1 According to the Lyon hypothesis (Chapter 4) all but one X chromosome in multi-X individuals degenerate and form sex chromatin bodies. How many sex chromatin bodies would be expected to occur in a cell from a person with (a) the Turner syndrome, (b) trisomic X, and (c) the Down syndrome?

10.2 How can human trisomy be explained?

10.3 If the Down syndrome occurs in 1/700 births in the general population (a) what is the chance that two cases will be recorded in a city hospital in the same day? (b) If the number of live births for a given year in a country is 42 million, how many would be expected to have the Down syndrome? (c) If 40 percent of the babies with the Down syndrome are born to mothers over 40 years of age and mothers in this age group produce 4 percent of all children, what is the chance that a given woman in this age group would have a baby with the Down syndrome?

10.4 If nondisjunction of chromosomes No. 21 is known to have occurred in the division of a primary oöcyte in a particular woman and the two 21 chromosomes had remained together in the division of the secondary oöcytes, what is the chance that a mature egg arising from this secondary oöcyte will receive the two No. 21 chromosomes?

10.5 If the Down syndrome occurs in about 1/700 and the Turner syndrome occurs in about 1/5000 in the general population, and each is separately and randomly distributed in the population, what is the chance that a baby will be born with both of these abnormalities?

10.6 (a) If X chromosome trisomy occurs in 1/1000 of the general population and No. 18 trisomy occurs in 1/4000, and each is separately and randomly distributed, what is the chance that a baby, such as one described by Uchida, will be born with both abnormalities? (b) If the mother is over 40 years old and if the increased occurrence of nondisjunction makes mothers in this age group ten times more likely to have babies with each of these abnormalities, what is the probability of the two trisomies occurring in the same baby?

10.7 The Poinsettia type of Datura carries an extra member of the chromosome set $(2n + 1)$ in which the genes for purple (p^+) and white (p) flower color are located. From the following crosses, give the expected proportions of purple and white. (Female parent is always written first. Female gametes may carry either 1 or 2 chromosomes of this set, but viable pollen carries only a single chromosome.) (a) $p^+p^+p \times p^+p^+p$; (b) $p^+p^+p \times p^+p$; (c) $p^+pp \times p^+p^+p$; and (d) $p^+pp \times p^+p$. (e) How do trisomic ratios differ from the usual Mendelian ratios?

10.8 Triplo-IV fruit flies have an extra member of the fourth chromosome in which the gene *ey* is located. Give the expected results from a cross between triplo-IV flies of the genotype ey^+ey^+ey and diplo-IV, ey^+ey, flies in terms of (a) $2n$ and $2n + 1$; and (b) eyeless and normal eye phenotypes.

10.9 (a) What evidence concerning the influence of the Y chromosome on sex determination in man can be obtained by comparing the characteristics of XO, XXY, and XXX individuals? (b) Compare the influence of the human Y chromosome on sex determination with that of Drosophila and Melandrium Y chromosomes (see Chapter 4).

10.10 How can aneuploidy be used as a tool to identify genes with chromosomes?

10.11 Why are tetrasomics and nullisomics found in nature less frequently than trisomics?

10.12 What values, potential if not realized at present, could monoploids have for genetic studies?

10.13 Polyploidy is rare in animals, yet some tissues in the bodies of certain diploid animals show evidence of polyploidy. Why do numbers above $2n$ persist in somatic tissues when they do not occur in the whole animal?

10.14 Describe two methods by which polyploidy might occur in nature.

10.15 (a) How may autopolyploidy and allopolyploidy originate? (b) Evaluate the significance that each might have in evolution.

10.16 Why do tetraploids behave more regularly in meiosis than triploids, and perpeturate themselves more readily in populations?

10.17 Why is chromosome irregularity associated with low fertility in plants?

10.18 What is the significance of chromosome numbers in (a) taxonomy; and (b) studies of evolution in plants and animals?

10.19 A plant species A, which has seven chromosomes in its gametes, was crossed with a related species B, which has nine. Hybrids were produced but they were sterile. Microscopic observation of the pollen mother cells of F_1 showed no pairing of chromosomes. A section of the hybrid that grew vigorously was propagated vegetatively and a plant was produced with 32 chromosomes in its somatic cells. What steps might have been involved?

10.20 A plant species A $(n = 5)$ was crossed with a related species B with $n = 7$. Only a few pollen grains were produced by the F_1 hybrid. These were used to fertilize the ovules of species B. A few plants were produced with 19 chromosomes. They were highly sterile but following self-fertilization produced a few plants with 24 chromosomes. These plants were different in phenotype from the original parents and the progeny were fertile. What steps might have been involved?

10.21 How does colchicine treatment result in chromosome doubling in plants?

10.22 What (a) practical and (b) theoretical significance may be associated with colchicine-induced polyploidy?

10.23 How could polyploidy be a significant factor in the evolution of such plants as cotton and wheat?

10.24 How might new species be produced through a combination of polyploidy and hybridization?

10.25 Give a plausible explanation for the origin of (a) *Triticum spelta*; (b) *Raphanobrassica*; (c) *Spartina townsendii*; (d) *Primula kewensis*; and (e) New World cotton.

10.26 What chromosome arrangement is symbolized by each of the following: (a) n; (b) $2n$; (c) $2n + 1$; (d) $2n - 1$; (e) $2n + 2$; (f) $3n$; (g) $4n$?

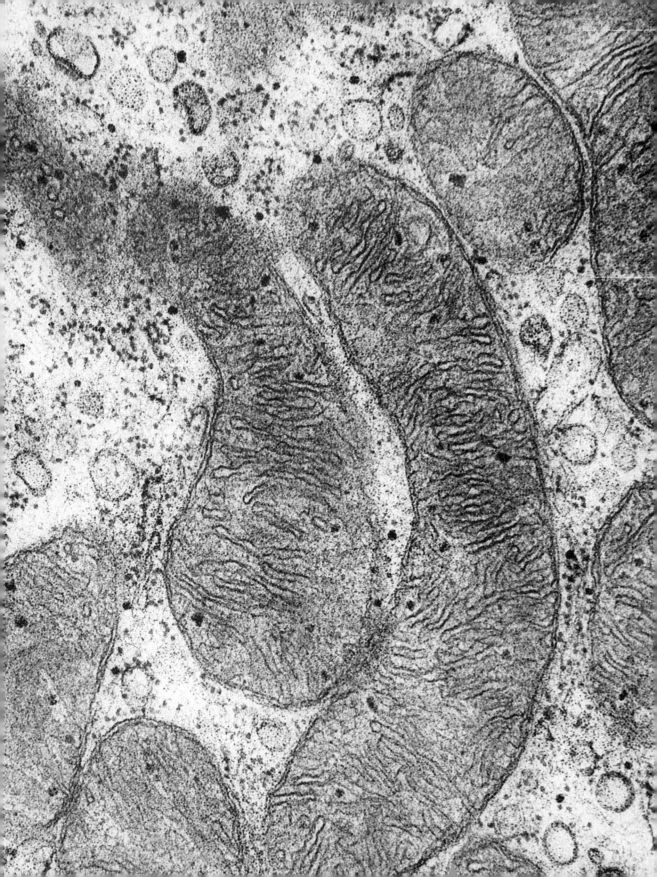

ELEVEN

Transmission of genetic information through nuclear genes and chromosomes has been amply demonstrated. Chromosomal DNA is the most important and very nearly the universal genetic material. Nevertheless, throughout the history of genetics, sporadic reports have indicated that extranuclear or cytoplasmic elements might act as agents for hereditary transmission. Most examples originally attributed to extranuclear inheritance have eventually been explained by other mechanisms. Some that appeared at first to depend on cytoplasmic factors and therefore were classified under maternal inheritance were shown by further investigations to be ultimately controlled by nuclear genes and were reclassified as maternal effects (Chapter 4). Others were associated with bacterial or viral infectious agents that are often difficult to distinguish from extrachromosomal genes. In some **extranuclear organelles** and **extracellular symbionts,** however, cytoplasmic DNA has been shown to exist and to function as an information-carrying genetic material. Some possible instances of **non-DNA inheritance** have also been recorded.

EXTRACHROMOSOMAL INHERITANCE

Mitochondria containing numerous cristae in a secreting stomach cell of a bullfrog. (89,300×). (Courtesy of Dr. Albert W. Sedar, Thomas Jefferson University, and the J. Biophys. Biochem. Cytol.)

DISTINGUISHING EXTRACHROMOSOMAL FROM CHROMOSOMAL INHERITANCE

What **criteria** might be followed in identifying and elucidating extrachromosomal inheritance mechanisms? (1) Differences in the results of **reciprocal crosses** would suggest a deviation from the pattern of autosomal gene transmission. If sex linkage were ruled out, such results would suggest that one gamete is exerting a greater influence than the other on the traits of the next generation. (2) If the egg, which usually carries a greater amount of cytoplasm and cytoplasmic organelles than the sperm, has more influence than the sperm, **maternal inheritance** is suggested. **Organelles** and **symbionts** in the cytoplasm might be isolated and analyzed for more specific evidence concerning maternal transmission in inheritance. (3) Chromosomal genes occupy particular loci and **map in certain places** with respect to other genes. This kind of information may rule out chromosomal inheritance if sufficient linkage data can be obtained. (4) **Lack of Mendelian segregation** and characteristic Mendelian ratios that depend on the chromosomal cycle in meiosis would suggest extrachromosomal transmission. (5) **Non-Mendelian segregation** different from that expected from chromosomal segregation may be recognized in the results of crosses. (6) Experimental **substitution of nuclei might** clarify the relative influence of nucleus and cytoplasm. (7) **Transmission of traits without transmission of nuclei** would suggest extrachromosomal inheritance. Genes and viruses have much in common and a fine line of distinction may be required to distinguish between infection and cytoplasmic inheritance. (8) When **extrachromosomal DNA** can be associated with the transmission of particular traits, the case for extranuclear inheritance is established.

NON-MENDELIAN INHERITANCE OF PLASTID CHARACTERISTICS

Carl Correns (in 1908) observed a **difference in the results of reciprocal crosses** and first described apparent deviations from Mendelian heredity. Different shades of color from white (albino) to dark green in the leaves of some plants were investigated. Instead of equal inheritance from the seed and pollen parent, as demonstrated by Mendel in garden peas, Correns showed in studies of four-o'clock *Mirabilis jalapa* plants that inheritance of certain traits came entirely from the seed parent. Color differences were related to plastids, cytoplasmic organelles in plant cells, most important of which are the chloroplastids or **chloroplasts** (Fig. 11.1), which carry chlorophyll. Chloroplasts arise from cytoplasmic particles called **proplastids** that contain DNA and duplicate themselves independently of other cell parts. They are distributed more or less equally during cell division. Although some proplastids are transmitted in the cytoplasm of the egg, few if any are transmitted in the pollen of most plants. Thus, some chloroplast characteristics are controlled by nonchromosomal inheritance independently of the nuclear genes and are inherited from the maternal or **seed parent cytoplasm.**

Nuclear genes, however, are involved in the overall presence or absence of chlorophyll in plants. This is illustrated by albino seedlings which occur when the plastids entirely lack chlorophyll. In barley, for example, the albino condition is dependent on a recessive nuclear gene *an,* which may be transmitted either through the egg or pollen. Many variations in plant colors depend directly or indirectly on nuclear genes. Phenotypes in which the leaves are mosaics of pale green or even white and dark green areas, however, are apparently dependent on nuclear genes but their occurrence in a particular plant is controlled by the cytoplasm of the maternal line. Some variations in chloroplasts can be traced to mutational changes in the DNA of the proplastids and are entirely independent of nuclear genes. Both spontaneous and induced mutations in these cytoplasmic bodies have been reported.

The most widely accepted hypothesis for persistently mottled or variegated leaves, branches, or whole plants is that **two kinds of chloroplasts,** normal and mutant, are present in the same plant. Normal green chloroplasts are produced in the cell through independent multiplication of pro-

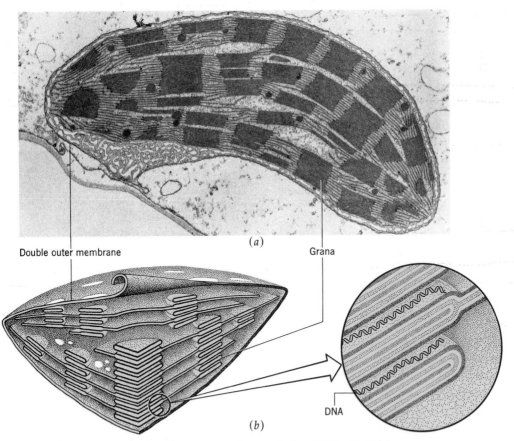

Double outer membrane   Grana

DNA

(a)

(b)

Figure 11.1 Chloroplast, cytoplasmic organelle containing DNA. (*a*) Electron micrograph of a chloroplast, Magnification $\times 15,200$, (*b*) enlarged diagram of grana, stacks of membrane sacs within chloroplast (Magnification $\times 78,500$) showing position of DNA (Photo courtesy T. Elliot Weier. From Weier, Stocking, Barbour, *Botany* (4th ed.), Wiley, 1970.

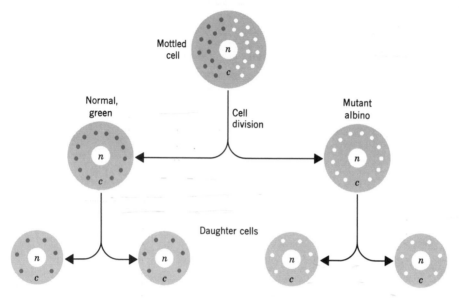

Figure 11.2 Diagrammatic illustration of the segregation of plastids in cell division: *n*, nucleus, and *c*, cytoplasm.

plastids. Mutant forms of proplastids give rise to abnormal plastids. Daughter cells ordinarily resemble parent cells from which they have arisen. Sometimes, in rapidly dividing cells, the multiplication of proplastids does not keep pace with cell division and the reduced numbers make chance distribution effective in changing the characteristics of daughter cells (Fig. 11.2). A cell with equal numbers of abnormal and normal chloroplasts may give rise to a green cell with mostly normal chloroplasts and a nearly colorless cell with all or nearly all abnormal plastids. Daughter cells with only abnormal plastids and those with only normal chloroplasts divide to form cells of their own kind.

Observations in plants such as maize, however, have not supported the hypothesis of mixed normal and abnormal plastids segregating during cell division. Furthermore, too many cell-division cycles seem to be required before phenotypic changes are observable, to account for color variations on the basis of segregation of mature chloroplasts. One hypothesis places the determining factor in the proplastid at the level of a maternally inherited **plasmagene**, a duplicating element that produces genetic effects of the type known to be produced by genetic material in chromosomes. To be effective, plasmagenes must segregate in cell division and give rise at a later time to the modifications that can be observed phenotypically. An alternative hypothesis assigns the basic abnormality to the cytoplasm itself rather than to individual proplastids. This postulates that **abnormal cytoplasm** affects the size and general efficiency of developing chloroplasts, resulting in gradations of chlorophyll production (and therefore color) in individual mature chloroplasts which are reflected in color patterns of whole plants or plant sectors.

Ovules as well as somatic cells of mottled plants (e.g., the four-o'clock, *Mirabilis jalapa*) may carry both abnormal, nearly colorless plastids and normal green chloro-

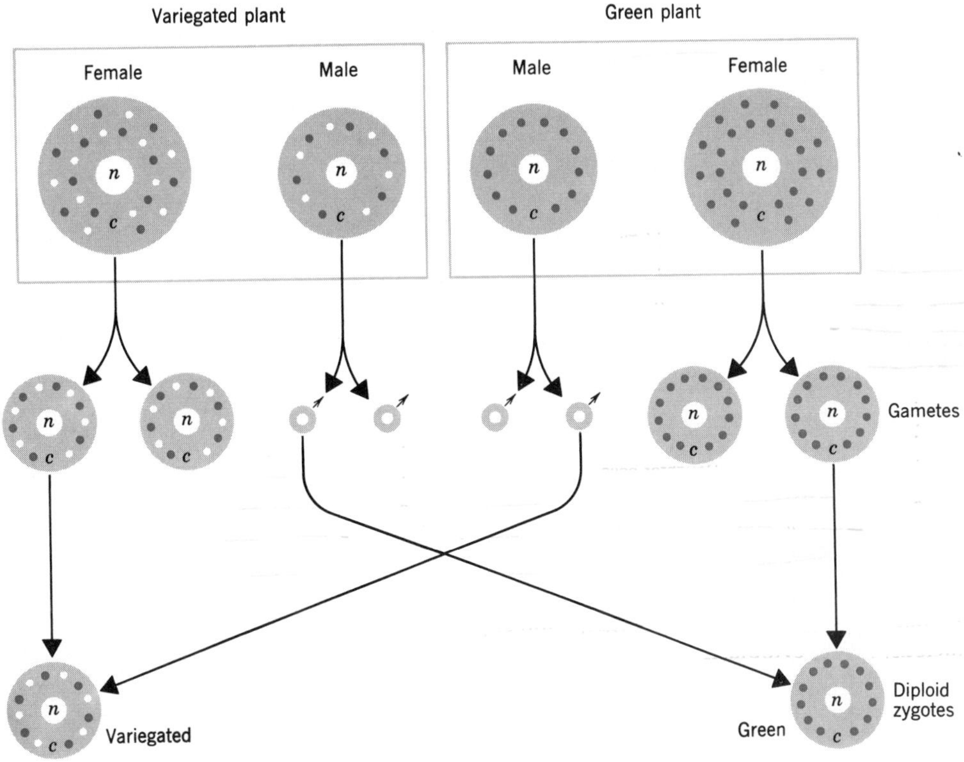

Figure 11.3 Diagrammatic illustration of maternal plastid inheritance in a diploid plant such as Mirabilis which has little or no cytoplasm in pollen gametes: n, nucleus, and c, cytoplasm.

plasts in their cytoplasm (Fig. 11.3). The mottled effect is transmitted through the maternal line, generation after generation. Because the **pollen** of the four-o'clock has **little if any cytoplasm,** its influence on the variegation is negligible. A single plant with green, white, and variegated branches or sectors may produce seed that perpetuates each of the three types. Seeds borne on white branches contain only primordia for colorless plastids, those on green branches only green, and those borne on variegated branches might contain either colorless or green chloroplasts or a combination of both.

In plants such as the primrose, *P. sinensis*, **chimeras** (sectors containing different plastid types) are sometimes formed with part of a plant containing chlorophyll and part not. The areas with abnormal plastids that lack chlorophyll can rely on the green parts of the plant for the products of photosynthesis and thus continue to live. Each part of the chimera may produce reproductive cells and thus transmit its type of plastids through female gametes.

DNA AND PLASTID CHARACTERISTICS

Both nuclear and chloroplast DNA have been demonstrated in Euglena and other plants. When nuclear DNA was removed from cells by treatment with **DNase,** the chloroplasts and chloroplast DNA were unaffected. Chloroplast DNA has a different buoyant density and a different base content than nuclear DNA. Three or four bands can be distinguished when Euglena are subjected to cesium chloride centrifugation. The first and largest band is presumed to be nuclear DNA. It contains about 50 percent GC base pairs. Other smaller bands represent satellite DNA with only about 28 percent GC pairs. **Chloroplast DNA** is estimated at 2 to 4 percent of the total DNA but chloroplasts contain all the enzymatic machinery required for protein synthesis. Although several arguments favor genetic continuity of chloroplast DNA, the possibility that chloroplast DNA is synthesized in the nucleus has not been entirely ruled out.

DNA IN MITOCHONDRIA

Mitochondria are usually small cytoplasmic organelles (Fig. 11.4) with distinct shelf-like internal layers or cristae that arise as invaginations from the inner mitochondrial membrane (Fig. 11.5). They occur in cells of eukaryotes but not in bacteria and viruses. Mitochondria provide higher animals and plants with life-sustaining cellular energy through the oxidative processes of the citric acid and the fatty acid cycles, and the coupled processes of oxidative phosphorylation and electron transport. They contain a small amount of **DNA** that has apparently remained autonomous, outside the nuclear genome, throughout the long evolutionary history of animals and plants.

In yeast cells, 10 to 20 percent of the cellular DNA is localized in the mi-

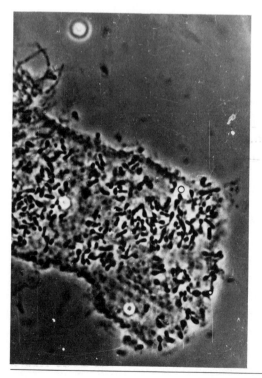

Figure 11.4 Mitochondria from wild-type cells of *Paramecium aurelia*. (Courtesy of Janine Beisson.)

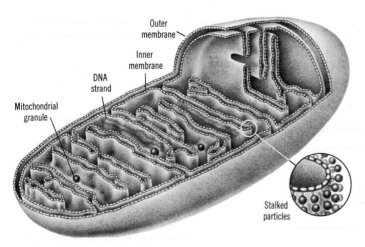

Figure 11.5 A mitochondrion showing smooth continuous outer membrane and periodically convoluted inner membrane with double membranes called cristae. Mitochondria are the principal energy source in all cells of eukaryotes. They contain a small amount of DNA, RNA polymerase, transfer RNAs and ribosomes which are presumably responsible for the extrachromosomal inheritance in mitochondria.

tochondrion. Mitochondrial DNA has properties different from those of nuclear DNA in density and proportion of **GC and AT base pairs.** One yeast study showed that mitochondrial DNA had a density of 6.683 gm/cm³ and a GC content of 21 percent, whereas nuclear DNA had a density of 1.699 gm/cm³ and a GC content of 40 percent.

The life cycle of normal baker's yeast *Saccharomyces cerevisiae* includes a haploid and a diploid phase. Mating normally occurs between vegetative haploid cells of opposite mating type (A or a). These cells fuse to form vegetative diploid cells that divide by mitosis. Cell division is usually unequal with a small "daughter" cell budding from a larger "mother." Both cells, however, are identical in nuclear composition. Vegetative diploid cells may undergo the complex process of sporulation in which meiosis occurs. The four resulting meiotic products are contained within an ascus. These ascospores are released and each divides to form a clone of vegetative haploid cells.

The first mutant found in yeast, a small colony type called "petite," has provided the best evidence that now exists for **mitochondrial mutations.** Petites proved to be defective in their ability to utilize oxygen in the metabolism of carbohydrates. When, for example, glucose is in the medium, petite yeast will grow to only small-sized colonies. Enzyme analyses indicated that the mitochondria lacked the respiratory enzyme cytochrome oxidase normally associated with mitochondria. Not only does this deficiency produce defective growth but it prevents petites from producing spores. Many petite strains have only a small proportion of G and C and a predominance of repetitive **AT base pairs.** This kind of DNA does not code meaningful biological information. Some petite strains are completely lacking in mitochondrial DNA. It is clear from this example that heritable alterations in mitochondrial DNA lead to heritable alterations in mitochondrial phenotypes.

On the other hand, yeast chromosomal genes must specify some activities carried out in mitochondria. Petite yeast strains with damaged or absent DNA continue to synthesize abnormal mitochondrial DNA. This indicates that the mechanism for mitochondrial DNA replication is not contained in mitochondrial DNA. Likewise, petite strains continue to synthesize the enzymes of the Krebs cycle that are located in the mitochondria. Control must come from chromosomal genes. It is more difficult to determine which genes are located in mitochondrial DNA because most petites have defective mitochondrial ribosomes resulting in **defective translation.** Defective ribosomes may be the cause of the absence of synthesis for some proteins in mitochondria.

Mutations other than petites can be induced in yeasts and transmitted by the cytoplasm. For example, resistance to the antibiotics chloramphenicol and erythromycin have been induced. These antibiotics have selective affinity for mitochondrial ribosomal proteins, suggesting that structural genes for some ribosomal proteins in yeast may be composed of mitochondrial DNA.

Recombination has been demonstrated in yeast mitochondria by crossing strains of yeast that differ in their resistance controlled by mitochondrial DNA to certain antibiotics: erythromycin, spiromycin, and paromomycin. Crosses between two strains, each carrying a particular resistance marker, produced clones that were resistant to neither or both drugs. Such clones must have arisen from recombination of genes within the mitochondria.

The amount of DNA in mitochondria of any given organism is sufficient to control only a few proteins. Nuclear genes have been shown to control the production of most of the enzymes associated with mitochondria. Since mitochondrial protein synthesis continues in the presence of antibiotics that interfere with cytoplasmic protein synthesis, **mitochondrial ribosomal proteins** must be coded by **mitochondrial genes.** Mitochondrial DNA must have a specific function because alterations cause severe changes as evidenced by the petite mutants of yeast. Presumably, if mitochondrial DNA was without function it would not be indefinitely retained in the living system. Mitochondria in presently living organisms arise only from preexisting mitochondria. Their growth and division are only partially controlled by nuclear DNA. They can incorporate amino acids into proteins. At least a few of the common proteins are produced within the mitochondria of all organisms from templates of mitochondrial DNA.

EXTRACHROMOSOMAL DNA IN GREEN ALGAE

When Ruth Sager placed green algae Chlamydomonas cells on a culture medium containing the antibiotic streptomycin, most of the cells were killed, but about one per million survived and multiplied, each to form a streptomycin-resistant colony. Mutants with resistance to streptomycin were being selected from the predominantly streptomycin-susceptible alga. About 90 percent of the mutants involved nuclear genes (sr-1). Subsequently, many mutations were identified in Chlamydomonas with most shown (by appropriate test crosses) to be changes in nuclear genes. Such mutations were merely being demonstrated by the antibiotic challenge. Approximately 10 percent of the mutations (sr-2), however, were **uniparental** and nonchromosomal. They were found to be induced by sublethal concentrations of the drug.

Eventually, nonchromosomal mutants were recovered from almost every colony. Streptomycin was not specific for a particular change but induced several kinds of mutational alterations in the treated colonies.

Nonchromosomal DNA changes gave the same phenotypes as chromosomal DNA changes, but their frequency of occurrence was much greater than the frequency of chromosomal gene changes. These nonchromosomal, uniparental genes are presumed to be located in the **chloroplasts**.

Reciprocal crosses (Fig. 11.6) demon-

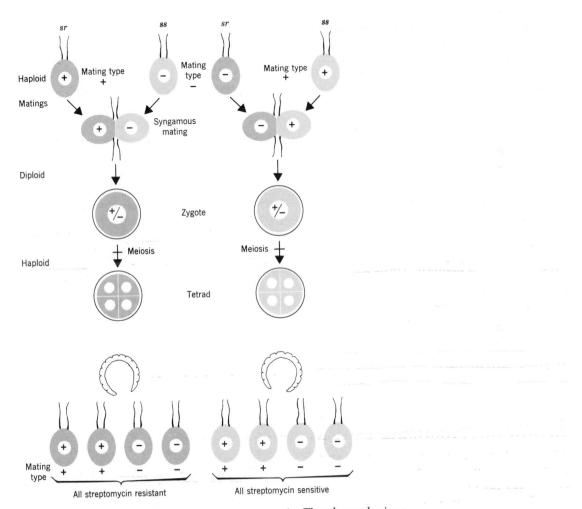

Figure 11.6 Inheritance of resistance to streptomycin. The plus and minus signs refer to mating type, which is inherited as a single gene difference. The progeny is with rare exceptions like the plus parent in its reaction to streptomycin, but in these crosses the mating type difference segregates 1:1 in every tetrad. (Based on Ruth Sager and F. J. Ryan, *Cell Heredity*, John Wiley & Sons, Inc., 1961.)

strated that antibiotic resistance, when controlled by nonchromosomal factors, was uniparental in inheritance. Mating types in this sexual unicellular alga were controlled by chromosomal genes, which were identified by the investigators as mt^+ and mt^- or simply + and − (instead of female and male). All progeny from each mating were like the + mating type with respect to relative streptomycin resistance. When the + mating type was resistant, all progeny were resistant; when the + mating type was nonresistant, all progeny were nonresistant. These results of reciprocal crosses demonstrated **non-Mendelian inheritance** and a single contrasting pair of traits. Other nonchromosomal genes, sr for streptomycin resistance and ss for streptomycin sensitive, were postulated to control these two alternative characteristics.

Another mutant, ac_2, which blocked photosynthetic activity was induced and a pair of nonchromosomal alleles, ac_1 and ac_2, was thus available for study in the same strain of chlamydomonas. The mutant required acetate in the medium for growth. With two pairs of nonchromosomal genes available, a dihybrid cross could be conducted in the same system to check for evidence of recombination. Crosses of the dihybrid type $ac_1 \, ss \times ac_2 \, sr$ were allowed to grow for a few vegetative multiplications. Each cell was then classified for its segregating markers, both nonchromosomal and chromosomal (that is, mating type and others known to be chromosomal). Both the ac_1/ac_2 and the sr/ss pairs of alleles were observed to segregate early but not always in the same division. After four or five mitotic doublings, equal numbers of parental ($ac_1 ss$ and $ac_2 sr$) combinations had been obtained. The results indicated **independent assortment**, suggesting that the two pairs of nonchromosomal genes were carried on different particles. Three and four point crosses and reciprocal crosses have been made with the addition of several mutants presumed to be carried in chloroplasts and mitochondria. A genetic map of **non-Mendelian genes** in Chlamydomonas has been constructed.

MALE STERILITY IN PLANTS

Another example of cytoplasmic inheritance is associated with pollen failure. This occurs in many flowering plants and results in male sterility. In maize, wheat, sugar beets, onions, and some other crop plants, fertility is controlled, at least in part, by cytoplasmic factors. In other plants, however, male sterility is controlled by nuclear genes. Critical observations and tests must be made in individual cases to determine the mechanism of inheritance. Male sterility has practical importance when crosses are made on a large scale to produce hybrid seed. Hybrid plants are produced commercially in maize, cucumbers, onions, sorghum and other plants for the purpose of obtaining hybrid vigor (Chapter 14).

MALE STERILITY IN A CROSS-POLLINATING PLANT

Classical examples of maternal inheritance mechanisms that transmit male sterility in maize (corn) were discovered and carefully analyzed by M. M. Rhoades. Pollen was aborted in the anthers of certain corn plants, causing them to be male sterile, but female structures and fertility were normal. Nuclear genes did not control this type of sterility. It was transmitted

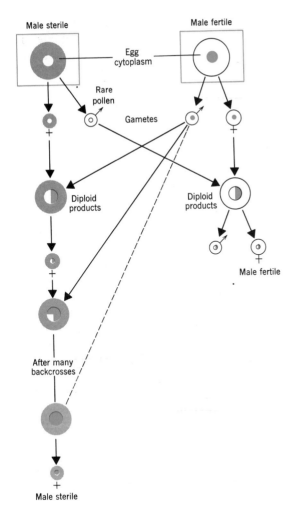

Male sterile

Male fertile

Egg
cytoplasm

Rare
pollen

Gametes

Diploid
products

Diploid
products

Male fertile

After many
backcrosses

Male sterile

Figure 11.7 Maternal inheritance of male sterility in maize. (After Rhoades.)

from generation to generation through the
egg cytoplasm.

A particular male-sterile variety produced
only male-sterile progeny when fertilized
with pollen from normal maize plants. The
male-sterile seed parent plants were then
backcrossed repeatedly with pollen-fertile
lines until all chromosomes from the male-
sterile line had been exchanged for those of
the male-fertile line (Fig. 11.7). In the genet-
ically-restored sterile line, male sterility
persisted, demonstrating that its inheritance was maternal and was not controlled
by chromosomal genes. As the investiga-

tions progressed, a small amount of pollen
was obtained from the male-sterile line
making reciprocal crosses possible. The
reciprocal crosses produced progeny from
the male-sterile seed plant line that were
always male-sterile, even when all chromosomes from the male fertile seed plant line
were present. Inheritance of male sterility
was maternal regardless of the direction in
which the cross was made. Male sterility in
this example was attributed to cytoplasmic
genes (plasmagenes) transmitted by female
gametes.

The cytoplasmic effect is not, however,

the only factor in male sterility. Specific nuclear genes are now known to suppress maternally inherited sterility in maize. A single dominant **chromosomal gene,** for example, **can restore** pollen **fertility** in the presence of cytoplasm that ordinarily would insure sterility. In one experiment, pollen abortion occurred only when a specific kind of cytoplasm was present along with a dominant gene for male sterility and the homozygous recessive allele was present at a suppressor locus.

Large scale use of male sterile maize for seed production brought disaster to the United States corn crop in 1970. Because of the desirability of uniformity in corn and the great advantage of male sterility in seed production, a single source of cytoplasm known as Texas (T), male-sterile cytoplasm had been used in producing seed for most of the corn hybrids planted that year. A new mutant strain of corn leaf blight was able to destroy some 15 percent of the crop. The lesson to be learned from the 1970 fiasco is to provide for more variation and avoid a single source of cytoplasm for a major commercial crop.

MALE STERILITY IN A SELF-POLLINATING PLANT

Following the successful applications of cytoplasmic male sterility to maize, experiments were designed to explore prospects for male sterility in self-pollinating plants. In these plants, the mechanism of normal seed production makes hybridization a tedious procedure. Enclosure of both male and female flower parts within the same protective glumes as in garden peas results in nearly complete self-fertilization. To produce single hybrid seeds, a series of delicate, exacting, properly timed hand operations must be carried out. The objective is to remove anthers from the seed parent before pollen is released and to in-

troduce pollen from the desired pollen parent at the appropriate time. Sorghum was the first normally self-pollinated field crop in which **male sterility** was used to produce commercial **hybrids.**

M. A. Overman and H. E. Warmke investigated the mechanism of male sterility in sorghum (*Sorghum bicolor* (L.) Moench). They found development of cytoplasmic male-sterile and male-fertile anthers to be similar until about the time of meiosis. After this time the tapetal layer of cells (nutritive cells around developing microspores in the anther) became thicker, more vacuolated and underwent cytoplasmic disorganization (Fig. 11.8) in some plants. In the normal male-fertile plants, these cells simply released their contents and disintegrated. These abnormalities of the tapeta within the anther resulted in pollen abortion and male sterility. When these experiments were concluded the explanation for the mechanism for male sterility in sorghum that had already been exploited empirically became established.

Cytoplasmically induced male sterility is now being developed in wheat. H. Kihara found cytoplasmic sterile plants in the progeny when "goat grass" was crossed with common bread wheat (*Aegilops caudata* × *Triticum aestivum*). Both cytoplasmic male-sterile and male-fertile lines were obtained. Numerous investigations are now devoted to the development of such lines for practical commercial use. Apparently in certain strains of wheat, **mutant nonchromosomal genes inhibit normal anther** and **pollen production** but have no adverse effect on the female parts (stigma and ovary) of the flower. Lines with these characteristics are being developed to become female parents for the final hybrid. In other lines, a dominant nuclear gene controlling a **pollen fertility restorer** is being developed for the male parent in the final cross to produce hybrid seed. In wheat, no

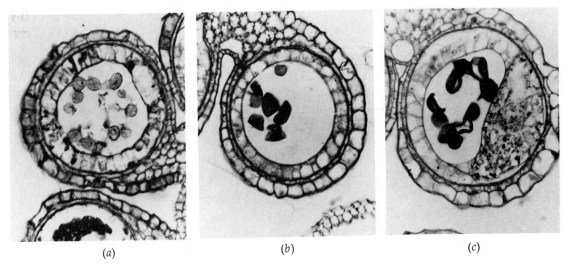

Figure 11.8 Cross sections of sterile anther locules in *Sorghum bicolor* exhibiting various tapetal irregularities. (*a*) Tapetum showing cytoplasmic disorganization and rupture of some radial walls shortly after microspore release. (*b*) Tapetum highly vacuolate and enlarged to fill most of locule. (*c*) Locule with intratapetal syncytium and heavy tapetal wall. These characteristics are associated with sterility. Magnification ×250. (Reprinted from Overman and Warmke, with permission from the *Journal of Heredity*.

simple, inexpensive method has been established for production of hybrid seed. Hybrid wheat will probably not have great commercial value as does hybrid corn. Other lines of investigation have resulted in wheat with desirable characteristics and high yield.

EXTRANUCLEAR INHERITANCE IN PARAMECIA

Paramecia are favorable material for genetic investigation. They are large, unicellular organisms with separate nuclei and reproduce by asexual and sexual mechanisms. Asexual reproduction occurs through cell fission and clones of genetically identical cells can be produced. They also conjugate periodically and transfer genetic material from one cell to another. It is possible in the laboratory to make sexual crosses through which nuclear DNA is transferred from a donor to a recipient resulting in heterozygous progeny, that is, $AA \times aa = Aa$. A process of self-fertilization, called **autogamy,** in heterozygous individuals results in homozygosis of the resulting progeny (Fig. 11.9). Following meiosis, the cells are haploid but through autogamy they become homozygous diploids. This provides a basis for comparing extranuclear and nuclear inheritance and thus for demonstrating that progeny can differ from wild type in traits controlled by both nuclear and extranuclear genes.

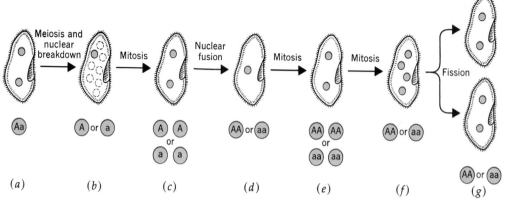

Figure 11.9 Autogamy in Paramecium resulting in homozygosis through the following steps: (*a–b*) meiosis of micronuclei resulting in one haploid product which divides mitotically (*c*) and unites to form a diploid product (*d*). Mitotic division of the nuclei (*e–f*) followed by fission of the cell results in two cells each homozygous (*g*).

G. H. Beale discovered that erythromycin resistance in Paramecium, like that in yeast, is due to **non-Mendelian inheritance**. A number of additional cytoplasmic and nuclear mutations affecting antibiotic resistance has been studied by both Beale and J. Beisson. These and other investigators made transfers of cytoplasm and also transfers of isolated mitochondria between strains of paramecia and showed that mitochondria (presumably **mitochondrial DNA**) are the basis for the trait. The studies have also shown that while some mitochondrial characters are determined by the mitochondria themselves, others are dependent on other elements in the nucleocytoplasmic environment.

KILLER BACTERIA IN PARAMECIUM CYTOPLASM

T. M. Sonneborn and others have investigated a persistent extranuclear effect in protozoa. Some stocks of *P. aurelia* produce a substance that has a lethal effect on members of other stocks of the same species. Paramecia from stocks capable of producing the substance were called "killers." When killers were subjected to low temperatures, their killing capacity gradually disappeared. Their toxic effect also decreased after repeated cell divisions. It was at first postulated that separate entities in the cytoplasm were responsible for the production of a toxic substance. From mathematical calculation, it was estimated that about 400 particles would be required to make killers effective. After the hypothetical calculations were completed, the killers were observed microscopically and "particles" called "kappa" were observed in about the expected numbers. These "particles" have been shown to be **symbiotic bacteria** and have been named *Caedobacter taeniospiralis* (the killer bacterium with the spiral ribbon).

A "toxic substance" produced by killer bacteria, is diffusible in the fluid medium (Fig. 11.10). When killers were allowed to remain in a medium for a time and were then replaced by sensitives, the sensitives were killed. The "toxic substance" had no

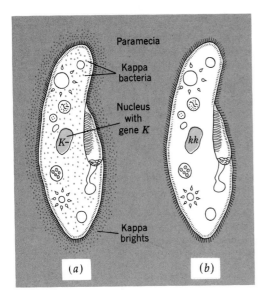

Figure 11.10 Kappa particles in *Paramecium aurelia*. (*a*) Killer with kappa particles inside and paramecin in the liquid medium outside the organism. Gene *K* is present in the nucleus. (*b*) Sensitive organism with no kappa particles, no paramecin, and genes *kk* in the nucleus. (After T. M. Sonneborn.)

effect on killers. The "toxic substance" has been shown to be associated with a particular kind of kappa, occurring in about 20 percent of a kappa population. These kappa bacteria possess a refractile protein—containing structure called an **R body.** Bacteria carrying R bodies are **brights** because they are infected with a virus that dictates the synthesis of viral protein as well as an R protein body in the kappa bacterium. The virus may act as the toxin in the killing response and the R body may aid the penetration of the toxin when the sensitive Paramecium is attacked. The virus may be in the provirus state in "nonbright" bacteria.

Kappa bacteria are perpetuated only in organisms carrying the dominant **nuclear gene** *K* which either provides directly for the production of the bacteria or establishes the kind of environment necessary for them to multiply. This can be demonstrated by allowing killers to conjugate with sensitives by appropriate techniques (to avoid killing the mate) and under conditions in which no cytoplasmic exchange occurs (Fig. 11.11). Two kinds of clones emerge; one from the original killer cell, which contains gene *K* and kappa bacteria, and the other from the original sensitive cell which lacks kappa. Following autogamy, half the progeny of the killers are killers and half are sensitive bacteria. All progeny of sensitives are sensitive. Since no cytoplasm was transferred in this conjugation only the cells from original killers inherit kappa bacteria. Kappa cannot reproduce in cells unless a *K* gene is present.

Under some conditions, conjugation persists much longer; a larger connection is established between conjugants, and cytoplasm as well as nuclear genes is exchanged (Fig. 11.12). Genes *K* and *k* are exchanged and both exconjugants are *Kk*. Cytoplasmic exchange has transferred kappa bacteria from the killer to the nonkiller cell. Kappa bacteria are maintained in the heterozygous cells carrying the *K* gene. Autogamy produces homozygous *KK* and *kk* cells each of which may produce clones of killers or nonkillers, respectively.

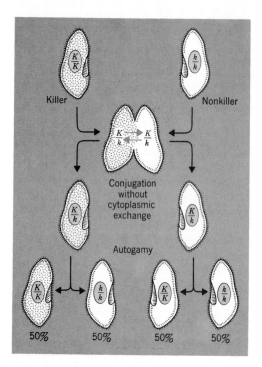

Figure 11.11 Conjugation between a killer Paramecium (stippled) and a sensitive Paramecium without cytoplasmic exchange but with an exchange of K and k genes. Autogamy that follows results in KK and kk homozygotes which may give rise to clones. (After T. M. Sonneborn.)

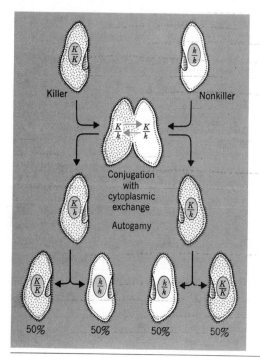

Figure 11.12 Conjugation as in Fig. 11.11 but with cytoplasmic exchange followed by autogamy. (After T. M. Sonneborn.)

CYTOPLASMIC DNA IN HIGHER ANIMALS

Mitochondria carrying DNA occur in cells of higher eukaryotes but phenotypic markers have not been detected in many cases to distinguish nuclear from mitochondrial inheritance. W. L. French has presented evidence that sterility in hybrid Culex mosquitoes is caused by interactions involving mitochondrial DNA. Several investigators have shown that mitochondrial DNA is inherited maternally in frogs. J. B. David has compared mitochondrial DNA in cell cultures of different mammals including the rat, mouse and man. He has also hybridized cells of different mammals in culture. In hybrid mouse and man cells, for example, he has shown that not only homogeneous mouse and homogeneous human mitochondrial DNA can be detected but also heterogeneous hybrid DNA. In one series of experiments, 20 percent of each circular DNA unit was mouse and 80 percent was human mitochondrial DNA. Heterogeneous DNA was shown to result from mitochondrial DNA recombination in the hybrids.

Extranuclear inheritance represents a small but persistent sector of the total inheritance of eukaryotes that has apparently remained **autonomous** throughout a long evolutionary period. What selective advantage could account for the separate maintenance of this small unit of perhaps 100 genes in the great germ plasm pool of eukaryotes?

All mitochondria have been shown to contain circular **DNA molecules like bacteria as well as ribosomes, tRNA and enzymes needed to make proteins.** Furthermore, mitochondrial ribosomes sediment at the 70S value which is characteristic of prokaryotic cells. This suggests that in early evolution mitochondria evolved from parasitic bacteria capable of growth within primitive eukaryotic cells. The efficiency of such bacteria in supplying ATP and energy to their hosts could have given this combination a selective advantage. Chloroplasts also carry DNA and prokaryotic ribosomes, suggesting that they might have developed from a type of prokaryote capable of photosynthesis. Reproduction of chloroplasts by increasing in length and dividing by fission is similar to that of bacteria. But chloroplasts have never been made to grow and divide outside of cells. This suggests that some vital **proteins** are coded by **nuclear DNA** with their synthesis occurring on **cytoplasmic (80S) ribosomes.** Experimental evidence supports this explanation for the origin of DNA in cytoplasmic organelles of eukaryotes. Some agents for extrachromosomal inheritance such as kappa bacteria have been shown to be symbionts that have become established in the cytoplasm of eukaryote host cells. Experimental evidence now indicates that **mitochondria** and **chloroplasts** may also be **prokaryote symbionts** with a long evolutionary history in the cells of eukaryotes.

CYTOPLASMIC INHERITANCE NOT INVOLVING DNA

Further examples of extrachromosomal inheritance have been cited such as stable modifications in gene action (serotypes in Paramecium) and cortical changes in ciliated protozoa that apparently do not involve DNA. The latter group includes traits in which pre-existing cellular structures influence the formation of newly developing structures. It is responsible for perpetuating a pattern in an organism. In nature, this type of system may have some advantages over the more basic genetic mechanism. Coordination in a section of a living organism may be more effective if a direct control can come from the system itself in preference to the segregating genes in the entire organism. Transmission of information by the usual genetic mechanisms may in certain situations be to the disadvantage of the organisms concerned. In bacteria, for example, it is important for parts that regenerate after injury to be coordinated with other parts of the organism. If the blueprints are maintained in a local area, restored parts can be made to fit and function with neighboring parts.

The flagellum of a bacterium can be dissolved experimentally and from the solution it may be precipitated out again, forming a new flagellum. This cannot be done, however, without a part of an organized flagellum that acts as a template. The template is apparently a non-DNA system that directs the molecules in such a way that they conform to a particular pattern. If a mutant type of flagellum that is curly is dissolved and precipitated again, the new flagellum will take the pattern of the "template" that provides the information for restoring the flagellum. When parts of wild-type, not curly, flagella are placed in the solution they induce wild-type flagella, regardless of the genetic makeup of the flagella that were originally dissolved. The immediate development of the flagellum, therefore, depends on the template that is in the vicinity where the precipitation is taking place and not on the genetic makeup of the organism.

A more complicated structure that behaves like the flagellum of a bacterium is the ciliate cortex. Each cilium originates in a basal body that is located in the cortex or surface of the organism. A unit cilium consists of nine pairs of microtubules around the edge and two pairs of microtubules in the center. Microtubules can be displaced or damaged and can be made to reassemble if a seed is present. It is possible to dissolve off an arm of a basal ciliary structure by adding magnesium. When a template is present the arm may be reformed. This process is called "heteronucleation." It can be accomplished experimentally in protozoa, and, as in the case of bacterial flagella, the final product depends on the seed. An organism may follow information other than that of its own genetic system and thus regenerate a part not like itself.

REFERENCES

Ashwell, M. and T. W. Work. 1970. "The biogenesis of mitochondria." *Ann. Rev. Biochem.* E. E. Snell (ed.), 39, 251–290. Annual Reviews, Inc., Stanford University Press, Palo Alto, Calif.

Duvick, D. N. 1965. "Cytoplasmic pollen sterility in corn." *Adv. in Genet.* E. W. Caspari and J. M. Thoday (eds.), 13, 2–56. Academic Press, New York.

Gibson, I. 1970. "Interacting genetic systems in Paramecium." *Adv. in Morphogen.* M. Abercrombin, J. Bracker, and T. J. King (eds.), 9, 159–208. Academic Press, New York.

Goodenough, U. W. and R. P. Levine. 1970. "The genetic activity of mitochondria and chloroplasts. "*Sci. Amer.* 223, 22–29.

Preer, J. P. Jr. 1971. Extrachromosomal inheritance: hereditary symbionts, mitochondria chloroplasts. H. L. Roman (ed.) *Ann. Rev. Genet.* 5, 361–406. Also, 1969. "Genetics of the protozoa." T. T. Chen (ed.) *Res. in Protozool.* 3: 130–278.

Overman, M. A. and H. E. Warmke. 1972. "Cytoplasmic male sterility in sorghum." *J. Heredity.* 63, 227–234.

Sager, R. 1972. *Cytoplasmic genes and organelles.* Academic Press, New York.

Sager, R. and Z. Raminis. 1970. "A genetic map of non-Mendelian genes in Chlamydomonas." *Proc. Nat'l Acad. Sci. U.S.* 65, 593–600.

Sonneborn, T. M. 1960. "The gene and cell differentiation." *Proc. Nat'l Acad. Sci.* 46, 149–165.

Swift, H. and D. R. Wolstenholme. 1970. "Mitochondria and chloroplasts: nucleic acids and problems of biogenesis." A Lima De Faria, (ed.) *Handbook Mol. Cytol.*, 972–1046.

PROBLEMS AND QUESTIONS

11.1 If a particular trait in a plant could be shown to be inherited solely through mitochondrial DNA would it be classified as a case of maternal inheritance or a maternal effect? Why?

11.2 In most animals, a larger amount of cytoplasm is carried by the egg than by the sperm. Likewise, the egg in plants carries more cytoplasm than the pollen. How could this difference affect the expression of inherited traits (a) dependent on chromosomal genes; and (b) dependent on nonchromosomal genes?

11.3 Reciprocal crosses sometimes give different results in the F_1. This may be due to (a) sex-linked inheritance; (b) cytoplasmic inheritance; or (c) maternal effects. If such a result were obtained, how could the investigator determine experimentally which category was involved?

11.4 Explain how single plants such as four-o'clocks could have green, pale green, and variegated sectors. If such sectors reached sexual maturity, what color characteristics would each type be expected to transmit through male or female gametes?

11.5 What practical applications could be made with male sterile lines of onions, wheat, and other crop plants?

11.6 How could kappa particles have become established in their host organism, Paramecium, in evolution?

11.7 O. Renner carried out reciprocal crosses between two types of the evening primrose, *Oenothera hookeri* and *O. muricata*, known to have the same chromosome constitution. When the seed parent was *O. hookeri* the plastids of the progeny were yellow but when the seed parent was *O. muricata* the plastids of the progeny were green. How might this difference in the results of reciprocal crosses be explained?

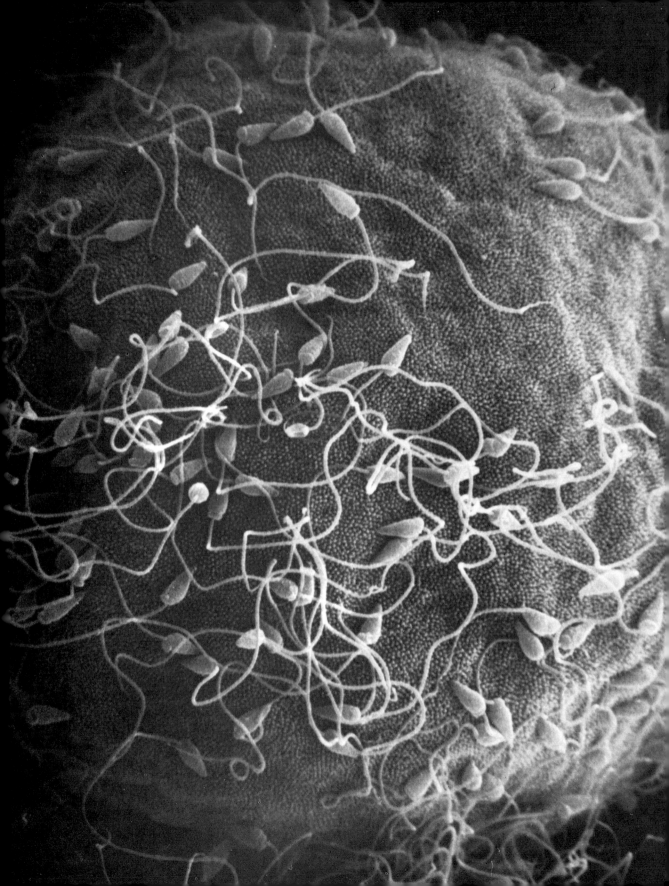

TWELVE

In 1760, Joseph Kölreuter reported but was unable to explain the results of crosses between tall and dwarf varieties of tobacco, Nicotiana. The F_1 plants were **intermediate** in size between the two parent varieties. The F_2 progeny showed a **continuous gradation** from the size of the dwarf to that of the tall parent. A normal distribution was obtained in the F_2, with the midpoint of the curve corresponding roughly with the midpoint of the F_1 curve and intermediate between the two original parents (P). Kölreuter was a careful experimenter and a good biologist, but he could not explain these results because the basic principles of genetics had not yet been established.

Only after Mendel's work was discovered, more than a hundred years after Kölreuter's experiments were completed, did we have an explanation for **discontinuous variation** (falling into distinct classes) based on a particulate mode of inheritance. (It should be noted that Mendel also described continuous variation. When he crossed white-flowered and purple-red flowered beans, an intermediate flower color was obtained in the F_1 progeny and

MULTIPLE GENE INHERITANCE

Sperm-egg interactions in sea urchin. (4,420×). (M. J. Tegner and D. Epel, *Science,* 179:687, February 16, 1973. (c) 1973 by the American Association for the Advancement of Science.)

399

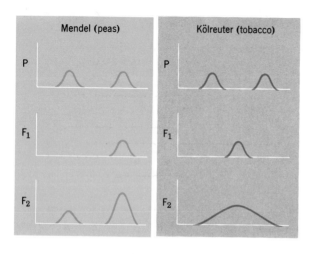

Figure 12.1 Curves representing the results of Mendel's experiments compared with those of Kölreuter. (Left) P, F_1, and F_2 from Mendel's crosses with garden peas; (right) P, F_1, and F_2 from Kölreuter's crosses on tobacco plants. The ordinate represents the number of plants and the abscissa the range in height.

a spread from white to red in the F_2.)

Mendel's results from garden peas could be analyzed in terms of simple frequencies that required only simple arithmetic. Bateson supported Mendel and strengthened the case for discontinuous variation. Later investigators reported simple ratios from crosses involving many traits in a wide variety of plants and animals. The more elusive and problematical results, which were not readily explained by Mendelian segregation, were pigeonholed or discarded. Several years elapsed after the discovery of Mendel's work before progress was made in the analysis of continuous variation.

The curves presented in Fig. 12.1 diagrammatically compare Mendel's results with those of Kölreuter. When Mendel crossed tall and dwarf varieties of peas, the F_1 progeny were all tall. Some of the F_2 plants were tall and some were dwarf, in the proportion of about 3 to 1. From these results, pairs of particulate elements were postulated, with one member of each pair being dominant over its allele. Distinct and clear-cut contrasting characters were observed and readily classified under such headings as tall or dwarf, yellow or green, and red or white. Kölreuter's results, on the other hand, showed continuous variation with no distinct class boundaries. The F_1 hybrids were intermediate between the parents, and the F_2 generation covered the entire range between the sizes of the parents.

QUANTITATIVE TRAITS

Between 1900 and 1910, many geneticists thought continuous variation reflected an entirely different mechanism of inheritance from that of discontinuous variation. A few keen investigators, however, began to envision a common basis for the results of Mendel and those of Kölreuter. Genes that had small but **cumulative effects** without dominance were postulated to behave in a Mendelian fashion. An explanation for continuous variation thus emerged in the form of the **multiple-gene hypothesis.** Experimental results and interpretations substantiating this hypothesis were obtained from the classical investigations of H. Nilsson-Ehle in Sweden and E. M. East in the United States during the period 1910 to 1913.

One of these studies was based on crosses between two varieties of wheat producing red and white kernels, respectively (Fig. 12.2). The F_1 seeds were intermediate in color between those of the two parents. They were lighter than those of the red parent but distinctly more colored than those of the white-parent variety. When the F_2 seeds were classified according to intensity of color, a continuous gradation was observed from red to white, and classes were more or less arbitrary. About 1/16 of the F_2 seeds were as red as those of the red parent, and about 1/16 were white. About 14/16 were intermediate, ranging between the color of the red and white original parents.

When the 14/16 of the F_2 seeds were classified further, on the basis of color intensity, it was shown that about 4/16 had more color than the F_1 intermediates, about 6/16 were intermediate like the F_1's and about 4/16 were lighter than the F_1's. This result suggested a segregation of two gene pairs and was explained on the basis of **duplicate genes** acting on the same character and producing a cumulative effect. When the results of this cross were examined critically, they were found to resemble those that Kölreuter had obtained many years before. A second cross between red-kernel and white-kernel varieties of wheat was carried to the F_2, and this time about 1/64 of the F_2 progeny produced white kernels and about 1/64 produced red kernels. Some 62/64 were intermediate in color, ranging between those of the two parental varieties. Again continuous gradations were recognized between the extremes of the parents. The results of this cross resembled those of a Mendelian trihybrid cross in that three independent gene pairs had to be postulated to explain this result in contrast to the two pairs for the previous cross. Evidently, one pair, which was segregating in the second cross, had been homozygous in both parents in the first cross.

The concept of multiple genes for **quanti-**

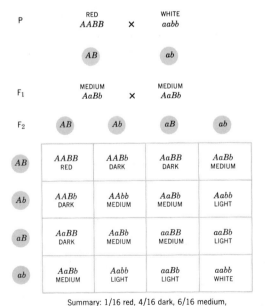

Figure 12.2 Cross between a wheat variety with red kernels and another variety with white kernels. This cross illustrates quantitative inheritance dependent on multiple genes.

tative inheritance is now one of the most important principles of genetics. It has been strengthened greatly by the use of statistical methods devised by R. A. Fisher in England, Sewall Wright in the United States, and others. The explanation is based on the action of **many genes** (polygenes) usually segregating independently but influencing the same phenotype in a cumulative fashion.

POLYGENE CONCEPT

A **polygene** is defined as a gene that individually exerts a slight effect on a phenotype but, in conjunction with a few or many other genes, controls a quantitative trait such as weight of an animal. Polygenic inheritance differs from the classical Mendelian pattern in that a **graded series** extends from one parental extreme to the other. Only averages and variances of populations are considered and not discrete values for individuals. Such factors as epistasis, cytoplasmic influences, interactions among genes and gene products, and interactions with the environment are reflected in the averages and variances. Polygenic inheritance is a statistical concept.

Because most characteristics of domestic plants and animals that have practical significance (including height, weight, time required to reach maturity, and qualities relevant to human nutrition) depend on polygenic inheritance, much attention has centered around this principle. If all of the practical genetic experimental projects that are now in progress at the various experiment stations throughout the world could be listed and classified, the results would probably indicate that some 80 to 90 percent involve polygenic or quantitative inheritance. Some human characteristics of interest and significance such as skin pigmentation also depend on multiple genes.

The inheritance of polygenic traits depends on the cumulative or **additive** action of several or many genes, each of which produces a small proportion of the total effect. This is in marked contrast to the inheritance of major gene traits which is an all or none phenomenon dependent on one gene or a few interacting genes. Polygenic traits can be measured and these measurements can be subjected to statistical treatment. Environmental factors also influence end products such as height, weight, and color intensity. The genetic component or heritability of the trait must therefore be separated from the environmental effect. Variation in such traits as kernel color in wheat and skin color in man are mostly hereditary under the conditions of the usual observation.

MEASUREMENT DATA

A classical study of quantitative inheritance, which did much to establish the multiple gene hypothesis, was made by R. A. Emerson and E. M. East on the inheritance of ear length in maize. A variety with ears averaging 6.6 cm in length was crossed with a variety having ears that averaged 16.8 cm. The F_1 progeny were intermediate in ear length, averaging 12.1 cm, and ranging from 9 to 15 cm. The F_2 represented a wider spread of variation than the F_1 with some of the ears as extreme in size as those of the original parents. This is the characteristic pattern for results from crosses involving polygenic traits dependent on only a few genes. When many genes are involved, the extremes represented by the parents occur infrequently in the F_2. The histograms in Fig. 12.3 represent the maize data and illustrate graphically the characteristic pattern of multiple gene inheritance. Since the parental lines were inbred and presumably homozygous, the variation within these lines and that shown for the fully heterozygous F_1 plants were environ-

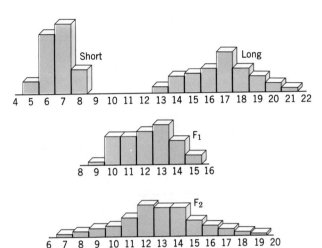

Figure 12.3 Histograms illustrating maize data from cross involving multiple gene inheritance. (Modified from Sturtevant and Beadle, based on data from Emerson and East.)

mental or a result of sequential development in time. The results of genetic segregation involving the genes from the two parents were illustrated by the wider variation in the F_2.

As in other examples involving polygenic inheritance, **environmental influences** are the most confusing factors for analysis. They must be considered and controlled as adequately as possible. The environment can produce results similar to those of genes with respect to size differences between large and small varieties. Plants raised in unfavorable environments—that is, without sufficient water, sunlight, or soil nutrients—will be smaller than others of the same genotype that enjoy a more satisfactory environment. On the other hand, under ideal environmental conditions, organisms with inferior genotypes (e.g., many lethals and mutational variants) may develop phenotypes equivalent or superior to those with better genotypes but inferior environments. Environmental variations can even affect the expression of polygenic traits. Potential environmental influences must be identified and controlled as completely as possible with experimental design or statistical adjustment.

In the maize example, let us assume that two gene pairs are active and the variation is all hereditary. Then, if each gene produces an equal effect on the size of the ear (beyond the size that is dependent on the residual genotype 6.6 cm), the individual contribution would be

$$\frac{\text{maximum height} - \text{minimum height}}{\#\ \text{of alleles}}$$

$$= \text{relative contribution of each allele}$$

$$= \frac{16.8 - 6.6}{4} = 2.55 \text{ cm per allele.}$$

Each active allele thus would produce 2.55 cm in addition to the 6.6 inherent in the residual genotype for the small variety. When the F_2 plants were classified according to phenotype, a ratio of $1:4:6:4:1$ was obtained. This is a modification of the $1:2:1:2:4:2:1:2:1$ ratio, which may be changed to $9:3:3:1$ by dominance. The analysis in Fig. 12.4 represents a model for these data based on the multiple gene hypothesis.

In other tests involving several thousand F_2 progeny from plant crosses, none even approached the parental phenotypes. Many gene pairs were therefore presumed to be

Individual contribution of genes:
  (A or B) 2.55 cm
Size produced by residual genotype: 6.6 cm
Genotypes of parents:

Black Mexican   Tom Thumb
    AABB   ×   aabb
        F₁ AaBb (12.1 cm)

F₂	Geno-type	Fre-quency	Phenotype (cm)	Phenotypic Ratio
	AABB	1	16.8	1
	AaBB	2	14.2	4
	AABb	2	14.2	
	AaBb	4	11.7	6
	aaBB	1	11.7	
	AAbb	1	11.7	
	aaBb	2	9.1	4
	Aabb	2	9.1	
	aabb	1	6.6	1

Figure 12.4 Analysis of Emerson's maize data based on the multiple gene hypothesis with two pairs of active alleles, each contributing equally to the phenotype.

involved. As many as 200 pairs of alleles were postulated to be active in some traits. E. M. East, for example, studied the corolla length in a cross between two varieties of tobacco, *Nicotiana longiflora*, and found that none of the several hundred F₂ plants resembled the parent (P). Many gene pairs were then postulated.

ESTIMATING THE NUMBER OF GENE DIFFERENCES

The contributions of individual genes to a particular characteristic can be evaluated roughly for some of the foregoing detailed crosses. In the maize data, for example, each extreme of the parents occurred in the proportion of about 1 in 16 in the F₂. Two pairs of alleles might therefore be assumed to be operating. Determining the number of genes involved in a given cross is usually more difficult, however, mainly because environmental as well as genetic variations are represented by the same measurements. In such cases, an estimate of the **degree** to which a trait is controlled by heredity (heritability, Chapter 14) can be useful in estimating the number of genes involved. A further complication is that the genes may not be equal in their influence on the phenotype. Models devised to estimate the number of genes are oversimplified if based on the assumption of equal effects but this is the assumption of choice when the degree of influence of individual genes is not known.

In dealing with quantitative traits in some organisms, a rough estimate has been made by determining the frequency of occurrence in the F₂ population of the **extremes** representing the parental phenotypes (Table 12.1). This is an inadequate approximation of the number of active genes. The method is only useful for preliminary genetic analysis. More complicated mathematical procedures are employed when sufficient F₁ and F₂ data are available for comparison. The simplified

TABLE 12.1
Probability of Occurrence of F_2 Individuals as Extreme as Either Parent

Pairs of Segregating Alleles	Fraction of F_2 as Extreme as Either Parent
1	$\frac{1}{4}$
2	$\frac{1}{16}$
3	$\frac{1}{64}$
4	$\frac{1}{256}$
5	$\frac{1}{1024}$

method presented here is based on the assumption that all genes produce comparable effects on the phenotype and that random assortment occurs among gene pairs. Gene combinations and interactions may become complex, and simple conclusions are not always possible.

TRANSGRESSIVE VARIATION

Extremes in some F_2 results have exceeded the corresponding parental (P) values. This pattern is called **transgressive variation** and its explanation is based on the hypothesis that the parents did not represent the extremes possible from the combined genotypes. If some genes for large size were lacking in the genotype of the large parent but were present in the genotype of the small parent, an F_2 individual might receive a combination of genes producing a **larger or smaller** size than that seen in either parent. For example, in a cross between a large Hamburgh chicken and a small Sebright Bantam, Punnett found that the F_1 chickens were intermediate between the two parents. While the F_2 progeny included some birds larger and some smaller than the parental varieties, most were intermediate between the original parents. Pun-

nett explained the result on the basis of a four-factor difference between the two parents. The Hamburghs were postulated to have the genotype $a^+a^+b^+b^+c^+c^+dd$ and the bantams $aabbccd^+d^+$. The F_1 birds would be uniformly heterozygous $a^+ab^+bc^+cd^+d$, accounting for the intermediate weight. Some F_2 individuals might have the genotype $a^+a^+b^+b^+c^+c^+d^+d^+$ and be heavier than the original Hamburgh parent, whereas others could have the genotype $aabbccdd$ and be smaller than the bantam parent.

Genes responsible for transgressive variation and other patterns of polygenic inheritance are considered to behave in the typical **Mendelian fashion.** Only their phenotypic results differ from the familiar ratios. Thus a $9:3:3:1$ ratio expected from a dihybrid cross involving complete dominance is modified to a $1:4:6:4:1$ ratio with continuous variation and no dominance (i.e., $AAbb = AaBb$ phenotypically). If three pairs of independent alleles were involved in a cross, the ratio of $1:6:15:20:15:6:1$ would replace the familiar $27:9:9:9:3:3:3:1$ ratio based on discontinuous variation and complete dominance. The **more genes** (polygenes) that are segregating, the **more continuous** is the phenotypic variation.

THE MECHANISM OF QUANTITATIVE INHERITANCE

The multiple-gene hypothesis was effectively illustrated by the kernel color of wheat experiment in which six genes with cumulative effects were postulated. Common bread wheat is a **polyploid** (a hexaploid, see Chapter 10) and has six chromosomes of a particular kind, each presumably carrying a gene for color. Segregation of the six genes postulated for kernel color might be attributable to multiple chromosome sets. Thus, the mode of inheritance would be more properly described on the basis of duplicate genes with a cumulative effect, than by polygenes. In animals, however, polyploidy is virtually nonexistent, but polygenic inheritance is very common. Polyploidy is also limited in applicability as an explanation for inheritance in plants, because the possible number of similar genes would depend on the number of chromosome sets that a plant could carry. The high number of genes postulated for some plants, therefore, could not be explained on this basis. *Crepis capillaris* ($n = 3$), for example, is **not a polyploid** but is reported to have a high proportion of **duplicate gene** or polygene inheritance.

Multiple genes might also arise through mutations in different chromosome areas producing similar effects, but it is difficult to visualize the occurrence of numerous mutations affecting the same trait. Duplicated segments of chromosomes (discussed in Chapter 9) or tandem duplications of genes might provide a partial explanation. Duplicate genes do exist and examples indicate that they are involved in quantitative inheritance, but the duplicate-gene concept cannot be used to interpret all cumulative effects of genes.

Modifier genes can influence size difference in animals and plants, but they do not always follow a uniform pattern. Some act as enhancers and some as inhibitors of a particular effect, and the end result is the average effect of many genes. Certain pleiotropic genes may influence both qualitative and quantitative traits. Furthermore, the effect of a particular gene substitution may vary with different genetic backgrounds.

The same phenotype may be produced by different genetic systems. Size in animals and plants, for example, is generally controlled by a polygenic system, but a single major gene may produce a dwarf animal or a dwarf plant such as among Mendel's peas (Chapter 2), thus accomplishing the same end result as that produced by multiple genes. In cases where only one or a few genes are involved, it is possible to develop homozygous, pure-breeding types in a few generations of inbreeding. This is not possible in systems of polygenes where many genes are involved; the genes cannot be individually recognized, and analytical techniques ordinarily applied to Mendelian inheritance are inadequate. New techniques making use of digital computers have been applied to various models based on different hypothetical patterns. It has been possible to demonstrate results comparable with those of natural systems of polygenes on the model of Mendelian segregation. These studies support the basic premise that polygenic inheritance is Mendelian. R. D. Milkman has studied a series of polygenes that controls the crossvein-making ability in the Drosophila wing. Some of these genes have been located by linkage studies in the three major chromosomes of *Drosophila melanogaster.*

Although there are many unresolved problems concerning the nature and action of polygenes associated with quantitative traits, the multiple-gene concept is a good working hypothesis. Because it is impossible at this time to identify particular polygenes or to obtain a detailed knowledge of their properties, polygenic systems must be considered in **statistical terms.** Statistical studies provide comprehensive averages in terms of **quantitative values.** In natural selection, the balance within the operative polygenic system is more important than the effect of individual genes. Polygenic inheritance fits the **Darwinian pattern** of gradual and continuous changes selected for by natural environments.

STATISTICAL METHODS

Data are usually numerous and complex in investigations of quantitative inheritance, and require analysis and organization before their significance can be fully appreciated. Large numbers of individuals usually typify populations to be studied and compared; therefore, sampling methods are used to facilitate comparison. One requirement that must be rigidly observed when sampling techniques are employed is that of **randomness.** A sufficiently **large sample,** taken at random without bias or favoritism, may adequately represent a complete population which could not be measured in its entirety. Most biological populations are so large that they are assumed to be infinite. In other cases such as occur in laboratory situations populations do not actually exist, but are only theoretical. Examples illustrating different kinds of populations will clarify the difference between total population and sample.

A study was made of the weight of deer presently living in the western part of the United States. Four hundred deer captured at random were weighed. The population consisted of the thousands of deer living in the area. A sample of 400, however, was deemed adequate to indicate the average weight of the population.

An animal nutritionist fed a special diet to 100 mice and then measured their increase in weight over a certain length of time. The sample consisted of the 100 mice from which the measurements were obtained. The population did not actually exist. It consisted of an infinite number of measurements which theoretically could be obtained if an infinite number of mice had been fed the diet.

Finally, a coin may be tossed 100 times and the number of heads and tails recorded. The sample here would consist of the 100 tosses, but the population would be the infinite number of tosses theoretically possible.

It is important to distinguish clearly between estimates based on samples and the actual values that would be obtained if it were possible to measure the entire population. Estimates are called **statistics,** and the true values based on entire populations are called **parameters.** In biological investigations, parameters are seldom known. If they were known, direct comparisons could be made between populations. Species A, for example, could be compared directly with species B in terms of the actual mean or degree of variability. When parameters are not known, statistics based on samples are used for comparisons. Measurements especially useful to the geneticist are the mean and the variance. The square root of the variance is the familiar statistic called the standard deviation.

Unfortunately, statisticians do not agree in the choice of symbols to represent statistical terms, but they generally use Latin letters for statistics, and Greek letters for parameters. In this discussion, the mean of a population (a parameter), will be represented by the Greek letter mu (μ), and the mean of a sample by $\bar{x}$. The variance of a population will be represented by sigma squared (σ^2) and the standard deviation by sigma (σ). Estimates of σ^2 and σ, calculated from samples, will be symbolized s^2 and s, respectively. The number of individuals in the sample will be symbolized by n and those of the entire population will be N.

A major reason for using statistical techniques is to estimate the properties of populations for which parameters are unknown. The objective in applying these techniques is to provide an unbiased estimate of a certain parameter. An unbiased estimate is as likely to be too low as too high. It is expected to more closely approximate the actual parameter as the sample size increases.

MEAN

The first statistic to be considered here is the mean, which is an estimate of **magnitude.** It can be defined as the sum of measurements divided by the total number of measurements. This can be symbolized by the following formula: $\bar{x} = \Sigma X/n$, where x is the sample mean, X the individual measurement, Σ the summation and n the number of individuals in the sample. The sample mean is an unbiased estimate of the population mean μ, which is given by $\mu = \Sigma X/N$, where X represents the individual measurements and N the total individuals in the entire population.

Data obtained from measuring the height of 122 guayule rubber plants, a random sample of a population, may be used in an illustration. The following measurements in inches rounded off to the nearest whole number were taken from plants growing in the field 107 days after planting.

12	12	11	11	12	15	12	11	10
13	14	12	13	13	10	13	14	13
11	12	12	10	15	16	13	11	9
12	13	11	13	14	11	13	8	10
11	13	14	13	12	14	13	10	10
11	11	10	12	10	12	11	12	11
12	11	13	12	9	10	14	11	13
13	11	12	11	11	13	11	12	13

10	10	11	11	12	16	16	12	12
14	12	13	10	9	12	13	12	12
17	13	13	10	11	10	13	12	11
14	11	11	11	10	11	14	11	14
8	10	10	11	13	10	11	12	12
13	12	15	15	15				

$$\bar{x} = \frac{\Sigma X}{n} = \frac{1458}{122} = 11.95$$

In analyses involving numerous measurements it is advantageous to begin by classifying or grouping the data. In the example of 122 guayule plants, the **interval** (or unit distance) between classes was 1 and the **range** was from 8 to 17 inches. To classify the data, intervals must be continuous and equal. Suppose that one plant measuring 19 inches was added to the sample. To keep the intervals equal it would be necessary to add another group for 18-inch plants (with frequency $f = 0$) and a group for 19-inch plants with $f = 1$.

If the range had been wider, for example from 2 to 36 inches, an interval of 2 might have been chosen for ease of computation and presentation. Under the plan, all measurements falling between 2 and 4 could arbitrarily be given the value of 3, which would be the class center. Other class centers would be 5, 7, 9, 11, 13 . . . 35. What if the range covered a much wider area, for example, from 1 to 100, and continuous variation was represented throughout the sample? Perhaps intervals of 10 would then be adequate. Class centers of 5, 15, 25, and so on, would be used for grouping the data. If the measurements covered a smaller range and were more precise, for example presented in terms of one or two decimal places, appropriate class centers, such as 70.5 to represent observations falling between 70 and 71, might be chosen. When the data are grouped, the mean is represented by the following formula:

$$\bar{x} = \frac{\Sigma f X}{n}$$

TABLE 12.2
The Steps in a Statistical Problem Involving the Height in Inches of 122 Guayule Seedlings Representing a Random Sample of Plants 107 Days After Planting

X	f	fX	$X - \bar{x}$	$(X - \bar{x})^2$	$f(X - \bar{x})^2$
8	2	16	-3.95	15.60	31.20
9	3	27	-2.95	8.70	26.10
10	18	180	-1.95	3.80	68.40
11	29	319	$-.95$	.90	26.10
12	27	324	.05	.00	.00
13	24	312	1.05	1.10	26.40
14	10	140	2.05	4.20	42.00
15	5	75	3.05	9.30	46.50
16	3	48	4.05	16.40	49.20
17	1	17	5.05	25.50	25.50
	122	1458			341.40

where f is the frequency of individuals having the value of X, and n the total number of individuals in the sample. In the example presented, the data are not extensive, ranging from 8 to 17, and may be arranged as shown in Table 12.2 The first two columns in the table show the various values of X and the frequency of measurements having these values, respectively. The classified data obtained from measurements of the sample may now be substituted into the formula and the calculations may be made as follows:

$$\bar{x} = \frac{\Sigma fX}{n} = \frac{1458}{122} = 11.95$$

The mean of 11.95 inches represents the average from the sample and an unbiased estimate of the population mean.

The mean is useful as an estimate of magnitude, but it does not provide all the information that may be desired from the sample. For example, in a certain experiment, one cross resulted in sample plants averaging 70 inches in height. An-other cross produced plants of approximately the same average. From the mean alone, the two samples seemed to be similar, but further analysis showed that the range of one was from 40 to 100, whereas that of the other was from 69 to 71. The variation, in addition to the average, is significant to the geneticist.

VARIANCE

One measure of **variation** is called **variance.** The population variance, σ^2, is represented by the following formula:

$$\sigma^2 = \frac{\Sigma(X - \mu)^2}{N}$$

where X represents the observed value of each individual measurement taken in the population, μ the actual, but usually unknown, population mean (parameter), and N the total number of individuals in the population. The population variance is the ideal value because it measures the actual population, but it is unrealistic because the true population mean μ is seldom known. The difficulty can be avoided by substituting the sample mean ($\bar{x}$) for the parameter mean μ, and the sample size n for the population size N. The sample variance is biased, however, because it tends to underestimate the true value of the population variance σ^2. On the average, the estimates would be too small. When $\bar{x}$ is used in place of μ in the formula, the number of independent measurements becomes one less than n. To use the sample mean $\bar{x}$ in the formula, a correction factor $n/(n-1)$ is introduced to overcome the bias. This modification results in the standard formula for the variance s^2, which is derived by multiplying the biased statistic $[\Sigma(X-\bar{x})^2]/n$ by the correction factor $n/(n-1)$ as follows.

$$\frac{\Sigma(X-\bar{x})^2}{n} \times \frac{n}{n-1} = \frac{\Sigma(X-\bar{x})^2}{n-1} = s^2$$

When this is done, s^2 becomes an unbiased estimate of the population variance σ^2. If the sample size is large, the difference between n and $n-1$ is small or negligible, but even for large samples the correct divisor for an unbiased estimate of the population parameter σ^2 is $n-1$.

When data are grouped, as in Table 12.2, the formula for the sample variance is

$$s^2 = \frac{\Sigma f(X-\bar{x})^2}{n-1}$$

The data from the sample of 122 guayule plants may now be substituted into the formula and the variance calculated as follows.

$$s^2 = \frac{\Sigma f(X-\bar{x})^2}{n-1} = \frac{341}{121} = 2.82$$

The variance $s^2 = 2.82$ is an unbiased estimate of the population variance σ^2. The accuracy of the estimate may be increased by enlarging the size of the sample or by taking the average of several samples of similar size.

STANDARD DEVIATION

The difficulty with using variance as a measure of variation is that it is given in terms of the units of measurements **squared.** Extracting the square root of the variance converts it to the same units in which the measurements were taken, and the statistic thus obtained is called the **standard deviation** s. Curiously, even though the sample variance s^2 is an unbiased estimate of the population variance σ^2, the sample standard deviation s is not an unbiased estimate of the population standard deviation. The bias is introduced by **extracting the square root.** When, however, the data are to be plotted on a curve and certain tests of significance are to be carried out, the estimate of variance must be expressed in the original units, for example, inches in the example from the guayule

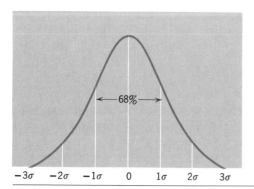

Figure 12.5 A symmetrical bell-shaped curve showing a normal distribution of individuals in a population.

seedlings. For such purposes, the standard deviation is employed despite its bias. From the sample of guayule seedlings the standard deviation s would be: $s = \sqrt{2.82} = 1.68$ inches.

The symmetrical bell-shaped curve presented in Fig. 12.5 represents a normal distribution of individual measurements. If one population standard deviation was plotted on either side of the mean, the arc would include about two-thirds (68.26) of the population represented. The shape of the curve is determined by the amount of variation, as indicated by the standard deviation. If the same area is maintained, a small s is associated with a high curve and a large s with a flat curve. Furthermore, if s is small and the curve is high, most of the observations are clustered around the mean. On the other hand, if s is large and the curve is flat, the observations are spread away from the mean.

STANDARD ERROR

Another useful statistical measurement is the standard deviation of means or the "standard error of the mean." This is a measure of how reliable is the sample mean $\bar{x}$ as an estimate of the population mean. As suggested by the name, the **standard error** is an estimate of the standard deviation of means of samples drawn from a single population. Means that represent values for groups of individuals, are expected to vary less than would individuals drawn at random from the population. It is not necessary to actually go through the process of taking several samples from the same population, since standard error is inversely proportional to the square root of the sample size n.

Two factors are involved in an evaluation of **reliability** based on a sample: the **size** of the sample, and the amount of **variation** in the population sampled. If great variation is present, a large sample will be required to represent the population adequately. If individuals are fairly uniform, a smaller sample may be satisfactory. These two factors are taken into consideration in the following formula for the standard error $s_{\bar{x}}$ of the sampling mean:

$$s_{\bar{x}} = \sqrt{\frac{s^2}{n}} = \frac{s}{\sqrt{n}}$$

Substituting the guayule data we have:

$$s_{\bar{x}} = \frac{s}{\sqrt{n}} = \frac{1.68}{11.05} = 0.15$$

The mean may now be written 11.95 ± 0.15. This standard error was based on one standard deviation which, plotted on the normal curve in a $+$ and $-$ position from the mean $\bar{x}$, includes about 68 percent of the population. Therefore, the investigator is about two-thirds or 68 percent confident that the true mean μ lies within the limits of $+$ one standard error and $-$ one standard error from the sample mean $\bar{x}$. Such proportions of the population on either side of the mean are "confidence intervals." The true mean μ is constant and does not possess a probability distribution. It either does or does not fall between established limits with reference to a given sample. If several samples are taken, and statements are made concerning each as to whether the population mean lies inside or outside the limits, two-thirds of these statements will be correct.

t TEST

Students' t test is a statistical method for determining the **significance** of the **difference** between two sample means. Suppose the investigator wishes to be 0.95 or 95 percent confident, that is, he wants to be able to say that the probability that μ will fall within certain limits on either side of $\bar{x}$ is 0.95. For this purpose, he may make use of the t-distribution concept. The t is defined as the difference between the sample mean and the true mean divided by $s_{\bar{x}}$ as shown by the formula

$$t = \frac{(\bar{x} - \mu)}{s_{\bar{x}}}$$

To solve for μ, the equation may be transposed as follows:

$$\mu = \bar{x} \pm t s_{\bar{x}}$$

The confidence interval may be obtained from the formula at a particular confidence level. One limit is on the $+$ side of the mean and the other on the $-$ side. Table 12.3 shows the distribution of t. At the left the degrees of freedom or n-1 are listed. The t values are given in the body of the table, and the probabilities of t values are given at the top. To use this device an appropriate t value from the table is substituted in the formula along with the sample mean $\bar{x}$ and its standard error $s_{\bar{x}}$. The equation is then solved for μ. At the 0.95 confidence level, the probability of being misled by sampling is 0.05. Lower (L_1) and upper (L_2) limits of the confidence interval may be established as follows:

$$L_1 = \bar{x} - t \text{ at } 0.05 \text{ level} \times s_{\bar{x}}$$
$$L_2 = \bar{x} + t \text{ at } 0.05 \text{ level} \times s_{\bar{x}}$$

These limits are random variables. If they are superimposed on the bell curve (Fig. 12.5) they will include the population mean μ in 95 percent of repetitions of the same experiment.

Now reading $t = 1.984$ from the table and 100 degrees of freedom, which is nearest to

TABLE 12.3
Probability (P) for Values of t (Vertical Columns) and Various Degrees of Freedom (d.f.)[a] (See Also Curves in Appendix A)

The degrees of freedom are one less than the number of classes.

d.f. \ P	0.5	0.4	0.3	0.2	0.1	0.05	0.01
1	1.000	1.376	1.963	3.078	6.314	12.706	63.657
2	0.816	1.061	1.386	1.886	2.920	4.303	9.925
3	0.765	0.978	1.250	1.638	2.353	3.182	5.841
4	0.741	0.941	1.190	1.533	2.132	2.776	4.604
5	0.727	0.920	1.156	1.476	2.015	2.571	4.032
6	0.718	0.906	1.134	1.440	1.943	2.447	3.707
7	0.711	0.896	1.119	1.415	1.895	2.365	3.499
8	0.706	0.889	1.108	1.397	1.860	2.306	3.355
9	0.703	0.883	1.100	1.383	1.833	2.262	3.250
10	0.700	0.879	1.093	1.372	1.812	2.228	3.169
15	0.691	0.866	1.074	1.341	1.753	2.131	2.947
20	0.687	0.860	1.064	1.325	1.725	2.086	2.845
25	0.684	0.856	1.058	1.316	1.708	2.060	2.787
30	0.683	0.854	1.055	1.310	1.697	2.042	2.750
50	0.680	0.849	1.047	1.299	1.676	2.008	2.678
100	0.677	0.846	1.042	1.290	1.661	1.984	2.626
∞	0.674	0.842	1.036	1.282	1.645	1.960	2.576

[a] Abridged from Table 4 of (14th Edn.) R. A. Fisher: *Statistical Methods for Research Workers*, (copyright © 1972 by Hafner Press.)

$n-1$ or 121, and substituting the values $\bar{x} = 11.95$ and $s_{\bar{x}} = 0.15$ for the guayule sample, we have:

$$L_1 = 11.95 - 1.984\,(0.15) = 11.65$$
$$L_2 = 11.95 + 1.984\,(0.15) = 12.25$$

These are the confidence limits. If statements based on samples are made repeatedly to the effect that the true mean μ falls within these limits, 95 percent of them will be correct. For confidence levels different from 0.95 and other levels given as headings in Table 12.3 where interpolation is required, the chart Appendix A will be useful.

Consequently, the standard error of the mean indicates the reliability with which the **sample mean $\bar{x}$ estimates the population mean μ.** The smaller the standard error, the more reliable is the estimate. It can be seen from the formula that, as the sample size n increases, the value of the standard error tends to decrease, and vice versa. This relation illustrates the advantage of large samples for obtaining the best possible estimates of parameters.

The application of statistical treatments to genetics has provided indirect as well as direct benefits. Statistical requirements have made it necessary for the geneticist to carefully design each experiment before he undertakes the actual investigation. He has thus developed a more critical attitude toward methods and interpretations of experimental results. A number of variations formerly attributed to genetic mechanisms have now been explained on the basis of sampling errors.

CONTROL OF ENVIRONMENTAL VARIATION

Some genetic patterns were not recognized in the past or were confused with the effects of environmental factors. The modern geneticist controls all aspects of his study that can be controlled and allows for the element of chance in sampling. The development of effective statistical methods has helped foster this trend. The value of the mean, variance, and standard error of the mean as basic statistical tools can be demonstrated by applying them directly to a problem. In cereal crops, the time required for the plants to mature depends to some extent on inheritance. Since **environmental factors** such as temperature are also involved, however, the plants to be compared must be grown in a single, controlled environment, with adequate checks for uncontrollable environmental factors.

In a particular experiment, seeds representing each of four types of wheat were planted in randomized field plots. The four types of wheat were: two inbred parent varieties (PA and PB) and F_1 and F_2 progeny from a cross between PA and PB. The seedlings were all raised in the same season and the randomized plots were designed to equalize minor environmental variations in soil, moisture, and other factors. The time required for maturing was recorded as days from the time of planting to the time when the heads were fully formed. A sample of 40 plants was taken at random from each population. The data representing the four samples of 40 individuals are given as follows.

PA

75	74	72	72	73	71	72	71
76	73	72	72	72	70	71	72
71	73	74	73	73	72	71	72
72	74	73	72	71	72	73	72
74	71	72	73	75	70	72	76

PB

58	55	56	56	53	55	55	57
54	55	56	55	58	57	55	56
55	57	55	57	56	57	55	55
56	57	55	54	59	57	55	55
58	56	57	54	53	56	58	56

TABLE 12.4

Data and Calculations from Samples of the Two Parents, F_1 and F_2, of Wheat Populations Classified According to Time in Days Required for Maturity; Forty Plants Are Included in Each Sample

	Days																								$\bar{x}$	$s_{\bar{x}}$	s^2	s
	53	54	55	56	57	58	59	60	61	62	63	64	65	66	67	68	69	70	71	72	73	74	75	76				
PA																		2	7	15	8	4	2	2	72.47±	0.23	2.05	1.43
PB	2	3	13	9	8	4	1																		55.85±	0.22	1.92	1.39
F_1							1	7	7	8	6	7	4												62.20±	0.27	2.88	1.70
F_2				1	1	1	1	3	2	5	7	6	4	2	1	1	1	1	1	1	1				63.72±	0.60	14.26	3.78

$F_1PA\ PB$

60	65	63	61	65	50	62	63
61	60	63	64	64	61	62	63
65	62	64	62	60	59	61	62
61	60	63	62	60	63	60	65
64	61	62	64	64	61	62	64

$F_2PA\ PB$

69	66	62	60	63	67	72	64
61	63	62	63	60	59	64	63
56	62	626	65	64	73	60	65
57	64	63	70	68	62	71	63
65	66	64	58	61	65	62	64

The data from each sample were classified and are summarized in Table 12.4. The means, variances, standard deviations, and standard errors of the means are also included in the table. Several facts can be derived from the data and calculations. First, each parent represents a population in regards to the length of time required for maturity. One population (PA) has a mean $(\bar{x}_A)$ of 72.47 and the other (PB) had a mean $(\bar{x}_B)$ of 55.85.

STANDARD ERROR OF A DIFFERENCE

The conclusion that the parents came from separate populations seems evident from the data. When the discontinuity is not obvious, however, it is necessary to devise a test to determine whether the difference between means indicates **different populations** sampled or merely **chance differences** in two different samples from the same population. The standard error of a difference (S_D) may be used for such a test. S_D is the square root of the sum of squares of the two standard errors calculated from the samples being compared, or $\sqrt{(S_{\bar{x}1})^2 + (S_{\bar{x}2})^2}$ where $S_{\bar{x}1}$ represents the standard error one sample and $S_{\bar{x}2}$ the standard error of the others. Substituting the

standard errors calculated for the two parents in the wheat we have:

$$S_D = \sqrt{(S_{\bar{x}1})^2 + (S_{\bar{x}2})^2}$$
$$= \sqrt{(0.23)^2 + (0.22)^2} = 0.318$$

The significance of the S_D of 0.318 can be appreciated by comparing it with the actual difference between the means of the two parents. This comparison can be shown by calculating t as follows:

$$t = \frac{\bar{x}_A - \bar{x}_B}{S_D} = \frac{72.47 - 55.85}{0.318} = 52.3$$

where t is the difference between the two means $\bar{x}_A - \bar{x}_B$ divided by the standard error of the differences S_D. If the actual difference is more than twice the standard error, it represents about the 5 percent level of probability and is considered to be significant. That is, it represents a real difference in addition to expected chance variations. More values of t in terms of probability are given in Table 12.3. The value of $t = 52.3$ with 78 degrees of freedom is off the table. Suffice it to say here that the probability that these two samples represent the same population is extremely low.

The next interpretation from the data and calculations concerns the mode of inheritance. The F_1 plants were intermediate between the two parents, and the F_2 were also generally intermediate but with a wider range than the F_1. This result suggests the polygenic pattern of inheritance.

Why are the variances for the PA, PB and F_1 about equal, whereas the F_2 variance is much greater? The inbred parents were highly homozygous. If they were completely homozygous, the variation would be entirely environmental. F_1 plants from homozygous parents are uniformly heterozygous, again all variation is environmental. Segregation expected in the F_2 provides for genetic variation in addition to that directly associated with environment.

STANDARD ERROR OF THE MEAN

The standard error of the F_2 mean (used as a measure of reliability) was larger than those of the parents and of the F_1 because of the increased variation in the F_2 and the comparable sample size. A large sample would provide a better estimate of the parameter (mean). For a sample of 80 with $s = 3.78$, the $s_{\bar{x}}$ would be 0.425. A sample of 200 with the same s would give $s_{\bar{x}} = 0.267$.

The number of gene pairs operating in the cross may be roughly estimated from the classified F_2 data. In the wheat crosses under consideration, 4 of the 40 F_2 plants, or about 1 in 10, occurred within the range of each parent. One representative of the parental type in every 16 would be theoretically expected if two independent gene pairs were segregating. This method of determining the number of active genes is too simplified to be of much value, however. If polygenes with small cumulative effects are involved in the quantitative traits being considered, it would take thousands of F_2 plants to obtain even a rough estimate of the number of genes present. Furthermore, the method just cited for estimating the number of active genes implies that all are additive and contribute equally to the end product.

Alternatively, an estimate of the number of active genes may be obtained mathematically by comparing the variances of the F_1 and F_2. Assuming that all variations being considered were hereditary, that only independent genes with cumulative effects were involved, and that the F_1 plants were fully heterozygous, the difference between the variances of the F_1 and F_2 should provide a minimum estimate of the number of genes acting. The larger the difference between the F_1 and F_2 variances, the fewer the number. Since the requirements for use of this method are seldom realized in ordinary genetic research, and standards are necessary to form a basis for comparison, the method will be left to the specialists. Instead, the simplified though imperfect procedure detailed earlier (which depends on the frequency with which the parental (P) phenotypes occur in the F_2) will be used here to estimate the number of segregating genes.

The premise of Nilsson-Ehle (which was later developed by Fisher, Wright, and others), that **polygenes behave in Mendelian fashion,** is accepted as generally correct. Furthermore, the effect of the individual genes in the polygenic system is cumulative and may operate in a plus or minus direction. The **cumulative** effect often seems to be geometrical rather than arithmetical. A logarithmic scale thus provides a more uniform pattern of the results of crosses than does a true scale developed in the units of actual measurements. So many factors are involved and so many assumptions are required that it is doubtful whether the number of genes operating in any but exceptionally simple cases can be estimated realistically by any current method.

CONTINUOUS VARIATION IN MAN

Most differences, in human populations follow the pattern of continuous variation. If, for example, 5000 men were arranged in order of height, each would differ only slightly from his neighbors on either side. Many polygenic traits could be identified by appropriate comparisons among individuals in a population. For most traits, however, only part of the continuous variation is inherited; the other part is due to environmental factors. In the example of stature in man, barring gross malnutrition or other severe environmental influence, about 90 percent of the variability would be hereditary. Measurements for other traits such as arterial pressures and pulse rates show a smaller genetic component. Hereditary continuous variation depends on the combined action of **many genes** (polygenes), each contributing to the same trait.

HEREDITARY AND ENVIRONMENTAL VARIATION

An important consideration in assessing relative contributions of heredity and environment is the requirement of one time and one place for the observations. When environmental conditions change the relative importance of heredity must usually also change. Susceptibility to pulmonary tuberculosis, for example, depends to some extent on hereditary predisposition. If the incidence of the disease decreases appreciably, the relative importance of the hereditary constitution will not remain the same. On the one hand, certain environmental conditions may offset high constitutional susceptibility. In such a case, the relative importance of heredity is increased among those who still suffer from the disease. Or, on the other hand, the genetic susceptibility may be overcome as environmental improvements take place, in which event the relative importance of heredity decreases.

Another equally important consideration is that each person, regardless of kinship, must have an equal chance of experiencing the various kinds of environmental factors; this could mimic the effect of heredity, and thus raise the measured resemblance. At least theoretically, the genes that make relatives alike and the shared environmental factors that can also make them alike can be distinguished. Genes in common decrease rapidly as relationships become more remote. Environmental similarities fall off less rapidly. A regression for uncles and aunts or for cousins that was higher than expected in proportion to that for sibs or parents would indicate shared **nongenetic factors.** Some data give indications of this, but the practical difficulties of making observations and measurements on any but the closest relatives are formidable. Better evidence for the relative influence of hereditary and environmental factors comes from studies of twins. **Identical twins** have all their genes in common. **Fraternal twins** are no more alike genetically than brothers and sisters born at different times (although they may share a more similar environment). Measures of resemblance between the two kinds of twins add useful information on the contributions of heredity and environment.

If both members of a pair of twins possess a particular trait or both are free from it, they are phenotypically similar or **concordant.** On the other hand, if only one member of the pair possesses the trait, the pair is phenotypically dissimilar or **discordant.** The relative degree of concordance for pairs of twins indicates the relative importance of hereditary and environment. In identical twins, eye color is 97 percent concordant and 28 percent in fraternal twins. Feeblemindedness is 94 percent concordant in identical twins and 47 percent concordant in fraternal twins.

REFERENCES

Dunn, L. C. and D. R. Charles. 1937. "Studies on spotting patterns." *Genetics* 22, 14–42.

East, E. M. 1910. "A Mendelian interpretation of variation that is apparently continuous." *Amer. Nat.* 44, 65–82.

East, E. M. 1916. "Studies on size inheritance in Nicotiana." *Genetics* 1, 164–176.

Emerson, R. A. and E. M. East. 1913. "The inheritance of quantitative characters in maize." *Nebraska Agric. Exp. Res. Bull.*

Fabian, Gy. 1969. *Phaenoanalysis and quantitative inheritance.* Adabe miai Kiado, Budapest.

Falconer, D. S. 1960. *Introduction to quantitative genetics.* Ronald Press, New York.

Falconer, D. S. 1967. "The inheritance of liability to disease with variable age of onset, with particular reference to diabetes mellitus." *Ann. Hum. Genet.* 31, 1–20.

Holt, S. B. 1968. *The genetics of dermal ridges.* Charles C. Thomas, Springfield, Ill.

Mather, K. 1949. *Biometrical genetics.* Dover Publ., New York.

Stern, C. 1970. "Model estimates of the number of gene pairs involved in pigmentation variability of the Negro-American." *Human Heredity* 20, 165–168.

Woolf, C. M. 1968. *Principles of biometry.* D. Van Nostrand Co., New York.

PROBLEMS AND QUESTIONS

12.1 Using the forked-line method, diagram the cross between a wheat variety with red kernels (*AABB*) and a variety with white kernels (*aabb*) in which the two pairs of genes have a cumulative effect. Classify the F_2 progeny under the headings, red, dark, medium, light, and white, and summarize the expected results.

12.2 From another cross between red and white kernel varieties, 1/64 of the F_2 plants had kernels as deeply colored as the red parent and 1/64 had white kernels. About 62/64 were between the extremes of the parents. How can the difference in F_2 results in Problems 12.1 and 12.2 be explained?

12.3 Different F_2 plants from the cross in Problem 12.2 were crossed with the white parent (*aabbcc*). Give the genotypes of the individual plants from which the following backcross results could have been obtained: (a) 1 colored, 1 white; (b) 3 colored, 1 white; and (c) 7 colored, 1 white.

12.4 Different F_2 plants from the cross in Problem 12.2 were selfed. Give the genotypes of the parents that could have produced the following results: (a) all white; (b) all colored; (c) 3 colored, 1 white; (d) 15 colored, 1 white; and (e) 63 colored, 1 white.

12.5 Different F_2 plants from the cross in Problem 12.2 all producing kernels with some red color, were crossed with each other. Give the genotypes of parents which could have produced the following results: (a) 7 colored, 1 white; (b) 31 colored, 1 white; and (c) 63 colored, 1 white.

12.6 Assume that two pairs of genes are involved in the inheritance of skin pigmentation. For the purpose of this problem, the genotype of a black person may be symbolized $AABB$ and that of the white $aabb$. What color might be expected in the F_1 from crosses between black and white people?

12.7 If mulattoes ($AaBb$) mated with other mulattoes of the same genotype, what results might be expected with reference to the intensity of skin pigmentation?

12.8 (a) If people with various degrees of skin pigmentation married only people known to have the genotype $aabb$, could they have "black" babies? Explain. (b) Could a "white" couple with black ancestry have a "black" baby? Explain.

12.9 If the number of gene pairs involved in skin pigmentation should actually be 4, 5, or 6, as indicated by Stern's model, (a) how would the expected results of crosses between white and black differ from those considered in the above problem? (b) What criteria and data would be necessary to determine the number of genes actually involved?

12.10 Two pairs of genes ($AABB$) with equal and additive effects are postulated to influence the size of corn in certain varieties. A tall variety averaging six feet was crossed with a dwarf averaging two feet. (a) If the size of the dwarf is attributed to the residual genotype ($aabb$), what is the effect of each gene that increases the size above two feet? (b) Diagram a cross between a large and small variety and classify the expected F_2 phenotypes.

12.11 The size of rabbits is presumably determined by genes with an equal and additive effect. From a total of 2012 F_2 progeny from crosses between large and small varieties, eight were as small as the average of the small parent variety and about eight were as large as the large parent variety. How many gene pairs were operating?

12.12 A sample of 20 plants from a certain population was measured in inches as follows: 18, 21, 20, 23, 20, 21, 20, 22, 19, 20, 17, 21, 20, 22, 20, 21, 20, 22, 19, and 23. Calculate (a) the mean, (b) the standard deviation, and (c) the standard error of the mean.

12.13 What is measured by (a) the mean, (b) the standard deviation, and (c) the standard error of the mean?

12.14 If the population sampled in Problem 12.12 were an F_2 involving parents from varieties averaging 7 inches and 33 inches, respectively, would you conclude that few or many gene pairs were involved? Why?

12.15 A sample of 20 plants from a certain population was measured in inches as follows: 7, 10, 12, 9, 10, 12, 10, 9, 10, 11, 8, 12,

10, 10, 9, 11, 10, 9, 10, and 11. Calculate (a) the mean, (b) the standard deviation, (c) the standard error of the mean, and (d) establish the 99 percent confidence limits.

12.16 If the population sampled in Problem 12.15 was an F_2 from parents averaging 7 and 12 inches, respectively, would you conclude that few or many gene pairs were involved? Why?

12.17 The width or spread of 122 guayule plants representing a random sample were measured 107 days after planting in inches as follows:

```
13 12 11 13 13 13 12 11 13 11 12 11 12 12 13 12 13
13 13 12 11 11 12 12 12 13 10 11 11 11 11 12 10 12
11 10  9 10 10 10 14 12 11 11  9 11 12 13 11 11 12
11 12 12 14 13 14 13 16 13 13 11 14 12 13 15 12 11
10 11 11 11 10 11 12 13 12 12 12 12 13 12 10 13 10
10 10 10 13 11 10 13 14 12  9 10 10 11 11 14 11  9
12 10 10 13 11 11 13 10 12 12 10  9 14 12 10 11 10
12 14 11
```

Calculate (a) the mean, (b) the standard deviation, (c) the standard error of the mean, and (d) establish the 95 percent confidence limits.

12.18 Seeds from four different types of wheat were planted and samples of each type were measured as to the time required for maturity. This is the number of days elapsing from the time of planting the seed until the heads of the grain appear. The following statistics were computed from the samples of different sizes:

	mean	standard deviation	standard error of the mean
Strain A	72.37	1.31	0.221
Strain B	55.85	0.93	0.179
$F_1(A \times B)$	61.40	1.35	0.350
$F_2(A \times B)$	63.84	3.45	0.330

Several of the F_2 plants were selfed. The F_3 seeds were planted in randomized plots and raised in the same season. The progeny thus produced were classified according to the time required for maturity. A sample of 24 plants from the seed of a single F_2 were recorded as follows:

```
56 55 54 56 57 56
55 56 57 56 57 56
55 57 56 55 56 58
56 55 57 56 59 58
```

Compute (a) the mean, (b) the standard deviation, (c) the standard error of the mean, and (d) establish the 95 percent confidence limits.

12.19 (a) Discuss briefly the biometrical significance of the sample in Problem 12.18. (b) From which part of the F_2 population was the F_2 parent likely obtained? (c) What suggestions can be made concerning the genotype of the F_2 parent?

12.20 Make a similar analysis of the following samples, each of which represents a random sample from the F_3 derived from the seed of a single F_2 plant:

(a)	73	72	70	74	76	71
	72	74	74	74	71	76
	73	72	77	73	78	75
	76	70	74	70	71	70
(b)	67	65	64	66	65	66
	68	64	65	66	69	71
	70	63	62	61	60	64
	63	65	64	63	68	67
(c)	65	64	66	67	65	64
	64	68	65	64	63	65
	65	64	66	65	67	66
	65	68	65	64	66	65

12.21 (a) What does it mean to be 90 percent confident? (b) What does a $s_{\bar{x}} = \pm 1.02$ cm tell the investigator? (c) What is the value of knowing the variance of a set of data?

12.22 What is the relation between (a) P, (b) confidence limits, and (c) acceptability on the part of the investigator.

12.23 If $P = 0.95$ what do the confidence limits $L_1 = 10$ cm and $L_2 = 12$ cm indicate?

12.24 Representatives of two different varieties of maize each averaging 68 inches in height were crossed. The F_1 also averaged 68 inches. In the F_2, however, there was considerable variation, ranging from 36 inches to 100 inches. Out of 2,016 F_2 plants, eight reached a height of 100 inches, and seven were as short as 36 inches. Assuming that the environment was held uniform for the entire experiment and that all genes had equal effects: (a) How many pairs of genes are involved? (b) How much of the height above that produced by the residual genotype (that of the 36 inch plants) is each gene symbolized by a capital letter responsible for in the F_2 plants reaching 100 inches? (c) Give probable genotypes of the parent plants. (d) Describe the type of variation exemplified in this problem.

12.25 Variety A of wheat required an average of 72 days from planting to heading in a particular environment. Variety B grown at the same time and in the same environment required an average of 56 days from planting to heading. Varieties A and B were crossed and the F_1 and the F_2 were grown in a comparable environment to that of the P. From a total of 5,504,000 F_2 progeny, 90 required the shortest time (50 days) and 86 required the longest time (90 days). Assuming that all factors contributed equally, how many pairs of genes contributed to the time required for heading?

THIRTEEN

Population genetics is the study of genes in populations. It is the branch of genetics that describes in mathematical terms the consequences of Mendelian inheritance at the population level. Population geneticists are concerned with the frequencies and interactions of genes in a **Mendelian population**, that is, an interbreeding group of organisms sharing a common **gene pool.** A gene pool is the total genetic information possessed by reproductive members in a population of sexually reproducing organisms. Genes in the pool have **dynamic relations** with other alleles and with the environment in which the organisms reside. Environmental factors such as selection tend to alter gene frequencies and thus to cause evolutionary changes in the population.

In 1908, an English mathematician, G. H. Hardy, and a German physician, W. Weinberg, independently discovered the principle concerned with the frequency of genes in a population. It is a theoretical statement called the Hardy-Weinberg equilibrium principle. It states that in an equilibrium

POPULATION GENETICS

Sheep herd. (H. W. Silvester/Rapho Guillumette.)

population both gene frequencies and genotype frequencies remain constant from generation to generation. It assumes that this occurs in a **large interbreeding population** in which mating **is random** and **no selection** or other factors for **changing** the gene frequency occur.

Mendelian segregation may be represented mathematically by the binomial expansion of $(a + b)^n$ where a is the probability that an event will occur and b, that it will not occur. The familiar ratio of $1 : 2 : 1$, representing the segregation of a single pair of alleles (Aa), in a monohybrid cross may be represented by the simplest expansion of $(a + b)^n = (A + a)^2 = 1AA + 2Aa + 1aa$. To express the relation in more general terms that will apply to any pair of alleles, the symbols p and q are introduced. At equilibrium the frequencies of the genotypic classes are $\mathbf{p^2(AA)}$, $\mathbf{2pq(Aa)}$ and $\mathbf{q^2(aa)}$. A frequency is the ratio of the actual number of individuals falling in a single class to the total number of individuals, and a probability represents the likelihood of occurrence of any particular form of an event. Chance is involved in combinations of alleles and genotypes in populations as well as in Mendelian segregation and independent assortment. Thus probability is the ratio of specified events to total events. Possible combinations of sperm and eggs from heterozygous individuals depicted in Table 13.1, illustrate the allele frequency relations following a mating of $Aa \times Aa$. In a larger population including AA and aa genotypes as well as Aa, an equilibrium is established for a single pair of alleles after one generation of random (panmictic) mating. The genetic proportions of an equilibrium population are entirely determined by its gene frequencies as illustrated in Table 13.2

When only two alleles are involved, $p + q = 1$. Since $p + q = 1$, $p = 1 - q$. Now, if $1 - q$ is substituted for p, all the relations in the formula can be represented in terms of q as follows: $(1 - q)^2 + 2q(1 - q) + q^2 = 1$. (This alteration provides a means for solving a problem with a single unknown, q. If allele A has a frequency of $1 - q$ and allele a has a frequency of q, the expected distribution of these alleles under conditions of random mating in succeeding generations may be calculated with the above formula. The formula may be applied to any pair of alleles, if the frequency of one member of the pair in the population can be determined. With no dominance, each member of the pair can be determined directly. With dominance, the decimal fraction representing the proportion of individuals with the homozygous recessive phenotype (q^2) is the index figure.

TABLE 13.1
Combinations of Sperm and Eggs from Heterozygous Individuals, Illustrating the Allele Frequency Relations that Form the Basis for the Hardy-Weinberg Equilibrium

Eggs	Sperm	
	$A(p)$	$a(q)$
$A(p)$	$AA(p^2)$	$Aa(pq)$
$a(q)$	$Aa(pq)$	$aa(q^2)$

Summary: $p^2(AA) + 2pq(Aa) + q^2(aa)$

TABLE 13.2

Algebraic Proof of the Maintenance of Genetic Equilibrium in a Randomly Mating Population. (It is assumed that only two alleles are present in the population (i.e., $p + q = 1$). Given any p and q such that their sum is one, it can be shown that only one generation of random mating is required to reach genetic equilibrium.)

Parental Matings	$AA \times AA$	$2(AA \times Aa)$	$2(AA \times aa)$	$Aa \times Aa$	$2(Aa \times aa)$	$aa \times aa$	Summations
Parental Mating Frequencies	$p^2 \times p^2$	$2(p^2 \times 2pq)$	$2(p^2 \times q^2)$	$2pq \times 2pq$	$2(2pq \times q^2)$	$q^2 \times q^2$	
Offspring Frequencies AA	p^4	$2p^3q$		p^2q^2			$\Sigma_{AA} = p^2$
Aa		$2p^3q$	$2p^2q^2$	$2p^2q^2$	$2pq^3$		$\Sigma_{Aa} = 2pq$
aa				p^2q^2	$2pq^3$	q^4	$\Sigma_{aa} = q^2$
Total Offspring Frequencies	p^4	$+\,4p^3q$	$+\,2p^2q^2$	$+\,4p^2q^2$	$+\,4pq^3$	$+\,q^4$	$\displaystyle\sum_{\text{total}} = 1$

$$
\begin{aligned}
&(p^4 + 2p^3q + p^2q^2) &&+\,(2p^3q + 4p^2q^2 + 2pq^3) &&+\,(p^2q^2 + 2pq^3 + q^4) \\
&\;p^2(p^2 + 2pq + q^2) &&+\,2pq(p^2 + 2pq + q^2) &&+\,q^2(p^2 + 2pq + q^2) \\
&\;p^2 &&+\,2pq &&+\,q^2 \\
&&&(p + q)^2
\end{aligned}
$$

$$= p^2 \qquad = 2pq \qquad = q^2 \qquad = 1$$

GENOTYPE EQUILIBRIUM

Genotypes in a population tend to establish an equilibrium with reference to each other, expressed as $p^2 : 2pq : q^2$. The absolute frequency of each genotype is thus seen to depend on the values of p and q. For example, if two alleles should occur in equal proportion in a large, isolated breeding population and neither has an advantage over the other, they would be expected to remain in equal proportion generation after generation. This would be a special case because alleles in natural populations seldom if ever occur in equal frequency. They may, however, be expected to maintain their relative frequency, whatever it is, subject only to change by such factors as **natural selection, differential mutation rates, migration, random genetic drift and meiotic drive,** all of which alter the level of the allele frequencies. An equilibrium in genotype frequencies is maintained through random mating, with absolute frequencies of genotypes being determined by the gene frequency. As long as the gene frequency does not change, genotypic proportions remain constant.

ALLELE FREQUENCY FOR INTERMEDIATE INHERITANCE

Two alleles for M and N blood antigens provide a model for the segregation of a single pair of autosomal alleles with intermediate inheritance in human populations.

To be consistent with symbols for alleles, the same letter should be used to represent the different alleles at the same locus. L^M and L^N will represent the two alleles in this discussion. These alleles are especially well adapted for a beginning study of gene frequency because dominance is not expressed. The M, N, and MN phenotypes are detectable by serological tests. Table 13.3 shows the relations between genotypes and phenotypes with the specific reactions of the blood cells to anti-M and anti-N sera. (Although no significant selective value has been associated with these blood traits, several investigators have suggested recently that a type of immunization is possible which may give these antigens significance in selection.) Numerous samples of blood representing different populations have been tested and the frequency of the L^M and L^N alleles has been calculated from the data obtained. Following are the proportions of the different phenotypes based on a sample of 6129 Caucasian people in the United States:

M	MN	N
1787	3039	1303

When the data are interpreted in terms of the frequency of genotypes ($L^M L^M + L^M L^N + L^N L^N$), the total numbers and frequencies of alleles are obtained, as shown in Table 13.4

Allele frequencies represented in decimal

TABLE 13.3
Detection of M, N and MN Phenotypes

Phenotype	Genotype	Anti-M	Anti-N
M	$L^M L^M$	+	○
MN	$L^M L^N$	+	+
N	$L^N L^N$	○	+

TABLE 13.4
Frequency of L^M and L^N Alleles

Allele	Proportion	Allele Frequency
$L^M = 2 \times 1787 + 3039 = 6613$	6,613/12,258	0.54
$L^N = 2 \times 1303 + 3039 = 5645$	5,645/12,258	0.46

fractions, as in this example, can be used directly as functions or probabilities. Thus, the probability is 0.54 that a member of the particular pair of chromosomes will carry L^M and the probability is 0.46 that a member of the pair will carry L^N. From such information, predictions can be made concerning the various combinations of the alleles in the population.

ALLELE FREQUENCY WHEN DOMINANCE IS INVOLVED

Dominance and recessiveness of alleles do not directly influence frequency; that is, dominance alone does not make an allele occur more frequently in the population than recessiveness. The same type of equilibrium is maintained for **autosomal** alleles showing dominance or recessiveness as is maintained for those with intermediate inheritance, such as the L^M and L^N alleles just described. However, it is true that dominant alleles express themselves in heterozygous combination and therefore are expressed more frequently than recessives. If one phenotype has a selective advantage over another, dominance could indirectly influence allele frequency.

An interesting pair of contrasting traits, which has been detected in human populations and has no known selective value is the ability or inability to taste phenylthiocarbamide (PTC). The difference in the ability of people to taste this chemical was discovered accidently by investigators in a university laboratory. A hereditary basis for the mechanism was postulated to account for tasters and nontasters. Inability to taste the chemical was found to be dependent on a single recessive allele (t).

From a group of 228 university students who were invited to taste the chemical, the following results were obtained: 160 tasters and 68 nontasters. What are the relative allele frequencies of T and t in the population? Since T is dominant over t, the 160 tasters include the genotypes TT and Tt. The 68 nontasters must have the genotype tt. Assuming equilibrium, the best approach to the Hardy-Weinberg formula is through the nontasters, all of whom can be assumed to have the same genotype. The 68 tt (q^2) individuals represent 0.30 of the population, $\hat{q} = \sqrt{0.30} = 0.55$. (The "hat" over a p or a q symbol indicates an equilibrium frequency.) Thus, the frequency of T in the sample is 0.45 and the frequency of t is 0.55. These allele frequencies are comparable with those obtained from larger samples of the Caucasian population in the United States. Predicted genotype frequencies based on random mating in the population are:

$$TT = p^2 = 0.45 \times 0.45 = 0.20$$
$$Tt = pq = 0.45 \times 0.55 = 0.25$$
$$tT = qp = 0.55 \times 0.45 = 0.25$$
$$tt = q^2 = 0.55 \times 0.55 = 0.30$$
$$p^2 = 0.20, \quad 2pq = 0.50, \quad q^2 = 0.30$$

When the allele frequencies are known, it is possible to predict the likelihood of occurrence of certain alleles and expressions of corresponding traits in populations. The probability of an expression of a trait dependent on a recessive allele may thus be calculated, even in family groups or other populations where no previous expression has occurred and no evidence is available to indicate which individuals carry the recessive allele in question.

FREQUENCY FOR MULTIPLE ALLELES

The equation $p + q = 1$ applies when only two autosomal alleles in the population occur at a given locus. If the system includes more alleles, more symbols must be added to the equation. The four human blood types, for example, are controlled by three alleles, A, A^B, and a. Both A and A^B are dominant over a but neither is dominant over the other, that is, both anti-A and anti-B sera will coagulate the cells from an individual with AA^B genotype. For substitution into the formula, the frequencies of A, A^B, and a may be symbolized p, q, and r, respectively. Since these three alleles represent all of the alleles involved, it follows that $p + q + r = 1$. Multiple alleles establish an equilibrium in the same way as the single pairs of alleles previously described.

The blood types of 173 students in genetics laboratory classes were determined as follows: O, 78; A, 71; B, 17; and AB, 7. Genotypes and frequencies in the population sampled are summarized in Table 13.5. From these data, the frequency of the various alleles may be calculated. The gene pool includes all of the alleles in this sample as well as all others in the population. Since r^2 (O) $= 78/173 = 0.45$, $r = \sqrt{0.45} = 0.67$. The Hardy-Weinberg formula may now be applied to two of the three alleles. The proportion of the A(p) and O(r) individuals in the population is represented by the equation:

$$O + A = r^2 + 2pr + p^2 = (r + p)^2.$$

Therefore,

$$r + p = \sqrt{O + A.} = \sqrt{78/173 + 71/173}$$
$$= \sqrt{0.45 + 0.41} = \sqrt{0.86} = 0.93$$

and

$$p = 0.93 - 0.67 = 0.26.$$

The frequencies of p and r have now been obtained and the next step is to calculate the frequency of q. Since $p + q + r = 1$, $0.26 + 0.67 = 1 - q$, or $q = 1 - (0.67 + 0.26) = 0.07$. Now summing up the allele frequencies calculated from the sample of 173 students, we have 0.26 (p) + 0.07

TABLE 13.5
Phenotypes, Frequencies, and Genotypes Represented by a Sample of 173 Genetics Students

Phenotypes	Phenotype Frequencies of Students	Genotypes	Genotypic Frequency	Sum of Frequencies of Genotypes with Similar Phenotypes
O	0.4509	aa	r^2	r^2
A	0.4104	AA Aa	p^2 $2pr$	$p^2 + 2pr$
B	0.0983	$A^B A^B$ $A^B a$	q^2 $2qr$	$q^2 + 2qr$
AB	0.0405	AA^B	$2pq$	$2pq$

$(q) + 0.67\ (r) = 1$. The value of 1 represents all the alleles considered in this example. As a result of dominance, allele distribution in the gene pool differs from the distribution of phenotypes resulting from combinations of alleles. Results from this sample of 173 students are comparable with those from larger samples in the general populations. The A and B from the AB blood group were not used during the calculations because the frequencies of alleles and not phenotypes were being determined. All three segregating alleles were included in the calculations.

FREQUENCY FOR SEX-LINKED ALLELES

Alleles in the sex chromosomes occur in a different frequency than those in autosomes because of the arrangements of **sex chromosomes** in the **two sexes**. In organisms such as Drosophila and man, with one heterogametic sex, there are 5 possible genotypes—AA, Aa, aa for the female and a, A for the male. Therefore, the total allele frequency in a mating pair is not one or one-half as for autosomal alleles but rather, one, two-thirds, one-third, or none.

$$A = p = \frac{1}{3}\ (p\,\male) + \frac{2}{3}\ (p\,\female)$$

$$a = q = \frac{1}{3}\ (q\,\male) + \frac{2}{3}\ (q\,\female)$$

Equilibrium genotypic values for sex-linked genes in organisms with male heterogametic sex determination are as follows:

1. For females $\underset{AA\quad Aa\quad aa}{\overline{p^2\ +\ 2pq\ +\ q^2}}$

2. For males $\underset{A\quad a}{\overline{p\ +\ q}}$

If, for example, a population includes only color-blind men carrying the allele rg (in the X chromosome and no allele in the Y) and normal women who are homozygous for rg^+, the expected allelic frequency in the population of progenies will be 1/3 rg and 2/3 rg^+ as shown in Table 13.6a. These frequencies would remain constant in large random mating populations if both males and females were considered together. The frequency of sex-linked alleles in families

TABLE 13.6
Frequency for Sex-linked Gene in Reciprocal Crosses (a) and (b) under XY Chromosome Pattern.

	Mating	Frequency	Number rg^+ Alleles	Number rg Alleles
	P $\frac{rg^+}{rg^+} \times \frac{rg}{Y}$		2	1
(a)	F₁ $\frac{rg^+}{rg} \times \frac{rg^+}{Y}$		2	1
	rg^+ allele	2/3		
	rg allele	1/3		
	P $\frac{rg}{rg} \times \frac{rg^+}{Y}$		1	2
(b)	F₁ $\frac{rg}{rg^+} \times \frac{rg}{Y}$		1	2
	rg^+ allele	1/3		
	rg allele	2/3		

TABLE 13.7

Calculation for Attainment of Equilibrium for Gametes in a Population with Two Separate Gene Pairs Aa **and** Bb **with Initial Gene Frequencies of** $A = B = 0.6$ **and** $a = b = 0.4$ **and Initial Genotypic Frequencies of** $AABB = AAbb = aaBB = 0.30$ **and** $aabb = 0.10$. **The difference (d) between repulsion and coupling products is:** $d = [(Ab) (aB)] - [(AB) (aB)] = [(0.3) (0.3)] - [(0.3) (0.1)] = 0.06$. **Generation 11 Approaches Equilibrium but Further Generations Are Required for Attainment (This Calculation Violates the Principle of Significant Figures but Small Increments Are Lost by Rounding Off).**

Generation	Proportion of $d = 0.06$ Added (AB,ab) or Subtracted (Ab,aB)	Gametes			
		AB	Ab	aB	ab
1.		0.3	0.3	0.3	0.1
2.	0.5d	0.33	0.27	0.27	0.13
3.	0.75d	0.345	0.255	0.255	0.145
4.	0.875d	0.3525	0.2475	0.2475	0.1525
5.	0.9375d	0.35625	0.24375	0.24375	0.15625
.					
.					
11.	0.99902342752d	0.35994140625	0.2400585937	0.2400585937	0.15994140625
.					
equilibrium	d	0.36	0.24	0.24	0.16

depends on the underlined allele arrangements in the initial matings. If the women had been color blind rg/rg and the men normal rg^+/Y the frequency would have been 1/3 rg^+ and 2/3 rg (Table 13.6b).

EQUILIBRIUM AT MORE THAN ONE LOCUS

When two gene-pair differences (e.g., Aa and Bb) are considered simultaneously the number of possible genotypes is 3^2 (i.e., $AABB$, $AABb$, $AaBB$, $AaBb$, $aaBB$, $aaBb$, $AAbb$, $Aabb$ and $aabb$). Let p, q, r and s represent the gene frequencies of A, a, B and b, respectively. The equilibrium formula depends on the terms pr, ps, qr and qs which are the equilibrium frequencies of the gametes AB, Ab, aB and ab, respectively. The equilibrium ratios of the genotypes are: $(pr + ps + qr + qs)^2$ or p^2r^2 ($AABB$) $2p^2rs$ ($AABb$) $2pqr^2$ ($AaBB$) $4pqrs$ ($AaBb$) q^2r^2 ($aaBB$) $2q^2rs$ ($aaBb$) p^2s^2 ($AAbb$) $2pqs^2$ ($Aabb$) and q^2s^2 ($aabb$). When the gametic equilibrium frequencies have been reached the genotypic frequencies will also be at equilibrium. When more than one gene-pair difference is considered simultaneously, the number of possible genotypes is 3^n (where n is the number of gene pairs). In such cases a multinomial rather than a binomial expansion is required to obtain equilibrium ratios.

How long will it take for gametic frequencies to reach equilibrium? The simplest case is a population of heterozygotes ($AaBb \times AaBb$) in which all genes have the same frequency, that is, $p = q = r = s = 0.5$. All four gametes (AB, Ab, aB and ab) are produced at equilibrium and genotypic equilibrium is attained in one generation. This is a rare situation. Usually initial gene frequencies are not alike and sometimes the initial population includes only homozygotes (e.g., $AABB$ and $aabb$) which produce only a part of the gametes required for equilibrium (namely the heterozygote $AaBb$). In such cases several or many generations may be required to attain equilibrium.

At equilibrium, frequencies of genes in gametes would be the same in coupling (AB and ab) as in repulsion (Ab and aB) (i.e., $(AB) \times (ab) = (Ab) \times (aB)$. If coupling and repulsion products are not equal in the initial population, the difference (d) may be used to calculate the time required to attain equilibrium. One half the difference from equilibrium is reduced in each generation of random mating. If, for example, the frequencies of A and B for each 0.6 and the frequencies of a and b are each 0.4, the equilibrium frequencies = $(Ab)(aB) = (0.24)(0.24) = (AB)(ab) = (0.36)(0.16) = 0.0576$. The difference in coupling and repulsion products represents the change in gametic frequencies required to attain equilibrium. If the initial genotypic frequencies are $AABB = AAbb = aaBB = 0.30$ and $aabb = 10$, the initial frequencies of gametes are $(AB) = 0.3$, $Ab = 0.3$, $aB = 0.3$, $ab = 0.1$ and $d = (Ab)(aB)$ for repulsion $- (AB)(ab)$ for coupling = $(0.3)(0.3) - (0.3)(0.1) = 0.06$. Calculations at random mating required to reach equilibrium with given initial gene frequencies and genotype frequencies are summarized in Table 13.7.

FACTORS INFLUENCING ALLELE FREQUENCY

The Hardy and Weinberg explanation for equilibrium in a population required three assumptions: (1) individuals of each genotype must be as reproductively fit as those of any other genotype in the population, (2) the population must consist of an infinitely large number of individuals; and (3) random mating must occur throughout the population. The Hardy-Weinberg theorem with its assumptions does not account for any change in allele frequency within populations. That is just what Hardy and Weinberg intended, because their formula described the **statics** of a Mendelian population. Something more was required to formulate a mathematical explanation of naturally occurring population change or **dynamics** in terms of allele frequencies. This need was filled by Fisher, Wright, and J. B. S. Haldane, who conceived additional theoretical models and superimposed mechanisms for change in allele frequencies upon the Hardy-Weinberg equilibrium. Population statics thus became population dynamics.

Now that mathematical tools have been established, the dynamic relations that exist between individuals representing particular genotypes and their variable environments can be determined. Population genetics seeks an explanation for the genetic architectures of different populations. This discipline is concerned with the origins of adaptive norms or arrays of genotypes as they become consonant with the demands of the environments. The genetic architecture of a population is the manner in which the genetic material is arranged and the frequency with which each allele occurs. In ascertaining the genetic architecture of a population, questions must be answered concerning allele frequency, the frequency of chromosome structural changes such as inversions and translocations, and chromosome numbers. Changes in allele frequency depend on selection, mutation, chance (random genetic drift), differential migration and meiotic drive.

SELECTION

Selection is the nonrandom differential reproduction of genotypes. Frequencies established by alleles in a population at equilibrium are dynamic and subject to change. If, for example, allele A makes the organism more efficient in reproduction (fit) than a, A is expected to increase generation after generation at the expense of a. Continued selection of this kind would tend to decrease the proportion of a in favor of A.

Observations made in natural populations, such as one involving moths in England, have provided examples of selection occurring in nature. In 1850, surveys indicated that most moths in various nonindustrialized communities were light in color, whereas a very few, less than one percent, were dark. As these cities became industrialized and factory smoke darkened the buildings and countryside, a parallel change occurred among the moths. Later surveys showed that 80 to 90 percent of the moths in certain industrial areas were dark. L. Doncaster, R. Goldschmidt, A. Kühn, and E. B. Ford showed, from experimental breeding, that dark color was inherited and controlled by one or two dominant genes. Presumably, those moths that matched their environment best, lived, reproduced, and transmitted their genes for dark color to their progeny, whereas the more conspicuous light-colored moths on a dark background (Fig. 13.1) were often caught by predators. The alternative explanation, that in some way the direct influence of the environment changed the genetic color-producing mechanism, was considered but not substantiated. Observations have now borne out the explanation that the change is based on selection favoring moths that most closely blend with their surroundings.

In a particular study conducted by H. B. D. Kettlewell, 447 black moths, *Biston betularia*, were collected elsewhere and released in smoky Birmingham. At the same time 137 white moths of a type not already present in the vicinity were also released. After a period of time, moths in the area

Figure 13.1 Dark and light forms of the peppered moth photographed on the trunk of an oak tree blackened by the polluted air of the English industrial city of Birmingham. The light form *Biston betularia* is clearly visible; the dark form (carbonaria) is well camouflaged. (H. B. D. Kettlewell, *Sci. Amer.* 200:48–53, 1956. From the experiments of H. B. D. Kettlewell, University of Oxford.)

were trapped and classified. Only 18 (13 percent) of the original 137 white moths and 123 (27.5 percent) of the blacks could be found. The collections were extensive and it was presumed that virtually all of the released moths that were still alive were captured. The blacks were favored 2 : 1 over the whites in survival. In Dorset, which is not industrialized, 398 black and 376 white moths were released and retrapped in a similar way. This time 14.6 percent of the whites but only 4.7 percent of the blacks survived. The moths that blended with their surroundings best were concealed from predators and therefore survived, whereas the conspicuous dark moths were more readily seen and devoured by birds.

The implication in the theory of natural selection is that certain genotypes in a species endow their carriers with advantages in survival and reproduction. In dealing with "fitness" traits that are less conspicuous than such traits as color in moths, it is difficult to evaluate the effects of individual alleles on survival and reproduction because it is whole organisms with their full complements of genes that survive or fail to survive. Usually only slight differences in survival and fitness characterize different genotypes. Nevertheless, single gene pairs contribute to total fitness in particular environments and it is appropriate to approach a complex problem through a consideration of its component parts.

The process of change in allele frequency

through selection can be illustrated with a large, random-mating population that has a single pair of alleles, A and a, segregating. Let $1 - q$ represent the frequency of A, which in this example is completely dominant, and q the frequency of the recessive allele a. A constant selective advantage is assumed for the dominant phenotype over the others. Reproductive rates for genotypes AA, Aa and aa would be $1 : 1 : 1 - s$, respectively. The selection coefficient, s, is a measure of the advantage or disadvantage of an organism relative to a particular reference genotype under selection and thus represents the intensity of selection. It expresses the fitness of one genotype compared with the other. It has a positive value if selection favors A over aa. If selection is complete and in favor of A (aa is lethal), $s = 1$. Change per generation in allele frequency can be determined when q and s are known.

For example, in a population with three-fourths (0.75) of the individuals expressing the phenotype of $A -$ (dominant allele) and one-fourth (0.25) the phenotype of aa, the relations may be summarized as shown in Table 13.8. Out of the total gene frequency of $1 - sq^2$, the AA parents would contribute proportionally $(1 - q)^2/1 - sq^2$ A gametes to the gene pool, and the Aa parents would contribute $[q(1 - q)]/1 - sq^2$ A gametes and $[q(1 - q)]/1 - sq^2$ a gametes. The aa individuals would contribute proportionally $q^2(1 - s)/1 - sq^2$ a gametes. Therefore, the frequency of a would be

$$\frac{q(1 - q) + q^2(1 - s)}{1 - sq^2}$$

or

$$\frac{(q - sq^2)}{1 - sq^2}$$

which $= q_1 =$ frequency of a in F_1.

The change (Δ) in frequency of $a(q)$ from the parental (P) to the filial (F_1) generation under selection pressure would be

$$\Delta q_s = q_1 - q \text{ or } \frac{q - sq^2}{1 - sq^2} - q$$

$$= \frac{-sq^2(1 - q)}{1 - sq^2}$$

This measures the amount of **change in q per generation of selection.** If s is very small, as is usually the case in nature, Δq_s can be approximated as $-sq^2(1 - q)$, since $1 - sq^2 \cong 1$.

Different values of s would indicate different trends in the population. Under conditions of complete selection ($s = 1$) against aa (homozygous lethal or completely sterile) in a particular environment, the parents for producing the next generation would be reduced to two genotypes: AA and Aa. If the initial allele frequencies before

TABLE 13.8
Relations Among Genotypes After Selection

	Genotypes			
	AA	Aa	aa	Total
Relative proportion before selection	$(1 - q)^2$	$2q(1 - q)$	q^2	$= 1$
Genotypic frequencies	0.25	0.50	0.25	$= 1$
Selective value	1	1	$1 - s$	
Relative proportion after selection	$(1 - q)^2$	$2q(1 - q)$	$q^2(1 - s)$	$= 1 - sq^2$

the selection factor was introduced were $q = 1 - q = 0.5$ the proportion of genotypes after selection would be $0.25 \ (1 - q)^2 \ (AA)$: $0.50 \ 2q \ (1 - q) \ (Aa)$. In the subsequent breeding population, they would represent $0.25/(0.25 + 0.50) = 0.33 \ (AA)$ and $0.50/(0.25 + 0.50) = 0.66 \ (Aa)$. If the AA and Aa individuals mate at random, the following progeny would be expected if $s = 1$:

type of mating	frequency	AA	Aa	aa
$AA \times AA$	$(0.33)^2 = 0.11$	0.11		
$AA \times Aa$	$2(0.33)(0.66) = 0.44$	0.22	0.22	
$Aa \times Aa$	$(0.66)^2 = 0.44$	0.11	0.22	0.11

The genotypes would again be at equilibrium but at a new level. The genotypic and phenotypic frequencies would have favored A– considerably under the condition of complete selection against one phenotype, that of aa. Frequencies of alleles under different fitness values are given in Table 13.9.

The efficiency of selection depends on the gene frequency. For example, selection increases with increasing gene frequency but decreases with decreasing gene frequency as shown in Table 13.10. Progress of selection is rapid at first when the gene is frequent enough to produce a significant number of recessive homozygotes but decreases as gene frequency declines. When q approaches 0 no rapid change occurs because most recessive alleles are tied up in heterozygotes and therefore are not exposed to selection. Recessive alleles cannot be entirely eliminated by selection. When q is small enough to allow homozygotes to occur very infrequently, recessive alleles are still carried in heterozygotes.

Thalassemia (or Cooley's anemia) provides an example of selection in human populations. It occurs mostly in children and is nearly 100 percent fatal. The disease is controlled by an allele (c), which, in homozygous condition (cc), produces the severe thalassemia major. The same allele in heterozygous condition results in a mild form of the disease called thalassemia minor or microcythemia. People with the heterozygous combination may have no outward symptoms, but they can be identified with blood tests. The importance of the disease from a public health standpoint and the ease with which carriers can be detected have led to many allele frequency surveys in areas where the disease is prevalent.

The thalassemia allele is widespread, mostly in equatorial areas (as shown in Fig. 13.2). It is rarely noted in the United States, but is especially frequent in the Mediterranean region, particularly in Italy, Greece,

TABLE 13.9
Frequencies p of Gene A after One Generation from an Initial Population Where Genes A and a Are Equally Frequent ($p = q = 0.5$), Selective Value of the Dominants (AA and Aa) Is 1 and That of the Recessives (aa) Is 0 (a Recessive Lethal), 0.4 (a Semilethal), 0.9, 0.99 (Subvital) or 1.5 (Supervital)

Selective value	0	0.4	0.9	0.99	1.5
Selection coefficient (s)	1.0	0.6	0.1	0.01	−0.5
Frequency p after one generation of selection	0.67	0.59	0.5128	0.5012	0.444
Increment of gene frequency (Δp)	+0.17	+0.09	+0.0128	+0.0012	−0.056

TABLE 13.10
Gene Frequency and Efficiency of Selection. A Recessive Gene q for an Undesirable Trait Is Present in the Initial Population with a Frequency of 0.5 and Homozygous Recessives Are Eliminated in Every Generation ($s = 1$). Gene Frequencies Are Given for Successive Generations

Generation	Frequency
0	0.500
1	0.333
2	0.250
3	0.200
4	0.167
8	0.100
10	0.083
20	0.045
30	0.031
40	0.024
50	0.020
100	0.010
200	0.005
1000	0.001

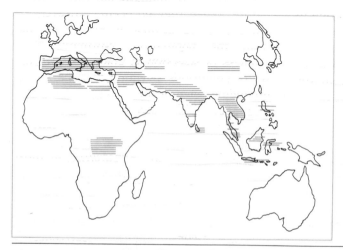

Figure 13.2 Map showing world distribution of thalassemia. This blood disease is concentrated mostly in equatorial regions. (Reprinted from *Human Biology*, 32:29–62, 1960 by A. Motulsky by permission of the Wayne State University Press, © 1960.)

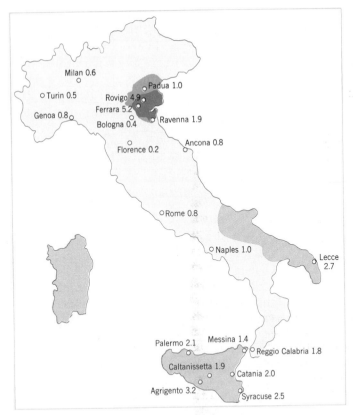

Figure 13.3 Map of Italy and Sicily illustrating the distribution of the gene for thalassemia. Percentage frequencies of allele c are given for several cities where samples have been taken. (From Bianco, Montalenti, Silvestroni and Siniscalco, *Annals of Eugenics* 16:299–315, 1952.)

and Syria. Some localities may have as many as five percent carriers (Fig. 13.3). In these areas the chance of the severe form (which requires the homozygous condition) occurring by random mating is about one in every 1600. Since the homozygous form is lethal and heterozygotes can be identified early in life, the allele frequency could be greatly reduced or even removed from the population if heterozygotes would not have children.

In a similar situation of complete dominance of allele A but with complete selection against AA and Aa genotypes (A-phenotype) in a particular environment, A could be completely eliminated in one generation. This allele would be lost from the population and would be restored only by new mutation.

Examples such as the two above involving complete dominance and complete selection against aa and A-phenotypes, respectively, are dramatic but infrequent in nature. When they do occur, however, the allele selected against is eventually lost from the population. Normally effectiveness of selection is graded. Under one set of conditions selection may be strong, producing significant changes in allele frequency within only a few generations. Under other conditions selection might be weak, resulting in only slight changes over long periods of time. Apparently, in most common situations in nature, one genotype is only **slightly** less efficient than another.

Mechanisms for selection have been identified in many higher forms, but questions have arisen concerning the effectiveness of selection in viruses. S. Spiegelman and his associates have demonstrated that nucleic acids produce phenotypes that are subject to selection and can evolve to lesser or greater complexity. QB, an RNA bacteriophage, can replicate in a reaction mixture consisting of an RNA template, precursors for RNA synthesis, and the replicase enzyme. The investigators selected for QB RNA molecules with a high rate of replication in this *in vitro* system, and showed that considerable variation can occur in a population of RNA molecules. Progress was made in a series of experiments toward a molecule with a replication rate that was 15-fold greater than the normal rate. This corresponded with the elimination of 83 percent of the original genome, a significant change in base composition, and an increased efficiency of interaction of the genome with replicase. Further studies, with suboptimal concentrations of riboside triphosphates and inhibitory base analogs, showed that selection can occur among mutant viruses in a selective environment.

A large number of certain polymer sequences (composed of few nucleotides) occurred in both strands of a double-stranded RNA virus. Complements of these sequences also appeared in both strands. Polymer sequences common to both strands paired with their complementary parts. Studies of sequences and of the chemical properties of the QB RNA molecule indicated extensive base pairing between complementary regions of the same strand, and a resultant extensive secondary and tertiary structure to the RNA molecule. This added new dimensions to the selective mechanisms and allowed a distinction between the genotype (primary structure) and phenotype (secondary and tertiary structures) of a simple molecular species. Results indicated that when a molecule exists outside of a cell as a highly evolved self-replicating structure, the structural "phenotype" of the molecule may be acted upon by a selective agent.

These experiments along with those of J. H. Campbell, and others showed that selection occurs at the molecular level and that it depends on interactions between genetically determined phenotypes and environmental agents. In this way the selection process resembles that in higher forms. Selection is a mechanism for **changing gene frequency** in virtually all forms of life.

MUTATION

Recurrent or "one-way" mutations that change one allele to another, such as A to a, tend to alter the relative frequency of members of that pair of alleles by increasing the proportion of a at the expense of A. If mutations occur only in one direction, that is, A to a, eventually only a alleles will be expected in the population. Because mutation rates for particular genes are low, a long period of time is required for any appreciable change in the relative frequency of alleles. In fact, mutation rates for most genes in higher organisms are so low that it is doubtful that mutation alone is an important direct factor for allele frequency changes. In addition, however, to the direct effect of mutations in changing the relative proportion of alleles, mutations have a more general and indirect influence on allele frequency changes in populations. They represent an original **source for variation and provide alternative genes with their corresponding traits.** This variation is basic for the operation of segregation, recombination and selection.

Assume a population that is very large so sampling errors may be ignored. Consider a single genetic locus at which two alleles, A and a, are segregating. The population will be composed of three genotypes, AA, Aa, and aa. Suppose that all of these reproduce with equal frequency (i.e., there is no selection). Let the experiment begin by counting the number of individuals of each of the three genotypes in the population. From these counts we may compute

$$A = p = p_o \text{ (}p \text{ in the initial generation)}$$

as well as

$$a = 1 - p_o = q_o \text{ (}q \text{ in the initial generation)}$$

In each generation, a certain fraction, u, of the A alleles will change by mutation to a alleles. If, for example, initially $A = p_o = 0.9$ and one in each million A alleles mutates to a in each generation, what will be the frequency of A after 1000 generations? The mathematics are involved but it can be shown that after 1000 generations p is reduced from 0.9 to 0.899. The influence of a recurring mutation in altering the frequency of a particular allele in a population of organisms is called **mutation pressure.** This example shows that mutation pressure acting alone over a short range **does not** make much difference in allele frequency. Over a very long range one-way mutation pressure may result in the **near elimination** of an allele.

Mutations may be **reversible** as well as recurrent. If the mutation of A to a is reversible (from a to A), the proportion of A will increase and the relative rates of the two kinds of mutations will determine the proportion of alleles at any given time.

When the proportions become established, the population will be stabilized with respect to the two alleles that undergo recurrent and reverse mutations as illustrated in Fig. 13.4. Usually the two kinds of mutations are not equally frequent in occurrence and a mutational trend is established in one direction. If mutations A to a occur at rate u and a to A mutations occur at rate v, the forward and reverse mutations may be presented according to the following scheme: $A \underset{v}{\overset{u}{\rightleftharpoons}} a$. The frequency of A and a will be influenced by the relative rates u and v. If $1 - q$ represents the frequency of A, and q represents that of a, the population allele frequencies will be stable with reference to these alleles when $(1 - q)u = qv$. Solving for q, the equilibrium value of $\hat{q}$ is, $\hat{q} = u/(u + v)$. This value of $\hat{q}$ is approached but never reached as shown in Fig. 13.4.

Recombination of alleles already present in the population is much more significant for changing the gene frequency of higher organisms than original spontaneous mutations. In higher organisms, recombination (Chapters 5 and 6) provides the **immediate source** upon which natural selection can act.

MUTATION AND SELECTION OPERATING TOGETHER

When both mutation pressure and selection pressure can be estimated, corollaries

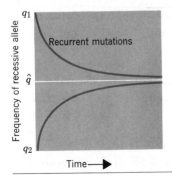

Figure 13.4 Time scale for change in gene frequency of a recessive allele subject only to mutation pressure; $q1$ falls exponentially toward $\hat{q}$ but never reaches equilibrium while $q2$ rises exponentially toward $\hat{q}$ but never reaches equilibrium.

to the Hardy-Weinberg formula may be introduced to describe mathematically the trend in the population. Change in allele frequency because of mutation pressure (m) would be:

$$\Delta q_m = (1 - q)\, u \text{ (gain)} - qv \text{ (loss)}$$

If we assume that a is a completely recessive, rare allele and that v is much smaller than u, that is, q is approximately 0 and $1 - q$ is approximately 1, then $\Delta q_m \cong (1 - q)u \cong u$. For change in allele frequency due to selection, $\Delta q_s = [-q^2 s(1 - q)]/(1 - q^2 s) \cong -sq^2$

thus

$$\Delta q_m \approx u \quad \Delta q_s \approx -sq^2$$

and

$$\Delta q = \Delta q_s + \Delta q_m$$

or

$$\Delta q = -sq^2 + u$$

$$\Delta \hat{q} = 0$$

at equilibrium

$$\therefore u - s\hat{q}^2 = 0$$

$$s\hat{q}^2 = u$$

$$\hat{q} = \sqrt{\frac{u}{s}}$$

If u is 0.00001 and s is 0.001, the numerical values of $\hat{q}$ and $1 - \hat{q}$ at equilibrium will be

$$\hat{q} = \sqrt{\frac{u}{s}} = \sqrt{\frac{0.00001}{0.001}} = \sqrt{0.01} = 0.1$$

$$1 - \hat{q} = 0.9$$

Thus we have the frequency of $a = \hat{q} = 0.1$ and the frequency of $A = 1 - \hat{q} = 0.9$ at equilibrium.

Observations in nature indicate that both mutation and selection operate in natural populations, and both must be recognized in defining population trends. In a natural environment, mutant organisms usually are at a disadvantage when compared with unmutated forms. A few changes might be inconsequential or neutral, but the chance of making an already well-adjusted organism better able to meet the conditions of an environment by random change is remote indeed. How often would the engine of an automobile, or any other well-adjusted machine, be improved by random changes? The random origin of newness that maintains population fitness is purchased at the cost of a genetic burden or **genetic load**. Genetic load is the average number of lethal genes or equivalents per individual in a population. It may be defined in terms of reduced fitness:

$$\frac{W_{max} - \overline{W}}{W_{max}}$$

where W_{max} is the fitness of the best genotype and $\overline{W}$ is the average fitness of the entire population. When H. J. Muller introduced the idea of the genetic load in 1950, he was concerned especially with the mutations caused in human beings by radiation. When such a mutation occurs, it almost always confers a lower fitness, at least in the homozygous condition. Lower fitnesses of the mutations are too often produced by hereditary diseases that cripple and prematurely kill their carriers. But this is only part of the story. The genetic load is also based on differences that do not harm the carriers in overt, physical ways. If one genotype produces an average of three offspring per generation while other genotypes produce only two, this contributes enormously to the genetic load of the population. The same will be true if one genotype is, say, twice as likely to take advantage of newly opened habitats as the other genotypes.

In organisms such as fungi and bacteria, which reproduce primarily by rapid, asexual means, mutations associated with selection may abruptly change the trend in the population. If, for example, a single bacterium *Staphylococcus aureus*, in an environment where penicillin is present, should undergo a spontaneous mutation and become resistant to penicillin, it might give rise to an entire population of **penicillin-resistant** bacteria in a short period of time. In this example, penicillin is not a mutagen but a selective agent. Animals as far up the phylogenetic ladder as scale insects and Australian rabbits have possessed chance mutations that made them more fit for their current environment. These mutants may rapidly give rise to whole populations of individuals favored in that environment. Organisms becoming **resistant** to a particular insecticide, for example, would be favored in an orchard sprayed with that insecticide. They and their descendants could take immediate advantage of the manmade environment from which competing insects had been eliminated. In the citrus groves of California, for example, whole populations of resistant scale insects have developed within a single season. Again, the insecticide is not a mutagen but a **selective** agent.

Mutations occur as frequently in higher, sexually reproducing organisms as in microörganisms, but the numbers in a population are too few to allow many mutations to become established even in long periods of geologic time. A mutation with a frequency of one per 100 million (10^8) germ cells might occur a few times in the population of human beings on the earth. A virus population of 10^8 units, on the other hand, may occur in a single plate culture. The F_2 virus, for example may produce 20,000 to 40,000 viral particles in a single bacterial cell.

Under most circumstances of recurrent mutation and selection, a dynamic equilibrium is reached with regard to genotype frequencies. As indicated previously, recombination (through meiosis and fertilization) of genes already present in a population of higher organisms provides a more effective source of variation than mutation. Recombination also occurs through crossing over (Chapter 5), with the rate depending on the strength of linkage. In sexually reproducing organisms, when random matings occur in a population, the genotype of each individual incorporates genetic contributions from many preexisting members of the species. As time goes on, mutant genes thus become **widely dispersed** throughout the entire population.

RANDOM GENETIC DRIFT

Random fluctuation in allele frequencies, called **random genetic drift**, also occurs in breeding populations. The effect of genetic drift is negligible in large populations, but in a **small effective breeding population**, the limited numbers of progeny might all become the same type with respect to certain gene pairs because of genetic drift. Should this happen, **fixation** or homozygosity will occur at the locus concerned. Fixation is defined in terms of gene frequency reaching $p = 1.00$ or $q = 1.00$. Random genetic drift may or may not lead to fixation, but fixation is much more likely in very small than in large effective breeding populations. The effective breeding populations may be much smaller than the actual population because of nonreproducing individuals, isolation into breeding units (demes) and limited sampling time. Distribution of gene frequencies in populations of different sizes is given in Fig. 13.5. Random genetic drift is a mathematical necessity because population samples are finite.

If a pair of alleles *Aa* should be present in all members of a small breeding population, normally one-fourth of the progeny would be *AA*, one-half *Aa* and one-fourth *aa*. If

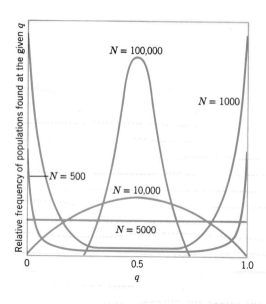

Figure 13.5 Distribution of gene frequencies for populations of different sizes. Populations of relatively small size ($N = 500$, $N = 1000$) show a considerable amount of random drift, many reaching elimination ($q = 0.0$) or fixation ($q = 1.0$). Only populations of large size ($N = 10,000$, $N = 100,000$) maintain the initial gene frequency $q = 0.5$ in appreciable proportions. (After Wright.)

TABLE 13.11
Simulation Experiment for Random Genetic Drift with Random Numbers

Generation	Random Number	Rule for Transforming Random Numbers into Genes	Corresponding Genes	Percentage of A Genes
0	—	—	—	50
1	9452274358	0,1,2,3,4 = A; 5,6,7,8,9 = a	aAaAAaAAaa	50
2	4262686819	0,1,2,3,4 = A; 5,6,7,8,9 = a	AAaAaaaaAa	40
3	1605133763	0,1,2,3 = A; 4,5,6,7,8,9 = a	AaAaAAAaaA	60
4	0824427647	0,1,2,3,4,5 = A; 6,7,8,9 = a	AaAAAAaaAa	60
5	5949704392	0,1,2,3,4,5 = A; 6,7,8,9 = a	AaAaaAAAaA	60
6	9715513428	0,1,2,3,4,5 = A; 6,7,8,9 = a	aaAAAAAAAa	70
7	9840966162	0,1,2,3,4,5,6 = A; 7,8,9 = a	aaAAAaAAAA	70
8	4547684882	0,1,2,3,4,5,6 = A; 7,8,9 = a	AAAaAaAaaA	60
9	8930069700	0,1,2,3,4,5 = A; 6,7,8,9 = a	aaAAAaaaAA	50
10	5005195137	0,1,2,3,4 = A; 5,6,7,8,9 = a	aAAaAaaAAa	50
11	3175385178	0,1,2,3,4 = A; 5,6,7,8,9 = a	AAaaAaaAaa	40
12	7915253829	0,1,2,3 = A; 4,5,6,7,8,9 = a	aaAaAaAaAa	40
13	4456038750	0,1,2,3 = A; 4,5,6,7,8,9 = a	aaaaAAaaaA	30
14	6832883378	0,1,2 = A; 3,4,5,6,7,8,9 = a	aaaAaaaaaa	10
15	4693938689	0 = A; 1,2,3,4,5,6,7,8,9 = a	aaaaaaaaaa	0

chance decrees that all progeny are either *AA* or *aa*, fixation would have occurred. No further genetic fluctuation would be expected at that locus without mutation. In small breeding populations, genetic drift will act on all loci represented by two or more alleles, but the direction and magnitude of the effect may be different at each locus. The chance factor accounts for an appreciable amount of variation in small populations of cross-mating organisms.

L. L. Cavalli-Sforza and associates have designed and conducted sampling experiments to simulate random genetic drift. In simulation studies, the procedure of random sampling is imitated by using random numbers. These are numbers of one or more digits in which each digit has the same probability of occurring, and no correlation exists between successive digits or numbers. For example, series of 10 random numbers simulating 5 (50 percent) *A* and 5 (50 percent) *a* genes were taken from a population of random numbers (Table 13.11). The first step was to rule that random digits between 0 and 4 inclusive are *A* "genes" and those between 5 and 9 inclusive are *a* genes. In the first generation, 5 (50 percent) are *A* and 5 (50 percent) are *a*. The proportion of first generation "progeny" is the probability for *A* genes in the parents for the second generation. From 5*A*: 5*a* "parents" the ruling again is that digits 0,1,2,3,4 are *A* genes and that all other digits (5,6,7,8,9) are *a* genes. In the second generation 40 percent are *A*. This procedure is followed until the fifteenth generation (Table 13.11) when all 10 alleles were *a*, and *A* became extinct.

The results of the simulation experiment outlined above are shown graphically in Fig. 13.6. Allele frequencies may fluctuate about their mean from generation to generation. If an effective population is large, the numerical fluctuations are small and random sampling of parents and thus gametes from the parental gene pool has little or no effect on succeeding generations. If, on the other hand, the effective population is small, random fluctuations may lead to complete fixation of one allele or another.

Cavalli-Sforza and his associates have investigated actual gene samplings with elaborate computerized experiments. In one series of experiments, some 20 alleles of blood groups, particularly ABO, Rh and MN, were studied in different world populations. These computerized studies have indicated that random genetic drift has been an important factor in the variations

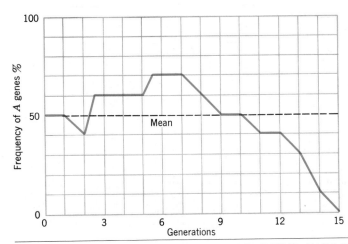

Figure 13.6. Random variation in frequency of a gene around the mean as shown in simulation experiment on genetic drift. (Data in Table 13.11.)

that have developed in different populations. The establishment of small but relatively stable agricultural communities has in the past provided a favorable situation for diversity among communities.

For further examples of probable genetic drift, an Irish colony in Liverpool, which originated from a few founders, was observed to have a different frequency for blood-group alleles than the Irish in Ireland. Again, some small Swiss isolates have a high proportion of A and others a high proportion of O-type blood; yet all these populations came from the same ancestry. Chance fluctuations in allele frequency presumably caused these changes.

MIGRATION

Allele frequencies may be altered among local units of a species by an exchange of genes with other breeding units. This exchange effectively modifies allele frequencies if the breeding populations have been partially or completely separated for enough time to have developed markedly different frequencies for the same alleles. Physical features in the environment such as rivers and islands provide models for a degree of isolation preceding migration among higher animals. The **significance** of migration in changing allele frequencies depends on the **degree of isolation** of the subpopulations involved. Migration is a source of variation similar to mutation in that alleles can be recurrently added and lost from a population but migration may be more effective in changing **allele frequency.**

Together with natural selection, the swiftest way by which gene frequencies can conceivably be altered is by introducing into the population groups of genetically

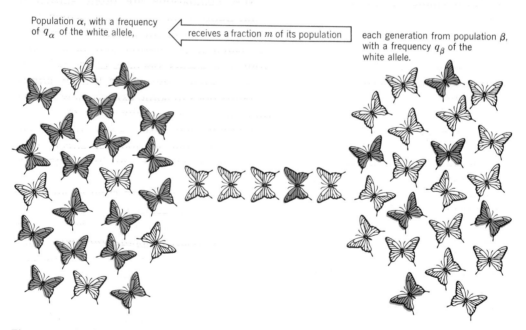

Population α, with a frequency of q_α of the white allele, receives a fraction m of its population each generation from population β, with a frequency q_β of the white allele.

Figure 13.7 Evolution by gene flow. Population α, with a frequency of $q\alpha$ of alleles for white butterflies, receives a fraction (m) of its individuals each generation from population β, which has a frequency $q\beta$ of alleles for white butterflies.

different individuals. Migration tends to **enhance** the effect of natural selection or to **blur** the effect of selection by replacing genes removed by selection. Let population *a* of butterflies containing a frequency q_a of a certain allele for white color, receive some fraction *m* of its individuals in the next generation from a second population *b* with a frequency q_b of the same allele (Fig. 13.7). The frequency (q_a) of the white allele in population *a* is altered to the frequency of the allele in the nonmigrant part of the population of *a*, times the portion of individuals that are not migrants $(1 - m)$, plus the frequency of the same allele among the immigrants q_b, times the proportion of individuals in the population that are new immigrants (m). The altered frequency (q_a') is thus

$$q_a' = (1 - m)q_a + mq_b$$

and the amount of change in one generation is

$$\Delta q = q_a' - q_a = -m(q_a - q_b)$$

The fractional change *m* in the size of the population is greatly influenced by *M* the number of migrants

$$m = \frac{M}{M + N}$$

where *N* is the number of individuals of a particular type already in the isolated population. If the frequency of migrants is low, the average change in *m* is equal to the genetic load. A population isolated for a prolonged period in the same environment has become specialized and has achieved **optimum fitness.** With the onset of migration the population could become extinct because of the disruption of optimum fitness.

Only a small difference in gene frequencies (of the magnitude that often separates populations), together with a moderate migration coefficient (*m*), is needed to effect a significant evolutionary change. The phenomenon is referred to as **gene flow** or migration pressure. Two categories can be distinguished: intraspecific gene flow between geographically separate populations of the same species; and interspecific hybridization. The former occurs constantly within many plant and animal species and is a major determinant of the patterns of geographic variation. Interspecific hybridization occurs during breakdowns of normal species-isolating barriers. Ordinarily it is temporary, or at least rapidly shifting in nature. Although much less common than gene flow within species, it has a greater effect because of the larger number of gene differences that normally separate species.

THE FOUNDER PRINCIPLE

New populations are often started by small numbers of individuals, which carry only a fraction of the genetic variability of the parental population and hence differ from it. If **chance** operates in the selection of the **founder individuals** (and it almost certainly does to some extent), new populations will tend to differ from the parent population and from each other. The founder principle (or founder effect), is of potential importance in the origin of species.

An isolated population such as an island population with restricted variability and genotype that migrates to another area occupied by related populations will most likely become extinct because of the broken population boundary that was established by selection pressure. If it does not become extinct, genetic drift will cause the small remaining population to become very divergent from the original population. Divergence is further enhanced because the different evolutionary pressures in the different areas occupied by the parent and

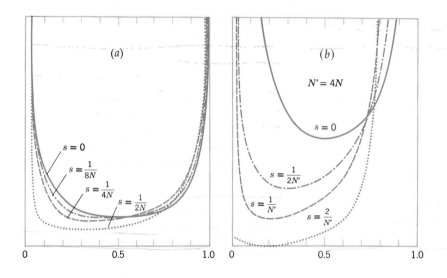

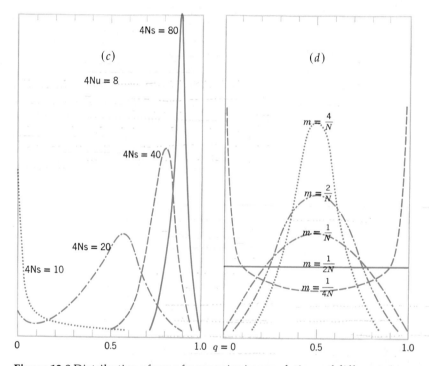

Figure 13.8 Distribution of gene frequencies in populations of different size under different selection, mutation and migration pressures. The abscissae represent the gene frequencies from 0 (loss) to 1 (fixation). The ordinates represent the frequencies of the different genes in a population. (*a*) Small populations (population = N) with U-shaped gene frequency curves. Selection coefficients (*s*) are small and selection pressure has little effect even when s changes from $s = 0$ to $s = 1/2N$. Most genes are either fixed or lost by

daughter populations will be operating on different **gene pools.** This may be the beginning of race and eventually species formation.

INTERACTIONS OF GENETIC DRIFT, SELECTION, MUTATION AND MIGRATION

Wright's curves in Fig. 13.8 illustrate the effects on gene frequencies of selection, mutation and migration pressures on populations of different sizes. The smaller the population size, the greater are random variations in gene frequencies, and the less effective become weak selection pressures. In small populations, alleles favored by selection may be lost and less favored ones may reach fixation. In large populations even very small selective advantages and disadvantages will eventually be effective, but a more rigorous selection is needed to overcome the random drift in small populations. In small populations (Fig. 13.8a) the gene frequency curves are U-shaped. A majority of variable genes are either fixed or lost most of the time. With small selection

coefficients, the shape of the curves is little modified. Genes are lost or fixed at random, with little reference to the selection pressure. Figure 13.8b represents the action of a selection of the same intensity as in Fig. 13.8a, but in a population that is four times larger $(N' = 4N)$. Here a selection of the order $s = 2/N'$ is rather effective; the curve no longer is U-shaped but has a maximum at the right, indicating that the gene alleles favored by selection largely supplant the less favored ones.

If the selection pressure is of the same order as the mutation rate, and both are small, the random variations of the gene frequencies in small populations become important. The curve is U-shaped, indicating that gene alleles reach fixation or loss largely irrespective of the mutation and selection pressures. In Figs. 13.8(a), (b), and (c), it was assumed that the colonies into which a species is subdivided are completely isolated from each other. In reality this rarely if ever happens. Even populations of oceanic islands receive occasional migrants from the mainland or from other islands. Fig. 13.8(d) shows the effects of migration. If the migration coefficient is $m =$

random fluctuations. (b) Selection pressure has the same intensity as in a, but the population is larger $(N' = 4N)$. Selection becomes more important in changing gene frequency as the population becomes larger. The curves are higher at the right indicating that the alleles favored by selection have replaced those less favored. (c) Population of intermediate size with a moderate mutation rate $(4Nu = 8)$ opposed by selection pressures of varying intensity: $4Ns = 10$, $4Ns = 20$, $4Ns = 40$ and $4Ns = 80$. At $4Ns = 10$ the selection pressure is near the mutation pressure and mutation pressure largely determines allele frequency. As selection pressure becomes greater, variance is restricted and the gene frequencies are kept within narrow limits. At $4Ns = 80$ the variation is greatly restricted. This situation is favorable for the differentiation of isolated populations into local races. (d) Effect of migration on gene frequency. When migration pressure is $m = 1/4N$ (1 migrant in 4 generations) the curve is U-shaped and the change is not significant. As migration increases, the effect of migration pressure becomes more important. In these curves the least variance occurs at $m = 4/N$. (From Wright and Dobzhansky.)

1/4N (one migrant individual on the average in four generations), the isolation is effective and the curve of gene frequencies becomes U-shaped. With $m = 1/2N$ (one migrant in alternate generations) the fixation of alleles is slowed down. With more frequent migration the distribution of the gene frequencies shows less and less variance.

MEIOTIC DRIVE

Another factor that may alter allele frequencies in a population is any irregularity in the mechanics of the meiotic divisions, called meiotic drive. Ordinarily, heterozygous Aa individuals produce A and a gametes in equal proportions and these gametes have equal probabilities of fertilization and development. For many years, cases of preferential segregation have been reported but most of these have been sporadic and nonhereditary. Now genetically based examples of systematic deviations from Mendelian ratios are on record for both male and female parents.

The SD (segregation distorter) locus in chromosome II of *Drosophila melanogaster*, for example, has two known alleles, a wild-type and a distorter of the wild-type allele. In the presence of the homozygous wild-type allele, the chromosomes segregate normally. Heterozygous males that carry the mutant allele show a marked departure from the 1:1 ratio under certain environmental conditions. The mutant allele apparently interacts with its wild-type homologue causing it to behave irregularly during spermatogenesis. As a result, only

a few sperm contain the normal allele.

Likewise in Drosophila, females that have a pair of structurally different chromosomes may undergo preferential segregation. One member of the pair may be systematically retained in the egg and the other segregated to the nonfunctional polar body. If the two homologues are of unequal length, the shorter one is usually included in the egg nucleus. Chromosomes that have undergone structural changes are often extruded. Any persistent factor disrupting a particular chromosome would result in preferential rather than random segregation.

Sex ratios are expected to be altered by irregularities in the segregation of X and Y chromosomes. Examples have been reported in certain races of *Drosophila pseudoobscura* where abnormal spermatogenesis resulted in failure of the X and Y chromosomes to pair. The Y degenerates and the X undergoes an extra division. As a result all sperm carry X chromosomes. When these males, which carry only X chromosomes in their sperm, are mated, only female progeny are produced. In other examples, all X chromosomes have systematically degenerated and only males have been observed in the next generation.

Meiotic drive could be a significant factor in evolution. Even the most favorable genes would not be perpetuated if the chromosomes in which they were carried were systematically excluded from gametes. The importance of meiotic drive in nature depends on the extent and persistence of preferential segregation at meiosis which is largely unknown.

GENETIC LOAD IN HUMAN POPULATIONS

Many recessive genes with detrimental effects are known to be segregating freely in the human population. This "genetic load" is the reduction in **average fitness** in a population which results from a population containing genotypes less fit than the **maximum genotype** (W_{max}). Several kinds of genetic load have been distinguished; two important ones are: (1) segregational or "balanced" load: the genetic disability sustained by a population due to genes segregating from advantageous heterozygotes to unfit homozygotes; and (2) mutational load: the amount of reduction of fitness in a population as compared with the fitness that would be expected if no mutations occurred.

The balanced load depends on heterozygous genotypes that are superior to both types of homozygotes. For a gene with two alleles, this load is $p^2 s_A$ for the AA homozygotes plus $q^2 s_a$ for the aa homozygotes as shown from the following calculations.

	AA	Aa	aa
frequency at fertilization	p^2	$2pq$	q^2
relative adaptive value	$1 - s_A$	1	$1 - s_a$
frequency after selection	$p^2 - p^2 s_A$	$2pq$	$q^2 - q^2 s_a$
reduction in frequency	$p^2 s_A$		$q^2 s_a$

At equilibrium

$$\hat{p} = \left(\frac{s_a}{s_A + s_a}\right) \text{ and } \hat{q}\left(\frac{s_A}{s_A + s_a}\right)$$

and be substituting equilibrium frequencies for p and q

into $p^2 s_A + q^2 s_a$ we obtain

$$\left(\frac{s_a}{s_A + s_a}\right)^2 s_A + \left(\frac{s_A}{s_A + s_a}\right)^2 s_a =$$

$$\frac{s_A s_a^2 + s_a s_A^2}{s_A + s_a} = \frac{s_A s_a (s_A + s_a)}{(s_A + s_a)^2} = \frac{s_A s_a}{s_A + s_a}$$

If s_A and s_a are both 0.1, $(0.1 \times 0.1)/(0.1 + 0.1) = 0.01/0.2 = 0.05$. The **segregational load** will be 0.05. This is the average number of lethal equivalents per individual in the population.

Mutational load results from the accumulation of deleterious alleles in a population. Naturally occurring mutations (without a known cause) and those produced by environmental agents such as radiation are usually deleterious to their carriers. Random changes in a highly integrated, adaptive system are expected to be disadvantageous in the environment to which the unchanged organisms are well adjusted.

Some mutations are lethal, that is, the organism in which one occurs cannot reproduce, or survive. The result is a **genetic death** or extinction of a gene lineage through premature death or reduced fertility of the individual carrying the gene. Most mutations are not immediately lethal. They may persist in a population generation after generation reducing the life span of individuals carrying them and, in human populations, increasing the level of physical and mental illness in the population.

POLYMORPHISM

Polymorphism is the existence of two or more genetically different classes in the same interbreeding population. In contrast, examples of interactions in populations considered thus far have involved systems of two competing alleles in the same locus. In a constant environment one allele eliminates the other, or else the frequencies of the two are carried to some intermediate equilibrium point between zero and one by the countervailing forces of selection, gene flow, and mutation pressure (with meiotic drive a possible fourth factor). However, even in the absence of a balance between the primary evolutionary agents, two or more alleles can be maintained together in the same panmictic population for indefinite periods of time. This condition is called balanced polymorphism. It is the preservation of genetic variability through selection. Such a balance can be achieved by several means. It can occur, for example, through frequency-dependent selection when the fitnesses of the two alleles are not constant, but change with their frequency. If one allele has a lower fitness than the other while it is at higher frequencies, but gains the advantage when its frequency descends to a certain level, the frequency will tend to stabilize at about that level.

A second condition in natural populations, is heterozygote superiority or heterosis. If a heterozygote Aa is superior to both homozygotes AA and aa, neither allele can eliminate the other. The frequency of a (q) and the frequency of A (p) are expected to stabilize at some intermediate frequency between 0 and 1. As long as heterozygosis is maintained, $\hat{p} = \hat{q}$ will tend to remain constant. The equilibrium value $\hat{q}$, (at which selection against the two homozygotes is in balance) and $\Delta q = 0$, (Fig. 13.9), is obtained from the following relations.

$$\hat{q} = \frac{s_A}{s_A + s_a}$$

A real case of balanced polymorphism has occurred in the sickle-cell trait (Chapter 8) that is common in the human population in Africa and parts of the Middle East. Sickle-cell anemia apparently became established in human populations while the way of life in parts of Africa was changing from hunting and gathering to a more stable agricultural pattern. The environment in agricultural areas was favorable for mosquitoes and some mosquitoes carried malaria. Both sickle-cell anemia and one type of malaria *Plasmodium falciparum* are diseases of human red blood corpuscles. Patients with the sickle-cell disease have collapsed, sickle-shaped corpuscles that clog the capillaries, thus interfering with circulation and depriving the cells in the body of oxygen. People with the gene for sickle-cell anemia carry an immunity to falciparum malaria. The allele Si^s controls the sickle-cell trait which causes the red blood cells to assume

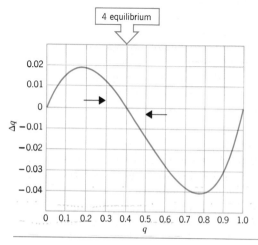

Figure 13.9 Selection favoring heterozygotes. Change in the frequency of a (Δq) when the adaptive values are $AA = 0.80$, $Aa = 1.00$, $aa = 0.70$ and population size is infinite. These values provide a stable balanced polymorphism ($\hat{q} = 0$) at $q = 0.4$. That is, Δq is positive (q increases) if q is less than 0.4 and negative (q decreases) if q is more than 0.4. If one allele is accidentally eliminated, (i.e., $q = 0$ and $p = 1$, Δq is zero, but polymorphism is lost. (After Crow, Li and Strickberger.)

a sickle-like shape when they are exposed to low oxygen tension outside the body. Si^A designates normal red corpuscles. People in the heterozygous condition (Si^s/Si^A), show the trait in less than one percent of their red corpuscles. The hemoglobin is slightly abnormal and the disease symptoms are slight but usually sufficient to make possible the recognition of heterozygous individuals. A simple blood test can verify carriers. About 40 percent of the people in some tribes in parts of Africa carry the allele. Homozygotes (Si^s/Si^s) display the trait in a large percentage of their red blood cells and they suffer from sickle-cell anemia that usually proves fatal in childhood.

Malaria is common in the same areas where sickle-cell anemia is prevalent, but the greatest death rate from malaria occurs among natives with normal hemoglobin in their red corpuscles. Those homozygous for the sickle-cell trait (Si^s/Si^s) usually die early from anemia, while those homozygous for normal blood hemoglobin (Si^A/Si^A) suffer from malaria. The heterozygotes (Si^s/Si^A) do not contract malaria, but perpetuate the sickle-cell gene in their progeny. Si^s/Si^A $W = 1$; $Si^s/Si^s = 1 - S_a \therefore W = 0$; $Si^A/Si^A = 1 - S_A$; where $S_A = 1 \therefore W = 0$.

Selective forces in a malaria-plagued environment and among people whose agricultural mode of life favors exposure to the malaria-bearing mosquito have apparently resulted in two different genotypes being maintained in the population. By effective control of mosquitoes and other public health measures, malaria is being irradicated in many parts of the world. With this change in the environment, the polymorphism is expected to disappear and the selection factor favoring sickle-cell anemia will be removed. The population in previously malarial areas is expected to become homozygous for Si^A except for rare mutations from Si^A to Si^s. Therefore the relative fitness of the heterozygote will no longer = 1 and p and q will no longer be stable. The balanced polymorphism will no longer exist.

GEOGRAPHIC POLYMORPHISM

Another major type of polymorphism is geographic polymorphism in which the classes are located in different regions. The ABO blood alleles are present in different frequencies in populations originating in different geographical areas of the world. When the hereditary antigens that help define the different blood profiles are studied individually in various populations, it is apparent that their distribution varies among different races. Thus, traits that have a known mode of inheritance allow us to determine how the frequency of the given allele in a given population compares with the frequency of that allele in another population. This objective approach to racial classification offers a number of advantages over morphological characteristics that are much more difficult to analyze genetically.

More data are available on the world distribution of alleles for the ABO system than for any other blood system and for any other set of human traits. Comprehensive reports of data on the ABO blood groups have been published independently by W. C. Boyd, A. E. Mourant, and other investigators. Native populations of aborigines have been studied extensively.

Among American Indians, the Utes from Utah, the Navajos from New Mexico (Table 13.12), and most other tribes in North, Central, and South America have a high incidence of O, comparatively little A, and no B or AB. On the other hand, Blackfeet Indians (Fig. 13.10) of Montana have less O and a high proportion of A (76.5 percent in the sample listed). The Polynesians sampled have 60.8 percent A and very little B and AB. Australian aborigines have a high A (57.4 percent in the sample cited) and no

TABLE 13.12
Frequencies of Blood Groups O, A, B, and AB in Samples of World Populations[a]

Population	Place	Number Tested	Percent of Phenotypes				Gene Frequency		
			O	A	B	AB	r	p	q
American Indians (Utes)[b]	Utah	138	97.4	2.6	0.0	0.0	0.987	0.013	0.000
American Indians (Blackfeet)[b]	Montana	115	23.5	76.5	0.0	0.0	0.485	0.515	0.000
American Indians (Navahos)	Northern Mexico	359	77.7	22.5	0.0	0.0	0.875	0.125	0.000
Caucasians[b]	Montana	291	42.3	44.7	10.3	2.7	0.650	0.257	0.053
Polynesians	Hawaii	413	36.5	60.8	2.2	0.5	0.604	0.382	0.018
Australian Aborigines	South Australia	54	42.6	57.4	0.0	0.0	0.654	0.346	0.000
Basques	San Sebastian	91	57.2	41.7	1.1	0.0	0.756	0.239	0.008
Eskimos	Cape Farewell	484	41.1	53.8	3.5	1.4	0.642	0.333	0.027
Buriats	Siberia	1320	32.4	20.2	39.2	8.2	0.570	0.156	0.277
Chinese	Peking	1000	30.7	25.1	34.2	10.0	0.554	0.193	0.250
Pygmies	Belgian Congo	132	30.6	30.3	29.1	10.0	0.554	0.227	0.219
Asiatic Indians	Southwest India	400	29.2	26.8	34.0	10.0	0.540	0.208	0.254
Asiatic Indians	Bengal	160	32.5	20.0	39.4	8.1	0.571	0.154	0.278
Siamese	Bangkok	213	37.1	17.8	35.2	9.9	0.595	0.148	0.257
Japanese	Tokyo	29,799	30.1	38.4	21.9	9.7	0.549	0.279	0.172
English	London	422	47.9	42.4	8.3	1.4	0.692	0.250	0.050
Germans	Berlin	39,174	36.5	42.5	14.5	6.5	0.604	0.285	0.110

[a] Boyd, W. C., *Genetics and the Races of Man.* Boston, D. C. Heath and Company, pp. 223–225.
[b] Matson, G. A., "Hereditary Blood factors Among American Indians." *Fifth Intn'l. Congress of Blood Trans. Reports and Comm.,* 274–283.

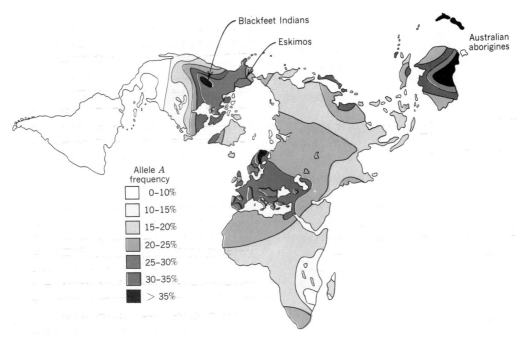

Figure 13.10 World map showing percentage frequencies of blood-group allele *A*. High frequencies occur among the Blackfeet Indians and Australian aborigines as also shown in Table 13.12 (Data from Mourant, Kopeć, Domaniewska-Sobczak, Boyd, and Matson.)

B or AB. Basques, who are believed to be similar to the ancestors of present-day European populations, have a high incidence of A, moderate B, and virtually no AB. Siamese and Japanese, on the other hand, have a high proportion of A and B. Western European peoples and Caucasians of the United States have a fairly high proportion of A and some B and AB. In none of the extensive samples, however, was the frequency of A as high as among the Blackfeet and related tribes of Indians.

Frequencies for the L^M, L^N alleles show less geographical variation than do those for the ABO group. American Indians and Australian aborigines, however, differ widely in this respect. Groups *M* and *MN* are higher than *N* among American Indians, but *M* and *MN* are low among the Australian aborigines on the other side of the world.

Peoples of western Europe have a more even distribution of the *MN* phenotypes. The highest frequency for L^N and the lowest for L^M are found among the Papuans in New Guinea, whereas the highest L^M and lowest L^N occur among the American Indians.

The value of the *M* and *N* antigens has been greatly increased with the discovery of a pair of antigens, *S* and *s*, which are closely associated with the *M* and *N* groups. It is now possible to postulate four gene combinations: $L^M S$, $L^M s$, $L^N S$, and $L^N s$ and to distinguish sharply between the peoples of New Guinea and the Australian aborigines on these factors alone. The antigen *S* is present in New Guinea and absent in native peoples of Australia.

The Rh system is the most useful of the blood-group systems employed in human population studies. Most population surveys have been carried out with four antisera called anti-C, anti-D, anti-E, and anti-c, plus certain special sera for detecting the variants of the antigen. If the rare anti-d and anti-e antisera are included, 27 different phenotypes can be distinguished. With only the four common sera, however, 12 phenotypes can be distinguished. Particular allele frequencies have been associated with geographical areas and populations. The blood group known as V or ce^s, for example, has a frequency among American blacks of about 27 percent and in West African blacks of about 40 percent. Very little V occurs among American Indians. Only two of 444 New York Caucasians were V-positive and these were Puerto Ricans.

The P_1 blood group is more prevalent among blacks than among Caucasians. It is infrequent among Chinese. The Diego factor (Di^a) is prevalent among Eskimos, Chinese, Japanese, and Koreans. The Duffy bloodgroups system, determined by genes Fy^a and Fy^b, and by the rare allele Fy^x, is found in different frequency in Caucasian and black populations. In England the gene Fy^a is present in about 0.4 percent of the people, but in Lapland and Asia it is more frequent. One or both Duffy factors can be found in the blood of all Caucasians. Only Negroes are likely to lack both Fy^a and Fy^b, and this is true of 68 percent of the New York blacks sampled and in nearly 90 percent of the West African black samples.

When the serum of a Caucasian male patient previously transfused with "black" blood was studied, an antibody was discovered which reacted to a blood group apparently present only in blacks. This blood group is called Kell (JS^a). The antibody reagent failed to react with any specimen of blood from 500 Caucasian donors, but it did react strongly with 20 percent of random blood specimens from American blacks. A total of about 77 percent of blood samples from American Negroes can be unmistaka-

bly identified as such by V and Duffy blood grouping alone. In West Africa, over 90 percent of all random samples would so label themselves. Twenty of the remaining 23 percent of blood samples from American blacks could be identified as being "black" blood by using anti-JS^a regents. Other genetic factors useful for identifying races are the secretor alleles, the haptoglobin, gammaglobulin, and hemoglobin types.

CHROMOSOMAL POLYMORPHISM

Different types of chromosomes as well as different types of alleles may become established in populations. Chromosome polymorphism is the inclusion in the population of two or more alternative structural forms of chromosomes. The structurally changed chromosomes are the result of rearrangements (Chapter 9). On the basis of limited observations, chromosome inversions would seem to be more common in animals and translocations more common in plants. The rearrangements may be fixed as homozygotes and heterozygotes in a certain percentage of the population. This is called **balanced chromosomal polymorphism.** Chromosomal structural types then fluctuate around the mean in roughly constant proportions from generation to generation and possess adaptive value. They behave as homologous chromosomes in their ability to mutually replace the alternative structural form in the karyotype. Chromosomal polymorphism may be eliminated if one of the structural variants is superior under prevailing environmental conditions. On the other hand, its proliferation may be strongly enhanced by natural selection. A classical investigation of chromosomal polymorphism has been conducted over the years by Dobzhansky and his associates.

Several structural variations of the third chromosome, which were associated with different gene arrangements in natural populations of *D. pseudoobscura*, formed the basis of Dobzhansky's study. Appropriate comparisons showed chromosome inversions to be mainly responsible for the different chromosome types. The standard (ST) chromosome type, for example, differed from a type symbolized CH on the basis of certain areas of the third chromosome of ST that were inverted within CH. Chromosomes carrying inversions behaved essentially like other chromosomes in the population. Some individual flies were found to be homozygous and others heterozygous for a given inversion.

Natural populations were usually not identifiable by the universal presence or absence of a particular chromosome type. Quantitative, rather than qualitative, differences were found to be characteristic of different geographical populations of *D. pseudoobscura*. The relative frequency of different alleles or of particular chromosome arrangements formed the main criterion for distinguishing one population from another. Homologous chromosomes, distributed through random mating, tend to establish an equilibrium in the same way as do the alleles described by the Hardy-Weinberg theorem. Wide differences were observed for the relative frequencies of different chromosome types in geographic races of *D. pseudoobscura* in California, Arizona, and New Mexico. Seasonal variations were also detected when the frequencies of chromosome types were compared at different times of the year. Sampling at Piñon Flats on Mount San Jacinto in Southern California in March of one season, for example, showed that 53 percent of the third chromosomes were ST and 23 percent were CH. When certain laboratory conditions (environments) were established to compare the reactions of different populations, it became evident that at particular seasons of the year flies carrying certain chromosome types were favored. Flies with those chromosome types were apparently better fitted to particular temperature and nutritional conditions. The effectiveness of such selection was indicated by the variations in the relative frequencies of chromosome types between populations and between seasons within a given population. Field observations were tested in the laboratory by subjecting mixed populations of flies to different temperatures. Flies from a warm locality were favored in the laboratory at a warm temperature, while those from a colder region were adapted to a lower temperature. A laboratory model was thus created that illustrated one mechanism by which natural selection influences chromosome and gene frequencies in populations.

In some laboratory experiments, the weaker competitors were not eliminated completely, but instead an equilibrium was established between the frequencies of the more favorable and less favorable chromosome types. Dobzhansky explained the survival of the less favorable on the basis of the superiority of the heterozygous combination (i.e., heterosis, Chapter 14). The heterozygous arrangement, which in itself was considered advantageous, contained both competing chromosome types and thus perpetuated the weaker competitor, which would otherwise have been eliminated.

REFERENCES

Bajema, C. V. (ed) 1971. *Natural selection in human populations.* John Wiley, New York.

Boyd, W. C. 1950. *Genetics and the races of man*. D. C. Heath, Boston.

Cavalli-Sforza, L. L. and W. F. Bodmer. 1971. *The genetics of human populations*. W. H. Freeman, San Francisco.

Crow, J. F., and M. Kimura. 1970. *An introduction to population genetics theory*. Harper and Row, New York.

Darwin, C. 1951. *The origin of species by means of natural selection, or the preservation of favoured races in the struggle for life*. Philosophical Library, New York. (Reprint of first edition, published November 24, 1859.)

Dobzhansky, T. 1951. *Genetics and the origin of species*, 3rd ed. Columbia University Press, New York.

Haldane, J. B. S. 1961. "Natural selection in man" in *Progress in medical genetics*, Vol. 1, Chapter 2. A. G. Steinberg (ed.) Grune and Stratton, New York.

Hardy, G. 1908. "Mendelian proportions in a mixed population." *Science* 28, 49–50.

Kettlewell, H. B. D. 1959. "Darwin's missing evidence." *Sci. Amer.* 200(3), 48–53.

Lewontin, R. C. 1967. "Population genetics." *Ann. Rev. Genet.* H. L. Roman, L. M. Sandler, and G. Stent (eds.), 1, 37–70.

Li, C. C. 1955. *Population genetics*. University Chicago Press, Chicago.

Matson, G. A., H. E. Sutton, J. Swanson, A. R. Robinson, and A. Santiana. 1966. "Distribution of heredity blood groups among Indians in South America." *Amer. J. Phys. Anthro.* 24, 51–70.

Stern, C. 1962. "Wilhelm Weinberg." *Genetics* 47, 1–5.

Wallace, B. 1970. *Genetic load, its biological and conceptual aspects*. Prentice-Hall, Englewood Cliffs, N.J.

Wright, S. 1931. "Evolution in Mendelian populations." *Genetics* 16, 97–159.

Wright, S. 1956. "Modes of selection." *Amer. Nat.* 90, 5–24.

PROBLEMS AND QUESTIONS

13.1 (a) Do dominant genes spread more readily in a population than recessives? (b) What factors influence the relative frequency of genes in a population? (c) When the relative frequency of alleles in a population is altered, how can equilibrium be maintained?

13.2 The frequency of children homozygous for a recessive lethal gene is about 1 in 25,000. What is the proportion of carriers (heterozygotes)?

13.3 The following MN blood types were obtained from the entire populations of an isolated American Indian village and an isolated village of Central American Indians.

group	population size	M	N	MN
Central American Indian	86	53	4	29
American Indian	278	78	139	61

Calculate the allele frequency of L^M and L^N for (a) the Central American Indian and (b) the American Indian populations. List the results in tabular form. (c) Discuss possible reasons for the differences in allele frequency.

13.4 (a) Among 205 American blacks the frequencies of the L^M and L^N alleles were 0.78 and 0.22 respectively. Calculate the percentage of individuals with M, MN, and N type blood. (b) A group of 212 college students was invited to taste PTC. There were 149 tasters and 63 nontasters. Calculate the gene frequency of T and t.

13.5 Among 798 students, 70.2 percent were tasters. (a) What proportion of the students were TT, Tt, and tt? (b) What proportion of the tasters who marry nontasters might expect only taster children in their families? (c) What proportion might expect some nontaster children?

13.6 Among 11,335 people, the following blood types were obtained: 5150 O, 4791 A, 1032 B, and 362 AB. Calculate the gene frequencies of the three alleles A, A^B, and a.

13.7 Among 237 Indians, the gene frequencies of a, A, and A^B blood alleles were 0.96, 0.03, and 0.01, respectively. Calculate the percentage of individuals with O, A, B, and AB type blood.

13.8 Blood samples from 999 (883 male and 116 female) students were typed as follows:

	O	A	B	AB
Male	419	371	68	25
Female	68	38	7	3

(a) Calculate the gene frequency for males and females separately. (b) Determine the proportion of heterozygotes (Aa, A^Ba, and AA^B) and the proportion of homozygotes (aa, AA, A^BA^B) for males and females in the sample. (c) Give the genotypic frequencies (r^2, p^2, $2pr$, q^2, $2qr$ and $2pq$) for the entire sample (males and females combined).

13.9 When the blood samples from 999 students were tested with anti-Rh serum, 74.9 percent were positive and 25.1 percent were negative. Assuming a single pair of alleles R and r (for this problem only) what proportion of the students would be expected to be RR, Rr, and rr?

13.10 In a species of animals with heterogametic (XY) males, alleles A and a may occur in the X chromosome and no allele is in the Y chromosome. Females have genotypes AA, Aa and aa. Give the equilibrium ratios and the time required to reach equilibrium if the initial frequency for A is 0.8 in males and 0.5 in females (assume number of ♂♂ = number of ♀♀).

13.11 In a large population of random-mating animals, 0.84 of the individuals express the phenotype of the dominant allele $(A-)$ and 0.16 express the phenotype of the recessive (aa). Calculate the amount of change in gene frequency in the first generation under 0.05 selection pressure against the aa phenotype.

13.12 In a large population of random-mating animals, 0.84 of the individuals express the phenotype of the dominant allele $(A-)$ and 0.16 express the phenotype of the recessive (aa). Under complete selection $(s = 1)$ against the $A-$ phenotype, what proportion of AA, Aa, and aa would be expected in the next generation?

13.13 In a large random-mating population in which A is completely dominant, the gene frequency of A $(1 - q) = 0.7$ and a $(q) = 0.3$. Calculate the rate of change of q per generation if selection is against aa with $s = 0.005$.

13.14 In a large random-mating population, if the mutation rate of A to a is 0.00001 and the reverse mutation rate a to A is 0.000001 and neither A nor a has a selective advantage, at what gene frequency level would A and a be at equilibrium?

13.15 A mutation has been found in a strain of bacteria that is a change in a single gene. It can only be detected by such sensitive techniques as antigen-antibody reactions in rabbit's blood. The rate of mutation in large populations has been found to be $u = 5.2 \times 10^{-3}$. If the bacteria had no mutations initially and then due to the introduction of a mutagen mutated at a rate of 5.2×10^{-3}, after two generations what would be the frequency of the mutant (q)? If the mutation was reversible and the reversible rate $(v) = 8.8 \times 10^{-4}$, what would be the equilibrium value of the mutant (q)?

13.16 In a large random-mating population, if an unfavorable recessive recurrent mutation A to a should occur at a net rate (u) of 0.000001 and selection against the phenotype aa is $s = 0.01$, what would be the frequency of a at equilibrium?

13.17 In a large random-mating population, an unfavorable recessive, recurrent mutation A to a occurs at a rate of $u = 0.000001$ and $s = 0.01$. What will be the frequency of A at equilibrium?

13.18 Why is random genetic drift effective in altering gene frequencies only in small breeding populations?

13.19 Why are the mechanisms of meiotic drive different in males and females?

13.20 Why has the polymorphism between Si^A and Si^s (for sickle-cell anemia) been maintained in African populations?

13.21 In a small African village the frequency (q) of the sickle-cell gene (Si^s) is 0.10. What does this fact suggest about the incidence of falciparum malaria in this area?

13.22 If the frequency of the Si^s gene is 0.10 and there is no more malaria ($S_A = O$) what will the frequency of Si^s be in two generations?

13.23 How is genetic isolation of a population related to race and species formation?

13.24 A blood stain associated with an accident was tested for blood antigens and the results were recorded as follows: A_2, MN, Hu, He, p^1, $V^{su}m$ JS^a, and Duffy negative. Does this information give a clue that could be useful in an investigation?

13.25 A letter sealed with blood was known to have originated either in a remote area of New Guinea or a remote area of Australia. Antigens A, N, and S were detected in a blood sample. Does this give a clue that might help in tracing its origin?

13.26 The label was accidentally removed from a blood sample in a blood bank which had originated from a particular hospital ward. At the time the blood samples were taken the ward was occupied by three men of different ancestry, (1) Caucasian, (2) Japanese and (3) Australian aborigine. The blood was A_1B, MS, Di^a, and Fy^a. (a) Would this information help in identifying the patient? (b) Which patient was most likely the donor of the blood?

13.27 On the oceanic island of St. Helena in the South Atlantic, where winds are strong and nearly constant, a number of species of wingless insects have become established. (a) How might the wingless insects have developed? Teissier carried out the following experiment with Drosophila in a windy part of the Mediterranean region: equal numbers of flies with normal wings and those with vestigial wings were placed in open but protected dishes well supplied with food on the roof of a building. At the end of several generations most of the flies in the dishes had vestigial wings. (b) How could this change in proportion of long-wing and vestigial-winged flies be explained?

13.28 Discuss why two linked genes do not attain equilibrium as rapidly as two unlinked genes.

13.29 Explain how the onset of migration into a previously isolated population can cause extinction of that population.

FOURTEEN

The Hardy-Weinberg equilibrium and population changes, discussed in previous chapters, are based on random mating. Two systems of mating are encountered in nature and in man-controlled populations: **inbreeding** and **outbreeding.** Inbreeding refers to the production of offspring by closely related parents, while outbreeding means the production of offspring by unrelated parents. The distinction between the main categories is not hard and fast; gradations occur. It is often helpful to think in terms of degree of inbreeding. Self-fertilizing plants typify extreme inbreeding whereas cross-fertilizing plants and animals can evidence varying degrees of inbreeding and outbreeding depending on such factors as compatibility, motility, and proximity.

A model (Table 14.1) involving a very small population of mice (two families), can illustrate the effects of a particular mating system on allele frequency. In family I, one parent is white (cc) and the other parent is black (CC). Their progeny, four females and four males are all black and hetero-

SYSTEMS OF MATING

Hybrid Corn. (Grant
Heilman.)

zygous (Cc). In family II, both parents and all eight (four females and four males) progeny are black (CC). In the total population (parents and progeny), there are 10c and 30C alleles. If only the F_1 of the two families were considered as the population, there are 8c and 24C.

When the entire population is considered, the same proportion of c and C alleles is maintained whether inbreeding or outbreeding is practiced. The system of mating in itself does not influence the proportion of alleles. A population with a frequency of 0.9 for allele A and 0.1 for a will maintain these frequencies, whether close relatives or unrelated individuals are mated, unless some other factor changes the proportions.

One difference between inbreeding and outbreeding is that more recessive alleles express themselves when the mating is between related individuals. **Although the total allele frequency remained the same in the model, the proportion of phenotypes was different.** Two of the sixteen progeny resulting from inbreeding expressed the trait (white) controlled by the recessive allele, but none of those resulting from

outbreeding were white. Also, more homozygous blacks (CC) resulted from inbreeding. If the environment should be less favorable to whites than to blacks and if the population were large, the allele frequency might be changed through selection over a period of time. If the population should be very small, random genetic drift might influence the trend in allele frequency, but the type of mating alone has no effect.

Continued mating among close relatives tends to **eliminate heterozygotes and** to produce **homozygotes** or distinct (pure) types within the population. Although the total proportion of different alleles in the entire population remains the same, a number of subgroups of more or less pure lines may develop. Again, if a certain trait or type is better adapted than another, or operates to the advantage of man, inbreeding is a way to get the necessary alleles in homozygous condition. If, on the other hand, a certain uncommon recessive allele is disadvantageous, outbreeding can hide it (by heterozygosity) from the forces of selection, while inbreeding could allow it to be eliminated.

TABLE 14.1

Expected Distribution of Genes in Families of Normal Mice of the Same Strain from Inbreeding and Outbreeding When the Initial Proportion of Genes Is the Same in the Parents and All Progenies Are Equal in Size (4)

	Family	Cross	Progeny
Inbreeding	I	Cc × Cc	1CC 2Cc 1cc
	I	Cc × Cc	1CC 2Cc 1cc
	II	CC × CC	4CC
	II	CC × CC	4CC
	Total genes		20C 4C, 4c 4c = 24C, 8c
Outbreeding	I × II	Cc × CC	2CC 2Cc
	I × II	Cc × CC	2CC 2Cc
	II × I	CC × Cc	2CC 2Cc
	II × I	CC × Cc	2CC 2Cc
	Total genes		16C 8C, 8c = 24C, 8c

INBREEDING

Extreme inbreeding occurs naturally in self-fertilizing plants such as peas and beans. This was a real advantage to Mendel in his studies. It meant pure lines of peas were available to hybridize, and it also facilitated his experimental procedure (Chapter 2). In 1903 Johannsen recognized the uniformity that characterized self-fertilizing plants grown in the same environment. On this basis he postulated pure lines: populations that breed true without appreciable genetic variation. Johannsen considered the pure lines resulting from inbreeding to depend on the similarity of alleles in the various individuals making up the group. Completely pure lines, in which all allelic pairs are homozygous, would represent the ultimate result of inbreeding. It is doubtful that complete **homozygosity** ever actually occurs because of new mutations and the tendency of nature to maintain a few heterozygotes in the system, but for all practical purposes lines subjected to inbreeding over long periods of time are pure.

DEVELOPMENT OF PURE LINES IN PLANTS

How do pure lines develop in nature? If tall heterozygous (*Dd*) garden peas and their descendants were allowed to self-fertilize over a period of time and neither tall nor dwarf was favored by selection, what would be expected in five or ten generations? In the first generation, the proportion would be 1 *DD* : 2 *Dd* : 1 *dd*. In terms of fractions of the total population, the proportion would be $\frac{1}{4}$ *DD* : $\frac{1}{2}$ *Dd* : $\frac{1}{4}$ *dd*. The heterozygotes represent only 50 percent of the population instead of 100 percent as in the beginning. If, for simplicity, the population at this point is considered to consist of 4 and each plant should produce 4 progeny in the next generation, the *DD* would produce 4 *DD* and the *dd* would produce 4 *dd*, but the 2 *Dd* plants would each produce 4 distributed as 1 *DD* : 2*Dd* : 1 *dd*. The total would be 6 *DD* : 4 *Dd* : 6 *dd*. Now 25 percent are heterozygotes and 37.5 percent represent each of the two homozygotes (Fig. 14.1). This trend would go on generation after generation, as illustrated in Table 14.2. In the fifth generation the proportion in whole numbers would be about 48 percent *DD*, 4 percent *Dd*, and 48 percent *dd*. In the tenth generation, only about one in a thousand would be heterozygous. Most of the plants would eventually be *DD* or *dd* with a smaller and smaller proportion of heterozygotes. The population of peas would eventually consist of two types, tall and dwarf, in essentially equal proportion. In contrast, if the plants

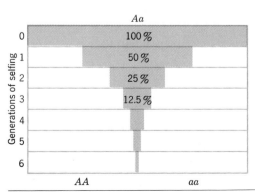

Figure 14.1 Consequences of self-fertilization in the various generations of selfing. In each generation of selfing, heterozygous individuals are reduced by one-half.

TABLE 14.2
Expected Progeny in Generations Following Self-Fertilization of an Annual Plant with Genotype *Dd,* **for Progenies of Equal Size (4) (After Gregor Mendel)**

Generation	Progeny Representing Different Genotypes Based on Families of 4			Percent of Each Genotype		
	DD	*Dd*	*dd*	*DD*	*Dd*	*dd*
I	1	2	1	25	50	25
II	6	4	6	37.50	25	37.50
III	28	8	28	43.75	12.50	43.75
IV	120	16	120	46.875	6.25	46.875
V	496	32	496	48.4375	3.125	48.4375

had been cross-fertilized and mated at random, the proportion of 1 *DD* : 2 *Dd* : 1 *dd* would have continued generation after generation.

Mendel worked this out with a model including 1600 plants. He asked what would be the proportion of the three genotypes in five or ten generations if the original population consisted of 1600 heterozygous (*Dd*) plants and each produced only one offspring. In the first generation, the proportion would be 1 *DD* : 2 *Dd* : 1 *dd*, or 400 *DD* : 800 *Dd* : 400 *dd*. In the second generation, there would be 600 *DD* : 400 *Dd* : 600 *dd*, and so on. The **heterozygotes** would be **decreased by half in each generation.**

Now suppose that the hybrid peas had been heterozygous for flower color alleles (*Rr*) as well as plant height in the beginning. About half of the plants over a period of time would be tall and half dwarf. They would be expected to occur in the following (about equal) proportions: *DDRR, DDrr, ddrr,* and *ddrr.* Four "pure" phenotypic lines: tall, red; tall, white; dwarf, red; and dwarf, white would have developed. When other genes were included they would behave in the same way, and a greater variety

of pure lines would be expected. The entire population would eventually be made up of **distinct types** or races. This has actually occurred in many self-fertilizing plants in nature.

Self-fertilizing plants that undergo inbreeding and selection for long periods of time, can virtually eliminate undesirable recessive alleles from their breeding populations. Further improvements, possible by selection alone, are limited because of the small amount of genetic variation. Pure lines may therefore be developed and tested under different environmental conditions. Hybridization may be used to bring together desirable traits, and new combinations may be established as pure lines. It is thus possible to select desirable varieties and develop lines that have practical advantages and from which further breeding experiments can be conducted.

DEVELOPMENT OF HOMOZYGOSITY UNDER DIFFERENT DEGREES OF INBREEDING

Self-fertilization occurs only among a limited group of plants and is nonexistent among the higher animals. In experimental animals such as mice, therefore, homozygosis can be approached most rapidly and conveniently by brother and sister matings. Backcrosses between progeny and one parent are efficient, especially when sex-linked genes are involved, but technical factors such as the age difference between parents and progeny, make this system less practical than matings between litter mates. Another simple method of inbreeding in mice involves double first cousins, that is progeny of parents that are brothers or sisters of each other. Sewall Wright has contributed greatly to the theoretical aspects of homozygosis with his studies on the effect of inbreeding in guinea pigs. His series of papers on "Systems of Mating" (see the references at the end of this chapter) represents a landmark in the field of population genetics.

Wright's **coefficient of inbreeding** (F) can measure the proportion by which inbreeding has **reduced heterozygosity** in a given situation. It is the probability that both of two allelic genes united in a zygote are descended from the same gene found in an ancestor common to both parents. When $F = O$, no inbreeding is occurring and the values for allele frequencies are those of the Hardy-Weinberg formula. When $F = 1$, the population is completely homozygous. By inbreeding, the proportion of heterozygotes is reduced from $2pq$ to $2pq(1 - F)$. In a population with an inbreeding coefficient of F, the proportion of heterozygous loci is reduced by the factor F compared with what it would be if mating had been random.

If frequencies of two alleles a_1 and a_2 are in the proportions p and q, the probability that two uniting gametes will contain identical alleles from the same ancestor is F. Thus an individual has a probability (F) of being homozygous for such an allele as a_1 or a_2. He also has a probability $(1 - F)$ that the two alleles did not come from the same gene in a common ancestor. If assortative matings were complete (i.e. no random mating) an equilibrium would be established between the two homozygous states. Since p and q represent the relative frequencies of the two alleles, the inbred individual has a probability (pF) of being a_1/a_1 and (qF) of being a_2/a_2. If assortative mating were not complete, some heterozygotes would be expected. The chance of being homozygous for any alleles in a population with some random mating is $p^2(1 - F)$ for a_1/a_1 and $q^2 (1 - F)$ for a_2/a_2. Table 14.3 shows the frequencies of three genotypes that would occur in a population under random mating as compared with those of a population with a coefficient of inbreeding equal to F.

The problem is to determine the value of F in a given situation. F values may be de-

TABLE 14.3
Frequencies of Three Genotypes Represented by One Pair of Alleles Under Random Mating and Inbreeding with Coefficient of Inbreeding Equal to F

Genotype	Frequency	
	Random Mating	Inbreeding
a_1/a_1	p^2	$p^2(1 - F) + pF$
a_1/a_2	$2pq$	$2pq(1 - F)$
a_2/a_2	q^2	$q^2(1 - F) + qF$

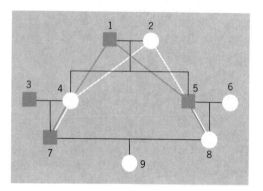

Figure 14.2 Pedigree of a man (No. 7) and woman (No. 8) from common ancestry, illustrating a first cousin marriage. The probability that any gene carried by either the common grandfather (No. 1) or common grandmother (No. 2) to both No. 7 and No. 8 would be the sum of the probabilities that the man and the woman would receive this gene from each grandparent. The probability for each parent of the child (No. 9) would be the product of the separate probabilities (each one half) that the particular allele and not the alternative allele would be transmitted through each of the five steps in the pathway.

termined from pedigree studies. In the pedigree shown in Fig. 14.2, for example, representing a marriage between first cousins, the probability that any allele (for example, a_1 or a_2) carried by either of the common ancestors would be present in both parents of the child would be the **product** of the probabilities that this particular allele was transmitted through each step in the pathway. The probability that an allele such as a_2 came from an ancestor common to the father and also to the mother of the child would be 1/2 for 4 (the father's mother) × 1/2 for 2 (the father's grandmother) × 1/2 for 5 (the mother's father) × 1/2 for 2 (the mother's grandmother) making a total of four genetic segregations in a line by which the father and mother are related through their common grandmother. A similar pathway for four steps would represent the probability that the same allele came to both parents from the common grandfather. The chance that both father and mother would carry the same allele (such as a_2) from either

common ancestor would be the *sum* of the probabilities for the two pathways from the two common ancestors and would represent the **relationship** (R) between father and mother.

$$R = \left(\frac{1}{2}\right)^4 + \left(\frac{1}{2}\right)^4 = \frac{1}{8}$$

The inbreeding coefficient (F) of the offspring equals 1/2 the relationship coefficient between its parents, so $F = 1/2 \times 1/8 = 1/16$. If, in the general population, a particular trait such as albinism dependent on a recessive allele (a_2) in homozygous condition occurs in about one person in 40,000, how much is the probability of expression enhanced if the parents are cousins?

$$q^2 = \frac{1}{40,000}, \qquad q = \sqrt{\frac{1}{40,000}} = \frac{1}{200}$$

The proportion of homozygous recessives in the population when $F = 1/16$ (first cousins) would be:

$q^2 \left(\text{after inbreeding with } F = \frac{1}{16}\right)$

$$= q^2 (1 - F) + qF = \left(\frac{1}{40,000} \times \frac{15}{16}\right)$$

$$+ \left(\frac{1}{200} \times \frac{1}{16}\right) = \frac{13.4}{40,000} = 0.000335$$

therefore, the probability of an expression of this trait would be about 13.4 times higher if the parents were cousins than if the parents were unrelated.

Percentages of homozygosis compared with the original degree of heterozygosis have been calculated by Wright and others for successive generations under different systems of inbreeding. Results for simple combinations are shown graphically in Fig. 14.3. As shown in the graph, self-fertilization produces homozygosis most rapidly. The 90 percent mark is passed in the third generation and 95 percent in the fourth. After eight generations, nearly all the genes that were heterozygous before inbreeding began would theoretically be homozygous. Brother and sister matings are somewhat less efficient than selfing in producing homozygosis, but these and parent-offspring matings are the most intense form of in-

breeding possible among nonself-fertilizing organisms. Under this system, 95 percent of the genes that were originally heterozygous would become homozygous in eleven generations of inbreeding, and 5 percent would remain heterozygous. Following continued brother and sister matings, homozygosity is closely approached in twenty generations. Double first cousins would yield about 84 percent homozygosity in the fourteenth generation. It might be expected, on the basis of superficial observations, that any system of inbreeding followed consistently would lead to complete homozygosis of all the originally heterozygous pairs. This is not true for matings between individuals in a large population who are less related than first cousins. Continued matings between half first cousins will result in a rise of only from 50 to 52 percent after an infinite number of generations. Continued matings of second cousins would cause a rise in homozygosis only from 50 to 51 percent.

Other factors such as **linkage,** the common procedure of **selecting** the more vigorous animals for mating, and spontaneous **mutations** interfere with the development of completely homozygous strains. Genes located near each other in the same

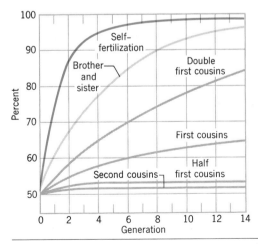

Figure 14.3 Graph representing percentage of homozygotes in successive generation under different systems of inbreeding. (After Sewall Wright.)

chromosome pair cross over infrequently and the investigator must often wait a long time for appropriate crossovers to occur and thus to make possible a high degree of homozygosity. Selection of the more vigorous animals in a population for mating tends to delay homozygosity because vigor is associated with heterozygosity. Spontaneous mutations may occur at any time and decrease homozygosity. To obtain and maintain homozygous strains in higher animals, it is necessary, therefore, to continue the inbreeding over long periods of time and to constantly watch for new variations in the inbred stocks, even after homozygosity has become well established.

Wright's calculations have established a background for theoretical as well as practical aspects of inbreeding. They have been employed extensively by breeders of domesticated animals. Wright contributed further to this aspect of genetics through his later studies on polydactyly and otocephaly in guinea pigs. These traits were found to follow the pattern **of multiple-gene inheritance.** The mathematics was somewhat more complicated but the polygenes responsible for these traits were also shown to become **homozygous through inbreeding.**

INBRED MICE AND CANCER STUDIES

In the early part of this century, C. C. Little recognized a practical application of inbreeding in connection with studies on the genetics of cancer. He and others had been studying the inheritance of tumors in ordinary laboratory mice, with little success. The possibilities of greater success with **homozygous strains** led to an extensive inbreeding program which continued for several years. After many generations of inbreeding and selection, strains highly susceptible to specific types of cancer were produced. The inbred lines have lived up to expectations and proved to be a most valuable material for investigating the genetics of different kinds of tumors as well as other aspects of genetics. Some inbred mouse strains had a high incidence of mammary tumors; others were high in the incidence

TABLE 14.4
Incidence of Mammary Tumors, Lung Tumors, and Leukemia in Several Inbred Strains of Mice (After W. E. Heston)

Strain	Mammary Tumor (Percent)	Lung Tumor (Percent)	Leukemia (Percent)
dba	55–75	low	30–40
A	70–85	80–90	low
C$_3$H	75–100	5–10	low
C57 black	low	low	20
C57 brown	low	low	Most common neoplasm in strain
C57 leaden	low	low	low
C58	low	low	90
Ak	low	low	60–80

of lung tumors or leukemia, and still others were relatively free from all types of tumors. The cancer incidence in some well-known inbred strains is summarized in Table 14.4. Little started the well-known dba strain in 1909 by selecting a stock he was then using for a coat-color experiment. He selected three recessive coat color alleles, *d* for dilute, *b* for brown, and *a* for nonagouti, from which the strain (dba) obtained its name. The original progenitors presumably carried alleles favorable for mammary gland tumors which were eventually **fixed** in the stock by **inbreeding.**

These studies have shown that tumors do not usually follow a simple Mendelian pattern in inheritance in mice. Apparently, either combinations of interacting genes or polygenes with cumulative effects are involved in the hereditary mechanism. Furthermore, the different types of cancer affecting different sites are tissue specific and genetically independent. W. E. Heston has suggested that inbred strains of mice also be used in studies of heart disease and tooth decay and of physiological characters that occur in mice as well as men. Besides their

help in clarifying the genetics of cancer and their potential value for studies of other diseases, inbred strains of mice have contributed much background knowledge concerning the nature and effect of inbreeding itself such as the relation between the time required for homozygosity to develop, and the system of inbreeding.

PRACTICAL APPLICATIONS OF INBREEDING IN DOMESTIC ANIMALS

By controlling the matings of animals within his herd or flock and then selecting for certain traits, the breeder has a powerful method for developing a desirable genotype. He can restrict matings to a small circle of animals and through inbreeding prevent the introduction of alleles from outsiders. This tends to decrease variation within the group and to stabilize type. Limiting the size of his breeding unit may also give the breeder access to the effects of random sampling or genetic drift.

Inbreeding combined with selection over periods of time has produced many valuable

Figure 14.4 Result of breeding Merino sheep. Left, modern Merino ram; right, direct descendant of the original strain from which the present-day Merino was developed. (Australian News and Information Bureau.)

breeds of domestic animals. Merino sheep, for example, which are widely known as fine wool producers, are the result of about 200 years of breeding. This breed was developed in Spain in the seventeenth century by stock raisers. Ancestors of the present-day Merino sheep had two different coats of wool, one composed of long, coarse fibers arising from primary follicles, and a second coat composed of short, fine wool arising from clusters of secondary follicles. Intensive selection was maintained for animals with more uniform production of fine wool and a lesser amount of coarse wool. For a time, Spain had a monopoly on the valuable Merino sheep. However, when France invaded Spain, Merinos were removed to France where they were maintained and eventually distributed to other parts of the world. Merinos were also taken to South Africa and in 1796 were introduced into Australia, which has since become the world's largest producer of fine wool. A modern Merino is shown in Fig. 14.4 along with a descendant of the original unselected strain from which the present-day Merino was developed.

EFFECTS OF INTENSE INBREEDING

Intense inbreeding has led to an unfortunate situation in certain breeds of cattle. Recessive genes for undesirable characteristics have become concentrated in the descendants of certain common ancestors. Hereford beef cattle have been inbred for type over a period of many generations. Along with the desirable traits, which have made the Hereford strain the best range breed in western United States, expressions of **insidious recessives** have crept in with increasing frequency. Records indicate that only a few sires have produced most of the present population. It is likely that the defective alleles have been present for a long time, but until recently they have nearly always been hidden by the dominant alleles.

Dwarf calves have demoralized some otherwise enthusiastic breeders of Angus and Shorthorn beef cattle. Five of a total of 36 calves in a recent spring crop of one herd were dwarfs. One bull and several cows in the herd were found to carry a recessive allele for dwarfism. A dwarf ten months old is shown with a normal animal of the same age in Fig. 14.5. From the side view, the characteristic head and body features of the dwarf can be compared with those of the normal animal. This type of dwarf is called brachycephalic dwarf; the name was suggested by the characteristically short, broad head. The lower jaw is extra long, the forehead is bulging, the abdomen is out of proportion, and the legs are short. More critical observations of the separate bones have shown other anatomical differences. Breeding data indicate that a basic recessive allele is necessary for dwarfing, but additional modifiers have been postulated to account for the different types of dwarfs. Those encountered in the Angus and Shorthorn breeds may depend on the same allele, but different modifiers have developed in the different breeds. Dwarfs in any breed are of little economic value and are avoided by stock raisers whenever carriers can be detected.

LINE BREEDING

Line breeding is the mating of animals to maintain a close relationship to an unusually desirable individual. All matings may be to a **sire** with particularly valuable qualities with the objective of perpetuating his **type.** This time-honored form of inbreeding used to maintain a desirable type has yielded good results in quarter horses and other types of livestock. Since it is a form of inbreeding, the limitations men-

(a)

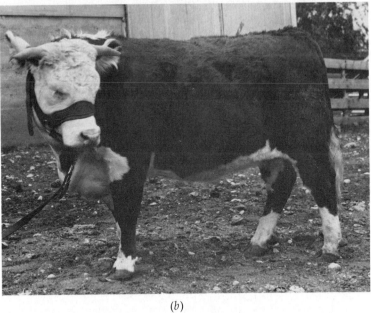

(b)

Figure 14.5 Hereford calves: (*a*) dwarf calf compared with a normal calf of the same age (*b*) older calf showing normal head and body features. The dwarf shows characteristic head and body features of brachycephalic dwarfism. (Photographs by A. L. Hansen.)

Figure 14.6 Quarter horse. (Walter Chandoha.)

tioned previously must be considered. As the individuals involved become more homozygous, their progeny are more likely to suffer from the expression of undesirable recessives. This has caused many breeders to be fearful of continued line breeding as well as other forms of close inbreeding.

Unfortunately, the ill effects of inbreeding have been overemphasized by some practical breeders. Some breeders have avoided line breeding when it might have been used to good advantage to fix a type. It is the only method by which an unusually good stock may be maintained. Good combinations of alleles have too often been lost by outcrossing when some form of inbreeding might have maintained them intact. The advantages of perpetuating something exceptional must not be lost because

of the risk of having to cull animals with unfortunate gene combinations. Rewards for ruthless culling over a prolonged period of time are often great. Breeders of quarter horses (Fig. 14.6), for example, have accomplished remarkable results through **line breeding and culling.**

Animal breeders speak of a quality that they call **prepotency,** the ability of animals to perpetuate characteristics in their offspring. Obviously, if the progeny perpetuate the characteristics of an ancestor, they will be uniform among themselves. Homozygosity and dominance provide the main underlying genetic bases for prepotency. A breeder interested in greater prepotency must promote **homozygosity** in his breeding stock and he can best do this by inbreeding.

HERITABILITY

The effectiveness of selecting for a particular characteristic depends upon the relative importance of heredity and environment in

the development of that trait. In practice, breeding animals are selected on the basis of visible or measurable traits (phenotypes) displayed by themselves or their relatives. Relative agreement between phenotype and genotype is measured by the **coefficient of heritability** (h^2) and is of utmost importance to the animal breeder. Heritability of a trait (narrow sense) is the ratio of its additive genetic variance to total variance as shown by the following expressions: $h^2 = V(G)/V(P)$ where $V(G)$ is the variance due to the additive effects of genes and $V(P)$ is the total phenotypic variance. Heritability is a measure of the degree to which a phenotype is genetically influenced and, therefore, can be modified by phenotypic selection.

The simplest method of selection used by animal breeders is phenotypic or mass selection. This is systematic choosing of the part of the population in which the desired qualities are most strongly developed and using the individuals chosen as parents of the next generation. In farm animals, this procedure involves evaluation of production records or other evidence of desired traits. Rate of progress depends on the ability of the breeder to choose individuals that are not only phenotypically but also genotypically superior. **Mass selection** is most efficient when heritability is high. Progeny tests, family selection, and pedigrees, however, can be used to supplement information pertaining to the individual himself when heritability of the trait is low. It follows that the individual's own phenotype is not an accurate indicator of his genotype when heritability is low.

Three different approaches have been used for estimating heritability. The first is the time-honored practice of trial and error combined with good husbandmanship that began in prehistoric time. The actual gain or success is realized through selection. The estimate of heritability derived by this method is the degree of progress related to the effort expended in selection as measured by the **selection differential** (superiority of selected parents).

The second approach is based on how much more alike in phenotype are related individuals having somewhat similar genotypes than are unrelated or less closely related individuals. In **isogenic stocks** made up of individuals with exactly the same gene component, all variation is **environmental** as shown by the relation: $h^2 = V(P) - V(E)/V(P)$; where $V(P)$ is the variance among individuals in the general population made up of genetic and environmental differences, and $V(E)$ (environmental variance) is the variance among individuals within the same isogenic line. Comparisons of variation in populations that are otherwise similar to isogenic stocks but undergo random breeding rather than inbreeding can provide estimates of heritability. This method can be readily accomplished with fruit flies and other laboratory animals but among farm animals the only isogenic individuals are identical twins. These are infrequent and difficult to identify.

Another approach that is suitable for some laboratory organisms but not practical for farm animals is to select in both directions and thus develop high and low lines with reference to a particular trait. The extent to which the lines differ in response to selection provides an **estimate of heritability.** Such an experiment has been performed in honey bees as illustrated in Fig. 15.5.

The most widely used procedure for estimating heritability for farm animals is to compare the resemblance between relatives. Greater average resemblance between parents and progeny of half or full sib groups is associated with heritability. When little correlation exists between the degree of expression of a trait by members within these related groups, the greater part of the observed variation in that trait is environmental.

When heritability of a trait is known, the progress to be expected from a generation of

TABLE 14.5
Heritability Coefficients

Trait and Animal	h^2
Egg production in chickens	0.05–0.15
Back fat in swine	0.50–0.70
Milk yield in cattle	0.2–0.4
Daily weight gain in cattle	0.3–0.5
Fleece weight in sheep	0.3–0.6

selection can be predicted. The coefficient of heritability is not a constant, but only indicates the proportion of variance caused by differences in additive gene effects in a particular population at a particular time. Some examples of heritability coefficients are listed in Table 14.5. Egg production in chickens has a low heritability coefficient, hence little progress would be expected per generation of mass selection. Improvement of the environment, or selection on some basis other than mass selection or the use of breeding systems that increase genetic variation, is suggested. Rate of weight gain in cattle, on the other hand, has a moderately high heritability coefficient, and considerable progress is possible and, indeed, has been made through mass selection.

Assume that in a cattle population the average daily gain is 2.4 pounds, and the bulls selected to be sires of the next genera-tion average 3.4 pounds and the cows 3.0, giving a selection differential of $[(3.4 + 3.0)/2] - 2.4 = 0.8$ pounds. If heritability is 0.4 or 40 percent, the expected gain will be 40 percent of the selection differential or 40 percent of 0.8 pounds, giving 0.32 pounds. The offspring would then be expected to average $2.4 + 0.32 = 2.72$ pounds as compared to 2.4 pounds for the entire population from which the parents were selected. If heritability had been only 10 percent, as it is for egg production, the gain would have then been 0.08 (10 percent of 0.8) and the average for the next generation would be only 2.48. The generation interval in beef cattle is 4½ to 5 years so the expected progress of 0.32 pounds per generation would represent an annual increase of about 0.06 to 0.07 pounds. This points out the value of reducing generation interval in livestock.

OUTBREEDING

The most valuable result of outbreeding is increased vigor of the hybrids as compared with inbreds. This **heterosis** has not been exploited in animals to the same extent as in plants, but is now providing greater efficiency in the production of meat, milk, and eggs. Poultry raised for meat responds to outcrossing. Rabbits also show marked increases in birth weight and slaughter weight following outbreeding. Swine producers have for some time utilized the profits resulting from outbreeding.

A type of outbreeding practiced among Hereford cattle, of which one animal is illustrated in Fig. 14.5B has resulted in distinct advantages over the usual practice of inbreeding. When inbred bulls were mated with unrelated cows that had been sired by other inbred bulls, significant improvement was demonstrated. Hybrids from a large experiment averaged about 12 percent greater weight at weaning than did inbreds. An even more impressive advantage in favor of the hybrids was noted after the feeding period. In final weight, the hybrids average 65 pounds more than the inbreds, which were maintained under the same conditions. An investigation of rabbits has also demonstrated the value of outbreeding. The weights at birth and at 180 days of 915 rabbits of four inbred strains and of their F_1, F_2, and backcross generations substantiates heterosis effects. The crossbreds were all significantly heavier than the inbreds. The birth weights and the weights at 180 days of the F_1 crossbreds had the smallest variability, whereas those in the F_2 generation had the largest variability. All the crossbreds showed significant **heterosis effects** on their weights at 180 days.

Outbreeding has been employed effectively in the development of improved strains of sheep. Hybrid sheep have been bred successfully for meat and also for increased wool production. Crosses were conducted at the Western Sheep Breeding Laboratory at Dubois, Idaho, between the Rambouillet, a gregarious range animal with poor meat quality and fine, short wool, and the Lincoln, a solitary animal with good meat quality but long, coarse (braid) wool. The result was the Columbia breed of sheep which has an intermediate grade of wool that is considerably longer than that of the Rambouillet, good characteristics for the range and good meat quality.

Hybrids between black-faced Hampshire sheep and Rambouillets have yielded smutty-faced lambs of excellent meat quality. Since sheep breeders are interested in maintaining the registers of the standard types, they usually dispose of the hybrids early as lambs and maintain only standard breeds for mating purposes.

CROSSBREEDING

Mating of individuals from entirely different races or even from different species (when possible) is called **crossbreeding.** This represents the most extreme form of outbreeding that is possible among animals. The extent to which this pattern of mating can be carried out under natural conditions is limited by the ability of distantly related individuals to mate and produce fertile hybrids. Successful interspecific crosses are rare in domestic animals. Numerous breeds and races of different animals have been intercrossed with good results, but members of only a few different species have mated and produced fertile hybrids. The ringneck dove and tumbler pigeon can mate and produce progeny (Fig. 14.7), but the hybrids are all **sterile.**

The mule is the result of a successful cross between two different species, the horse and the donkey. In many ways the mule is superior to either parent. It is a larger, swifter, and stronger animal than the donkey, and more hardy, more resistant to disease, and more capable of prolonged work with short food rations than the horse. Mules have been produced commercially for a long time. They have served and continue to serve as valuable beasts of burden in many parts of the world. Qualities that make the mule valuable are attributed partly to heterosis. The mule is usually **sterile.** A few fertile mules have been reported, but the number is too small to have significance in practical breeding. Each new generation must be produced from new crosses between donkeys and horses.

The zebu, representing a geographical race of cattle native to India, has been crossed successfully with domestic cattle of European origin. Wide phenotypic differences separate these two races. They differ in type of horns, skull shape, dewlap, voice, size of digestive tract, behavioral

Figure 14.7 Ringneck dove, (top) baldhead tumbler pigeon, (middle) and sterile hybrid (bottom) resulting from interspecific hybridization. (From *American Naturalist* 84:275–308, 1950. Photo courtesy of Willard F. Hollander.)

traits, and disease resistance. One type of zebu breed called "Brahman" cattle (Fig. 14.8) has thus far been the most successfully used for outcrossing. The characteristics of **economic importance,** such as adaptability, meat production and milk production, are the qualities sought in the hybrid.

HETEROSIS IN MAIZE

In the second half of the last century it was shown by W. J. Beal and others that the products of varietal crosses were superior to the products of inbred varieties of maize (corn). G. H. Shull, working at Cold Spring Harbor (1908), E. M. East at the Connecticut Agricultural Experiment Station (1908), and later D. F. Jones, H. K. Hayes, and others continued the investigation and found that **hybrid corn** had more vigor than the inbred lines or the open pollinated varieties then in use. This observation has been elaborated and tremendous increases in production have been obtained. The fieldcorn acreage in the United States de-

Figure 14.8 Brahman bulls. (Dick Greenberg/Black Star.)

creased from slightly over 100 million acres in 1929 to less than 80 million acres in 1974, but the yield has increased by one third. In 1929 practically none of the acreage was hybrid; ten years later about 23 percent was planted to hybrid seed. Now 99 percent of the fieldcorn acreage in the United States is planted with hybrid seed. Hybrid corn is the most practical attainment of genetics thus far, but paradoxically, the early contributors were not located in the corn belt and apparently had no thought of making a practical contribution.

Both inbreeding and outbreeding are involved in the production of hybrid corn. Hybrid seed is produced by crossing inbred strains, which are developed by controlled pollination. This is done on a small scale by placing paper bags over the ears and tassels of the corn plants when these parts first develop. Pollination is accomplished by transferring the pollen from the tassel to the silk of a male sterile plant and protecting the silk from foreign pollen. Only the most desirable lines are kept during the several generations of inbreeding. The

inbred strains are then crossed to produce **hybrid seed.** To make the cross, a few rows of one type of inbred corn are planted near a few rows of another inbred variety in the direction of the prevailing wind. The tassels are removed from the seed parent plants and the silks are fertilized by the wind-borne pollen of the other inbred line (pollen parent). F_1 plants are larger and more productive than those from further generations of selfing, as illustrated diagrammatically in Fig. 14.9. Experimental results for the different generations are illustrated by sketches drawn to scale. Commercial hybrid seed corn is produced on a large scale in places where facilities are suitable.

Vigorous hybrids from two properly matched F_1 plants produce seed more efficiently and economically than do the weak inbred plants used in the single cross. Therefore, the double cross (Fig. 14.10) has been developed for hybrid seed corn production. This is accomplished by developing four inbred varieties, making parallel crosses between two pairs in the same season, and, in the next season, crossing the

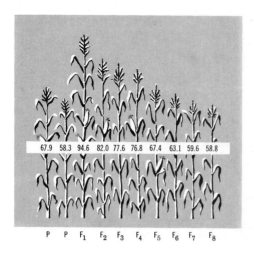

Figure 14.9 Heterosis in corn. The figures in the center represent the average height in inches for the different generations identified at the bottom. (Data from D. F. Jones.)

67.9 58.3 94.6 82.0 77.6 76.8 67.4 63.1 59.6 58.8

P   P F_1 F_2 F_3 F_4 F_5 F_6 F_7 F_8

hybrids. Inbred strains A and B, for example, are crossed together, and other inbred strains, C and D, are crossed with each other. F_1 plants from AB are then crossed with F_1 plants from CD. Seed from this cross is sold to the farmer. Double crosses do not actually improve the hybrid vigor above that of the single cross; they mainly provide large uniform vigorous plants for seed production and thus reduce the cost of commercial seed. The uniformity of the crop in height, yield, and ear characteristics can also be improved. Plant size and vigor, and particularly the size of the cob, determine the amount of seed that can be produced. A weak inbred plant must be the cob producer for F_1 and therefore the amount of seed is limited. On the other hand, if an F_1 plant can be used as a **seed parent,** a much **larger cob** and greater **plant vigor** are available.

One theory postulated to explain heterosis is based on **dominant alleles** for increased vigor. It requires that these alleles accumulate in inbred lines by random mutations and selection. Vigor depends on a large and efficient root system, well-developed leaves with a good supply of chlorophyll, firm supporting tissue, and other properties. Alleles for vigor are brought together in hybrids and, because of their dominance, produce maximum expression in the F_1. When all the many avenues through which plants could become better fitted to their environment are considered, different inbred lines may be expected to accumulate many different alleles for increased vigor. Dominant alleles postulated in this theory would become homozygous through continued inbreeding. Crosses between inbred lines would result in the heterozygous F_1 plants expressing the favorable characteristics from both inbred lines. Under this system, maximum vigor would be expected to occur in individuals having the loci with dominant favorable alleles.

Let us consider a simple model with only two pairs of genes involved; one inbred line will be considered to have become better able to meet the conditions of the environment through an improved root system controlled by allele A. Another inbred line is also improved because it has a better chlorophyll system controlled by allele B. Through inbreeding, each line has become homozygous for its dominant allele, that is Ab/Ab and aB/aB, respectively. A cross be-

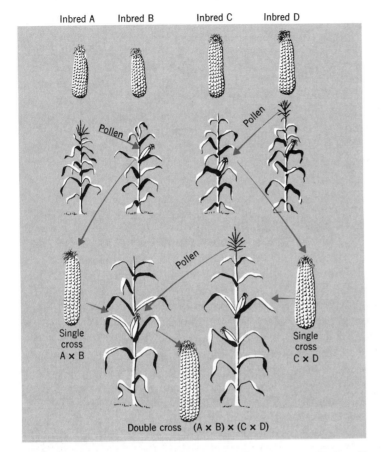

Inbred A   Inbred B   Inbred C   Inbred D

Pollen

Pollen

Single
cross
A × B

Pollen

Single
cross
C × D

Double cross   (A × B) × (C × D)

Figure 14.10 Double cross method for producing hybrid corn. (From Dobzhansky, *Evolution, Genetics, and Man*, John Wiley & Sons, Inc., New York,
1955.)

tween these inbred lines brings the good
qualities from both together in the hybrid
Ab/aB. The weak alleles *a* and *b* are also
present, but they are recessive and do not
influence the efficiency of the hybrid. If the
inbred parents were *AB/AB* and *ab/ab*, the
hybrid would be no better than the more
vigorous parent. Complete homozygosity
would be rare if many genes were involved.
The theory is dependent on dominant alleles controlling in separate ways the vigor
of progeny. Assuming that such alleles for
vigor are distributed between the two

inbred parents, the **hybrids** should be **better
than either parent.**

What happens when hybrids with a high
level of vigor are inbred? If the hybrids illustrated in the oversimplified model presented above were self-fertilized, that is,
Ab/aB × Ab/aB, nine combinations would
be expected, but only those with *A* and *B*
(that is, *AABB, AABb, AaBB,* and *AaBb*)
would be as vigorous as the F_1. Continued
inbreeding over many generations would
result in more or less pure lines than were
crossed to produce the hybrids. If fully ho-

mozygous dominant plants could be distinguished from the others, these should be as vigorous, but no more so, than the hybrid. The population as a whole would decrease in vigor and the plants would lose the uniformity that was conspicuous in the original hybrids. By the eighth generation, when the lines would be relatively pure, further inbreeding would have little effect.

On the basis of this explanation, it would seem to be possible to develop eventually inbred lines carrying the genes for increased vigor. This has been attempted many times without success. A possible explanation is that dominant alleles for vigor are **linked** with recessives for undesirable characteristics, which also accumulate through inbreeding and become **homozygous.** The deleterious recessives, which on this assumption come together in later generations, more than counteract the good effects observed in the F_1. The genetic basis of heterosis is probably more complex than that expressed in any single explanation offered thus far. Such genetic principles as (1) complementary gene action, (2) epistasis masking deleterious recessives, (3) effects of multiple alleles and (4) overdominance (selective advantage of heterozygote over both homozygous types) may be involved in the process. It now seems evident that only certain genes produce heterosis when they are in heterozygous condition. Favorable chemical combinations controlled by particular alleles may be responsible for the increased vigor.

ADVANTAGES AND DANGERS OF UNIFORMITY

In the United States alone, billions of dollars are saved annually by planting hybrid corn instead of inbred varieties. Genetics has played a key role in the development of hybrid corn. The plant breeder cannot rest easily, however, because of the danger of **epidemic diseases** in new crops.

Rusts and other fungi present a constant challenge to the breeder of cereal crops. New varieties of corn, like those of wheat, may become susceptible to fungus pests. The southern corn leaf blight epidemic in 1970 and 1971 (Chapter 11) reduced the 1970 corn crop by some 710 million bushels below the estimate for that year.

What caused this disaster to the corn crop? A new mutant of the fungus *Helminthosporium maydis* (Nisikado and Miyake) became a virulent pathogen on a particular kind of hybrid corn. It was especially destructive on corn with Texas (T) **male-sterile cytoplasm** which was extensively used in hybrid seed production. Pathologists and plant breeders met the epidemic by searching for corn varieties that were resistant to the T form of the fungus. Because of a previous less serious yellow leaf blight, some 1970 seed production had been shifted to corn without T-cytoplasm. This corn required manual detasseling but was used for winter planting in many 1971 seed fields. It also produced some resistant seed for immediate general farm use. Some growers preferred the predictable 20–30% loss of yield to the high risk of much larger losses from growing susceptible hybrids.

Most of the 1971 seed production was, therefore accomplished without the use of male sterility and T-cytoplasm. The T race of *H. maydis* was not serious in 1972. Still another race of *H. maydis* may appear or one of the other corn diseases could become a threat to the highly uniform hybrid corn with T malesterile cytoplasm. Several varieties of corn resistant to the existing T race of *H. maydis* have now been identified and are available for seed production.

This example illustrated the danger of germ plasm **uniformity** for a crop grown on a large scale. It also indicates that sustained research programs are essential in protecting food supplies from potential losses of catastropic magnitude.

HETEROSIS IN OTHER SPECIES

F_1 plants obtained by Mendel from his cross between tall and dwarf garden peas were larger than the tall parents and may represent an example of heterosis. E. M. East and H. K. Hayes, two of the foremost investigators on heterosis in maize, crossed tobacco strains A and B, 31 and 54 inches in average height, respectively, and obtained hybrids averaging 67 inches under the same environmental conditions. F_1 tomato hybrids are widely grown in home gardens. Other garden vegetables such as cucumbers, squash, cantaloupes, and sweet corn have been developed for hybrid vigor. Pine trees respond to appropriate cross-pollination and their progeny display hybrid vigor. This principle is also utilized with millet, Sudan grass, Bermuda grass, and several other grasses.

In onions, as in sugar beets, the procedure for producing hybrids has been markedly facilitated through the discovery of male sterile lines. Male sterility, for example, has been introduced into three varieties of onions — the crystal wax, the yellow Bermuda, and the sweet Spanish. With male sterile plants it is possible to make crosses without the laborious task of emasculating each individual flower. It should be pointed out that factors other than yield are important in hybrid plants. Ability to stand up in wind or rain, resistance to pests, uniformity of maturity, and high-quality seed are also of practical concern.

Since the successful development of hybrid corn, plant breeders have sought to utilize the principle in **self-pollinated crops** such as sorghum, wheat, rice and barley. These cereal crops present a much more difficult problem than did corn. Even before male sterile and fertility restorer systems were developed for corn (Chapter 11), enormous quantities of commercial hybrid corn seed was produced by hand detasseling of the female line and cross-pollinating from the male line. Artificial hybridization is much more **tedious and costly** in self-pollinating plants. Only one self-pollinating plant, sorghum, has been developed thus far for commercial productions of hybrid seed.

Another self-pollinating plant, wheat, has been studied extensively to access the prospects of producing hybrid seed commercially. Results of crosses among some **wheat** strains have demonstrated heterosis. Besides increased yield, the hybrids had acceptable milling and baking qualities and a high level of disease resistance. High cost of seed production, short life of wheat pollen, difficulty with male sterility and the restorer system and amount of seed required per acre (100 pounds compared with 20 for corn) now make commercial hybrid wheat doubtful. Development of **high yielding dwarf wheats** and successful interspecific **crosses** between tetraploid wheat and rye resulting in **triticale** have made the production of commercial hybrid wheat less urgent than it was a few years ago.

HETEROSIS IN A MENDELIAN POPULATION

Heterosis may be represented in a Mendelian population by the following relations: Where s_A is the selection coefficient for the AA genotype and s_a is the selection coefficient for the aa genotype. The rate of change (Chapter 13)

$$= \Delta q = \frac{pq(ps_A - qs_a)}{1 - p^2 s_A - q^2 s_a}$$

	AA	Aa	aa	Total
Proportion before selection	p^2	$2pq$	q^2	1
Fitness	$1 - s_A$	1	$1 - s_a$	
Proportion after selection	$p^2(1 - s_A)$	$2pq$	$q^2(1 - s_a)$	$1 - p^2 s_A - q^2 s_a$

In analyzing this equation:

1 $1(-p^2 s_A - q^2 s_a)$ is the genetic load

2 $\Delta q = 0$ when $p = 0$ or $q = 0$ (subject to change as mutations occur)

3 When $ps_A = qs_a$, $\hat{q} = \dfrac{s_A}{s_A + s_a}$ and

$$\hat{p} = \frac{s_a}{s_A + s_a}$$

Heterosis is associated with increased heterozygosity. It can be **disrupted** by **inbreeding** and **restored** by **interbreeding**. As long as heterozygotes maintain a fitness of 1 ($\overline{W} = 1$), $\hat{p}$ and $\hat{q}$ will tend to remain constant.

For example, a certain trait in corn is determined by two alleles, c_1 and c_2. One hundred individual corn stocks of the $c_1 c_1$ variety produce 300 ears of corn, the same number of $c_1 c_2$ variety plants produce 400 ears of corn and 200 ears are produced by the same number of $c_2 c_2$ individuals. The relative fitness ($\overline{W}$) of each variety may be determined:

ratios

$$\overline{W}_{c_1 c_1} = \frac{300}{100} = 3 \qquad 3/4 = 0.75$$

$$\overline{W}_{c_1 c_2} = \frac{400}{100} = 4 \qquad 4/4 = 1.00$$

$$\overline{W}_{c_2 c_2} = \frac{200}{100} = 2 \qquad 2/4 = 0.50$$

The selection coefficient of the homozygous groups may be determined by subtracting the fitness ratio (based on a $\overline{W} = 1$ for the heterozygotes) from 1 (i.e., $(1 - \overline{W}) = s$)

$$s_{c_1} = 1 - 0.75 = 0.25$$
$$s_{c_2} = 1 - 0.50 = 0.50$$

The equilibrium frequency ($\hat{q}$) of c_1

$$= \hat{q} = \frac{s_{c_1}}{s_{c_1} + s_{c_2}} = \frac{0.25}{0.50 + 0.25} = 0.33$$

and the equilibrium frequency ($\hat{p}$) of $c_2 = p = s_{c_2}/(s_{c_1} + s_{c_2}) = 0.5/(0.50 + 0.25) = 0.67$. These equilibrium values indicate that the frequency of the alleles c_1 and c_2 will remain constant at this level until altered by factors such as selection, mutation, migration and random genetic drift.

INBREEDING AND OUTBREEDING IN MAN

The present trend in human populations is toward outbreeding. Improved transportation and communication have resulted in greater and greater mixing of peoples. Recent studies show that inbreeding is now at a very low level in most parts of the world whereas some groups of people had considerable inbreeding in the past. The present trend keeps recessive genes heterozygous and thus protected from selection.

REFERENCES

Hayes, H. K. 1963. *A professor's study of hybrid corn.* Burgess Publishing Co., Minneapolis.

Heston, W. E. 1949. *Development of inbred strains in the mouse and their use in cancer research.* Roscoe B. Jackson Memorial Laboratory, Twentieth Commemoration. Maine, Bar Harbor.

Lush, J. L. 1945. *Animal breeding plans.* Collegiate Press, Ames, Iowa.

Müntzing, A. 1967. *Genetics: basic and applied.* Lts. Förlag, Stockholm, Sweden.

Pirchner, F. 1970. *Population genetics in animal breeding.* W. H. Freeman Co., San Francisco.

Schull, W. J. and J. V. Neel. 1965. *The effect of inbreeding on Japanese children.* Harper and Row, New York.

Tatum, L. A. 1971. "The southern corn leaf blight epidemic." *Science* 171, 1113–1116.

Wright, S. 1921. "Systems of mating, I, II, III, IV, V." *Genetics* 6, 111–178.

Wright, S. 1934. "On the genetics of subnormal development of the head (otocephaly) in the guinea pig, an analysis of variability in number of digits in an inbred strain of guinea pigs, the results of crosses between inbred strains of guinea pigs, differing in number of digits." *Genetics* 19, 471–551.

PROBLEMS AND QUESTIONS

14.1 How does inbreeding affect (a) gene frequency? (b) heterozygosity?

14.2 What are the general effects of inbreeding and outbreeding among plants or animals that are ordinarily cross-fertilized?

14.3 Why do self-fertilizing plants such as peas and beans not lose vigor through continued inbreeding?

14.4 If it was known that self-fertilization resulted in decreased vigor in a certain type of plant but the natural method of breeding was not known, what speculation might be drawn as to the natural system of mating?

14.5 How are inbreeding and outbreeding related to (a) natural selection? (b) methods of plant and animal breeding?

14.6 Why are pure lines valuable for (a) hybridization experiments? (b) experiments designed to compare genetic and environmental variation?

14.7 Garden peas heterozygous for three genes, *g* for seed color (yellow or green); *y* for pod color (green or yellow); and *c* for pod shape (inflated or constricted), are allowed to self-fertilize and their descendants are self-fertilized for many generations. What phenotypic pure lines would be expected to develop?

14.8 When pure lines are obtained, would the self-fertilizing individuals be expected to produce anything but homozygous progeny? Explain.

14.9 If 2000 garden peas heterozygous (Dd) for height were inbred for five generations, about how many plants would be expected to be heterozygous? (Assume no differential adaptive value and that each parent produced one offspring so that 2000 plants will be present in each generation.)

14.10 A large number of fruit flies, all heterozygous for gene a, were allowed to mate at random and their descendants also mated at random for ten generations in a population cage. What proportion of AA, Aa, and aa would be expected assuming no differential adaptive value?

14.11 Evaluate the following systems of mating with reference to their relative efficiency in producing homozygosis: self-fertilization, brother and sister mating, double first cousins, first cousins, second cousins, and random mating in a large population.

14.12 In some herds of beef cattle, dwarf calves occur. A recessive gene is responsible for the abnormality. (a) How could this condition have developed? (b) What measures might be taken to avoid the financial losses which come from the production of dwarfs?

14.13 If genes A, B, C, and D, produce vigor in maize, evaluate the efficiency with which superior hybrid seed could be produced from the following crosses and give reasons for evaluation: (a) F_1 from $abCD/abCD \times ABcd/ABcd$ (b) $F_1 \times F_1$, from a and (c) F_1 from $aBCD/aBCD \times AbCd/AbCD \times F_1$ from $ABcD/ABcD \times AbCd/ABCd$?

14.14 In certain breeding stocks of maize, assume that gene A for increased vigor is closely linked with a recessive gene w which makes the plant weak, and that gene B for another quality also resulting in increased vigor is closely linked with a recessive gene l for low viability. (a) What combination would be most efficient for production of vigor? (b) What kind of mating would produce the desired combination? (c) What would be the likelihood of obtaining a pure breeding strain with superior vigor?

14.15 To what extent does heterosis have practical significance in animals? Based on the studies of rabbits, mules, and sheep, what are the prospects for the future in exploiting heterosis in animals?

14.16 Describe the practical usefulness of inbred strains: (a) mice in cancer research; (b) cattle breeding; and (c) production of hybrid seed maize.

14.17 How can the high concentration of certain tumors in certain inbred strains of mice be explained?

14.18 What effects might be expected from (a) inbreeding, and (b) outbreeding in the present human population?

14.19 A trait dependent on a recessive gene (a) occurs in the general population at the rate of one in 10,000 people. In the following pedigree, (a) what would be the chance of it occurring in A's child if he married at random and if the trait was not known to occur in his family? (b) What is the inbreeding coefficient for C? (c) What is the chance that C will express the trait? (d) If A's grandmother was known from family history to be a carrier for *a* and A's grandfather was known not to carry the gene, what is the chance that C will express the trait? (e) If A's grandmother was known to have expressed the trait and his grandfather was known not to carry the gene, what is the chance that C will express the trait?

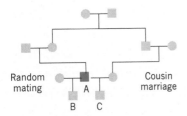

Random mating A Cousin marriage B C

14.20 What would be the inbreeding coefficient of a child whose parents were (a) second cousins (i.e., the children of first cousins)? (b) uncle and niece? (c) double first cousins?

14.21 The average height of two inbred corn varieties and their hybrids measured in inches are: $P_1 = 70$, $P_2 = 60$ inches, $F_1 = 90$ and $F_2 = 80$. Calculate the amount of heterosis in the (a) F_1 and (b) F_2 hybrids.

14.22 The gene r_1 when found in a homozygous state causes death of 7/8 of all offspring while the r_2 allele in a homozygous condition causes the death of 1/4 of all offspring. The heterozygotes are vigorous. If these trends continue, what will be the frequency of r_1 and r_2, in an equilibrium state?

14.23 The total genetic variance of 180 day body weight in a population of swine is 250 pounds. The phenotypic variance is 600 pounds. What is the heritability of the trait (narrow trait)?

14.24 Three hundred individual milo stocks of the m_1m_1 genotype produced 600 seeds; the same number of the m_1m_2 genotype produced 1200 seeds and 900 seeds were produced by the m_2m_2 genotype. What is the relative fitness ($\overline{W}$) of each genotype and what is the equilibrium frequency of alleles m_1 and m_2?

FIFTEEN

Genetic mechanisms associated with structurally and numerically classifiable traits have been investigated much more extensively and successfully than those associated with behavioral characteristics. This is true primarily because of the **complexity of behavioral traits** and the difficulty of studying them at the molecular level. Furthermore, behavioral characteristics of any animal develop under the joint, tightly entwined effects of heredity and **environment.** The DNA in the genome determines the individual's physiological, structural, and behavioral potentials, but not all of the potentials are inevitably realized in the developing individual. Behavior genetics is concerned with the effects of genotype on behavior and with the role that genetic differences play in determination of behavioral differences in a population.

A fundamental question in the study of the relation between genes and behavior is whether heredity directly affects behavior or merely defines the stage on which behavioral patterns may be molded by environmental fac-

GENETICS OF BEHAVIOR

Grasshoppers mating.
(Eric Haas/Rapho Guillu-
mette.)

tors. Biologists and psychologists have taken opposite views on this issue in past decades, but now the two groups recognize that both heredity and environment are relevant for virtually all behavior patterns. The question of direct versus indirect effects of heredity can be approached most meaningfully by recognizing the sequential levels of organization in developing animals. Environmental factors are **interwoven** with inheritance mechanisms at every point in the developmental process. The problem is to recognize the role of each and to evaluate their relative importance in specific situations.

In some cases, genetic programming simply provides that animals have specific learning capabilities at certain developmental periods. Zebra finches, for example, have inherited physical equipment for the learning process from their parents but when they are isolated from species members before they are 35 days old they can never learn to distinguish males and females of their own species. Other aspects of behavior such as calls of some birds are less learned and more inherited. An incubator-raised chicken, for example, having never heard the sound of a hen, will still mature and produce notes typical of other chickens. In other cases, it has been shown that the environment in which an animal matures can drastically affect certain aspects of its adult behavior. H. F. Harlow and his associates, for example, have shown that female Rhesus monkeys separated at birth from their own mothers and deprived of early interaction with other monkeys were deficient in basic patterns of maternal behavior, social play, and sexual activity. Again structural and physiological characteristics that provide the tangible framework for these behavior patterns must follow the **DNA blueprints** for the organism, but an important component of the behavior itself must be learned through **contact** with the environment.

Basically, then, it is **ranges of modifiability** that are inherited, with the segregation of genes and the forces of natural selection accounting for observable individual differences in behavior patterns within a given population.

GENETIC MECHANISMS

In contrast to the more common complex meshing of environment and heredity, the genetic mechanisms of some behavior traits have proved surprisingly simple. Some specific examples depend on a **few genes** and respond to a limited range of environmental stimuli.

W. C. Rothenbuhler, for example, has found evidence that an interesting behavior pattern in honey bees is controlled by two pairs of recessive genes in simple Mendelian fashion. Two different races of bees differ in their "hygienic" behavior. Worker bees from the Brown line, a hygienic race, open compartments in the hive that contain pupae dead from American foulbrood and remove the dead (Fig. 15.1a). Those from the Van Scoy line, a nonhygienic race, leave the dead pupae in the closed compartments and thus allow the infectious agent for American foulbrood *Bacillus larvae* to spread in the colony (Fig. 15.1b).

Rothenbuhler crossed the two races and obtained F_1 worker bees all of which were nonhygienic (Fig. 15.2). When F_1 drones were backcrossed to hygienic queens, four kinds of backcross colonies were obtained in about equal proportions: (1) hygienic bees, (2) bees that opened cells but did not remove the dead pupae, (3) bees that did not

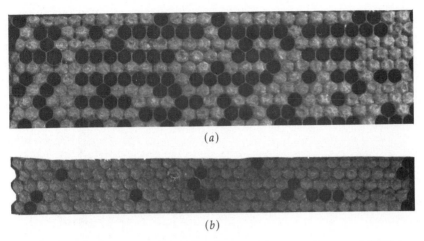

(a)

(b)

Figure 15.1 Combs of brood from (a) Brown, hygienic colony and (b) Van Scoy nonhygienic colony. In the Brown colony about two days before brood emergence many individuals were missing from the spore-inoculated rows but all brood remaining in the comb were found to be alive. In the Van Scoy colony most of the brood were present but when the cells were uncapped in the laboratory, many individuals in the spore inoculated rows were dead of American foulbrood. (From *American Zoologist,* 4:111–123, 1964. Photos courtesy of Walter C. Rothenbuhler.)

open cells but removed dead pupae when cells were opened by the beekeepers, and (4) nonhygienic bees. If a single recessive gene *u* controls the behavior pattern of uncapping cells and another single recessive gene *r* controls the behavior pattern for removal of dead pupae, the results obtained by Rothenbuhler can be explained by Mendelian independent combinations. The dihybrid cross is reconstructed in Fig. 15.3. Although many genes and environmental influences may be associated with the complicated neuronal mechanism underlying the behavior patterns of uncapping and removing, the response threshold is determined primarily by the **single alleles.**

In this example, two colony-behavior characteristics that appeared together in parental lines, separated in the backcross colonies. This manifestation of genetic segregation is expected among individuals, but not among colonies. It can occur only

when members of a colony are genetically similar. Colonial bees can be genetically similar when the mother (the queen) is highly inbred and has mated with a single haploid drone to produce the colony. An inbred queen—single drone mating performed by artificial insemination facilitates the study of colony behavior.

In the course of studying the 63 colonies of honey bees in their hygienic behavior experiment, Rothenbuhler and his associates made observations on several other behavioral characteristics of bees. Most obvious and impressive was stinging behavior. In the course of 98 visits to 7 Van Scoy colonies, the beekeepers were stung only once, whereas the same number of visits to 7 Brown colonies brought 143 stings. The first proposed explanation was that keeping the brood nest free of dead larvae and defending the colony against beekeepers were manifestations of the same general charac-

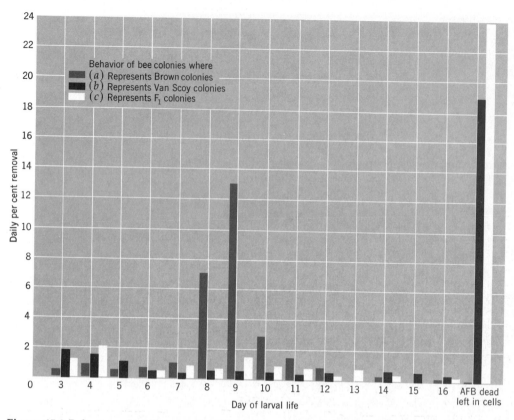

Figure 15.2 Behavior of (*a*) Brown, (*b*) Van Scoy and (*c*) F_1 colonies of bees with reference to hygienic measures for control of American foulbrood. (*a*) Brown colonies removed all American foulbrood, (*b*) Van Scoy colonies left most foulbrood-killed individuals in the comb and (*c*) F_1 colonies behaved like Van Scoy colonies. (Data from Walter C. Rothenbuhler.)

teristic—a high level of vigor in the worker bees. It was suggested that the same genes might explain the two behavior patterns. If this was the explanation, the hygienic colonies among the backcross colonies would also be the stingers. This was not the case. Only a very few of the bees in the hygienic backcross colonies were stingers. The stinging behavior in all of the 29 backcross colonies indicated that more than two pairs of genes were involved.

The genetic aspects of such observable elements of behavior are believed to be determined basically by the same mech-

anisms that function for the more tangible physical traits that Mendel and others have described. The genetic bases for behavior patterns in animals, however, are characteristically difficult to confirm through experimental procedures.

INHERITANCE AND LEARNING IN BEES

On the genetic side, W. P. Nye and O. Mackensen have shown that honeybee preference for alfalfa (*Medicago sativa*) pollen depends to a large extent on genetic de-

P   hygienic queen   nonhygienic drone
       (uncapping and
       removing)

$$uurr \qquad\qquad \times \qquad\qquad u^+r^+$$

F_1 nonhygienic

$$u^+ur^+r$$

The backcross of the four kinds of drones (u^+r^+, ur, u^+r, ur^+) with a queen for the hygienic race ($uurr$) resulted in the following:

B.C.	hygienic	uncapping no removal	no uncapping, removal	non-hygienic
	1:	1:	1:	1
	$uurr$	uur^+r	u^+urr	u^+ur^+r

Figure 15.3 Cross between hygienic and nonhygienic bees, and backcross of four kinds of drones with $uurr$ queens; u is the gene for uncapping, r is for removing.

terminers that respond to **selective breeding.** These investigators observed that some colonies of honeybees collected a much higher percentage of alfalfa pollen than did others. Separate inbred lines were developed from colonies showing either a high or a low preference for collecting alfalfa pollen. At the end of the eighth generation of inbreeding, the high and low lines had been completely separated (Fig. 15.4). Subsequent hybridization of the bees from the two lines produced bees that were intermediate between the high and low lines. Because the preference for alfalfa pollen can be changed markedly by selection, and the trait follows a predictable pattern based on Mendelian inheritance, preference for alfalfa pollen is presumed to have a high hereditary component.

Both inheritance and learning are involved in the complex social patterns of some Hymenoptera such as honeybees, that make use of chemical, optical, and aural signals. The effectiveness of the activities of bees in and around a hive as well as foraging areas, depends on the exchange of information among the individual bees. Communication symbols for **distance** and **direction** of food-source material are **mostly learned** by individuals. Even so, in some groups, this behavior is so stereotyped and well established that it can serve as well as a morphological characteristic for distinguishing species.

Although the basic communication system is the same for all *Apis mellifera* bees, different **dialects** have developed in different races. Members of an Italian race, for example, have a slower dancing rhythm than those of an Austrian race. When bees of these two races are mixed they misunderstand each other. An Austrian bee re-

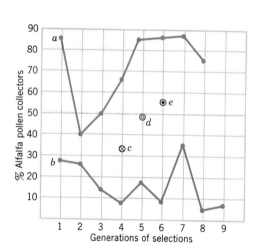

Figure 15.4 Results of selection experiments for high alfalfa pollen collectors (HAPC), low alfalfa pollen collectors (LAPC), and results from outcrosses and backcrosses. (*a*) HAPC queens and drones selected in each of 8 generations. (*b*) LAPC queens and drones selected in each of 9 generations. (*c*) Hybrid colonies resulting from crosses between LAPC queens and HAPC drones from generation 3. (*d*) Results from backcrosses between queens from *b* (hybrid colonies) and drones from generation 4 HAPC queens. (*e*) Results from outcrosses between HAPC queens from generation 5 and drones from a commercial stock. The dip from 85 percent to 40 percent between generation 1 and 2 in *a* is presumed to be the result of heterozygosity of the genes involved in the parents selected to initiate the HAPC line. Both HAPC and LAPC lines are expected to become more homozygous as inbreeding continues. Environmental as well as genetic factors are known to influence pollen collection. In the spring of the 7th year, frosts destroyed most of the early plants on which LAPC bees usually forage and they were forced to move to the more abundant alfalfa. (Data from William P. Nye.)

ceiving information from an Italian bee about food 100 meters from the nest will fly 120 meters because she interprets the "Italian dialect" on the basis of her Austrian knowledge. Conversely, the Italian bee will fly 80 meters when given the information for 100 meters by the Austrian forager.

Nevertheless, when Lindauer compared communication systems of three different species of Apis, he observed wider differences in behavior as well as in structural characteristics between the species than between races.

EXPERIMENTAL BEHAVIORAL GENETICS

Behavior genetics is a challenging field for experimental work. Several investigators, including T. Dobzhansky (Fig. 15.5), S. Benzer, M. Delbrück, J. Alder, C. Kung, M. Nirenberg, G. Stent, M. Levinthal, S. Brenner, and C. M. Woolf, who have distinguished themselves in other aspects of experimental genetics are now investigating behavior genetics. Each investigator has sought an ideal experimental material with properties such as large size, rapid reproduction, few neurons, and wide behavior patterns. Such organisms as *E. coli*, phycomycetes, Paramecium, nematodes, rotifers, fresh water crustacea (Daphnia), leaches, Drosophila and mice have been chosen for basic investigations on behavior. Examples from *E. coli*, Paramecium, and Drosophila will be cited here.

E. COLI CHEMOTAXIS

In the bacterium *E. coli*, behavior patterns are comparatively simple. **Responses** (taxis) to chemicals, light, gravity, and temperature have been studied by different investigators. J. Adler, for example, investigating chemotaxis, has shown that the mechanics include detection of the chemical (attractant) by a chemoreceptor and transmission to an effector which produces a flagellar response, the swimming of the organism (Fig. 15.6). By isolating mutants, genes responsible for each step in the process have been postulated. Three *che* mutants, for example, render the bacteria nonchemotactic to specific chemicals: galactose, aspartate, and serine. Eight *fla* (nonflagellar) genes interfere with swimming activity.

Movements of the bacteria are followed with a tracking microscope and the data are fed into a computer. When no chemical gradient is present in the medium, the movement is haphazard. A bacterium moves in one direction (run) for two to four seconds, tumbles around (twiddles), makes another short run, twiddles and so on without ap-

Figure 15.5 Theodosius Dobzhansky, eminent researcher, writer, and teacher in the fields of genetics and evolution. He has made basic contributions to population genetics as well as behavior genetics through his extensive studies on *Drosophila pseudoobscura.*

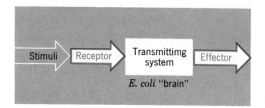

Figure 15.6 Receptor-effector system in *E. coli.*

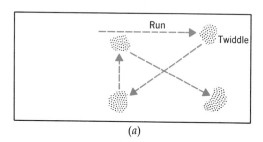

(*a*)

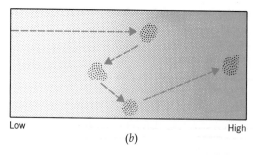

(*b*)

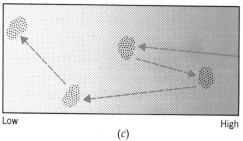

(*c*)

Figure 15.7 Movement of *E. coli* in a liquid medium. (*a*) No chemical gradient in medium, (*b*) gradient for attractant, (*c*) gradient for repellent.

preciably changing location (Fig. 15.7). If an attractant is present, the runs are longer in the direction of the gradient. At the end of a run the organism twiddles, takes a short run, and repeats the activity until the orientation is again in the direction of the gradient. At this time a long run occurs. If a repellent, instead of an attractant, is pres- ent, long runs, twiddles, and short runs occur but the direction is toward the low concentration of the chemical. Several theories have been advanced to explain the cause of twiddling: (1) a diffusible substance, (2) a change in membrane potential, (3) a change in membrane configuration, and (4) a gene-controlled enzyme that influences chemotaxis. If the fourth is the correct explanation, some **mutations** may be expected to block the twiddle and others to make bacteria twiddle all the time.

PARAMECIUM BEHAVIOR MUTANTS

The protozoan *P. aurelia*, a giant cell with a regulating, excitable membrane, has proved to be an excellent material for studies of Mendelian as well as cytoplasmic inheritance (Chapter 11). Its movements, accomplished by numerous cilia, are better coordinated than those of *E. coli*. *P. aurelia* moves in a helical direction. If it hits an object, it backs up and starts in a new direction.

Some 200 lines of behavioral mutants are available for study with several unlinked loci already identified. Among the behavior mutants are: fast 1, fast 2, paranoiacs (with a violent avoiding reaction), pawns (with a membrane mutation that makes it impossible to swim backwards, therefore without an avoiding reaction), jerkers (with high-frequency avoiding action), spinners (that turn violently), slows (that are slow swimmers), cilia reverse, chemotaxis, and galvanotaxis. Many **mutants** have a modified membrane physiology and are detected by studying the effects of potassium, sodium, barium, and calcium on the movement of the organisms.

DROSOPHILA BEHAVIOR GENETICS

Drosophila adults are much more complex than *E. coli* or Paramecium. The brain contains some 10^5 neurons arranged in nerve tracts. For such complex animals, Drosophila are small; they reproduce within a few days, live on simple food in the laboratory, and are well known genetically. Their chromosomes have already been extensively mapped and behavior genes have been detected in particular areas of the chromosomes. On the X chromosome of *D. melanogaster*, for example, S. Benzer and associates have located several behavior loci including visual receptors, stress receptors, and "wings-up" determiners with reference to other known genes. Developmental studies by these investigators have related cells in the blastocyst stage with imaginal disks that give rise to eyes, antennae, legs, and wings in the adult flies. Mosaics (gynandromorphs, Chapter 4) are being used to relate structural parts of the fly to cells in the blastoderm and to genes in the zygote. The wings-up mutation, for example, has been related to a developmental abnormality of the muscles. The "drop-dead" mutation is related to a defect in the brain. Sex and learning behavior are analyzed in component parts through mutant blocks and are then related to sequential steps in gene activity.

RESPONSE TO LIGHT AND GRAVITY IN DROSOPHILA

C. M. Woolf has carried out hybridization studies with strains of *D. pseudoobscura* that differ in phototactic (i.e., response to light) and geotactic (response to gravity) behavior. These behavior traits were measured by running virgin females and males separately through Hirsch-Hadler classification mazes (Fig. 15.8). Matings were then made among the flies that earned particular classifications, to obtain evidence with respect to **genetic mechanisms.**

Hirsch-Hadler mazes provide 15 downward or upward choices, that is, toward or away from light. Eventually a fly enters one of 16 different collecting tubes. Number 1 collecting tube is entered by flies making 15 choices upward or away from light. Number 16 collecting tube will be entered if the 15 opposite choices are made. If an

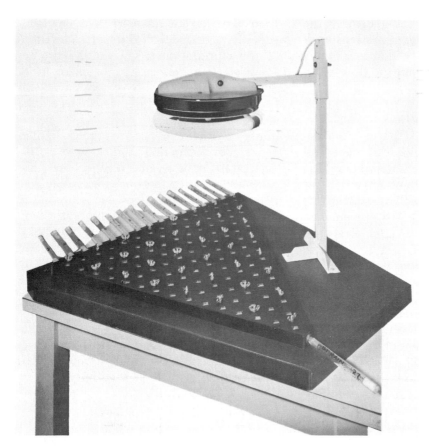

Figure 15.8 Hirsch-Hadler classification maze designed to select Drosophila for positive and negative geotactic and phototactic behavior. In each generation, virgin females and males are introduced separately at the maze entrance. The flies have a number of choices indicating positive or negative behavior in the maze. Those showing each response are collected at the exit and used for progenitors of the next generation. Beginning with strains that were geotactic and phototactic neutral, it was possible to obtain positive and negative strains after several generations of selection. (Courtesy of C. M. Woolf, Arizona State University.)

equal number of downward and upward choices are made, the fly enters number 8 or 9 collecting tube. The number of flies reaching each collecting tube can be used to calculate the reaction of particular flies to light. A completely neutral strain has an expected mean of 8.5. The strongest possible positive strain would have an expected mean of 16. The strongest possible negative strain would have an expected mean of 1.

The strains of flies used by Woolf had been built up through selection as a part of the extensive research program of T. Dobzhansky and his associates. Beginning with strains that were phototactic and geotactic neutral, the Dobzhansky group had selected flies with positive and negative response to light and gravity. Strain number 25, for

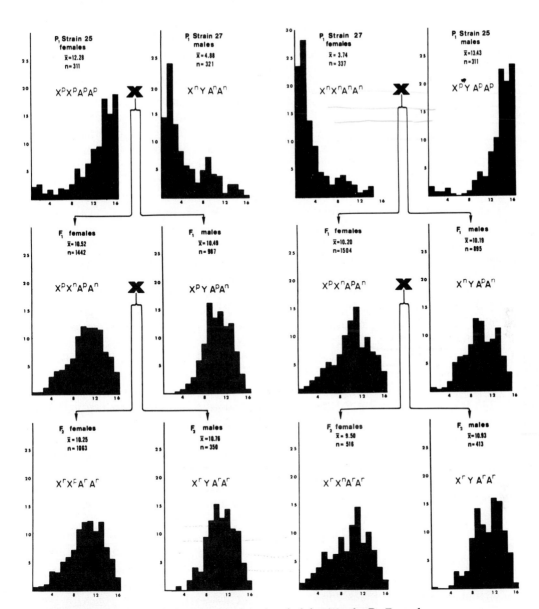

Figure 15.9 Distribution of phototactic scores (pooled data) in the P_1, F_1, and F_2 generations for the parental mating between (left), female strain 25 (positive) X male strain 27 (negative) and (right), female strain 27 (negative) X male strain 25 (positive). In this series of histograms, the area of each rectangle represents the percentage of flies occurring in each of the 16 collecting tubes of the Hirsch-Hadler classification maze. The X chromosome is symbolized by X, the Y chromosome by Y, and the set of autosomes by A. The p and n indicate whether the X chromosome and autosomes are from the positive or negative strain respectively. The r designates that the X chromosome and autosomes are recombinant chromosomes as a result of crossing over or segregation. (Courtesy of C. M. Woolf.)

example, had become strongly positive to phototaxis whereas strain 27 became strongly negative. Observations of this kind provided evidence for genetic variance in these strains of *D. pseudoobscura*. The rate of divergence under artificial selection indicated that the genetic component consisted of polygenes. When selection was relaxed, the positive and negative strains reverted to the neutral state, suggesting that the genes responsible for the positive and negative phototactic behavior were maintained in the heterozygous state in natural and laboratory populations by natural selection.

Woolf crossed females from strain 25 with males from strain 27 and also crossed females from 27 strain with males from strain 25. As indicated in Fig. 15.9, the 311 female flies from strain 25 had a mean of 12.28. The 321 male flies from strain 27 chosen for this same experiment had a mean of 4.88. The means of the F_1 progeny and of the F_2 (that were obtained from mating F_1 females with F_1 males) were intermediate between the parental strains. The particular arrangement of parental chromosomes in the F_1 and F_2 male and female progeny are indicated along with the histograms. Phototactic behavior scores are characterized as being relatively consistent from replication to replication and relatively similar in males and females of the F_1 and F_2 generations.

The position of the F_1 means between the values of the parental means indicates that **polygenic** inheritance (groups of genes controlling quantitative traits, Chapter 12) is involved. A consistent shift, however, was observed in the F_1 and F_2 means toward that of the parental positive strain (strain 25) suggesting that some type of nonadditive influence is present. The similarity of the means of F_1 males and F_1 females indicated that the X chromosome had little or no effect on phototactic behavior. Therefore, the responsible genes are largely located in the

autosomes. Further studies are in progress to localize the polygenes within the autosomes. The variances of the F_1 and F_2 scores are similar. This exception to classical polygenic inheritance, in which a wider spread is usually observed in the F_2 as compared with the F_1, apparently results from the **low heritability** (Chapter 14) of phototactic behavior in *D. pseudoobscura*, the lack of homozygosity of the parental strains and perhaps some unknown interaction between genotype and environment.

In other comparable experiments carried out in other types of mazes and designed to determine geotactic behavior, Woolf's results were quite different. As in the studies of phototactic behavior, the consistency of the replications indicated the importance of the genotype for these types of behavior under controlled laboratory conditions in spite of the low heritability. Female progeny in the F_1 generation were essentially intermediate between the two parents, but the F_1 males were strongly positive for geotactic behavior. This is the pattern of crisscross inheritance expected when the X chromosome carries the gene or genes involved in the transmission of the trait. This same pattern was held through the F_2, F_3, and backcross results, giving evidence that the **X chromosome** is strongly involved in the transmission of geotactic behavior. The evidence indicated further that the autosomes carried some genes for this trait but that the great proportion of genes for positive geotactic behavior were located in a region of the X chromosome not readily divisible by crossing over. Females homozygous or males hemizygous for this region tended to express positive geotactic behavior. Although geotactic behavior in Drosophila is not controlled by single major genes like that associated with hygienic behavior in bees, particular regions of DNA are involved in the mechanism of this behavior characteristic.

GENETIC AND ENVIRONMENTAL INTERACTIONS IN DOGS

Dogs, like human beings, have personality differences even within a breed. Some are timid and others are confident; some are gentle and others are aggressive. Those that are "socialized" early in life (allowed to interact with people) function as friendly and understanding companions of man, whereas others of the same breed (or even of the same litter) that are not given similar experiences while young may become fearful or even hostile toward people.

J. P. Scott and J. L. Fuller have made extensive observations on genetic and environmental factors involved in the building of **behavior patterns** in dogs. In some of their work, daily observations were initiated with newborn puppies of several diverse breeds and were continued until the dogs were 16 weeks of age. Every effort was made to observe the earliest manifestation of hereditary difference and to detect the effects of heredity before they could be contaminated by experience. During the first few days after birth, of course, there was very little behavior to observe in the pups.

As soon as recognizable behavior began to be apparent, interaction between hereditary and environmental influences was already present. The puppies changed markedly in reactions from day to day even though the genes in their chromosomes remained essentially constant throughout their life. Different determiners were evidently becoming active at the different **developmental stages.** Furthermore, the genes were acting on a very **different animal** at birth as compared with the same individual a few weeks later.

Results obtained by Scott and Fuller were quite unexpected and scientifically significant. During the very early stages of development when behavior is minimal, genetic difference had few opportunities to be expressed. When behavior patterns did appear, however, the evidence supported the conclusion that genetically determined differences in behavior do not appear all at once early in development, to be modified by later experience. Instead, they are themselves developed under the influence of environmental factors. Scott and Fuller concluded that the raising and lowering of response **thresholds** is one of the most important ways behavior in dogs is affected by heredity. The Scott and Fuller studies showed that while heredity is an important factor in dog behavior, details, and sometimes the actual appearance of specific patterns depend to varying degrees on the **individual's experience.** Furthermore, they demonstrated that at least some genetic difference in behavior can be measured and compared as validly as can hereditary physical differences.

Dog breeds generally have managed to retain a great deal of genetic flexibility despite man's intensive selection over time. This was borne out by the studies of Scott and Fuller in which 50 traits were examined in five pure breeds of dogs. Almost all the traits were significantly different between the breeds. But a very few of them were found to breed-true, as would be expected if they were controlled by single homozygous pairs of alleles. Additionally, lack of correlation was observed between behavior and phenotypic "type" within each breed. In their crossbreeding work with cocker spaniels and basenjis, the basenji personality was often seen in spaniel-appearing dogs and vice versa. Selection has, however, apparently produced near homozygosity for certain traits in particular breeds. Fighting behavior, for example, is almost nonexistent in the hound, but is well developed in the terrier. But

Figure 15.10 Golden retriever. This breed of dogs has been selected and trained for more than 100 years for the characteristics that make good retrievers. (Gordon Jenney/Black Star.)

such instances of near homozygosity are rare.

Through selection of genetic qualities and training, remarkable behavior patterns have been made available in some breeds of dogs. The basenji, for example, commonly known as the African nonbarking hound, is by nature a "scent" hunter. This dog is used in Africa to find and drive wild game. He is basically intelligent and can be taught such feats as the advanced American Kennel Club obedience program, which includes retrieving a dumbbell over a high jump. In this demonstration, the basenji's inherent intelligence is put to a relatively artificial use.

Retrieving dogs have been selected and trained for more than 100 years in England by enthusiasts in the waterfowling sport. Several breeds exhibited some of the characteristics necessary for an excellent retriever such as: strength, moderate size, endurance, enthusiasm, aquatic ability, keen scenting ability, courage, favorable temperament, and trainability. By intercrossing the most favorable breeds, Newfoundland, setter, and spaniel, and selecting progeny, a litter of four puppies with a favorable combination of traits was obtained in 1868. From this beginning the golden retriever stock was developed. These dogs (Fig. 15.10) are light yellow in coat color. They have aquatic ability, pleasing temperament, keen nose, tracking ability, and tenacity to retrieve under severe conditions. It is natural for a pup from this stock to want to retrieve. Training merely perfects and polishes the performance.

Breed differences in behavior are both real and important. The great variability available in dogs means that it is possible to modify a breed markedly within a few generations by careful selection. Through crossbreeding, entirely new and unique combinations of behavioral traits can be created and studied.

Human beings also show great variation in temperament and behavior as well as in structural features. Attempts to understand the bases for such differences in man led to much controversy. In the past, a rigid distinction was made between nature and nurture. Some human traits were considered to be hereditary and some acquired. More insight into the mechanisms actually involved has shown that all traits develop within limits set by genetic material under the influence of the environment. Both heredity and environment are thus involved in the development of any trait, but variations among individuals may depend more on one than on the other factor. The blood type expressed by an individual, for example, depends almost entirely on his genetic endowment. His ability to use language, however, depends on structural characteristics of his throat and mouth (developed according to genome information), combined with what he learns through experience in the environment. Other structural and physiological traits, such as body size and carbohydrate metabolism, are significantly and directly influenced by both genetic and environmental factors, and in turn **affect behavior.**

The **nervous system** and the **hormonal system** are of particular importance in setting behavior patterns, and both of these systems are ultimately dependent on genetic determiners. In higher animals, including man, hormones function in initiating and sustaining the reproductive drive, but are also important in determining reactions to stress. An individual's temperament or mood may be greatly influenced by the particular hormones that are circulating in his blood at a given time. Animals whose reproductive behavior is seasonal, experience periodic surges in the activity of certain hormones that render them sensitive to certain stimuli from the environment. At another time, the same stimuli might evoke no response on the part of the animal. In human beings, whose ability to reproduce is not seasonal, hormonal influence tends to be less obvious.

Many family studies have uncovered unit characters that are related indirectly to psychological patterns. By studying a family pedigree, it is sometimes possible to fit a hypothesis of dominant or recessive, single-gene inheritance to the pattern represented in the family. A progressive dementia called Huntington's chorea, for example, follows the pattern of a single dominant autosomal gene. Certain sensory anomalies, such as defective color vision and taste blindness are also controlled by single gene substitutions. The pedigree method is useful for determining the genetic patterns of conspicuous and rare traits that are determined by single genes and are not limited in expression by the environment.

Quantitative traits generally are influenced by the environment as well as by inheritance. Elaborate statistical methods have been used to control environmental influences in some experiments and thus to determine the degree of heritability of a trait, but these are extremely difficult to impose in human studies. The twin study method has become increasingly useful as a way to control the genetic component and thus to **compare environmental influences.** Because members of monozygotic or identical twin pairs have the same genotype, except for rare mutations and chromosome alterations, essentially all variation is environmental. Comparison of intrapair differences in monozygotic and dizygotic twins is a useful basis for evaluating the degree of inheritance of a particular trait.

MODEL FOR STUTTERING

Some human traits such as cleft lip and cleft palate, pyloric stenosis, and schizophrenia do not follow a Mendelian pattern and cannot be analyzed directly in terms of heritability because the environmental influence is not known. Since particular genes cannot be identified, Cavalli and others have developed models to account for the apparent inheritance patterns.

Stuttering, for example, does not show any obvious Mendelian behavior, but it does show a strong familial concentration. It occurs in males three to four times more frequently than in females. Female stutterers, however, have more affected relatives than do male stutterers. When models were prepared to account for the data, a single major locus model accounted more accurately for the data than a multifactorial model. Neither genetic model could be excluded, nor could the analyses prove the existence of a single major locus. This model, however, fits almost exactly the observed data ($\chi^2 = 2.68$).

This treatment of the data shows a large dominance variance associated with female to female relationships but not with male to male relationships. Stuttering is best explained by a single locus with two alleles. The predicted interaction of environmental factors with the genotype is as follows: homozygotes for the normal allele never stutter; homozygotes for the "stuttering allele" almost always stutter; about 25 percent of the male heterozygotes stutter, though only about 4 percent of the female heterozygotes stutter. Environmental factors thus are presumed to affect only heterozygotes, though differently in the two sexes.

Students of human behavior genetics are concerned with problems of social importance (such as intelligence and psychoses) but they are restricted in their experimental approach. Because human psychological traits have not yet been reduced to quantitative mechanisms, however, the investigator must either settle for data from animal studies, or confine his efforts to studies of inheritance of mental dysfunction and the determination of heritability of quantitative traits such as intelligence and personality.

GROUP BEHAVIOR PHENOMENA

Groups of animals and groups of human beings have behavior characteristics in common that apparently involve biological mechanisms with selective value. Adherence to a particular social order has value when it provides that the group may act as one in behalf of the individual. On the other hand, without a certain amount of social disorder or freedom, the strengths inherent in diversity are lost. A dynamic balance between order and disorder is a healthy condition within living populations.

One aspect of "order" requires respect for one another's "distance" needs. Different species have different individual distance requirements and differ in their **responses to crowding.** Black-headed gulls resting in a row will space themselves at about one foot apart. Flamingos maintain about twice that much space between neighbors, and swallows require about half as much space as the gulls. Some tortoises and hedgehogs will crowd together and make an animal pile. Inherent species differences control such behavior patterns. **Individual distance** or "personal space" requirements, however, may vary with individuals and with seasons. For example, virtually all bird species tend to gather in tight flocks during the winter, but as soon as the breeding season comes in the spring they disperse and strenuously defend their individual territories. **Social distance** is the farthest point an animal will stray from the group.

A baboon troop is dispersed widely while feeding in the daytime but at night the group will come together and sleep in a few adjacent trees. In this case the acceptable "social distance" varies with environmental conditions.

Overcrowding, whether among human beings or other animals, can lead to a breakdown in social structure and open the gates to overt aggressiveness. Innate aggressiveness, which has been necessary throughout evolution to help each species to survive, can take a destructive turn under conditions of overcrowding. In some cases the aggression is directed towards others, but it can also become manifest in ulcers, nervous disorders, and various psychosomatic maladies.

Fortunately, aggression, like many other behavior patterns, is subject to modification—although only within genetic bounds. **Aggression** can be increased or decreased by purposeful education. If mice or chickens are paired with others of their kind, fighting ensues and one becomes dominant and the other subordinate. If these encounters are arranged so that a particular animal always wins, that animal becomes more and more emphatically dominant over his fellows. In psychological terms, winning reinforces further aggression. On the other hand, repeated losses make an animal submissive.

One can also train animals to be nonaggressive in other ways. Scott prevented puppies from making playful attacks on their handlers by picking them up frequently, thus rendering them helpless with their feet off the ground. This process of "passive inhibition" produced nonaggressive adults. He obtained similar results with young male mice by repeatedly stroking them at an early age. Aggressiveness among nonhumans is a product of inheritance, maturation, various endogenous factors, and experience. Manifestations of aggression among nonhumans,

however, depend upon the presentation of proper external stimuli, usually specific **sign stimuli** from other individuals of the same species.

THE EVOLUTIONARY APPROACH

In trying to trace the evolution of behavior in man, researchers use the same general approaches that have proved productive with other species. It has been found (predictably) that some behavior patterns that serve the function of food intake are phylogenetically quite old—and the human infant shares them with many other mammals. One example is the rhythmic searching for the nipple.

The grasping reflex is also characteristic, perhaps originally serving to keep the baby attached to the mother. In premature babies it is especially strong and they are able to cling to an outstretched cord. Climbing movements can also be elicited in premature infants.

Observation of congenitally deaf-blind children (in effect, deprivation experiments) are producing insights into which human behavior patterns are truly unlearned. Expressions of anger (stamping of feet, facial contortions), high-intensity laughter, and the rejection of strangers seem innate and, therefore, probably have a genetic base. Similarly, components of facial expressions—such as raising the eyebrows when greeting someone—are common to all cultures that have been studied.

Inborn releasing mechanisms also seem to account for a remarkably universal desire for cover and unobstructed vision—as evidenced in modernday life by a preference for corner and wall tables in restaurants.

Culture imposes specific limits on individual-distance requirements, but the basic need for space seems inborn and probably evolved over time.

The tendency to seek membership in some sort of group and to accept the group-dictated exclusion of others also seem to be universal human behaviorisms—and are therefore likely evolutionary phenomena.

Obviously, human beings have many kinds of **behavior in common**—despite a heavy veneer of individually learned, culturally dictated modifications. It has also been proved (e.g., as in Scott and Fuller's work with puppies) that experiments with other animals can provide valid insights into human behavior. So it seems reasonable to expect that, as more is learned not only about behavior as it exists today but also about how it got that way, that the future evolution of man may be modified in the direction of true humaneness.

REFERENCES

Ardrey, R. 1970. *The social contact.* Atheneum, New York.

Benzer, S. 1973. "Genetic dissection of behavior." *Sci. Amer.* 229(6): 24–37.

Dobzhansky, T., B. Spassky, and F. Sved. 1969. "Effects of selection and migration on geotactic and phototactic behavior of Drosophila." II. *Proc. Roy. Soc.* B. 173, 191–207.

Ehrman, L., G. S. Omen, and E. Caspari. 1972. *Genetics, environment and behavior.* Academic Press, New York.

Eibl-Eiberfeldt, I. 1970. *Ethology, the biology of behavior.* Holt, Rinehart and Winston, New York.

Fuller, J. L. and M. W. Fox. 1968. *The behavior of domestic animals.* E. S. E. Hafex (ed.) Wm. Wilkins Co., Baltimore, Md.

Glass, D. C. (ed.) 1968. *Biology and behavior.* Rockefeller University and Russell Sage Found., New York.

Hirsch, J. (ed.) 1967. *Behavior-genetic analysis.* McGraw-Hill, New York.

McClearn, G. E. 1970. "Behavioral genetics." *Ann. Rev. Genet.* 437–468. H. L. Roman (ed.) Annual Reviews Inc., Palo Alto, Calif.

McGill, T. E. (ed.) 1973. *Readings in animal behavior,* 2nd ed. Holt, Rinehart and Winston, New York.

Nye, W. P. and O. Mackensen. 1970. "Selective breeding of honeybees for alfalfa pollen collection: with tests on high and low alfalfa collection regions." *J. Apicultural Res.* 9, 61–64.

Porter, J. H. and R. G. Skalko. 1973. *Heredity and society.* Academic Press, New York.

Rothenbuhler, W. C. 1968. "Bee genetics." *Ann. Rev. Genet.* 2, 413–437.

Scott, J. P. and J. L. Fuller. 1965. *Genetics and social behavior of the dog.* University of Chicago Press, Chicago.

PROBLEMS AND QUESTIONS

15.1 How is animal behavior related to genetics?

15.2 Why has the genetics of behavior patterns developed more slowly than the genetics of other characteristics such as size and color patterns?

15.3 Why do studies such as Rothenbuhler's hygienic and non-hygienic bees have particular significance in behavior genetics?

15.4 Why are comparative studies of behavior patterns especially useful for investigating the evolution of behavior?

15.5 What evidence suggests a genetic basis for the preference of some bees for alfalfa pollen?

15.6 (a) How may hormones influence behavior? (b) How and to what extent is hormone production under genetic control?

15.7 What conclusions may be drawn from the studies of Scott and Fuller on dogs concerning the relative influence of heredity and environment on behavior?

15.8 How have adjustments been made in Drosophila strains to compensate for low level courtship behavior in males and perpetuate a gene such as y for yellow body color?

15.9 Why are dogs highly reactive to selection?

15.10 Why was the dog one of the first animals to be domesticated and to come into close association with man?

15.11 How could natural or human selection account for non-barking in basenji dogs even though they are capable of barking?

15.12 What are potential dangers in trying to transfer behavioral insights that are gained with laboratory or wild populations of animals to man (or even between populations of people)?

15.13 If a behavioral trait has demonstrable survival value, is it more likely to be genetically controlled or a learned phenomenon? Why?

15.14 How does genetics modify behavioral potentials of quarter horses and thoroughbreds?

15.15 If two strains of rats seemed to differ in their ability to solve maze-running problems, what sorts of environmental factors should be considered before concluding that the difference was genetically controlled (whether through general physiology or brain capacity)?

15.16 How may aggression be reduced in animal and human populations?

15.17 Why are Drosophila more suitable than mice as experimental material for maze studies designed to determine genetic mechanisms?

15.18 How could it be determined whether sex-linked polygenes control geotactic behavior in Drosophila?

SIXTEEN

Human genetics is a highly significant area for genetic research and for potential as well as immediate applications of known genetic principles. As our understanding of the nature of genes and our control of gene activity increase, possibilities for improving the lot of man in his environment may become realities. At least the symptoms of many hereditary diseases and deficiencies may be alleviated or overcome after we thoroughly understand their biochemical cause and effect. Changes in genes themselves may eventually be accomplished in individuals, and changes in gene frequencies may be achieved in populations. These possibilities engender both enthusiasm for the prospects of such manipulations and fear that the wisdom of man and quality of governance in human populations may be inadequate when it comes to deciding what should be done and who should do it.

Five immediate problems face those concerned with the broad aspects of human genetics. (1) Do the basic principles of genetics (determined largely from experimental studies on microorganisms, plants, and animals) apply to

PRINCIPLES OF GENETICS APPLIED TO MAN

Human embryo 39 days old. ($3\frac{1}{2}\times$). (By courtesy of Carnegie Institute of Washington.)

man? This has been largely answered in the affirmative. Basic principles, as outlined in Chapter 1, apparently do apply to man. Indeed, examples from human genetics have illustrated principles in the preceding chapters of this book. (2) How freely may a physician or a genetic counselor apply the principles of genetics in preventing, diagnosing, and treating illness? (3) Who, if anyone, is accountable for the quality of the germ plasm going to future generations? (4) What moral and legal considerations must attend the genetic engineering that might be accomplished by applying expanding technical knowledge and skill? The potentials may seem both distant and nebulous, but they illustrate the depth and significance of already accomplished genetic research. The latter three questions will require the best thought and judgment that can come from all segments of human society. (5) Should the optimum quantity as well as quality of the human population be determined by genetic scientists? Any projections for mankind's future depend on the size as well as the genetic nature of the population.

The worldwide population increase has crowded other human problems into the background. Predictions made in the 1940s indicate that the population in the United States would be 150 million by about 1968 and about 153 million by about 1980. Population analysts of the 1940s were predicting that the population of the United States would stabilize shortly after the year 2000 and that it would remain more or less stationary after that time.

Actually, the population was approaching 200 million in the late 1960s. What happened? A tremendous increase in birth rate occurred after World War II. An increase was expected to make up for delays in reproduction during the war years, but its extent was not foreseen. Several factors now seem to have been involved.

1. A large deficit of marriages had accumulated by the end of World War II because of postponement during the war and its preceding depression years. Following the war, this deficit was rapidly reduced.

2. Many married couples postponed having children during the war years. The birth rate increased greatly after 1946.

3. The median age at first marriage declined from 21.5 years in 1940 to 20.3 in 1950 and 1960.

4. A change in spacing of children by married couples resulted in larger families at an earlier age of the mother. Bottle feeding of babies replacing breast feeding was a factor in the changed spacing.

The population increase in the United States has fluctuated considerably in the years since 1960 but the trend of the curve is distinctly downward. In 1973 the rate of increase was near the replacement level. This level is expected to be two per woman (couple) but considering the number of women who do not marry and the married couples who do not have children, the replacement level in the United States is about 2.4 per woman. The population increase, however, has developed its own momentum which makes it difficult to slow down. Even if immigration from abroad ceased and couples had only enough children to replace themselves, the U.S. population would continue to grow for about 70 years. The past rapid growth equates with many young couples today. To bring population growth to an immediate halt, therefore, the birth rate would have to drop by about 50 percent. Couples, on the average, would be limited to one child.

Throughout the entire world the population continues to rise at a rapid rate, while land and water limitations suggest that this trend cannot persist indefinitely without

TABLE 16.1
Growth Rate of World Population
(From John D. Durand, "The Modern
Expansion of World Problems,"
Proceedings of the American
Philosophical Society, **Vol. III, No. 3,**
1967)

World Population (Millions)		Average Annual Growth Rate (Percent)	
1750	791	1750–1800	0.4
1800	978	1800–1850	0.5
1850	1,262	1850–1900	0.5
1900	1,650	1900–1950	0.8
1950	2,515	1950–1965	1.8
1965	3,281		

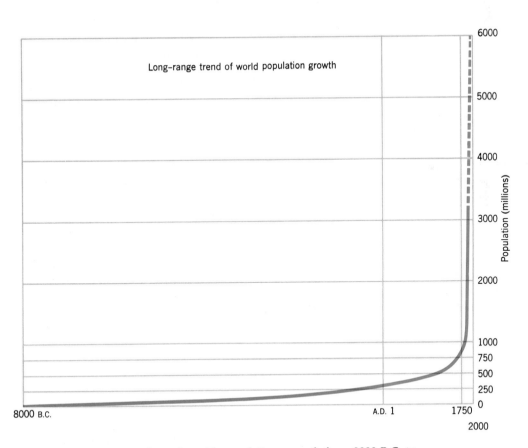

Figure 16.1 Long-range view of world population growth from 8000 B.C. to 2000 A.D. (John D. Durand, "The Modern Expansion of World Problems," *Proceeding of the American Philosophical Society*, Vol. III, No. 3, (1967): chart on page 139.)

dire results. Large groups of people already lack the basic necessities for life. Average annual growth rate has increased from 0.4 percent in the period 1750–1800 to 1.8 percent in the period 1950–1965 (Table 16.1). The world population is approaching four billion and according to estimates of the United Nations it is increasing at the rate of some 50 million per year.

The entire history of man on earth up to about 1830 had produced a population of one billion (Fig. 16.1). By 1930 the population doubled to about two billion. The third billion was added in about 30 years (1930–1960), and the fourth billion in 15 years (1960–1975). United Nations' statisticians estimate that, at the present rate, 4.47 billion will be reached by 1980. If the present trend continues, more than six billion people will be living on the earth before the end of the twentieth century. Opinions differ concerning the possibilities

for meeting the needs of so many people. But in a world with finite space and other essential resources, it seems inevitable that a time will come when no more people can be supported.

Even now many of the people on earth are hungry. A high proportion are poorly clothed and sheltered. Energy sources are in short supply. Large numbers of people in many countries are poorly educated. Moreover, in the increasingly complex culture of the most advanced countries, proportionately more adults are needed to properly educate the children and to provide adequately for the young and old. Already there is overcrowding in many places. The birth rate must be controlled and the world living standard increased if the future of man on earth is to be viable. The science of genetics may function actively in efforts to achieve that future.

HISTORICAL OBSERVATIONS OF HUMAN HEREDITY

Greek philosophers were keen in discerning similarities among related people. Hippocrates (460–370 B.C.), the father of medicine, called attention to the recurrence of human traits such as crossed eyes and baldheadedness in certain family groups. He also observed that certain disorders such as epilepsy and a particular eye disease causing blindness in older people occurred in some families and not in others. Disease syndromes were recognized by the Greeks, and the idea of a physical constitution providing immunity or susceptibility for disease became well established.

These conceptions of the Greeks were lost in the middle ages, however, and were not restored until modern times. Pierre Maupertuis, an eighteenth-century biologist, was one of the first to revive interest in human genetics. He collected pedigrees

of families in which polydactyly and albinism occurred and analyzed the results by applying the theory of probability. Because the requisite scientific background had not yet been established, the significance of his studies was not appreciated. In the latter part of the nineteenth century, Sir Francis Galton applied more sophisticated statistical tools to human problems and directed attention to the social implications of human genetics. Nevertheless, human genetics developed slowly until after the discovery of Mendelian inheritance, when it was soon shown that Mendel's principles were at least generally applicable to man.

During the first decade of the present century, considerable interest was shown in human genetics. Many traits were known to be more prevalent in certain families than in the general population. Some were initially explained in simple Mendelian terms. Feeblemindedness, for example, was considered to be a single entity and was interpreted on the basis of a single gene substitution. Now it is known that only a few conditions that result in feeblemindedness depend on single gene substitutions, while most other types are genetically and environmentally complex. Critical analyses have indicated that most of the early explanations for human traits were oversimplified.

After the first rush of interest in the early 1900s, human studies became rare because they were too imprecise. It was easier and more productive to apply the experimental method to other animals and to plants. Controlled matings among experimental organisms were preferred over uncertain human studies as the way to establish basic genetic principles. Fortunately, experiments with organisms that were more amenable to the experimental approach demonstrated apparently universal applicability of the basic principles of genetics and thereby indirectly furthered human genetics.

MAN AS AN OBJECT FOR STUDY

Many types of experiments simply cannot be performed on man, and objective data are elusive even when experiments are possible. Ideally, genetic investigations maintain standardized individuals in a controlled environment. Man is far from standardized genotypically and, to a large extent, he regulates his own environment. Circumstances at home and school, nutritional status, and numerous other factors influence the development of a child. It is difficult, therefore, to find anything in human society that resembles the ideal material required for objective experimental genetics. Identical twins and other types of multiple birth provide the only human units that approach a genotype standard. They occur infrequently, and usually the members of pairs or groups are raised in the same home, subject to essentially the same environmental conditions. When identical twins are separated early in life and experience different environmental situations, it is possible to compare the effects of different environments on similar genotypes.

The problem of human genetics has been further complicated by the long period (on the average, about 30 years) between generations. The life span of the investigator is no greater than that of his material for study. Only recently have methods been developed that allow the gathering of significant statistical data from one or two generations. Furthermore, large families are desirable for genetic studies, and even the largest human families fall short of the size necessary to establish genetic ratios and pursue orthodox statistical analyses. New statistical tools, however, now permit analysis of data from small families.

Incomplete knowledge of human cytology also has been limiting, although much is now known about the egg and sperm (Fig. 4.8) and their contents. Human chromosomes were observed several years ago and the sex chromosomes identified, but little other basic cytological knowledge was available until recently (Chapters 9 and 10). Present interest in counting human chromosomes, and the increased availability of human material for cytological studies, have greatly stimulated chromosome studies in man. At present, human chromosomes at mitotic metaphase in cultures are probably better understood than those of any other mammal. Human cytogenetics is now an active and progressive field of research.

Other difficulties encountered by the human geneticist have been associated with the genetic mechanism itself. The emphasis of necessity has been on phenotypes rather than genotypes. Obviously, a true genetic picture must be obtained before gene behavior can be properly analyzed and evaluated. Hereditary deafness, for example, may result from any one of several abnormalities in the ear, nerve tracts, or brain, with each abnormality controlled by a different combination of genes. On the other hand, different manifestations of a single gene may result in strikingly different phenotypes. These difficulties are being overcome by more precise information concerning gene action.

Many human traits result from the cumulative action of polygenes, (Chapter 12), which undergo segregation and crossing over but cannot be identified or localized individually. They resemble Mendelian genes in transmission but are dissimilar in action. Their individual effects are slight but cumulative. They play an important part in selection mainly because of their relatively small effect on viability. Most of the great variation known to occur in man is probably based on numerous polygene differences which cannot be studied individually.

Statistical approaches and refinements of twin studies perhaps hold the most promise for the immediate future. The application of statistical methods to human genetics in the last century made this discipline a quantitative science. Although statistical approaches are more comprehensive than Mendelian analyses and cover all variations of the trait, the individual units cannot be considered separately and are therefore lost in the analyses. The average effect of all genes acting in the group can be demonstrated by such methods, but not the effect of single genes.

Nevertheless, the classical analysis of single gene effects is still a valid and useful approach to human genetics. In his catalogue of human traits, McKusick (1971) has listed 944 phenotypes controlled by single dominant genes, 789 by recessives, and 149 by X-linked genes. With 1882 traits known to be controlled by single genes, conventional pedigree analyses are by no means obsolete. Modern computerized experiments can now help determine genetic mechanisms from family history data.

GENES IN MAN

As shown in Chapter 6, blood types can be genetically different and physiologically incompatible as compared with other types. Thousands of blood transfusions are now given each year, however, with comparatively few difficulties because of correct type matching. Blood banks have greatly facilitated the availability of the right kind of whole "typed" blood for patients requiring transfusions. In addition, blood plasma, without the corpuscles and their antigens, can be pooled with reduced anti-A and anti-B titers. Normal human plasma can be given without cross-matching. It can be transfused for immediate treatment of shock on the battle field or at the scene of an accident.

Blood groups are remarkably stable and free from any obscuring effects of age and from the influence of other genes in the body. On rare occasions, they are modified by disease. Several patients who previously had responded strongly to anti-A serum, had a weak or entirely absent antigenic reaction after the onset of leukemia. A weak anti-B response has sometimes been

TABLE 16.2

Genetic Map of Autosomes Giving Confirmed and Provisional Linkages (Based on the New Haven Conference, 1973. Details, References and Descriptions of Phenotypes Are Given in *Mendelian Inheritance in Man,* **3rd edition, by V. A. McKusick)**

A.   *Chromosomal assignments*

No. 1.
Adenylate kinase-2
Amylase, pancreatic
Amylase, salivary
Auriculo-osteodysplasia
Cataract, zonular pulverulent
*Colonic polyposis
*Duffy blood group
Elliptocytosis-1
Fumarate hydrate
Guanylate kinase
ᴾPeptidase C
ᴾPhosphoglucomutase-1
ᴾ6-phosphogluconate dehydrogenase
ᴾPhosphopyruvate hydratase
Rh blood group
5s RNA gene(s)
Uridyl diphosphate glucose pyrophosphorylase

No. 2.
ᴾAcid phosphatase
Galactose-1-phosphate uridyltransferase
ᵠHemoglobin alpha or beta
Interferon-1
Isocitrate dehydrogenase-1
Malate dehydrogenase-1
ᴾMNSs blood group

No. 3.
(no assignment)

No. 4 or 5.
Adenine B
Hemoglobin alpha or beta

No. 5.
Hexosaminidase B
Interferon-2

No. 6.
*Gm
*ᴾHagerman factor
HL-A histocompatibility region
Indophenoloxidase-B
Malic enzyme-1
Phosphoglucomutase-3

Table 16.2
Chromosomal assignments (Continued)

No. 7.
Hexosaminidase A
Mannosephosphate isomerase
Pyruvate kinase-3

No. 8.
(no assignment)

No. 9.
(no assignment)

No. 10.
Glutamate oxalacetic transaminase
Hexose kinase

No. 11.
Acid phosphatase, lysosomal
Esterase-A4
Killer antigen
Lactate dehydrogenase A

No. 12.
Citrate synthetase, mitochondrial
[p]Lactate dehydrogenase B
[*]Lactate dehydrogenase C
[q]Peptidase B
Serine hydroxymethylase
Triosephosphate isomerase

No. 13.
[q*]Retinoblastoma
[p]Ribosomal RNA

No. 14.
Nucleoside phosphorylase
[p]Ribosomal RNA

No. 15.
[*]ABO
[p]Ribosomal RNA

No. 16.
Adenine phosphoribosyltransferase
[*q]Haptoglobin, alpha

No. 17.
[p]Thymidine kinase

No. 18.
[*]IgA
Peptidase A

Table 16.2 (Continued)

No. 19.
Polio sensitivity
Phosphohexose isomerase

No. 20.
Adenosine deaminase, red cell

No. 21.
Anti-viral protein
Indophenoloxidase-A
Lipoprotein-Ag
^{q}Ribosomal RNA

No. 22.
*Pycnodysostosis
^{q}Ribosomal RNA

B. *Loci on the same chromosome but specific chromosome not yet verified.*
   a. Lutheran (*Lu*) blood group locus, secretor (*Se*) locus and myotonic dystrophy (*Dm*) locus
   b. ABO blood group locus, nail-patella (*Np*) locus, adenylate kinase (*AK*) locus, *Xeroderma pigmentosum
   c. Beta, delta and gamma hemoglobin loci
   d. *Am₂* immunoglobulin locus, *Gm* immunoglobulin region
   e. Transferrin (*Tf*) locus and pseudocholinesterase₁ (*E₁*) locus
   f. Albumin (*Alb*) locus and group-specific component (*Gc*) locus
   g. Pelger-Huët locus and unusual muscular dystrophy locus
   h. *MNSs* blood group locus and sclerotylosis (*Tys*) locus
   i. *HLA* region = HLA_{LA} and *HLA4*, phosphoglucomutase₃ (*PGM₃*) locus, adenosine deaminase locus, P blood group locus
   j. PTC (phenylthiocarbamide) taste and Kell blood group
   k. *GPT* (glutamate pyruvate transaminase) and epidermolysis bullose, Ogna type
   l. Lewis blood group and complement component-3
   m. *Ade B* (human complement for hamster adenine auxotroph B) and esterase activator gene
   n. IDH-1 and MDH-1
   o. Dombrock blood group and *MNSs* blood group

---

^{p}short arm
^{q}long arm
*not proved

---

seen in patients who should have tested as anti-A. Such cases prove that an antigenic change, ordinarily inherited, can be acquired. Nevertheless, blood serological traits can be used effectively to identify individuals and populations.

During World War II and since, the rapid expansion of blood transfusion services brought technical improvements in processing and increased numbers of blood samples for examination. With the increased volume of sampling and cross-matching, new alleles and new relations among existing alleles were discovered. In addition, several entirely new blood group systems were identified. Now some 14 systems including more than 60 blood group genes (Table 16.2) are known in man, and other more or less related systems are known in nonhuman organisms. Single alleles in the ABO system control the antibodies in the serum as well as the antigenic properties of the erythrocytes. Other tissues and body fluids besides the blood carry antigens. A "secretor" gene *Se* not linked to the ABO locus, permits the secretion of *A* and *B* antigens into the saliva and other body fluids such as tears, urine, semen, gastric juice, and milk.

SERUM PROTEINS

In addition to the genetic variations in blood group antigens, many inherited variations have been detected in serum or plasma proteins. Of the serum proteins, we will consider three types of particular genetic interest: (1) transferrins, (2) haptoglobins, and (3) immunoglobins. Each of these types is determined by a system of alleles at a single locus.

(1) Transferrins are beta globulins. They transport plasma iron to the bone marrow and tissue storage areas, and bind this metal in compounds such as hemoglobin, myoglobin, cytochrome, and several important enzymes. Transferrins can be detected by electrophoresis and are identified with

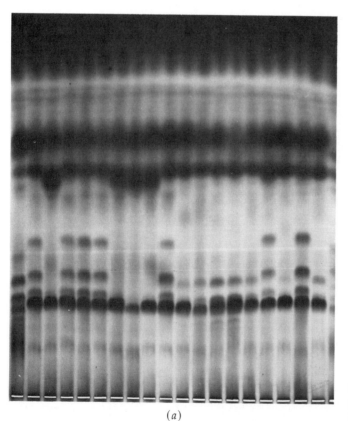

(*a*)

Figure 16.2 Electrophoresis experiments illustrating the separation of patterns of the three common haptoglobin types, Hp I-I, Hp 2-1, and Hp 2-2. (*a*) Photograph of a starch stained with amido-black 10B. (*b*) Photograph of a slice of the same gel stained (for hemoglobin) with benzidine and hydrogen peroxide showing examples of the pattern for the three common haptoglobin types Hp 2-1, Hp 1-1 and Hp 2-2. (*c*) Diagram of three tubes showing separation patterns for Hp 1-1, Hp 2-1 and Hp 2-2.

letters of the alphabet. The common form is designated as C. The 14 other transferrins that have been discovered are comparatively rare. All known transferrins seem equally able to bind and transport iron. The gene *Tf^c* has been associated with the C transferrin. Individuals homozygous for this gene (*Tf^cTf^c*) have a single electrophoretic band. The heterozygotes have two bands.

(2) Haptoglobins are alpha globulins that remove free hemoglobin (released when red blood cells are destroyed) from the blood

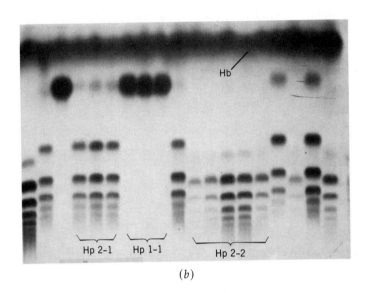

(b)

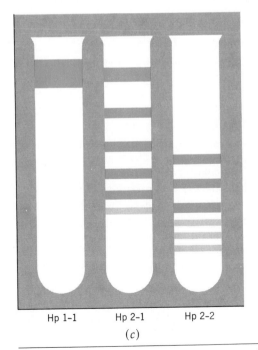

Hp 1-1   Hp 2-1   Hp 2-2

(c)

serum. O. Smithies separated three main types of haptoglobins, Hp1-1, Hp2-1, and Hp2-2, by starch-gel electrophoresis as shown diagrammatically in Fig. 16.2. The three types also differ from each other in the number of protein components present. In type Hp1-1, a single haptoglobin component was recognized. A component with the same mobility as that of Hp1-1 was detected in type Hp2-1 but not in 2-2. Both 2-1 and 2-2 types have several other components that migrate more slowly than that of 1-1. Components in type 2-1 differ in mobility from those in type 2-2.

Surveys of European populations showed that about 16 percent of the people were type 1-1, about 48 percent were type 2-1, and about 36 percent were type 2-2. Codominant, autosomal alleles, Hp^1 and Hp^2 were postulated by Smithies and others to account for the three phenotypes. Hp1-1 was attributed to the homozygous allele Hp^1 (Hp^1/Hp^1), Hp2-1 to the heterozygote (Hp^1/Hp^2), and Hp2-2 to the homozygous allele Hp^2 (Hp^2/Hp^2).

(3) Immunoglobulins are gamma globulins. They have been given phenotypic designations such as Gm^a and Gm^b and associated with a series of Gm alleles, more than 10 of which have been identified. These alleles are polymorphic in human populations and are thus useful in tracing the ancestral history of isolates and races of man. In one sample of 1284 Danish people, for example, 56 percent were Gm^a. Another phenotype, called "Gm-like," which is presumably produced by another Gm allele, does not occur in Caucasians but is found in Negroes. Gm alleles have been found in varying frequencies in different ethnic groups.

THE HUMAN GENE MAP

Much has been accomplished in mapping genes on human autosomes since the early 1970s. All human chromosomes were identified cytologically at the Paris conference in 1971 (Chapter 9). With the coming of somatic cell hybridization (Chapter 5), DNA-RNA hybridization (Chapter 7), and deductions from amino acid sequences in proteins (Chapter 3), many gene loci were placed on their chromosomes. Those either confirmed or provisionally accepted at the time of this writing are listed in Table 16.2.

EFFECT OF RECESSIVENESS AND DOMINANCE

If a recessive gene is rare in a population, it comes to expression very infrequently. In a series of albinos, for example, it could easily happen that no albinos would appear among the parents, grandparents, children, grandchildren, uncles, aunts and cousins. The resemblance of all these relatives to the albino subjects is zero. But the lack of resemblance is not genetically true. All the parents (Cc) of an albino child have one gene for albinism and are half like the subjects, who have two of these genes (cc). The genetic resemblances are simply hidden from expression by dominance.

With sibs the situation is different. However rare the recessive gene, one quarter of the sibs of an albino are expected to be albinos. In terms of genes, on the average, one out of every four sibs is an albino (cc) (wholly like the subject); one is a normal homozygote (CC), (wholly unlike the subject); and two are heterozygotes (Cc), (half like the subject). But owing to dominance, a quarter of the sibs are phenotypically alike (cc), while three-quarters are unlike the subject. When dominance is present, the resemblance between relatives thus depends on gene frequency. If the dominant gene is extremely rare, the values of regressions are those for genes in common, one-half for parent or sib and one-quarter for uncle or aunt. When the recessive

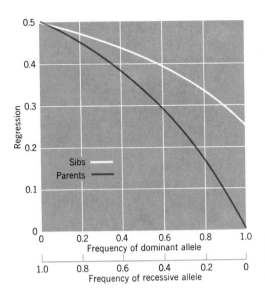

Figure 16.3 Effect of dominance on the regressions of parents and of sibs on subjects.

gene is extremely rare, regressions limit sibs to one-quarter and other relatives to zero. The decline from one extreme to the other is shown graphically in Fig. 16.3 for parents and sibs. The decline for other relatives is proportional to that for parents. For any gene frequency, the reduction for sibs from the value of one-half is half that for the parents.

CONSANGUINITY AND RECESSIVE GENES

The consequences of dominant and recessive genes in family groups and in populations (Chapters 2, 14) represent fundamental aspects of genetics. Data have been obtained from the accumulated results of many studies of a particular kind of religious community in the United States. Members of these communities are healthy and industrious. They live by a rigid moral code and give much attention to their families. Marriage generally involves members of the same sect who reside in the same or other similar communities. Much inbreeding has thus occurred. Several thou-

sands of people now constitute one big family. Genealogies are available and they are usually kept up to date. Family records are of considerable interest and the people are known for scrupulous record keeping. Indexes of family names have been prepared and some of these indexes are computerized.

Studies among these people have shown that rare recessive genes have become homozygous and have come to expression as expected in closed populations. Some genes have been traced for several generations. Such traits as albinism, phenylketonuria, (see p. 527) craniostenosis (narrowing of cranial sutures), adrenal hyperplasia (increase in amount of tissue resulting from increase in the number of cells), ataxia telangiectasia (frequent respiratory infection and abnormal immune mechanism), brittle hair, enlarged great toes, blood group genes, Mast syndrome (presenile dementia), Troyer's syndrome (spastic paraplegia), hemophilia B, limb girdle muscular dystrophy, Weill-Marchesani syndrome (dystrophy and dwarfism), symphalangy (fused joints in fingers), Byler's disease (jaundice), cartilage-hair hypoplasia and deafness have been studied and shown to follow patterns expected for recessive genes in closed populations.

MEDICAL GENETICS

In this discussion, *disease* will include any condition in which bodily or mental well-being is seriously impaired. J. Lederberg has estimated that one-quarter to one-half of all disease is of genetic origin.

In the general population about one in every ten human gametes carries a bit of misinformation introduced by mutation. Most of the cost of this "mutational load" is paid during the early stages of development, when it accounts for a fair part of all fetal losses. About two percent of newborns suffer from a recognizable genetic defect. A high proportion of the genetic illnesses that come to the attention of physicians are preventable if background mutation processes can be controlled. The most effective single avenue for preventing genetic disease is to prevent gene mutations. Some environmental mutagenic agents are irradiation, certain drugs, peroxycompounds in smog, and viral infections.

Somatic mutations have been associated with the cause of some cancer. Most chemical carcinogens are also mutagenic when properly tested. Induction of mutations in germ cells, and of cancer in somatic cells, are fundamentally similar processes. Many cases of cancer are of chemical-environmental origin, cigarette smoking being the best-known example.

CHANGING ROLE OF THE PHYSICIAN

Physicians will undoubtedly continue to make increasingly greater use of genetics, at least partly because of their greater awareness of the role that inheritance plays in human health and illness. Most medical students now include courses in human genetics in their program and some 19 medical colleges in the United States have their own departments of human genetics. Furthermore, a progressive change is occurring in the type of services a physician is called upon to perform. Childhood infections and adult diseases, in which heredity is relatively unimportant, occupied much of a physician's time just a few decades ago. Now these diseases are effectively controlled. Today, more people are reaching advanced ages during which degenerative diseases such as cancer, heart disease, and arthritis pose difficult problems that require basic biological and psychological insights. In these diseases, heredity is a factor of major importance.

Current progress in medical genetics is dramatic enough to be compared with the medical revolution of the last half of the nineteenth century that followed the establishment of the germ theory of disease. After years of basic accomplishments, the impact of genetics in medical practice is just beginning to be felt. This impact will be major, not because the incidence of genetic diseases is increasing but because vaccines, antibiotics, and sanitation have decreased infectious diseases and brought genetic diseases into new prominence. Infant mortality from congenital malformations is still roughly the same as in 1900, about 5 per 1000 live births. In 1900, however, the total infant mortality was 150 per 1000 live births compared with about 20 per 1000 in 1974. While the incidence of congenital abnormalities has remained constant, they now account for about 25 percent of all infant mortality compared with only 4 percent in 1900.

About one in every eight pediatric hospital beds in the United States is now occupied by a child with an illness in which genetic factors are involved. Not all such illnesses are inherited as entities, since both genetic and environmental factors are involved in some way in virtually all abnormalities. About 20 percent of all congenital defects are predominantly inherited, about 20 percent result from environmental factors such as mutagenic agents, bacterial and virus infections (e.g., rubella), and some 60 percent result from combinations of inherited and environmental factors.

Instead of a small number of infectious diseases that affect large numbers of people, the physician must learn to deal with approximately 1500 known genetic diseases, some of which have an incidence of one per 100,000 in general populations. A physician may only see a few patients with some genetic diseases in years of practice. Obviously, specialization and a system of counseling centers and referrals is necessary for coping with this situation. No medical center can afford to specialize in more than a few rare diseases. Cooperation among centers with expertise in particular diseases has proved effective in handling an otherwise formidable situation.

Furthermore, our complex civilization provides an increasing number of environments in which genes can act. Some 23,000 different occupations are now available to individuals. This is far more diversity than only a few years ago. People who work with tar products, for example, must take precautions to avoid skin lesions and cancer. Some 3000 occupational disorders, involving interactions between genes and particular environmental situations are now recognized. Physicians must deal with these and new occupational hazards as they appear. A dynamic place has thus been made for human genetics by industry, health education, and the health professions.

PRENATAL DIAGNOSIS OF GENETIC CONDITIONS

In the late 1950s several obstetricians were attempting to improve the treatment for erythroblastosis fetalis (Rh factor incompatibility; Chapter 6). In one approach, amniotic fluid was analyzed to determine whether the bilirubin content (a measure of red corpuscle destruction) was sufficient to justify premature delivery of the fetus. The obstetrician observed that the amniotic fluid also contained cells that belonged to the fetus and not the mother. Fritz F. Fuchs in Copenhagen began examining these cells for Barr bodies (Chapter 4), which appear in the cells of female but not in cells of male fetuses. His purpose was to determine whether the sex of the fetus could be determined prenatally. If Barr bodies are present, the fetus is female (except for some syndromes such as Klinefelter's XXY, Chapter 10) with virtually no chance of inheriting such male (X-linked) diseases as hemophilia. If, on the other hand, Barr bodies are not present in the cells, the fetus is male and the chance is one-half that a potentially present, male (X-linked) disease will be inherited.

DETECTING CHROMOSOME IRREGULARITIES

Prenatal diagnoses of some chromosomal diseases are now being conducted with virtually no risk to the mother (Chapter 10). This assumes, of course, that the diagnoses are performed by competent persons who have access to the necessary laboratory, technicians, and facilities. Tests of this kind are not simple, and they do not produce immediate definitive results.

Fortunately, amniotic fluid can be prepared and shipped to distant laboratories without deterioration. A sample is obtained by inserting a needle into the amniotic cavity, the fluid-filled sac in which the fetus develops. Cells from the sample must then be grown in tissue culture for ten days to three weeks. Then they can be processed and prepared for study during cell division, when the chromosomes are favorable for observation. Since the cells are from the fetus, any chromosome abnormality is indicative of abnormalities in the fetus.

The volume of amniotic fluid increases by about 25 cc's per week after the twelfth week. The fifteenth week is considered ideal by many physicians, if the particular disorder in question can be diagnosed that late in pregnancy. Diagnostic tests may be more satisfactory if the sample is taken even later in pregnancy, but the usefulness of the test is enhanced if it is made reasonably early.

IDENTIFICATION OF MALES WITH SEX-LINKED RECESSIVE GENES

Congenital hyperuricemia (Lesch-Nyhan syndrome), a serious kidney disease, is inherited through a sex-linked recessive gene. This means that the mother contributes the defective X chromosome to the male infant. One-half of the male children of carrier mothers may be expected to inherit the syndrome. Babies who have this disease appear normal at birth, but by about two months after birth they become abnormally irritable. By the second year of life, the nervous condition has progressed to a degree that self-mutilation occurs, manifested by lip-biting, finger-chewing, teeth-grinding, and marked swinging of the arms. Death, which is usually secondary to severe renal and neurological damage, normally occurs within a few years.

The ratio of two enzymes, inosinate pyrophosphate phosphoribosyl transferase (IMP phosphorylase) and adenylate pyrophosphate phosphoribosyl transferase (AMP phosphorylase) from fresh amniotic fluid of pregnant women has been used successfully in diagnosing the Lesch-Nyhan syndrome. IMP phosphorylase is virtually absent whereas AMP phosphorylase is not decreased by the disease.

Juvenile muscular dystrophy also depends on a sex-linked recessive gene. If the mother is known, either from her pedigree or through tests that are now becoming available, to be a carrier for this gene, about half of her male children are expected to inherit the disease. Male fetuses with a high risk can be identified by a chromosome study, as in the case of Lesch-Nyhan syndrome patients. This disease afflicts boys, usually before they reach their teens, with the muscular deterioration that progresses rapidly during the early teen years. Muscles of the legs and shoulders become stiff and the children usually become paralyzed and crippled during their middle or late teens. Virtually all die before age 21. All female children born to a carrier mother are expected to be normal, since the possibility for their being homozygous for a sex-linked recessive gene is virtually nonexistent.

Another severe disease following the pattern of sex-linked recessive inheritance is the Hunter syndrome. It is characterized by mental retardation, coarse feature, hirsutism (abnormal hairiness), and a characteristic facial appearance that includes a broad bridge of the nose and a large protruding tongue. Symptoms appear in early childhood. Beyond determining the sex of the fetus early in pregnancy, no diagnosis is possible on the basis of cell analysis. However, a chemical means of diagnosing this condition is being developed. Certain constituents in the amniotic fluid indicate the presence of this disease, which is associated with an abnormal processing of mucopolysaccharides in early pregnancy. Mucopolysaccharides also accumulate in skin cells of persons who are heterozygous for the gene for the Hunter syndrome. When amniotic or skin cells are grown in culture and stained with O-toluidine blue, any mucopolysaccharide cell inclusions will be stained pink. It is thus possible to identify heterozygous carriers of the gene as well as an affected fetus.

The pattern of mucopolysaccharide metabolism by Hunter cells is so strikingly different from the normal that it can be used along with chromosome analysis for sex determination in prenatal diagnosis, a situation in which clinical observation is obviously impossible. Of the many cell types originally present in amniotic fluid, fetal fibroblasts are the only ones to multiply in culture. Like fibroblasts from skin biopsies, they show an excessive accumulation of mucopolysaccharide or stainable cell inclusions if the fetus is affected with the Hunter syndrome.

DIAGNOSIS FOR METABOLIC DISORDERS

In general, metabolic disorders have not been as successfully diagnosed prenatally by amniocentesis as have chromosomal disorders. Much more uncertainty and risk are inherent in attempts to diagnose such diseases. The basic causes and conditions associated with a number of these diseases are being defined, however, and it seems likely that it will eventually be possible to diagnose metabolic disorders in the fetus. Inherited metabolic diseases generate considerable interest and research activity despite their relatively rare occurrence, because they afford a unique opportunity to combine genetic concepts with tools of biochemistry in studying the metabolism of man.

A severe and well-known metabolic disorder is the Hurler syndrome, named after Gertrud Hurler who described it in detail in 1919. This disease, characterized by the accumulation of monopolysaccharides in cells, is transmitted by an autosomal recessive gene. After several months of normal development, an infant with this syndrome deteriorates physically and mentally, and gradually acquires an extraordinary appearance. The head becomes abnormally large with a flat bridge of the nose, wide-set eyes, large lips, and coarse tongue. Other external and internal abnormalities are a part of the syndrome but mental retardation is most prominent. Affected children usually do not survive past the age of 20. The disease can be diagnosed prenatally by the presence of cell inclusions and cell cultures of skin fibroblasts.

In-utero diagnosis of Type II Glycogenosis (Pompe's disease), a glycogen storage abnormality, is being developed by H. L. Nadler and his group. In patients with Pompe's disease, the activity of the enzyme α-1, 4-glucosidase is deficient in amniotic fluid, amniotic fluid cells, and cultured amniotic fluid cells obtained between the fourteenth and sixteenth week of pregnancy. Affected fetuses lacked α-1, 4-glucosidase activity in all organs and cultured cells. This confirmed the in-utero diagnosis of Type II Pompe's disease. Inheritance of Pompe's disease depends upon an autosomal recessive gene. The disorder is characterized by intractable cardiac failure progressing to death within the first year of life. Activity of α-1, 4-glucosidase is deficient in the liver, leucocytes, and cultured fibroblasts of patients with this disorder.

Another prenatal diagnostic procedure for the future is "fetoscopy." This technique depends on a device called a fetoscope, a lighted needle designed to penetrate the uterus and facilitate the observation and photographing of the developing fetus. In experimental stages ultrasound is used at the time of insertion of the needle instead of the usual anesthetics. This method is also used for tracings to determine the position of the needle with respect to the uterine walls and other structures. Gross fetal parts such as knees, thighs and toes have been observed thus far. Both the physical apparatus and the technical skills are in the developmental stages. Ethical and moral issues associated with such a procedure must be considered in evaluating appropriate use of such a diagnostic technique.

POSTNATAL DIAGNOSIS AND TREATMENT OF BIOCHEMICAL GENETIC DISEASES

Cystic fibrosis is the most common genetic disease affecting Caucasians. In the United States it occurs once in every 2000 births. This hereditary disease, transmitted by an autosomal recessive gene, has been known for many years as a fatal disease of childhood. Cystic fibrosis involves abnormal functioning of several exocrine glands, including the pancreas, liver, and sweat glands. Nevertheless, early diagnosis and intensive, persistent treatment allow its victims to live fairly normal and active lives during childhood. The disease, however, is not susceptible to a single package treatment. Each malfunction of a particular gland must be treated separately.

The most conspicuous glandular defect associated with cystic fibrosis involves abnormally high concentrations of sodium chloride in the sweat. Individuals with the disease may sweat profusely over long periods of time and lose enough salt to upset the salt balance in the body. If the normal intake of salt is not supplemented, heat prostration may result. The cure for this particular symptom therefore is salt supplementation which may be accomplished by salt tablets taken orally.

Physiological problems occur when the mucous glands function abnormally. The mucous of a cystic fibrosis patient may be too viscous to flow properly in the small tubules of the lungs. It then becomes mixed with bacteria, serum fluids, and debris, and creates an unhealthy situation. Bronchial walls become infected and chronic bronchitis results. Infection of the respiratory tract is the most common cause of death among young cystic fibrosis victims. Treatment with antibiotics, sulfonamides, and pulmonary therapy designed to liquify the mucous and remove infected secretions has improved the health and extended the life expectancy of many patients.

The more severe glandular defects that threaten the life of the patient are associated with malfunctions of pancreas, liver, and intestines. These challenges can still be only partially resolved. Cystic fibrosis affects the pancreas by destroying the secreting tissue. The pancreatic enzymes can be supplied in the diet but the treatment must be constant. Some patients develop cirrhosis of the liver, which leads to an absence of fat-emulsifying factors needed during digestion. Bile salts may then be supplied in the diet. Along with the special treatments, a diet well balanced with fats, proteins, and carbohydrates, and supplemented with fat-soluble vitamins is prescribed for the cystic fibrosis patient. Much remains to be done before the symptoms of this disease will be conquered, but the consequences of the genotype are well known.

Staining cultured cells from cystic fibrosis patients has led to the identification of two and possibly three kinds of cystic fibrosis, differing on the cellular level. Hopefully a screening test will be developed for identifying affected fetuses as well as carrier mothers. Barbara Bowman has demonstrated that serum from patients afflicted with cystic fibrosis inhibits the movement of oyster gill cilia. Other materials such as rabbit trachea have also been used for the test. This test is useful for a postnatal diagnosis of the disease but the technique in its present state does not distinctly differentiate between carriers and those afflicted with the disease.

SCREENING FOR PHENYLKETONURIA

Phenylketonuria (PKU) is treatable but not detectable prenatally. Most states in the United States require routine screening at birth. The disease is transmitted by an autosomal recessive gene and it occurs once in about 18,000 live births. The test is accomplished by a simple color change in treated urine. Massachusetts, which tests 97 percent of the babies born in the state for PKU, identified eight infants in one year as having the disease. Infants giving a positive reaction are placed on a particular diet for the first few years of life. Care must be taken to avoid side effects and malnutrition of babies subjected to this diet. All eight of the infants detected in Massachusetts in that year were treated and all have apparently escaped the mental retardation associated with PKU.

The name phenylketonuria describes the excretion of phenylpyruvic acid (a phenyl ketone) in the urine. Normally, the amino acid phenylalanine is converted to tyrosine, which is oxidized further and has several other reactions of great physiological importance (Fig. 7.19). The conversion of phenylalanine to tyrosine is an oxidative step catalyzed by the liver enzyme phenylalanine hydroxylase. Phenylketonurics lack this enzyme and cannot make the conversion. The phenylalanine accumulates as protein is consumed and affected persons may have 50 to 100 times normal blood levels. These high levels induce side reactions that, under ordinary circumstances, are quantitatively unimportant. They are responsible for the formation of phenylpyruvic acid, o-hydroxyphenylactic acid, and other substances characteristic of PKU. Some of the products of these side reactions, however, are toxic to the central nervous sytem and produce irreversible brain damage.

Man cannot make phenylalanine and must rely on dietary protein as a source. If phenylalanine could be removed from the diet, affected persons should lose the PKU symptoms. The difficulty is that phenylalanine is required as a building block for body proteins, so that complete elimination would be very detrimental. The amount consumed can be controlled, however, and good results have been obtained in preventing mental defect when proper diets are instituted. Delay in starting diet control leads to irreversible changes in the nervous system.

Massachusetts also requires a test on umbilical cord blood for the level of galactose to detect galactosemia in newborn infants. This is a disease of considerable importance that is controlled by an autosomal recessive gene. In addition, each mother is requested to supply a sample of urine from her baby three or four weeks after they leave the hospital. A kit containing a strip of filter paper is given to the mother along with a preaddressed envelope to the state laboratory. Instructions are given for impregnating the filter paper with urine and mailing it. At the laboratory chromatographic tests are performed to detect enzyme abnormalities associated with histidinemia, hyperlysinemia, cystinuria, and several other genetic disorders.

PHARMACOGENETICS

Pharmacogenetics is the area of biochemical genetics dealing with inherited variations in sensitivity in response to drugs. The enzyme glucose-6-phosphate dehydrogenase (G6PD), for example, is widespread in tissues of normal people and is important in glucose metabolism. Some people inherit a form of the enzyme that is less stable than the common form. People with this defect are entirely normal except when challenged with certain drugs or when they eat fava beans. Those who have the G6PD deficiency appear to be protected from falciparum malaria but they are susceptible to the antimalarial drug, primaquine, which first led to the recognition of the condition. Other toxic drugs include a variety of more common substances such as napthalene mothballs. When exposed to these substances, G6PD-deficient red cells are unable to cope with the chemical stress and break open. The release of hemoglobin into the plasma may have serious consequences, including death. The genetic locus for G6PD is on the X chromosome. Males, with one X chromosome, are clearly either affected or nonaffected. Females may have one or both G6PD deficiency genes and show intermediated or severe effects.

Another drug metabolized differently by different persons is isoniazid, used primarily in treatment of tuberculosis. Some persons inactivate the drug rapidly, others slowly, depending on the gene combinations at a specific locus.

An important area still largely unexplored, concerns inherited variations in resistance to infection. Work with experimental animals indicates clearly that inbred strains of mice have innate differences in ability to cope with infections.

The most suggestive studies in human beings concern leprosy, to which only a portion of the population appears to be susceptible. Significant advances can be anticipated through efforts to define genetic influence on resistance to infection. The process should help clarify and evaluate the interrelations of heredity and environment as both contribute to the final makeup of the individual.

GENETIC COUNSELING

Human genetics as a science has become complex and has gained a high level of maturity, largely through team efforts. Areas of specialization include biochemical genetics with special tools such as electrophoresis for enzyme assay. Cytogenetics has defined a wide range of chromosome irregularities that have practical significance for diagnosis of human disease. Cell hybridization has brought a rebirth in linkage studies and promises to produce basic knowledge about differentiation and the origin of cancer in the cells of the body. Immunology has much to offer for human welfare. Population genetics with its dynamic relations involving changes in gene frequencies and interactions with environments provides a broad theoretical base for clinical medicine. All this suggests that help is needed for people trying to deal with facts and fears of inheritance. Genetic counseling provides and interprets medical information based on expanding knowledge of human genetics.

The family physician is the foremost genetic counselor for his patients. He now has available for his guidance much information and consulting assistance from many centers of genetic and medical research. Problems of social needs as well as individual predisposition to health and disease are becoming increasingly important. Appropriate safeguards for individual patients and society as a whole, will gain in significance as the present trends continue. Although the family physician will retain his place as a counselor, and nonmedical human geneticists will continue to give advice about genetic problems, the main counseling effort for difficult and involved problems will undoubtedly be concentrated in large medical centers where teams of specialists are available and appropriate tests can be made as the basis for counseling.

QUALITY OF GERM PLASM

If decisions are to be made that will affect quality of germ plasm in future generations, who should make them? Environmental hazards such as pollution, and particularly irradiation, some drugs and some viral infections are known to break chromosomes and induce mutations. The most effective way to prevent transmission of genetic diseases to future generations is to prevent mutations. For serious DNA lesions already in the gene pool, detection and human containment might be accomplished. Defective genes that produce serious mental and physical malformations will probably decrease the quality of the germ plasm in the environment of future generations, but the social problems of how the decisions are made and who makes them remain with mankind.

IMPROVEMENT OF GENE POOL

This line of thought inevitably leads to the practical issue of what Sir Francis Galton called "eugenics," social behavior designed to improve the genotype of the human population. The first requisites for such behavior are: (1) continued study of human genetics and other disciplines designed to clarify knowledge of man, and (2) public education to bring about widespread enlightened opinion based on sound facts. Eugenics, of all endeavors, must be based on accurate information. Research and education, however, will be fulfilled only when appropriate action is taken to improve mankind, with due consideration for moral and ethical principles.

In theory, the procedure is simple: those with superior genotypes should produce more and those with inferior, less (or none) of the next generation. But in practice, superior wisdom and courage are required to determine precisely who should and who should not have children. It is equally difficult to decide how and by whom such decisions should be made. Genotypes are, for the most part, not known, and criteria for selecting types best fitted for future environments are not established. Moreover, reproduction cannot be entirely controlled at will. Sterility may interfere with the best planning. Sir Francis Galton, the founder of eugenics, for example, who was unusually gifted and devoted to the principle that the better qualified people should produce at least their share of the children for the next generation, died childless.

Voluntary adherence to any eugenic movement in a free society requires a highly enlightened population that is deeply concerned for the quality of the gene pool. This ideal has been and probably will continue to be very difficult to attain. Most people would agree that severely disabled hereditary defectives should not reproduce, and usually nature takes care of this situation. But the larger segments of the population, those that carry defective genes but fall within the near-normal phenotypic range for basically important traits such as intelligence, ordinarily make their own decisions and naturally cherish that right. One puzzling feature of the statistics, particularly since World War II, is that the average intelligence of the population in the United States has not decreased even though many seemingly less-qualified parents are producing a larger share of the children.

In their extensive study of mental retardation, Reed and Reed concluded:

"There is a large negative correlation (about −0.3) between the number of children in a family and their average intelligence as measured by any test; this should result in a decrease in the intelligence of the population in each generation—but it doesn't. The explanation of the failure of the intelligence of the population to decline is that while a few of the retarded produce exuberantly large families of low average intelligence, most of the retarded produce only one child or no children at all. The persons at the upper end of the curve of intelligence are consistent in their production of smaller families of more intelligent children; thus the children of the smaller, more intelligent families balance, or perhaps outnumber, the children of the larger less intelligent families. The net result has been an increase in intelligence to the present level."

Reed and Reed also reported that only 0.5 percent of the children from normal parents with normal siblings were retarded. On the other hand, five-sixths of the retarded in-dividuals, representing 2.5 percent of the entire population in the United States, have at least one parent or close relative (uncle or aunt) who is mentally retarded. Although environmental as well as hereditary factors are involved in mental retardation of individuals, averages in the population indicate that heredity is of major importance.

ARTIFICIAL INSEMINATION AND SPERM BANKS

The late H. J. Muller had for many years articulated the position that the gifted, more able and socially minded people should raise large families while those less capable and relatively antisocial should produce none or less than their full proportion of progeny. Muller would rely mainly on education, enlightened public opinion, and voluntary choice to bring about the desired strengthening in the quality of germ plasm for the next generation. To facilitate such a plan, artificial insemination is suggested for cases in which a qualified woman is unmarried or married to a sterile or nonqualified man. Sperm banks with frozen sperm properly protected from radiation and other environmental hazards would eventually be employed. Again, the questions about who would make the decisions and how the quality of germ plasm would be defined are central, controversial issues.

GENETIC ENGINEERING

Although procedures for genetic engineering are only now being developed, the concept goes back to 1908, when Sir Archibald Garrod, physician to the British royal family, described several diseases as "inborn errors of metabolism." Among these diseases were alcaptonuria, cistinuria, porphyria, and albinism. Far ahead of his time, Garrod described chemical individuality for traits. This led to concepts of biochemical evolution. He foreshadowed the "1 gene–1 enzyme" theory and proposed that several biochemical diseases were dependent on recessive genes. Based on Garrod's discovery, it was possible before 1910 to predict the probability of expression of some hereditary

TABLE 16.3
Conditions for Which Enzyme Deficiencies or Malfunctions Are Known (Based on McKusick, 1969 and Levitan and Montagu, 1971)

Condition	Enzyme
acatalasia	catalase
adrenogenital syndrome, severe salt-losing form	3-β-hydroxy steroid dehydrogenase
adrenogenital syndrome, uncomplicated, probably two forms	steroid C-21β-hydroxylase
adrenogenital syndrome with high blood pressure	steroid C-11 β-hydroxylase
alactasia, two forms: infant and adult lactose intolerance	intestinal lactase (may be two forms)
albinism — complete oculocutaneous form	tyrosinase
alkaptonuria	homogentisic acid oxidase
argininosuccinicacidurea	argininosuccinase (ASAase)
carnosinemia	carnosinase
cataract	galactokinase
central-nervous-system disease with cataracts	sulfite oxidase
chronic granulomatous disease	leukocyte $NADH_2$-oxidase
citrullinemia	argininosuccinic acid synthetase
Crigler-Najjar syndrome; also Arias type hyperbilirubinemia?	glucuronyl transferase
cystathioninuria	cystathionine cleaving enzyme
disaccharide malabsorption	maltase 3,4 (or 3,4, and 5), possibly also palestinase
essential fructosuria	hepatic fructokinase
Fabry's disease	ceramide trihexosidase
familial nonhemolytic jaundice	glucuronyl transferase
favism (see primaquine sensitivity)	
formiminoglutamic aciduria with mental retardation	formimino transferase
fructose intolerance	fructose-1-diphosphate aldolase
fructosuria	fructokinase
galactokinase deficiency	galactokinase
galactosemia	galactose-1-phosphate-uridyl transferase
Gaucher's disease infantile (and adult?) form	glucocerebrosidase
generalized gangliosidosis	beta-galactosidase
glucose-galactose malabsorption	intestinal monosaccharidase

Table 16.3 (*Continued*)

Condition	Enzyme
1-glyceric aciduria	D-glyceric dehydrogenase
glycogen storage disease Cori type I (von Gierke disease)	glucose-6-phosphatase
glycogen storage disease Cori type II (Pompe disease)	α-1,4-glucosidase (acid maltase)
glycogen storage disease Cori type III (Forbes disease)	amylo-1, 6 glucosidase (debrancher enzyme)
glycogen storage disease Cori type IV (Andersen disease)	amylo-1,4 $\longrightarrow$ 1,6-transglucosidase (brancher enzyme)
glycogen storage disease Cori type V (McArdle-Schmid-Pearson disease)	muscle glycogen phosphorylase
glycogen storage disease type VI (Hers disease)	liver glycogen phosphorylase
glycogen storage disease XI	muscle phosphofructokinase
goitrous cretinism, type IIA (organification defect I)	iodide peroxidase
goitrous cretinism, type IV	iodotyrosine deiodinase
hemolytic anemia	glucose-6-phosphate dehydrogenase
hemolytic anemia	pyruvate kinase
hemolytic anemia	triose-phosphate-isomerase
hemolytic anemia	2,3-diphosphoglycerate mutase
hemolytic anemia	glutathione synthetase
hemolytic anemia	hexokinase
histidenemia	histidase
homocystinuria	cystathionine synthetase
hydroxykynureninuria	kynureninase
hydroxyprolinemia	hydroxyproline oxidase
hyperammonemia	hepatic orinthine transcarboxylase
hyper-β-alaninemia	β-alanine-γ-ketoglutarate transaminase
hyperoxaluria	2-oxo-glutarate glycoxalate carboligase
hyperprolinemia, type I	proline oxidase
hyperprolinemia, type II	Δ'-pyrrolline-5-carboxylate dehydrogenase
hyperuricemia	hypoxanthine-guanine phosphoribosyl transferase
hypervalinemia (see valinemia)	
hypoglycemia	hepatic glycogen synthetase
hypogonadism with mineralocorticoid excess	steroid C-17α-hydroxylase

Table 16.3 (*Continued*)

Condition	Enzyme
hypophosphatasia	alkaline phosphatase
infantile amaurotic idiocy (Tay-Sachs)	B-D-N-acetylhexosaminidase
isovaleric-acidemia	isovaleric-coenzyme A dehydrogenase
Lesch-Nyhan compulsive disorder	hypoxanthine-guanine phosphoribosyl transferase
maple sugar urine disease	branched-chain-keto acid decarboxylase
metachromatic leukodystrophy	arylsulfatase A (also B and C?)
methemoglobinemia with normal hemoglobin	diaphorase I
methylmalonic aciduria	methylmalonyl-coenzyme A isomerase
multiple disaccharide intolerance, may be several forms	sucrase, maltases, isomaltase probably a proenzyme common to all)
Niemann-Pick diseases	sphingomyelinase
nonspherocytic hemolytic anemia	ATPase
nonspherocytic hemolytic anemia	2-3-diphosphoglycerate mutase
nonspherocytic hemolytic anemia	glutathione reductase
nonspherocytic hemolytic anemia	hexokinase
nonspherocytic hemolytic anemia	pyruvate kinase
nonspherocytic hemolytic anemia	reduced glutathione (coenzyme)
nonspherocytic hemolytic anemia	triosephosphate isomerase
oroticaciduria	orotidylic pyrophosphorylase and/or orotidylic decarboxylase
oxalosis (primary hyperoxaluria)	glycine transaminase
Pendred syndrome	an iodine transferase
pentosuria (L-xyloketosuria)	NADP-xylitol (or L-xylulose dehydrogenase)
phenylketonuria	phenylalanine hydroxylase
primaquine sensitivity	glucose-6-phosphate dehydrogenase
Refsum disease	phytanic acid oxidase
spherocytosis, hereditary	red cell aldolase
Sidbury syndrome	green acyl dehydrogenase
suxamethonium sensitivity	pseudocholinesterase
tyrosinemia	p-hydroxyphenylpyruvic acid oxidase
tyrosinosis	p-oxyphenyl pyruvic acid hydroxylase or tyrosine transaminase
valinemia	valine transaminase
Wilson's disease	ceruloplasmin
xanthinuria	xanthine oxidase

diseases for which methods of early diagnosis, prevention, and treatment are only now being developed. The new techniques had to await the discoveries in molecular biology. Although the molecular basis of most genetic diseases is not known, some 92 human disorders are identified with genetically determined specific enzyme deficiency. Some of these are listed in Table 16.3. Curing single-gene diseases such as phenylketonuria by replacing genetic material would seem more feasible than altering behavioral traits that depend on many genes.

Enough is now known about gene action and genetic control to make some past speculations near realities. A real problem associated with altering and ultimately directing gene action involves ethical and moral issues. Even such a simple procedure as genetic counseling raises ethical questions. Reproduction by cloning of cells, sperm banks, predetermination of sex, genetic surgery, and selective reproduction pose complex issues. Scientists as well as laymen hesitate to make value judgments that could change gene frequencies and alter the human species.

INDUCING ENZYME ACTIVITY IN CELLS

On the mechanical level, genetic engineering involves basic cell biology and ways of adding, removing, or altering enzyme action within a cell. Enzymes that operate within a cell are made in that cell. Chance mutations can create new alleles that synthesize new enzymes. Thus far, such changes have not been induced or prohibited deliberately by genetic intervention.

Since viruses carry DNA (or RNA) into their host cells, they are potential agents with which to modify cellular enzymes. When DNA virus replaces to the genetic machinery of the host cell (i.e., transduction), enzymes are subsequently produced according to the specification of the virus. The objective of some research is to induce the production of a particular enzyme in the cells of an individual who may need that particular protein for health. If defective DNA could be replaced directly with good DNA (i.e., transformation), people suffering from genetic defects might be cured. In such gene therapy or gene surgery, comparatively minor genetic engineering could improve the health of individual patients. This level of genetic engineering is far more practicable than efforts to modify the gene pool of all mankind.

Several investigators have transferred genetic material to an animal cell through the agency of a virus. W. Munyon and his colleagues, for example, have infected mutant L cultured cells from a mouse that lacked the enzyme thymidine kinase, with the animal virus, herpes simplex. Some of the infected L cells were transformed into stable cells with thymidine kinase activity and were maintained in culture for eight months. Control L cells that were uninfected and those infected with a herpes simplex mutant virus that did not induce thymidine kinase showed no activity.

C. R. Merril and associates have shown that a bacteriophage can introduce a particular gene into human cells. Lambda phage carrying the *E. coli* galactose operon was introduced into cultured human fibroblast cells from the skin of a patient with galactosemia. The gene for the enzyme α-D-galactose-1-phosphate uridyl (GPU) which is part of the galactose operon was transferred by lambda phage to the fibroblasts and GPU activity persisted for more than 40 days, during which 8 doublings of the cells occurred. Control infections with normal lambda and with a mutant that inactivates its transferase resulted in the production of lambda-specific RNA but no transferase activity occurred.

Eventually, artificial viruses may be tailored to correct certain specific enzyme deficiencies associated with genetic diseases.

CLONAL REPRODUCTION

An extreme form of genetic engineering and one that may be a considerable distance away in time, if indeed, it is ever accomplished in man, is that of producing twins through a form of asexual reproduction. In this discussion we will consider the principles of genetics as applied to clonal reproduction with a minimum of speculation concerning the problems that would be raised in society.

A clone is a group of genetically identical, asexual cells all descended from a single ancestral cell by mitosis. A colony of bacteria descended from a single bacterium and cultured on a Petri dish is a clone. Several years ago J. B. Gurdon produced an African clawed frog by means other than gamete formation and fertilization. Gurdon obtained an unfertilized egg cell from a frog and destroyed its nucleus by radiation (in other experiments the nucleus was removed by microsurgery). He then replaced the haploid egg nucleus with the diploid nucleus from a cell in the intestine of a tadpole.

The egg with the acquired diploid chromosome set, characteristic of a fertilized egg, began to divide as if it had been fertilized, and proceeded through the embryological stages which culminate in an adult frog. Instead of producing a frog with combined genetic material from two parents, and therefore uniquely different from either parent, this cell received all of its genetic material from the tadpole used as the nuclear donor. It gave rise to a frog that was in effect a twin of the donor tadpole that had been hatched some months before.

Gurdon's clonal frog demonstrated that a nucleus from a highly differentiated cell could direct the development of a completely normal frog. This suggested that a tadpole (or frog) donor of a single nucleus could be the genetic parent of many thousands of identical progeny.

Is clonal reproduction possible for other differentiated cells? Evidence indicates that cells other than intestinal cells of a frog have nuclei that can furnish the genetic material for other twin frogs, if properly transplanted into enucleated eggs. Furthermore, the same results are expected eventually from most higher plants and animals, including man. The principle has been demonstrated in plants such as carrots.

Experimenting with clonal reproduction and transplanted nuclei is much more difficult in mammals, however, than in amphibia. Development occurs in the body of the mother and not in open water or a laboratory dish. The small eggs are difficult to find and to maintain *in vitro* for experimentation. Removing egg nuclei and replacing them with diploid nuclei from somatic cells requires great skill and superb precision in instrumentation. Nevertheless, considerable progress has been made with mammalian eggs.

Mouse eggs, for example, have been isolated, fertilized *in vitro*, maintained in a test-tube through the 64-cell (blastocyst) stage and implanted in uteri of living mice where they remained until normal birth. The egg nucleus can be removed by treatment with the mitotic poison, colchicine. A diploid nucleus can be introduced by fusion of cells in culture using Sendai virus (Chapter 5). Somatic cells fused with eggs divide several times, but do not develop into blastocysts, presumably because of the protective covering of the egg (zona pellucida). When a technique is perfected for fusing cells with the zona pellucida or for permitting unprotected cells to develop, clonal mammals may be a reality.

Clonal domestic animals and man may seem a long way ahead but considering the momentum of experimental biology, especially in England, and the interests and pressures for research in the areas of human reproduction and population control, clonal reproduction in man may be technically possible in the next decade. Already R. G. Edwards and P. C. Steptoe in England are maintaining human eggs in culture, observing fertilization with human sperm, and following the zygotes through several cleavage stages. If these investigations are as successful as the mouse studies, implantation of blastocysts may soon be technically possible in women. Women who are infertile but whose ovaries produce normal eggs could thus have children. They would contribute their own genetic material but other women would receive the implant of the blastocyst and carry the baby to normal birth.

The possibility of clonal reproduction stimulates the imagination about man's future. Could a great artist or scientist or political leader have a clonal twin or many twins and thus perpetuate his genetic endowment and accomplishments, even to a degree of physical immortality? Could many twins be produced and thus provide a community or a nation of identical people? What would be the result in a world where diversity has been the norm? Obviously many ethical, moral, and legal as well as biological issues would be inevitable.

REFERENCES

Durand, J. D. 1967. "The modern expansion of world population." *Amer. Phil. Soc., Proc.* 3, (3) 136–159.

Frantantoni, J. C., C. W. Hall, and E. F. Neufeld. 1969. "The defect in Hurler and Hunter syndromes, II. Deficiency of specific factors involved in mucopolysaccharide degradation." *Nat'l Acad. of Sci.* 64 (1), 360–366.

Harris, H. 1970. *Advances in human genetics.* Vol. 1. North-Holland Publishing Co., Amsterdam.

Harris, H. and K. Hirschhorn. (eds.) 1972. *Advances in human genetics.* Vol. 3. Plenum Press, New York.

Levitan, M. and A. Montagu. 1971. *Textbook of human genetics,* 2nd ed. Oxford University Press, New York.

McKusick, V. A. 1969. *Human genetics.* Prentice-Hall, Englewood Cliffs, N. J.

————, 1971. *Mendelian inheritance in man,* 3rd ed. The Johns Hopkins Press, Baltimore.

————, 1972. *Heritable disorders of connective tissue,* 4th ed. C. V. Mosby Co., Saint Louis.

Mertens, T. R. and S. K. Robinson. 1973. *Human Genetics and social problems.* MSS Information Corporation, New York.

Nadler, H. L. and A. B. Gerbie. 1970. "Role of amniocentesis in the intrauterine detection of genetic disorders." *New Engl. J. of Med.* 282, 596–599.

Stanbury, J. B., J. B. Wyngaarden, and D. S. Fredrickson (eds.) 1972. *The metabolic basis of inherited disease,* 3rd ed. McGraw-Hill, New York.

Stern, C. 1973. *Principles of human genetics,* 3rd ed. W. H. Freeman and Co., San Francisco.

Thompson, J. S. and M. W. Thompson. 1973. *Genetics in medicine.* W. B. Saunders Co., Philadelphia.

PROBLEMS AND QUESTIONS

16.1 Why has human genetics been slow in developing, in spite of its great importance and early beginnings?

16.2 Why are human traits dependent on dominant genes better known than those dependent on recessive genes?

16.3 What are the relative values and limitations of (a) pedigree analyses, (b) statistical methods, and (c) twin studies for studying human genetics?

16.4 How can the statements that human hereditary diseases are curable and are incurable be explained?

16.5 What effect from widespread medical "cures" of hereditary diseases might be expected on gene frequencies in the population?

16.6 What is the objective of eugenics?

16.7 Why are the prospects for genetic engineering desired and feared?

16.8 What special problems are involved when eugenic measures are developed in society?

16.9 Why will physicians of the future require more specialization and be expected to devote more time to counseling than those of 50 years ago?

16.10 Why has amniocentesis been more successful thus far for prenatal diagnosis of chromosome irregularities than metabolic disorders?

16.11 Why is sex determination of the fetus significant in predicting the Lesch-Nyhan and the Hunter syndrome?

16.12 Cystic fibrosis, a hereditary disease which occurs once in every 2000 births, is transmitted by an autosomal recessive gene. Would the frequency of this gene be decreased significantly by sterilization of diseased individuals?

16.13 Why are environmental factors especially important to patients with cystic fibrosis?

16.14 Why are genetics and basic cell biology now more important in medical research than in the late nineteenth century?

APPENDIX

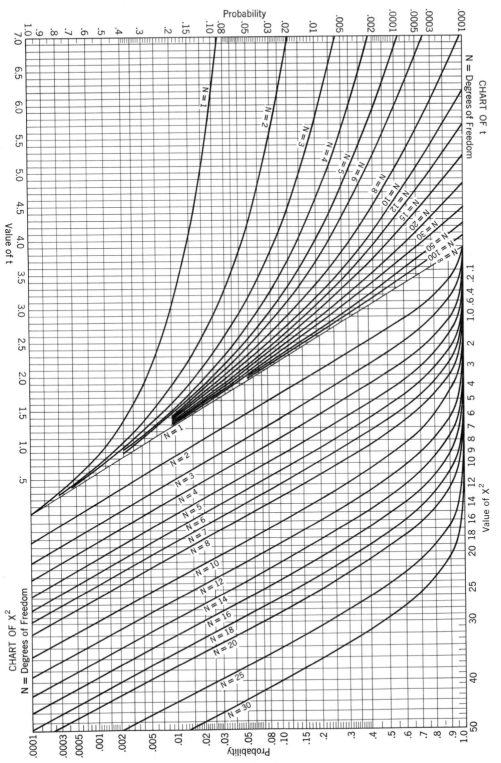

Curves showing probability (P) values for X^2 and t. When X^2 or t has been computed and the degrees of freedom are determined the P value can be read directly from the chart. For X^2, find the calculated value at the right of the chart, read left to the curve for the number of degrees of freedom and find P at the bottom of the chart. (Courtesy of James F. Crow.)

GLOSSARY

This glossary provides and introduction to the terms employed in the text and an aid in using them. Names of chemical compounds such as amino acids that are identified in the text are omitted in the glossary.

Acentric chromosome. Chromosome fragment lacking a centromere.

Acquired character. A modification impressed on an organism by environmental influences during development.

Acrocentric. A modifying term for a chromosome or chromatid that has its centromere near the end.

Acrosome. An apical organelle in the head of the sperm.

Adaptation. Adjustment of an organism or a population to an environment.

Adenine. A purine base found in RNA and DNA.

Agglutinin. An antibody in blood plasma that brings about clumping (agglutination) of blood cells carrying an incompatible agglutinogen.

Agglutinogen. An antigen carried in red blood cells that reacts with a specific agglutinin in the plasma and causes clumping of the cells. When a specific antigen is injected into an animal body it stimulates the production of a corresponding antibody.

Agouti. A grizzled color of the fur of animals resulting from alternating light and dark bands on the individual hairs.

Albinism. Absence of pigment in skin, hair, and eyes of an animal. Absence of chlorophyll in plants.

Alcaptonuria. An inherited metabolic disorder. Alcaptonurics excrete excessive amounts of homogentisic acid (alcapton) in the urine.

Aleurone. The protein matter occurring in the form of minute grains in the endosperm of ripe seeds.

Aleurone layer. The outer differentiated layer of cells of the endosperm (cells in this layer are filled with aleurone grains.)

Allele (allelomorph), adj. allelic (allelomorphic). One of a pair, or series of alternative alleles that occur at a given locus in a chromosome; one contrasting form of a gene. Alleles are symbolized with the same basic symbol (e.g., D for tall peas and d for dwarf), *see* Multiple alleles.

Allopolyploid. A polyploid having chromosome sets from different sources, such as different species. A polyploid containing genetically different chromosome sets derived from two or more species.

Allotetraploid. An organism with four genomes derived from hybridization of different species. Usually, in forms that become established, two of the four genomes are from one species and two are from another species.

Amino acid. Any one of a class of organic compounds containing the amino (NH$_2$) group and the carboxyl (COOH) group. Amino acids are building blocks of proteins. Alanine, proline, threonine, histidine, lysine, glutamine, phenylalanine, tryptophan, valine, arginine, tyrosine, and leucine are among the common amino acids.

Amniocentesis. A procedure for diagnosing genetic abnormalities in utero. Amniotic fluid is taken from a pregnant woman. Chemical contents of the fluid are studied directly for the diagnosis of some diseases. Cells are cultured and metaphase chromosomes are studied for detection of chromosomal irregularities (e.g., the Down syndrome).

Amnion. The inner fluid-filled sac in which the embryo develops in higher vertebrates.

Amniotic fluid. Liquid contents of the amniotic sac of higher vertebrates containing cells with the chromosomal content of the embryo (not that of the mother). Both fluid and cells are used for diagnosis of genetic abnormalities of the embryo or fetus.

Amorph. Mutant allele that has little or no effect on the expression of a trait.

Amphidiploid. A species or type of plant derived from doubling the chromosomes in the F$_1$ hybrid of two species; an allopolyploid. In an amphidiploid the two species are definitely known whereas in other allopolyploids they may not be known.

Anaphase. The stage of nuclear division during which the daughter chromosomes pass from the equatorial plate toward opposite poles of the cell (toward the ends of the spindle). Anaphase follows metaphase and precedes telophase.

Androgen. A substance with male sex hormone activity in vertebrate animals.

Anemia. Abnormal condition characterized by pallor, weakness, and breathlessness, resulting from a deficiency of hemoglobin or a reduced number of erythrocytes.

Aneuploid or heteroploid. An organism or cell having a chromosome number that is not an exact multiple of the monoploid (n) or basic number, hyperploid, higher (e.g., $4n + 1$); hypoploid, lower (e.g., $4n - 1$).

Anomalous gametes. Irregular and usually incompatable gametes with chromosome numbers different from those normally produced by members of the species.

Anther. Male part of a plant flower in which pollen is produced.

Anthocyanin. Glucoside pigment in plants. Anthocyanins range in color from red to blue.

Anthoxanthine. Any of a group of yellow pigments found especially in plants.

Antibody. Substance in a tissue or fluid of the body that acts in antagonism to a foreign substance (antigen).

Anticodon. Three bases in a transfer RNA molecule that are complementary to the three bases of a specific codon in messenger RNA.

Antigen. A substance, usually a protein that stimulates the production of antibodies when introduced into a living organism.

Antihemophilia globulin. Blood globulin that reduces the clotting time of hemophilic blood.

Apomixis. Development of an individual plant from an egg without fertilization or fusion with pollen nucleus. The egg may be normally reduced (haploid) or, more commonly, through failure of reduction division, may remain diploid.

Ascospore. One of the spores contained in the ascus of certain fungi such as Neurospora. Following the meiotic sequence, each ascus or sac contains eight ascospores.

Ascus, pl. asci. Reproductive sac in the sexual stage of a type of fungi (Ascomycetes) in which ascospores are produced.

Asexual reproduction. Any process of reproduction that does not involve the formation and union of gametes from the two sexes.

Asynapsis. The failure or partial failure of pairing of homologous chromosomes during the meiotic prophase; failure of chiasma formation resulting in a high frequency of univalents.

Atavism. *See* Reversion.

ATP. Adenosine triphosphate; an energy-rich compound that promotes certain activities in the cell.

Atrophy. Decrease in size or wasting away of an organ or tissue.

Autopolyploid. A polyploid that has multiple and identical or nearly identical sets of chromosomes (genomes). A polyploid species with genomes derived from the same original species.

Autoradiograph. A record or photograph prepared by labeling a substance such as DNA with a radioactive material such as tritiated thymidine and allowing the image produced by decay radiations to develop on a film over a period of time.

Autosexing. A method of distinguishing the sex of young chickens by introducing marker genes into the breeding stock that produce a conspicuous phenotype (e.g. delayed feathering) on the male or female progeny at an early age.

Autosome. Any chromosome that is not a sex chromosome.

Auxotroph. A mutant organism (bacterium) that will not grow on a minimal medium but requires the addition of some growth factor.

Axoneme. Central thread or bundle of fibrils in a cilium or a flagellum.

Backcross. The cross of a hybrid to one of the parental types. The offspring of such a cross are referred to as the backcross generation or backcross progeny (*See* Test cross).

Bacteriophage. Virus that attacks bacteria. Such viruses are called bacteriophages because they destroy their bacterial host.

Balanced lethal. Lethal genes on the same pair of chromosomes that remain in repulsion because of close linkage or crossover supresion. Only heterozygotes for both gene pairs survive.

Balanced polymorphism. Two or more types of individuals maintained in the same breeding population.

Basal body. Small granule to which a cilium or a flagellum is attached.

Binomial expansion. Exponential multiplication of an expression consisting of two terms connected by a $+$ or $-$ such as $(a + b)^n$.

Biometry. Application of statistical methods to the study of biological problems.

Biotype. Distinct physiological race or strain within morphological species. A population of individuals with identical genetic constitution. A biotype may be made up of homozygotes or of heterozygotes, of which only the former would be expected to breed true.

Bipartite structure (chromosome). One having two corresponding parts.

Bivalent. A pair of synapsed or associated homologous chromosomes that may or may not have undergone the duplication process to form a group of four chromatids.

Blastomere. Any one of the cells formed from the first few cleavages in animal embryology.

Blastula. A form of early animal embryo following the morula stage; typically, a single layered, hollow ball stage.

Blended inheritance. Inheritance in which traits of two dissimilar parents appear to be blended in the offspring, and segregation fails to occur in later generations. A preMendelian concept of genetics.

Bud sport or chimera. A branch, flower, or fruit that differs genetically from the remainder of the plant.

Carcinogen. A substance capable of inducing cancer in an organism.

Carotenoid. Any of a group of yellow and red pigments found in plants and in the fat of animals.

Carrier. An individual which carries a recessive gene that is not expressed (i.e., obscured by a dominant allele).

Centriole. Central granule in many animal cells that appears to be the active principle of the centrosome and which undergoes duplication preceding the division of the centrosome proper.

Centromere, kinomere, or kinetochore. Spindle-fiber attachment region of a chromosome.

Centrosome. A self-propagating cytoplasmic body usually present in animal cells and those of some lower plants, but not present in flowering plants; consisting of a centriole and sometimes a centrosphere (when inactive) or astral rays (when active); located at each pole of the spindle during the process of nuclear division (mitosis).

Chalcones. Any of a group of yellow pigments in plants.

Character (contraction of the word characteristic). One of the many details of structure, form, substance, or function that make up an individual organism. The Mendelian characters represent the end products of development, during which the entire complex of genes interacts within itself and with the environment.

Chiasma, pl. chiasmata. A visible change of partners or crossover in two of a group of four chromatids during the first meiotic prophase. In the diplotene stage of meiosis the 4 chromatids of a bivalent are associated in pairs but in such a way that in one part of their length two chromatids are associated, but in the remainder of their length each is associated with one of the other two chromatids. This point of "change of partner" is the chiasma.

Chimera. A mixture of tissues of genetically different constitution in the same organism. It may result from mutation, irregular mitosis, somatic crossing over, or artificial fusion (grafting); may be: periclinal with parallel layers of genetically different tissues; or sectorial.

Chloroplastid. Green structure in plant cytoplasm that contains cholorphyll and in which starch is synthesized. A mode of cytoplasmic inheritance independent of nuclear genes has been associated with these cytoplasmic structures.

Chondriosomes. *See* Mitochondria.

Chondrodystrophic children. Children with a hereditary abnormality of the bones.

Chondrodystrophy. Trait in man characterized by abnormal growth of cartilage at ends and along shafts of long bones.

Chromatid. One of the two identical strands resulting from self-duplication of a chromosome during mitosis or meiosis. One of the four strands making up a bivalent during late meiotic prophase.

Chromatin. The nuclear substance that takes basic stain and becomes incorporated in the chromosomes, so called because of the readiness with which it becomes stained with certain dyes (chromaticity).

Chromatography. A method for separating and identifying the components from mixtures of molecules having similar chemical and physical properties.

Chromocenter. Body produced by fusion of the heterochromatin regions of the autosomes and Y chromosome in salivary gland preparations of certain Diptera.

Chromomeres. Small bodies described by

Belling which he identified by their characteristic size and linear arrangement on the chromosome thread.

Chromonema, pl. chromonemata. An optically single thread forming an axial structure within each chromosome.

Chromosome aberration. Abnormal arrangement of parts of a chromosome caused by chromosomal breakage and reunion.

Chromosomes. Microscopically observable nucleoprotein bodies, dark-staining with basic dyes, in the cell during cell division. They carry the genes that are arranged in linear order. Each species has a characteristic chromosome number.

Cilium; pl. cilia; adj. ciliate. Hairlike locomotor structure on certain cells; a locomotor structure on a ciliate protozoan.

Cis. *See* Coupling.

Cistron. A unit of function in a DNA system; a working definition of a gene. One cistron in the DNA specifies one polypeptide in protein synthesis.

Clone. All the individuals derived by vegetative propagation from a single original individual.

Codominant genes. Alleles, each of which produces an independent effect when heterozygous.

Codon. A set of three adjacent nucleotides that will code one amino acid (or chain termination).

Coenzyme. A substance necessary for the activity of an enzyme.

Coincidence. The ratio of observed double crossovers to expected doubles calculated on the basis of probability for independent occurrence and expressed as a decimal fraction.

Colchicine. An alkaloid derived from the autumn crocus that is used as an agent to arrest spindle formation and interrupt mitosis.

Colinearity. The sequence of nucleotides in a cistron corresponds with the order of the polypeptide it specifies.

Competence. Ability of a bacterial cell to incorporate DNA and become genetically transformed.

Complementary genes. Genes that are similar in phenotypic effect when present separately, but which together interact to produce a different trait. If two such genes are complementary for a dominant effect, a $9:7$ ratio results in F_2; if two are complementary for a recessive effect, a $15:1$ ratio results in F_2.

Complementation map. An illustration of a series of mutants in a chromosome segment showing mutually complementing mutants as nonoverlapping lines and noncomplementing mutants as overlapping, continuous lines.

Complementation test (cis-trans test). Introduction of two mutant chromosomes into the same cell to determine whether the two mutations occurred in the same gene. If the phenotype is wild type, each chromosome complements the defect of the other and the genes may be written in the trans arrangement $(a+/+b)$. If the mutations are allelic, a mutant phenotype will be expressed and the genotype of the hybrid may be written $ab/++$.

Complementation unit. A subunit of the complementation map of a cistron as determined by a complementation test.

Concordance. Identity of matched pairs or groups for a given trait, e.g., identical twins both expressing the Down syndrome.

Conidium, pl. conidia. An asexual spore produced by a specialized hypha in certain fungi.

Conjugation. Union of sex cells (gametes) or unicellular organisms during fertilization; in *E. coli*, a one-way transfer of genetic material from a donor ("male") to a recipient ("female").

Constitutive enzyme. An enzyme that is produced in fixed quantities regardless of need (c.f. inducible and repressible enzymes).

Continuous variation. Variation not represented by distinct classes. Individuals grade into each other and measurement data are required for analysis, c.f. discontinous variation. Multiple genes or polygenes are usually responsible for this type of variation.

Coordinate repression. Control of structural genes in an operon by a single operator gene.

Copolymers. Mixtures consisting of more than one polymer. *Example:* polymers of two kinds or organic bases such as uracil and cytosine (poly-UC) have been combined for studies of genetic coding.

"Copy choice." An explanation for crossing over first suggested by J. Belling in 1930, which assumes that crossing over occurs during the process of chromosome duplication. Duplication or "copying" proceeds partially along one homologue and partially along the other.

Coupling or cis-arrangement. The condition in linked inheritance in which an individual heterozygous for two pairs of genes received the two dominant members from one parent and the two recessives from the other parent, (e.g., *AABB × aabb*). Compare repulsion.

Covalent bond. A bond in which an electron pair is equally shared by protons in two adjacent atoms.

Cross breeding. Mating between members of different races or species.

Crossing over. A process inferred genetically by new association of linked genes and demonstrated cytologically from new associations of parts of chromosomes. It results in an exchange of genes and therefore produces combinations differing from those characteristic of the parents. The term "genetic crossover" may be applied to the new gene combinations (*See* Recombination).

Crossover unit. A frequency of exchange of 1 percent between two pairs of linked genes; 1 percent of crossing over is equal to 1 unit on a linkage map.

Cytogenetics. Area of biology concerned with chromosomes and their implications in genetics.

Cytokinesis. Cytoplasmic division and other changes exclusive of nuclear division that are a part of mitosis or meiosis.

Cytology. The study of the structure and function of the cell.

Cytoplasm. The protoplasm of a cell outside the nucleus in which cell organelles (mitochondria, plastids, etc.) are located. All living parts of the cell except the nucleus.

Cytoplasmic inheritance. Hereditary transmission dependent on the cytoplasm or structures in the cytoplasm rather than the nuclear genes; extrachromosomal inheritance. *Example:* Plastid characteristics in plants may be inherited by a mechanism independent of nuclear genes.

Cytosine. A pyrimidine base found in RNA and DNA.

Deficiency (deletion). Absence of a segment of a chromosome involving one or more genes.

Deme. A local interbreeding group of plants or animals.

Determination. Process by which embryonic parts become capable of developing into only one kind of adult tissue or organ.

Deviation. As used in statistics, a variation from an expected number.

Diakinesis. A stage of meiosis just before metaphase I in which the bivalents are shortened and thickened.

Dicentric chromosome. One having two centromeres.

Differentiation. Modification of different parts of the body for particular functions during development of the organism.

Dihybrid. An individual that is heterozygous with respect to two pairs of alleles. The product of a cross between homozygous parents differing in two respects.

Dimer. A compound having the same percentage composition as another but twice the molecular weight; one formed by polymerization.

Dimorphism. Two different forms in a group as determined by such characteristics as sex, size, or coloration.

Dioecious plant. A unisexual plant; each plant is either a male or a female (c.f. Monoecious).

Diploid. An organism or cell with two sets of chromosomes ($2n$) or two genomes. Somatic tissues of higher plants and animals are ordinarily diploid in chromosome constitution in contrast with the haploid (monoploid) gametes.

Diplonema, adj. diplotene. That stage in prophase of meiosis following the pachytene stage, but preceding diakinesis, in which the chromosomes are visibly double; stage characterized by centromere repulsion of bivalents resulting in the formation of loops.

Discontinuous variation. Distinct classes such as red vs. white, tall vs. dwarf, c.f. continuous variation.

Discordance. Dissimilarity of matched pairs or groups for a given trait, e.g., identical twins with different eye color.

Disjunction. Separation of homologous chromosomes during anaphase of mitosis or meiotic divisions (*See* Nondisjunction).

Disome. *See* Monosomic.

Distinguishable hybrid. A hybrid in which intermediate inheritance is expressed, (i.e., the heterozygous combination is distinguishable by a phenotype).

Ditype. A tetrad that contains two kinds of meiotic products. e.g. $4AB$ and $4ab$.

Dizygotic twins. Two-egg or fraternal twins.

DNA. Deoxyribonucleic acid, the chemical material of which the information-carrying material or genes are composed.

DNase. Any enzyme that hydrolyzes DNA.

Dominance. Applied to one member of an allelic pair of genes that has the ability to manifest itself wholly or largely at the exclusion of the expression of the other member. An inherited trait expressed when the controlling gene is either homozygous or heterozygous.

Drift (*See* Random genetic drift).

Duplication. The occurrence of a segment more than once in the same chromosome or genome.

Dysgenic. A situation that tends to be harmful to the hereditary qualities of future generations. (c.f., eugenic).

Ectoderm. Outside cellular layer of an early animal embryo that gives rise to the outer skin and nervous system.

Egg (ovum). A germ cell produced by a female organism.

Electrophoresis. The migration of suspended particles in an electric field.

Embryo. A young organism in the early stages of development; in man, first period in uterus.

Embryo sac. A large thin-walled space within the ovule of the seed plant in which the egg and, after fertilization, the embryo develop; the mature female gametophyte in higher plants.

Endoderm. Inner layer of an early animal embryo that gives rise to the lining of the digestive tract.

Endogenote. The part of the bacterial chromosome that is homologous to a genome fragment (exogenote) transferred from the donor to the recipient cell in the formation of a merozygote.

Endomitosis. Duplication of chromosomes without division of the nucleus resulting in increased chromosome number within cells or endopolyploidy. Chromosome strands separate but the cells do not divide.

Endonuclease. An enzyme that breaks

strands of DNA and thus is involved in replication and recombination of DNA.

Endoplasmic reticulum. Network in the cytoplasm to which ribosomes adhere.

Endopolyploidy. Occurrence of cells in diploid organisms containing multiples of the $2n$ genomes (i.e., $4n$, $8n$, etc.)

Endosperm. Nutritive tissue arising in the embryo sac of most angiosperms. It usually follows the fertilization of the two fused primary endosperm nuclei of the embryo sac by one of the two male gametes. In most diploid organisms the endosperm is triploid $(3n)$ but in some, for example, lily, the endosperm is $5n$.

Enhancer. A substance or object that increases a chemical activity or a physiological process, a major or modifier gene that increases a physiological process.

Environment. The aggregate of all the external conditions and influences affecting the life and development of the organism.

Enzyme. A protein that accelerates a specific chemical reaction in a living system.

Epigenesist. One who visualized embryological development as a step-by-step process from a relatively undifferentiated zygote to a complex adult.

Episome. A genetic element that may be present or absent in different cells associated with a chromosome or independent in the cytoplasm. *Example:* Fertility factor (F) in *E. coli.*

Epistasis. The suppression of the action of a gene or genes by a gene or genes not allelomorphic to those suppressed. Those suppressed are said to be hypostatic. Distinguished from dominance that refers to the members of one allelomorphic pair.

Equational or homotypic division. Mitotic-type division that is usually the second division in the meiotic sequence; somatic mitosis and the nonreductional division of meiosis.

Equatorial plate. The figure formed by the chromosomes in the center (equatorial plane) of the spindle in mitosis.

Estrogen. Female hormone or estrus-producing compound.

Euchromatin. Parts of chromosomes that carry Mendelian genes and have characteristic staining properties. Most of the armlike parts of Drosophila salivary chromosomes are euchromatin. (c.f., heterochromatin).

Eugenic. A situation that tends toward improvement in the hereditary qualities of future generations of mankind. (c.f., dysgenic).

Eugenics. The science of improving the qualities of the human race; the application of the principles of genetics to the improvement of mankind.

Eukaryote. A member of the large group of higher organisms composed of cells with true nuclei that are enclosed in nuclear envelopes. These organisms undergo meiosis (c.f., prokaryote).

Euploid. An organism or cell having a chromosome number that is an exact multiple of the monoploid (n) or haploid number. Terms used to identify different levels in an euploid series are: diploid, triploid, tetraploid, etc. (c.f., heteroploid and aneuploid).

Exogenote. Chromosomal fragment homologous to an endogenote and donotated to a merozygote.

Exonuclease. An enzyme that digests DNA, beginning at the ends of strands.

Expressivity. Degree of expression of a trait controlled by a gene. A particular gene may produce varying degrees of expression in different individuals.

Extrachromosomal. Structures that are not a part of the chromosomes; DNA units in the cytoplasm that control cytoplasmic inheritance.

F_1. The first filial generation. The first generation of descent from a given mating.

F_2. The second filial generation. produced by crossing *inter se* or by self-pollinating the F_1. The inbred "grandchildren" of a given mating. The term is loosely used to indicate any second generation progeny from a given mating, but in controlled genetic experimentation, inbreeding of the F_1 (or equivalent) is implied.

F factor. Fertility factor in a bacterium. In the presence of F^+ a bacterial cell functions as a male.

Fertility. Ability to produce offspring.

Fertilization. The fusion of a male gamete (sperm) with a female gamete (egg) to form a zygote.

Fetal hydrops. A form of dropsy in the newborn caused by incompatability between an Rh-negative mother and her Rh-positive fetus during pregnancy.

Fetus. Prenatal stage of a viviparous animal between the embryonic stage and the time of birth. In man, the final seven months before birth.

Filial. *See* F_1 and F_2.

Fitness. The number of offspring left by an individual as compared with the average of the population, or compared to individuals of different genotype.

Flagellum; pl. flagella; adj. flagellate. A whiplike organelle of locomotion in certain cells; locomotor structures in flagellate protozoa.

Founder principle. A newly isolated population soon diverges from the parent population because of sampling errors. Parent and daughter populations operate on different gene pools.

Freemartin. A sexually underdeveloped female calf born twined with a male.

Gamete. A mature male or female reproductive cell (sperm or egg).

Gametogenesis. The formation of the gametes.

Gametophyte. That phase of the plant life cycle that bears the gametes; cells have n chromosomes.

Gastrula. An early animal embryo consisting of two layers of cells; an embryo-logical stage following the blastula.

Gene. A particulate hereditary determiner; a unit of inheritance; a unit of DNA; located in a fixed location in the chromosome.

Gene flow. The spread of genes from one breeding population to another by migration which may result in changes in gene frequency.

Gene frequency. The proportion of one allele as represented in a breeding population.

Gene pool. Sum total of all different alleles or genetic information in the breeding members of a population at a given time.

Genetic drift. *See* Random genetic drift.

Genetic equilibrium. Condition in a group of interbreeding organisms in which particular gene frequencies remain constant throughout succeeding generations.

Genetic load. The proportion by which the fitness of the optimum genotype is decreased by deleterious genes; expressed in lethal equivalents or "genetic deaths." In the human population the genetic load was estimated by Muller to be 4 lethal equivalent or recessive genes that are lethal as homozygotes.

Genetics. The science of heredity and variation.

Genome. A complete set of chromosomes (hence of genes), inherited as a unit from one parent.

Genotype. The genetic constitution (gene makeup), expressed and latent of an organism (i.e., *Dd* or *dd*.) Individuals of the same genotype breed alike. (c.f., phenotype).

Germ cell. A reproductive cell capable when mature of being fertilized and reproducing an entire organism.

Germ plasm. The germinal material or physical basis of heredity. The sum total of the DNA.

Globulins. Common proteins found in the blood, that are insoluble in water and soluble in salt solutions. Alpha, beta, and gamma globulins can be distinguished in human-blood serum. Gamma globulins are important in developing immunity to diseases.

Gonad. A sexual gland, (i.e., ovary or testis).

Guanine. A purine base found in DNA and RNA.

Gynandromorph. An individual in which one part of the body is female and another part is male; a sex mosaic.

Haploid or monoploid. An organism or cell having only one complete set (n) of chromosomes or one genome.

Haptoglobin. A serum protein, alpha globulin in the blood.

Hardy-Weinberg equilibrium. Mathematical relation between gene frequencies and genotype frequencies within populations. At equilibrium frequencies of genotypic classes are p^2 (AA), $2pq$(Aa) and q^2(aa).

Helix. Any structure with a spiral shape. The Watson and Crick model of DNA is in the form of a double helix.

Hemizygous. The condition in which only one allele of a pair is present, as in sex linkage or resulting from deletion.

Hemoglobin. Conjugated protein compound containing iron, located in erythrocytes of vertebrates; important in the transportation of oxygen to the cells of the body.

Hemolymph. The mixture of blood and other fluids in the body cavity of an invertebrate.

Hemophilia. A bleeder's disease; tendency to bleed freely from even a slight wound; hereditary condition dependent on a sex-linked recessive gene.

Heredity. Resemblance among individuals related by descent; transmission of traits from parents to offspring.

Heritability. Degree to which a given trait is controlled by inheritance.

Hermaphrodite. An individual with both male and female reproductive organs. *See* Monoecious plant.

Heterocaryon. A fungus hypha with two nuclei of different genotypes; the nuclei do not fuse but divide independently and simultaneously as new cells are formed.

Heterochromatin. Chromatin staining differently and functioning differently than the euchromatin that contains the Mendelian genes. In Drosophila salivary preparations the heterochromatin including the Y chromosome is mostly in the chromocenter. The nucleolar organizing region, the site of ribosomal RNA synthesis is flanked on either side by heterochromatin.

Heterochromatinization. Cytological transformation of euchromatin to the heterochromatin state.

Heterogametic sex. Producing unlike gametes, particularly with regard to the sex chromosome. In species in which the male is "XY", the male is heterogametic, the female (XX), homogametic.

Heteronucleation. Control of development of a body part by a template or seed superimposed on a basic genetic pattern. Dissolved microtubules of a cilium, for example, can be reassembled experimentally but will conform to the pattern of the seed that is present.

Heteroploid or aneuploid. An organism characterized by a chromosome number other than the true haploid (monoploid) or diploid number ($2n + 1$ or $2n - 1$). *See* Euploid.

Heteropyknosis, adj, heteropyknotic. Property of certain chromosomes or of their parts to remain more dense and stain more intensely than other chromosomes or parts during the nuclear cycle.

Heterosis. Superiority of heterozygous genotypes in respect to one or more traits in comparison with corresponding homozygotes.

Heterozygote, adj. heterozygous. An organism with unlike members of any given pair or series of alleles, that consequently produces unlike gametes.

Hfr. High frequency recombination strain of "*Escherichia coli*" the F episome is integrated into the bacterial chromosome.

Histone. Groups of proteins rich in basic amino acids. They may function in coiling of DNA in chromosomes and regulating gene activity.

"Holandric" gene. A gene carried on the Y chromosome and therefore transmitted from father to son.

Homogametic sex. Producing like gametes (c.f., heterogametic sex.)

Homologous chromosomes. Chromosomes that occur in pairs and are generally similar in size and shape, one having come from the male and one from the female parent.

Homozygote, adj. homozygous. An organism whose chromosomes carry identical members of any given pair of alleles. The gametes are therefore all alike with respect to this locus and the individual will breed true.

Hormone or internal secretion. An organic product of cells of one part of the body that is transported by the body fluids to another part where it influences activity or serves as a coordinating agent.

Hybrid. An offspring of homozygous parents differing in one or more genes.

Hybridization. Interbreeding of species, races, varieties, etc. among plants or animals. A process of forming a hybrid by cross pollination of plants or by mating animals of different types.

Hybrid vigor or heterosis. Unusual growth, strength, and health of hybrids from two less vigorous parents.

Hydrocephalus. Abnormal increase in the amount of cerebral fluid resulting in enlargement of the head and other symptoms.

Hypha, pl. hyphae. A branched filament of a fungus.

Hypostasis. *See* Epistasis.

Idiogram. A diagrammatic representation of the chromosomes of an individual illustrating their relative sizes and appearance.

Immunize. To induce a resistance to a parasite or foreign substance. Noun: immunization.

Immunoglobulin. *See* Globulin.

Inbreeding. Matings among related individuals.

Incomplete dominance. Expression of heterozygous alleles different from those of the parents, producing distinguishable hybrids.

Independent combinations or independent assortment. The random distribution of genes to the gametes that occurs when genes are located in different chromosomes. The distribution of one pair of genes is not controlled by other genes located in nonhomologous chromosomes.

Inducer. A substance of low molecular weight that increases the proportion of a repressor and thus decreases the repression of enzyme synthesis.

Inducible enzyme. An enzyme that is synthesized only in the presence of the substratum that acts as an inducer.

Inhibitor. Any substance or object that retards a chemical reaction; a major or modifier gene that interferes with a reaction.

Interference. Crossing over at one point reduces the chance of another crossover in adjacent regions. Detected by studying crossovers of three or more linked genes.

Intermediate inheritance. An alternative to dominance in which the heterozygotes are distinguishable from both homozygotes.

Interphase. The stage in the cell cycle when the cell is not dividing; the metabolic stage; the stage following telophase of one division and extending to the beginning of prophase in the next division.

Intersex. An organism displaying secondary

sexual characters intermediate between male and female; a type that shows some phenotypic characteristics of both males and females.

Inversion. A rearrangement of a group of genes in a chromosome in such a way that their order in the chromosome is reversed.

Isoagglutinogen. An antigen such as A or B blood-type factor that occurs normally (i.e. in an individual, without artificial stimulation.)

Isogenic stocks. Strains of organisms that are genetically uniform; completely homozygous.

Kappa particles. DNA containing, self reproducing cytoplasmic particles in certain strains of *Paramecium aurelia*. They control a toxic substance *paramecin* that is released into the culture medium and kills sensitive paramecia. Nuclear gene *K* is required for maintenance of kappa in the cytoplasm of killers.

Karyotype. The appearance of the metaphase chromosomes of an individual or species; comparative size, shape, and morphology of the different chromosomes.

Kinetochore. *See* Centromere.

Kinetosome. Granular body at the base of a flagellum or a cilium.

Kinomere. *See* Centromere.

Kynurenine. A compound derived from tryptophan metabolism that occurs in the urine of rabbits and under certain conditions in the urine of other animals.

Lampbrush chromosomes. Greatly enlarged chromosomes in the oöcytes of amphibians. They have a main axis and side loops suggesting the name "lampbrush."

Leptonema, adj. leptotene. Stage in meiosis immediately preceding synapsis in which the chromosomes appear as single fine threadlike structures.

Lethal gene. A gene that renders inviable an organism or a cell possessing it in proper arrangement for expression.

Ligase. An enzyme that joins two segments of a broken strand of double stranded DNA.

Line breeding. Mating of selected members of successive generations among themselves to fix desirable characteristics.

Linkage. Association of genes that are physically located in the same chromosome. Such a group of linked genes is called a linkage group.

Linkage map. Genes of a given species listed in linear order to show their relative positions on the chromosomes.

Locus, pl. loci. A fixed position on a chromosome occupied by a given gene or one of its alleles.

Lysis. Bursting of a bacterial cell by the destruction of the cell membrane following infection by bacteriophage.

Lysogenic bacteria. Those harboring temperate bacteriophages.

Macromolecule. A large molecule; term used to identify molecules of proteins, nucleic acids and other large molecules.

Map units. One percent of crossing over represents one unit on a linkage map.

Maternal effect. Predetermination by the genes of the mother.

Maternal inheritance. Inheritance controlled by extrachromosomal (e.g., cytoplasmic) factors.

Maturation. The formation of gametes or spores.

Mean. The arithmetic average; the sum of all measurements or values of a group of objects divided by the number of objects.

Megaspore (macrospore). A plant spore having the property of giving rise to a gametophyte (embryo sac) bearing only a female gamete. One of the four cells produced by two meiotic divisions of the megaspore-mother-cell called a megasporocyte.

Megaspore mother cell or megasporocyte. The cell that undergoes two meiotic divisions to produce four megaspores.

Megasporogenesis. Process of production of megaspores. *See* Megaspore.

Meiosis. The process by which the chromosome number of a reproductive cell becomes reduced to half the diploid ($2n$) or somatic number; results in the formation of gametes in animals, or of spores in plants; important source of variability through recombination.

Meiotic drive. Any mechanism that results in unequal gametes produced by a heterozygote and thus affects the genetic composition of a population.

Melanin. Brown or black pigment of animal origin.

Mendelian population. A natural interbreeding unit of sexually reproducing plants or animals sharing a common gene pool.

Merozygote. Partial zygote produced by a process of partial genetic exchange such as transformation in bacteria. An exogenote may be introduced into a bacterial cell in the formation of a merozygote.

Mesoderm. The middle germ layer that forms in the early animal embryo that gives rise to such parts as bone and connective tissue.

Messenger RNA. A particular kind of RNA that carries information necessary for protein synthesis from the DNA to the ribosomes.

Metabolic cell. A cell that is not dividing.

Metabolism. Sum total of all chemical processes in living cells by which energy is provided and used.

Metacentric chromosome. One with the centromere in the middle and two arms of about equal length.

Metafemale or superfemales. In Drosophila, abnormal type of female usually sterile with an overbalance of X chromosomes with respect to autosomes.

Metamorphosis. Change of form, structure, or substance.

Metaphase. That stage of cell division in which the chromosomes are most discrete and arranged in an equatorial plate; following prophase and preceding anaphase.

Microspore. One of the four cells produced by the two meiotic divisions of the microspore-mother-cell or microsporocyte. A spore having the property of giving rise to a gametophyte bearing only male gametes.

Microspore mother cell. *See* Pollen mother cell.

Microsporogenesis. Process of production of microspores. *See* Microspore.

Microtubules. Hollow filaments in the cytoplasm making up a part of the locomotor apparatus of many motile cells.

Migration. Movement of individuals or groups from one population to another resulting in transfer of genetic material which may change gene frequencies in the population to which the migrants move.

Mitochondria. Small DNA containing bodies in the cytoplasm of most plant and animal cells.

Mitosis. Cell division in which there is first a duplication of chromosomes followed by migration of chromosomes to the ends of the spindle and dividing of the cytoplasm.

Modifier or modifying gene. A gene that affects the expression of another gene.

Modulation. The more frequent translation of particular sequences of messenger RNA; those sequences at the beginning of the genetic transcription are more likely to be read.

Monoecious plant. Plant with separate staminate (male) and pistillate (female) flowers on the same plant.

Monohybrid. An offspring of two homozygous parents differing from one another by alleles at only one gene locus.

Monohybrid cross. A cross between parents

differing in only one trait or in which only one trait is being considered.

Monomer. A simple molecule of a compound of relatively low molecular weight.

Monoploid or haploid. Individual having a single set of chromosomes or one genome (*n*)

Monosomic. A diploid organism lacking one chromosome of its proper complement ($2n - 1$); an aneuploid. Monosome refers to the single chromosome, disome to two chromosomes of a kind, trisome to three chromosomes of a kind.

Monozygotic twins. One-egg or identical twins.

Morphology. Study of the form of an organism. Developmental history of visible structures and the comparative relation of similar structures in different organisms.

Morula. A mass of cells formed by repeated cleavage in early animal embryology.

Mosaic. An organism part of which is made up of tissue genetically different from the remaining part.

Mosaic egg. A fertilized egg (zygote) in which a high degree of organization has already occurred. Parts develop according to a predetermined plan.

Multiple alleles. Three or more alternative alleles representing the same locus in a given pair of chromosomes.

Mutable genes. Those with an unusually high mutation rate (unstable genes).

Mutagen. An environmental agent, either physical or chemical that is capable of inducing mutations.

Mutant. A cell or individual organism that shows a change brought about by a mutation. A changed gene.

Mutation. A change in the DNA at a particular locus in an organism. The term is used loosely to include point mutations involving a single gene, and chromosomal changes. *Recurrent m.* One way mutations, e.g., $A \to a$.

Mutation Pressure. A constant mutation rate that adds mutant genes to a population.

Muton. The smallest unit of DNA that can undergo change resulting in a mutation.

Mycelium, pl. mycelia. Threadlike filament making up the vegetative portion of thallus fungi.

Natural selection. Natural processes favoring individuals that are better adapted and tending to eliminate those unfitted to their environment.

Nebenkern. A structure derived from clumped mitochondria in developing sperm that becomes a two-stranded helical body surrounding the proximal region of the tail filament of the mature sperm.

Nondisjunction. Failure of disjunction or separation of homologous chromosomes in meiosis, resulting in too many chromosomes in some daughter cells and two few in others.

Nucleic acid. An acid composed of phosphoric acid, pentose sugar, and organic bases. DNA and RNA are nucleic acids.

Nucleolar organizer. A chromosomal segment that controls the synthesis of ribosomal RNA.

Nucleolus. Structure within the nucleus of some metabolic cells; storage area for ribosomal RNA.

Nucleoprotein. Conjugated protein composed of nucleic acid and protein, and making up the chromosomes.

Nucleotide. A unit of the DNA molecule containing a phosphate, a sugar, and an organic base.

Nucleus. Part of a cell containing genes and surrounded by cytoplasm.

Nullisomic. An otherwise diploid cell or organism lacking both members of a chromosome pair (chromosome formula: $2n - 2$).

Octoploid. Cell or organism with eight genomes or monoploid sets of chromosomes; a polyploid.

Ommochrome. A product of tryptophan metabolism that gives rise to pigments, particularly eye pigments in animals.

Ontogeny. The complete developmental history of an organism from egg, spore, bud, etc., to the adult individual.

Oöcyte. The egg-mother cell; the cell that undergoes two meiotic divisions (oögenesis) to form the egg cell. Primary oöcyte—before completion of the first meiotic division; secondary oöcyte—after completion of the first meiotic division.

Oögenesis. The formation of the egg or ovum in animals.

Oögonium, pl. oögonia. A germ cell of the female animal before meiosis begins.

Operator gene. A part of an operon that controls the activity of one or more structural genes.

Operon. A group of genes making up a regulatory or control unit. The unit includes an operator and structural genes.

Organelle. Specialized part of a cell with a particular function or functions (e.g., cilium of a protozoan).

Organizer. An inductor; a chemical substance in a living system that determines the fate in development of certain cells or groups of cells.

Otocephaly. Abnormal development of the head of a mammalian fetus.

Outbreeding. Mating of unrelated individuals or of those not closely related.

Ovary. The swollen part of the pistil of a plant flower that contains the ovules. The female reproductive gland or gonad in animals.

Ovule. The macrosporangium of a flowering plant that becomes the seed. It includes the nucellus and the integuments.

P. Symbolizes the parental generation or parents of a given individual.

Pachynema, adj. pachytene. A midprophase stage in meiosis immediately following zygonema and preceding diplonema. In favorable microscope preparations, the chromosomes are visible as long, paired threads. Sometimes four chromatids are observed.

Panmictic population. A population in which mating occurs at random.

Panmixis. Random mating in a population in contrast to nonrandom mating.

Paracentric inversion. An inversion that is entirely within one arm of a chromosome, does not include the centromere.

Parameter. A value or constant based on an entire population (c.f., statistic).

Parthenogenesis. The development of a new individual from an egg without fertilization.

Paternal. Pertaining to the father; set of chromosomes derived from the sperm in animals or pollen in plants.

Pathogen. An organism or virus that causes a disease.

Pedigree. A table, chart, or diagram representing the ancestral history of an individual.

Penetrance. The proportion (in percent) of individuals with a particular gene combination that express the corresponding trait.

Peptide. A compound containing amino acids. A breakdown or buildup unit in protein metabolism.

Peptide bond. A chemical bond holding amino acid subunits together.

Pericentric inversion. An inversion including the centromere, hence involving both arms of a chromosome.

Petites. Colonies of yeast cells that grow slowly because they lack respiratory enzymes ordinarily produced in mitochondria; dwarfs.

Phage. *See* Bacteriophage.

Phenocopy. An organism whose phenotype (but not genotype) has been changed by the environment to resemble the phenotype of a different (mutant) organism.

Phenogroup. A term used to describe a large number of antigenic responses associated with a particular locus. In cattle a large number of antigenic responses presumably determined by alleles has been associated with the *B* locus.

Phenotype. Characteristic of an individual observed or discernable by other means (i.e., tallness in garden peas; color blindness or blood type in man). Individuals of the same phenotype may appear alike but may not breed alike.

Phenylalanine. *See* amino acid.

Phenylketonuria. Metabolic disorder resulting in mental retardation; transmitted as a Mendelian recessive and treated in early childhood by special diet.

Pistil. The female part of a flower consisting of ovary, style, and stigma.

Plaque. Clear area on an otherwise opaque culture plate of bacteria where the bacteria have been killed by a virus.

Plasmagenes. Self-replicating cytoplasmic particles capable of transmitting traits in inheritance. Units believed to be responsible for some extranuclear inheritance.

Plastid. A cytoplasmic body found in the cells of plants and some protozoans. Chloroplastids produce chlorophyll which is involved in photosynthesis.

Pleiotropy, adj. pleiotropic. Condition in which a single gene influences more than one trait.

Polar bodies. In female animals, the smaller cells produced at meiosis that do not develop into egg cells. The first polar body is produced at division I and may not go through division II. The second polar body is produced at division II.

Polarity mutation. Gene mutation that influences the functioning of other more distal genes in the same operon.

Pollen mother cell, microsporocyte or microspore-mother-cell. The plant cell that undergoes the meiotic sequence and produces four microspores.

Pollen parent. The male plant that produces the pollen. The term is used to designate the pollen producing parent in a cross.

Polydactyly. The occurrence of more than the usual number of fingers or toes.

Polygene. One of a series of multiple genes involved in quantitative inheritance.

Polymer. A compound composed of two or more units of the same substance; results from a process of polymerization.

Polymerase. DNA p specific enzyme which catalyzes the synthesis of DNA. **RNA p** specific enzyme which catalyzes the formation of ribonucleic acid.

Polymerization. Chemical union of two or more molecules of the same kind to form a new compound having the same elements in the same proportions, but a higher molecular weight and different physical properties.

Polymorphism. Two or more kinds of individuals maintained in a breeding population.

Polynucleotide. A linear sequence of joined nucleotides in DNA or RNA.

Polypeptide. A compound containing two or more amino acids and one or more peptide groups. They are called dipeptides, tripeptides, etc., according to the number of amino acids contained.

Polyploid. An organism with more than two sets of chromosomes or genomes (e.g., triploid (3n), tetraploid (4n), pentaploid (5n), hexaploid (6n), heptaploid (7n), octaploid (8n).

Polysaccharide capsules. Carbohydrate coverings with antigenic specificity on some types of bacteria.

Population. Entire group of organisms of one kind; an interbreeding group of plants or animals. The infinite group from which a sample might be taken.

Population (effective). Breeding members of the population.

Population genetics. The branch of genetics that deals with frequencies of alleles and genotypes in breeding populations.

Position effect. A difference in phenotype that is dependent on the position of a gene or group of genes or heterochromatin in relation to other genes.

Prepotency. Ability of an individual to transmit particular characteristics to the offspring.

Primary oöcyte. *See* Oöcyte.

Primary spermatocyte. *See* Spermatocyte.

Probability. Likelihood of occurrence.

Progeny. Offspring of animals or plants; individuals resulting from a particular mating.

Prokaryote. A member of the large group of lower organisms including viruses, bacteria, and blue-green algae that lack well-defined nuclei. They do not undergo meiosis. (c.f., eukaryote).

Prophage. Noninfectious stage of a temperate phage in a bacterial cell.

Prophase. The stages of mitosis or meiosis from the appearance of chromosomes following interphase to metaphase.

Prosthetic group. An organic component of conjugated proteins. Conjugated proteins are a combination of simple protein and another substance called the prosthetic group. *Example:* Nucleoproteins have as prosthetic groups nucleic acids.

Protoplast. A unit body of protoplasm, the mass of living material within a single cell, including cytoplasm and nucleus.

Prototroph. A wild-type organism (bacterium) that will grow on a minimal medium.

Pseudoalleles. Closely linked genes that behave ordinarily as if they were alleles but which have been shown by extensive experiments to be separable by crossing over.

Pseudodominance. Apparent dominance of a recessive allele in the area opposite a chromosome deficiency. A recessive gene may come to expression because its dominant allele is absent.

Pure line. A strain of organisms that is comparatively pure genetically (homozygous) because of continued inbreeding or through other means.

Quadripartite structure (chromosome ring). One having four corresponding parts.

Quadrivalent. A group of 4 chromosomes of the same kind in a cell. They may be united by chiasmata in the first division prophase of meiosis. Quadrivalents result from chromosome translocations.

Race. A genetically distinct segment of a species. Usually a race is geographically isolated from other divisions of the species.

Random genetic drift. Changes in gene frequency in small breeding populations due to chance fluctuations. Called the "Sewall Wright Effect."

Recessive. Applied to one member of an allelic pair lacking the ability to manifest itself when the other or dominant member is present. An inherited trait expressed only when the controlling gene is homozygous.

Reciprocal crosses. A second cross involving the same strains but carried by sexes opposite to those in the first cross, e.g., a female from strain A X a male from strain B and a male from A X female from B are reciprocal crosses.

Recombination. The observed new combinations of traits different from those combinations exhibited by the parents. Percentage of recombination equals percentage of crossing over only when the genes are relatively close together. Cytological chiasma refers to the observed change of partners among chromatids whereas recombination or genetic crossing over refers to the observed genetic result. Random recombination occurs in meiosis.

Recon. The smallest unit of DNA capable of recombination; in bacteria, the smallest unit capable of being integrated or replaced in a host chromosome subjected to the transformation process.

Reduction division or heterotypic division. Phase of meiosis in which the maternal and paternal elements of the bivalent separate, (c.f., equational or homotypic division).

Regression. The correlation between parents and offspring or other related individuals when used as a measure of inheritance.

Regulator gene. A gene that controls the rate of production of another gene or genes. Example: the operon involved in lactose production in *E. coli* has a regulator, an operator and structural genes.

Replication. A duplication process that is accomplished by copying from a template (e.g. reproduction at the level of DNA).

Replicon. A unit of replication. In bacteria replicons are associated with segments of the cell membrane controlling replication and coordinating it with cell division.

Repressible enzyme. Enzyme produced in a cell when its end product is not present. It is repressed as the end product increases.

Repulsion or trans arrangement. The condition in linked inheritance in which an individual heterozygous for two pairs of linked genes received the dominant member of one pair and the recessive member of the other pair from one parent and the reverse arrangement from the other parent, (e.g., *AAbb × aaBB = AaBb*). (c.f. coupling).

Restitution nucleus. A nucleus with unreduced or doubled chromosome number resulting from the failure of a meiotic or mitotic division.

Reticulocyte. A young red blood cell.

Reversion. Appearance of a trait expressed by a remote ancestor; a throwback; atavism.

Ribonucleic acid. *See* RNA.

Ribosome. Cytoplasmic structure in which proteins are synthesized.

RNA Ribonucleic acid. The information carrying material in plant viruses. Certain kinds of RNA are involved in the transcription of genetic information from DNA (i.e., mRNA); transfer of amino acids to the ribosomes for incorporation into proteins (i.e., tRNA); and the makeup of the ribosome (i.e. rRNA).

Roentgen (r). Unit of ionizing radiation.

Secondary oöcyte. *See* Oöcyte.

Secondary spermatocyte. *See* Spermatocyte.

Secretor. A person with a water soluble form of antigen A or B. In such a person the antigen may be detected in body fluids (e.g., saliva) as well as on the erythrocytes.

Seed. The enlarged and matured ovule of a plant embryo in a dormant stage of development.

Seed parent. A female parent in a cross between two plants.

Segregation. The separation of the paternal from maternal chromosomes at meiosis, and the consequent separation of alleles and their phenotypic differences as observed in the offspring. Mendel's first principle of inheritance.

Selection. Differential reproduction of different genotypes. The most important of factors that change the frequencies of alleles and genotypes in a population and thus influence evolutionary change.

Selection pressure. Effectiveness of the environment in changing the frequency of alleles in a population of individuals.

Self-fertilization. Pollen of a given plant fertilizes the ovules of the same plant. Plants fertilized in this way are said to be selfed.

Semisterility. A condition of only partial fertility in plant zygotes (e.g., maize), usually associated with translocation.

Serology, adj. serological. The study of interactions between antigens and antibodies.

Sex chromosomes. Chromosomes that are particularly connected with the determination of sex.

Sexduction. The incorporation of bacterial genes by sex factors and their subsequent transfer by conjugation to a recipient cell.

Sex factor. In bacteria, an episome (F$^+$ in *E. coli*) which enables a bacterial cell to be a donor of genetic material. The episome may be transferred in the cytoplasm during conjugation or it may be integrated into the bacterial chromosome.

Sex-influenced dominance. The dominant expression depends on the sex of the individual (e.g., horns in sheep are dominant in males and recessive in females).

Sex-limited. Expression of a trait in only one sex. Examples: milk production in mammals, horns in Rambouillet sheep, egg production in chickens.

Sex linkage. Association or linkage of a hereditary character with sex; the gene is in a sex chromosome.

Sex mosaic. *See* Gynandromorph.

Sex reversal. A change in the characteristics of an individual from male to female and *vice versa.*

Sexual reproduction. Reproduction involving the formation of mature germ cells (i.e., eggs and sperm).

Sib-mating, crossing of siblings. Matings involving two or more individuals of the same parentage, brother-sister mating.

Sickle-cell anemia. An inherited blood abnormality produced by defective hemoglobin. Red blood cells become sickle shaped under reduced oxygen tension.

Sire. The male parent in mating.

Soma cells. The cells that make up the body in contrast with germ cells that are capable of reproducing the organism.

Somatic cells. Referring to body tissues; having two sets of chromosomes, one set normally coming from the female parent and one from the male, as contrasted with germinal tissue that will give rise to germ cells.

Somatoplasm. The nonreproductive material making up the body of the organism in contrast to germ plasm.

Species. Interbreeding, natural populations that are reproductively isolated from other such groups.

Sperm, abb. of spermatozoön, pl. spermatozoa. A mature male germ cell. Also pl.

Spermatids. The four cells formed by the meiotic divisions in spermatogenesis. Spermatids become mature spermatozoa or sperm.

Spermatocyte or **sperm mother cell.** The cell that undergoes two meiotic divisions (spermatogenesis) to form four spermatids; primary spermatocyte before completion of the first meiotic division; secondary spermatocyte, after completion of the first meiotic division.

Spermatogenesis. The process by which maturation of the gametes (sperm) of the male takes place.

Spermatogonium, pl. spermatogonia. Primordial male germ cell that may divide by mitosis to produce more spermatogonia. A spermatogonium may enter a growth phase and give rise to a primary spermatocyte.

Spermiogenesis. Formation of sperm from spermatids; the part of spermatogenesis that follows the meiotic divisions of spermatocytes.

Spore. A unit of protoplasm capable of developing asexually into a new individual; in higher plants, the haploid product of meiosis that gives rise to male or female gametes.

Sporocyte. The spore mother cell of a plant.

Sporogenesis. Formation of the spore or reproductive element in plants.

Sporophyte. The spore-forming generation in the life cycle of plants, normally diploid.

Stamen. Male part of the flower which includes the pollen-producing anthers and the filament.

Standard deviation. A measure of variability in a population of individuals.

Standard error. A measure of variation of a population of means; used to indicate how well samples represent populations or parameters.

Statistic. A value based on a sample or samples of a population from which estimates of a population value or parameter may be obtained.

Step allelism. The concept of series of alleles with graded effects on the same trait.

Sterility. Inability to produce offspring.

Stigma. Female part of the flower which receives pollen.

Structural gene. A gene that controls actual protein production by determining the amino acid sequence (c.f. operator and regulator genes.)

Style. Part of the pistil between the stigma and the ovary in a flower through which the pollen tube grows.

Sublethal gene. A lethal gene with delayed effect. The gene in proper combination kills its possessor in infancy, childhood, or adulthood.

Symbiont. An organism living in intimate association with another dissimilar organism.

Synapsis. The pairing of homologous chromosomes in the meiotic prophase.

Syndrome. A group of symptoms that occur together and represent a particular disease.

Syngamy. Union of the gametes in fertilization.

Telophase. The last stage in each mitotic or meiotic division in which the chromosomes are assembled at the poles of the division spindle.

Temperate phage. A phage (virus) that invades but does not destroy (lyse) the host (bacterial cell) (c.f. virulent phage).

Template. A pattern or mold. DNA stores coded information and acts as a model or template from which information is taken by messenger RNA.

Terminalization. Repelling movement of the centromeres of bivalents in the diplotene stages of the meiotic prophase, that tends to move the visible chiasmata toward the ends of the bivalents. The point where the exchange of chromatids or the chiasma occurs is believed to be the point of a genetic crossover, but the chiasmata appear to slip toward the ends of the bivalents, and therefore after the diplotene looping begins there is no longer a relation between the chiasma and the point of crossing over.

Test cross. Backcross to the recessive parental type or a cross between genetically unknown individuals with a fully recessive tester to determine whether an individual in question is heterozygous or homozygous for a certain allele. Also used as a test for linkage.

Tetrad. The four cells arising from the second meiotic division in plants (i.e., pollen tetrads). The term is also used to identify the quadruple group of chromatids formed by the association of split homologous chromosomes during meiosis, but the term bivalent is preferred in this usage.

Tetraploid. An organism whose cells contain four haploid (4n) sets of chromosomes or genomes.

Tetrasomic, noun: tetrasome. Pertaining to a nucleus or an organism having four members of one of its chromosomes, the remainder of the chromosomes being normally diploid. (Chromosome formula: 2n + 2).

Tetratype. A tetrad in which the four meiotic products are different, e.g. *AB*, *aB*, *Ab* and *ab*. Crossing over has occurred in such a tetrad.

Thymine. A pyrimidine base found in

DNA. The other three organic bases, adenine, cytosine and guanine are found in both RNA and DNA but in RNA thymine is replaced uracil.

Totipotent cell. An undifferentiated cell such as a blastomere which when isolated develops into a complete embryo.

Trans arrangement. *See* Repulsion.

Transduction. Genetic recombination in bacteria mediated by bacteriophage.

    Abortive t. Bacterial DNA is injected by a phage into a bacterium but it does not replicate. **Generalized t.** Any bacterial gene may be transferred by a phage to a recipient bacterium. **Restricted t.** Transfer of bacterial DNA by a temperate phage is restricted to only one site on the bacterial chromosome.

Transferrin. Blood serum protein, beta globulin. *See* Globulin.

Transfer RNA. A particular kind of RNA that transports amino acids to the ribosome where they are assembled into proteins.

Transformation. Genetic recombination in bacteria brought about by adding foreign DNA to a culture.

Transgressive variation. The appearance in the F_2 (or later) generations of individuals showing a more extreme development of a character than either parent. Assumed to be due to cumulative and complementary effects of genes contributed by the parents of the original hybrid. Adequate testing of variation in the parents is required to establish its occurrence.

Transition. A mutation caused by the substitution of one purine by another purine or one pyrimidine by another pyrimidine in DNA or RNA.

Translocation. Change in position of a segment of a chromosome to another part of the same chromosome or to a different chromosomes.

Transversion. A mutation caused by the substitution of purine for a pyrimidine or a pyrimidine for a purine in DNA or RNA.

Trihybrid. The offspring from homozygous parents differing in three pairs of genes.

Trisomic. An otherwise diploid cell or organism which has an extra chromosome of one pair (chromosome formula: $2n + 1$).

Tryptophan. *See* Amino acids.

Twins. Two individuals from the same birth. Identical twins—from the same fertilized egg. Fraternal twins—from different fertilized eggs.

Tyrosine. *See* Amino acids.

Unipartite structures (chromosomes). Single units.

Univalent. A chromosome unpaired at meiosis.

Uracil. A pyrimidine base found in RNA but not in DNA. In DNA uracil is replaced by thymine.

Variance. The square of the standard deviation. A measure of variation.

Variation. In biology, the occurrence of differences among the individuals of the same species.

Viability. Degree of capability to live and develop normally.

Virulent phage. A phage (virus) that destroys the host (bacterial) cell (see temperate phage).

Wild type. The customary phenotype or standard for comparison.

X chromosome. A chromosome associated with sex determination. In most animals the female has two and the male one.

Xenia. Immediate effect of pollen on the endosperm, due to the phenomenon of double fertilization in the seed plants.

Y chromosome. The mate to the X chromosome in the male of most animal species, usually carries few genes; in Drosophila composed mostly of heterochromatin. In man the Y chromosome carries genes which influence maleness.

Zygonema, adj. zygotene. Stage in meiosis during which synapsis occurs; after leptotene stage and before the pachytene stage in the meiotic prophase.

Zygote. The cell produced by the union of two mature sex cells (gametes) in reproduction: also used in genetics to designate the individual developing from such a cell.

ANSWERS TO PROBLEMS

CHAPTER 2

2.1 (a) All tall; (b) 3/4 tall, 1/4 dwarf; (c) all tall; (d) 1/2 tall, 1/2 dwarf.

2.2 (a) $WW \times ww$; (b) Ⓦ and ⓦ; (c) Ww; (d) $Ww \times Ww$; (e) Ⓦ, ⓦ and Ⓦ, ⓦ

(f)

phenotypes	genotypes	genotypic frequency	phenotypic ratio
round	WW	1	3
	Ww	2	
wrinkled	ww	1	1

2.3 (a) The 3:1 ratio suggests a single pair of genes with the gene for color dominant over that for white.

(b) $CC \times cc$ P

Ⓒ ⓒ   gametes

Cc F_1

$Cc \times Cc$ $F_1 \times F_1$

Ⓒⓒ Ⓒⓒ F_1 gametes

$1CC : 2Cc : 1cc$ F_2

phenotypes	observed	calculated	deviation
colored	198	202.5	−4.5
white	72	67.5	4.5

2.4 (a) Woman, Pp; her father, Pp; her mother, pp. (b) Half of her children are expected to have ptosis.

2.5 (a) 3:1; (b) 3:1; (c) 1:1:2; (d) 2:1:1; (e) 1:1:1:1

2.6 $CCBB \times ccbb$ P

ⒸⒷ ⓒⓑ   gametes

$CcBb \times CcBb$ $F_1 \times F_1$

gametes	ⒸⒷ	Ⓒⓑ	ⓒⒷ	ⓒⓑ
ⒸⒷ	$CCBB$	$CCBb$	$CcBB$	$CcBb$
Ⓒⓑ	$CCBb$	$CCbb$	$CcBb$	$Ccbb$
ⓒⒷ	$CcBB$	$CcBb$	$ccBB$	$ccBb$
ⓒⓑ	$CcBb$	$Ccbb$	$ccBb$	$ccbb$

Summary of F_2:

phenotypes	genotypes	genotypic frequency	phenotypic ratio
checkered red	$CCBB$	1	9
	$CCBb$	2	
	$CcBB$	2	
	$CcBb$	4	
checkered brown	$CCbb$	1	3
	$Ccbb$	2	
plain red	$ccBB$	1	3
	$ccBb$	2	
plain brown	$ccbb$	1	1

2.7 (a) *CCVv × ccVv*; (b) *CcVv × CcVv*; (c) *CcVv × ccvv*.

2.8 (a) *BBRR × bbrr*   P

⟨BR⟩   ⟨br⟩   gametes

BbRr
BbRr × bbrr   F₁ × F₁

gametes	⟨BR⟩	⟨Br⟩	⟨bR⟩	⟨br⟩
⟨BR⟩	*BBRR*	*BBRr*	*BbRR*	*BbRr*
⟨Br⟩	*BBRr*	*BBrr*	*BbRr*	*Bbrr*
⟨bR⟩	*BbRR*	*BbRr*	*bbRR*	*bbRr*
⟨br⟩	*BbRr*	*Bbrr*	*bbRr*	*bbrr*

Summary of F₂ phenotypes: 9 black, long: 3 black, rex: 3 brown, long: 1 brown, rex.
(b) 1/9
(c) *BbRr × bbrr*

gametes	⟨BR⟩	⟨Br⟩	⟨bR⟩	⟨br⟩
⟨br⟩	*BbRr*	*Bbrr*	*bbRr*	*bbrr*

Summary of back cross results: 1 black, long: 1 black, rex: 1 brown, long: 1 brown, rex.

2.9 (a) All red; (b) 1/2 red: 1/2 roan; (c) all roan; (d) 1/4 red: 1/2 roan: 1/4 white; (e) 1/2 roan: 1/2 white; (f) all white; (g) red. Mate red × red and all progeny will be red. If roan animals are mated together red and white as well as roan will be produced.

2.10 (a) 1/2; (b) 1/2; (c) 3/8.

2.11 (a) 81/256; (b) 108/256; (c) 54/256; (d) 12/256.

2.12 (a)   *Dd × Dd*   P

⟨D⟩ ⟨d⟩ ⟨D⟩ ⟨d⟩   gametes

DD 2*Dd*   *dd*   progeny
dies dichaete wild type
Summary: 2 dichaete: 1 wild

(b)   *Dd × dd*   P

⟨D⟩⟨d⟩   ⟨d⟩   gametes

Dd   *dd*   progeny
Summary: 1 dichaete: 1 wild

2.13 (a) 1/4; (b) 1/4; (c) 1/4; (d) 1/4.

2.14 (a) Single autosomal recessive gene (b) 1. parents, *aa × Aa*, progeny, *Aa, aa, Aa, aa*; 2. parents, *Aa × Aa*, progeny *Aa* or *AA, aa, Aa* or *AA, aa*; 3. parents, *Aa × Aa*, progeny, *aa, Aa* or *AA, Aa* or *AA, Aa* or *AA, aa*; 4. parents, *aa × aa*, progeny, *aa, aa, aa, aa*.

2.15

(a) *DDGGWW × ddggww*   P

⟨DGW⟩   ⟨dgw⟩   gametes

DdGgWw   F₁

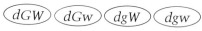
⟨DGW⟩⟨DGw⟩⟨DgW⟩⟨Dgw⟩

F₁ gametes

⟨dGW⟩⟨dGw⟩⟨dgW⟩⟨dgw⟩

(b) *DdGgWw × DdGgWw*

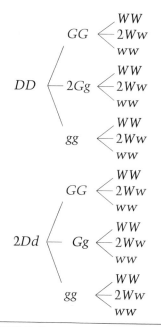

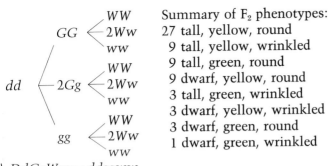

Summary of F₂ phenotypes:
27 tall, yellow, round
9 tall, yellow, wrinkled
9 tall, green, round
9 dwarf, yellow, round
3 tall, green, wrinkled
3 dwarf, yellow, wrinkled
3 dwarf, green, round
1 dwarf, green, wrinkled

(c) $DdGgWw = ddggww$

Summary of backcross results:

	pheno-types	genotypes	genotypic frequency	phenotypic ratio
Dd — *Gg* — Ww	tall, yellow, round	*DdGgWw*	1	1
	ww	tall, yellow, wrinkled	*DdGgww*	1	1
gg — Ww	tall, green, round	*DdggWw*	1	1
	ww	tall, green, wrinkled	*Ddggww*	1	1
dd — *Gg* — Ww	dwarf, yellow, round	*ddGgWw*	1	1
	ww	dwarf, yellow, wrinkled	*ddGgww*	1	1
gg — Ww	dwarf, green, round	*ddggWw*	1	1
	ww	dwarf, green, wrinkled	*ddggww*	1	1

2.16

P cross	$AA \times aa$	$AABB \times aabb$	$AABBCC \times aabbcc$	general formula
F₁ gametes	2	4	8	2^{n*}
F₂ genotypes	3	9	27	3^{n}
F₂ phenotypes**	2	4	8	2^{n}

* n = number of segregating pairs of alleles
** under complete dominance of A, B and C

2.17 The Punnett Square method is illustrated as follows:

$$RRLL \times R'R'L'L' \qquad P$$

$$\widehat{RL} \qquad \widehat{R'L'} \qquad \text{gametes}$$

$$RR'\,LL' \qquad F_1$$

$$RR'LL' \times RR'LL' \qquad F_1 \times F_1$$

gametes	$\widehat{RL}$	$\widehat{RL'}$	$\widehat{R'L}$	$\widehat{R'L'}$
RL	$RRLL$	$RRLL'$	$RR'LL$	$RR'LL'$
$\widehat{RL'}$	$RRLL'$	$RRL'L'$	$RR'LL'$	$RR'L'L'$
$\widehat{R'L}$	$RR'LL$	$RR'LL'$	$R'R'LL$	$R'R'LL'$
$\widehat{R'L'}$	$RR'LL'$	$RR'L'L'$	$R'R'LL'$	$R'R'L'L'$

phenotypes	genotypes	genotypic frequency	phenotypic ratio
red, long	$RRLL$	1	1
red, oval	$RRLL'$	2	2
red, round	$RRL'L'$	1	1
purple, long	$RR'LL$	2	2
purple, oval	$RR'LL'$	4	4
purple, round	$RR'L'L'$	2	2
white, long	$R'R'LL$	1	1
white, oval	$R'R'LL'$	2	2
white, round	$R'R'L'L'$	1	1

2.18 (a) 3 walnut: 1 rose; (b) 1 walnut: 1 pea; (c) 3 walnut: 3 rose: 1 pea: 1 single; (d) 1 rose: 1 single.

2.19 $Rrpp \times RrPp$.

2.20 12 white: 3 yellow: 1 green.

2.21 13 white: 3 colored.

2.22 (a) $\chi^2 = 0.75$, $P = 0.20 - 0.50$, not significant; (b) $\chi^2 = 8.39$; $P = 0.004$ significant

2.23 (a) $\chi^2 = 0.264$, $P = 0.80 - 0.50$, not significant; (b) $\chi^2 = 0.390$, $P = 0.80 - 0.50$, not significant; (c) $\chi^2 = 0.450$, $P = 0.80 - 0.50$, not significant; (d) $\chi^2 = 0.618$, $P = 0.95 - 0.80$, not significant; (e) $\chi^2 = 0.530$, $P = 0.95 - 0.80$, not significant;

2.24 (a) 1/16; (b) 1/16; (c) 4/16; (d) 4/16; (e) 6/16.

2.25 (a) 6/16 or 3/8; (b) 1/16; (c) 2 boys and 2 girls, because more combinations of the four independent events result in 2 and 2 than in any other arrangement; (d) 1/2.

2.26 (a) 6/64; (b) 20/64; (c) 1/64.

2.27 (a) 0; (b) 1/2.

2.28 (a) $1/2 \times 1 \times 1/4 = 1/8$; (b) 1/16; (c) 1/6; (d) 1/24.

2.29 (a) 243/1024; (b) 405/1024; (c) 270/1024; (d) 90/1024; (e) 15/1024; (f) 1/1024.

CHAPTER 3

3.1 (a) Griffith carried out *in vivo* experiments and demonstrated the transformation principle. Avery and others, after long and painstaking *in vitro* investigations, demonstrated beyond question that the transforming principle was DNA. (b) Griffith showed that a transforming substance existed, Avery et al. defined it as DNA. (c) Griffith was only able to show that some chemical carried into live mice could change the genetic material. Avery et al. refined the studies and isolated DNA as the transforming substance.

3.2 (a) All other materials except DNA were successfully ruled out and DNA was established as the genetic material. (b) Transformation has been demonstrated in several bacteria besides *E. coli*.

3.3 The chemical DNA extracted, purified, and introduced separately as a nonliving chemical entity caused transformation. Transformation did not occur in the presence of DNase which degrades DNA but did occur in the presence of RNase and protease.

3.4 To be effective, the DNA must be in a

living system. Combinations of chemicals have not been shown to have a comparable reaction when mixed in a test tube. Transmission of the newly acquired genetic material in later generations could be demonstrated.

3.5 (a) The objective was to determine whether the genetic material was DNA or protein. (b) By labeling phosphorus, a constituent of DNA, and sulfur, a constituent of protein in a virus, it was possible to demonstrate that only the labeled phosphorus was incorporated in the host cell during the cell's reproductive cycle. This was enough to transduce new phages. (c) Therefore DNA, not protein, is the genetic material.

3.6 (a) The ladderlike pattern was known from X-ray diffraction studies. Chemical analyses had shown that a 1:1 relation existed between the organic bases adenine and thymine and between cytosine and guanine. Physical data concerning the length of each spiral and the stacking of bases were also available. (b) Watson and Crick developed the model of a double helix with rigid strands of sugar and phosphorus forming spirals around an axis and hydrogen bonds connecting the complementary base pairs.

3.7 (a) A spiral pattern was interpreted from X-ray diffraction studies of DNA. Double strands were required to fit the chemical data of pairing complementary bases and to fit the requirements for replication and stability of the genetic material. (b) Hydrogen bonds were known to hold complementary bases together, to separate readily during replication, and to fit the chemical requirements of the substances and conditions involved.

3.8 (a) 400,000; (b) 20,000; (c) 400,000; (d) 680,000 Å.

3.9 (a) (1) 1/2 N^{15}, 1/2 N^{14}, each molecule, 2 heavy or 2 light strands; (2) 1/2 N^{15}, 1/2 N^{14}, each molecule, 1 heavy and 1 light;

(b) (1) one molecule 2 heavy to 3 molecules each with 2 light; (2) half of the molecules would contain only light and half would contain 1 heavy and 1 light.

3.10 C A G T A C T G

3.11 (a) DNA has one atom less of oxygen than RNA in the sugar part of the molecule. In DNA, thymine replaces the uracil that is present in RNA. (b) The main function of DNA is to carry information from cell to cell. RNA may also act as an information carrier in some viruses that have no DNA. In cells with both DNA and RNA, mRNA acts as an intermediary in transcribing the information carried by DNA to the ribosomes, where it is used in protein synthesis; tRNA carries amino acids to the ribosomes, where they are assembled into polypeptides; and rRNA makes up an essential part of each ribosome. (c) DNA is located mostly in the chromosomes, whereas RNA is in the cytoplasm, nucleolus, and also in the chromosomes at stages when they are microscopically visible in the cell.

3.12 A C G U C U G U

3.13 G A C T A

3.14 An enzyme, RNA polymerase, was found to catalyze a reaction through which mRNA is synthesized in the cell. With this basis to work from, mRNA was synthesized and prepared in a cell-free system. It was found to have the properties necessary to act as a transcribing agent for the DNA coded message.

3.15 Proteins are long chainlike molecules made of amino acids joined together by peptide linkages. They compose the total nitrogenous substance in vegetable or animal material. Proteins are compounds of carbon, hydrogen, nitrogen, oxygen, and usually sulfur. They provide the structure and enzymatic capacity for the living system. DNA is composed of phosphate, pentose sugar, and four organic bases (adenine, cytosine, guanine, and thymine). It is the information carrier of a living organism. The synthesis of proteins is of concern to the geneticist because a particular protein is coded for by a particular unit of DNA.

3.16 Protein synthesis occurs mainly in the ribosomes. Some protein synthesis apparently occurs in the nucleus, and the mitochondria are involved to a lesser extent.

3.17 Ribosomes are from 100 to 200 Å in diameter. They are located in the cytoplasm of the cells of higher organisms. In bacteria and other organisms that do not have a clear separation between nuclear and cytoplasmic material, as well as in higher organisms, they are associated with the endoplasmic reticulum of the cell. Functionally, they are the main centers for protein synthesis. Chemically, they are composed of protein and rRNA.

3.18 Size and particle weight can be determined by centrifugation of cell extracts. Single ribosomes can be separated from cell extracts by chemicals such as phenol and by precipitation at the proper pH. They can be identified chemically by chromatography.

3.19 (a) Nucleus; (b) cytoplasm

3.20 Messenger RNA is comparatively large and single-stranded. It transcribes information from DNA and carries it to the ribosomes. Transfer RNA is a small single-stranded agent that brings amino acids to the ribosomes for incorporation into proteins. Each tRNA is specific for a particular amino acid. Ribosomal RNA is nonspecific and makes up an integral part of the ribosome.

3.21 (a) Ribosomes held together by mRNA form a polysome. (b) Ribosomal RNA is nonspecific, whereas mRNA and tRNA are specific for complementary elements. (c) The tRNA molecule is much smaller than that of DNA and mRNA. It is single-stranded but folded in such a way that it gives the appearance of being double-stranded.

3.22 A reaction catalyzed by a specific enzyme combines an amino acid with energy rich ATP and forms AA-AMP. The enzyme then transfers AA-AMP to a tRNA molecule forming AA-tRNA and free AMP.

3.23 (a) A synthetic RNA (polyuridylic acid) containing only one base (uracil) was prepared. It was added to a cell-free extract from *E. coli* together with a mixture of 20 amino acids. A small protein-like molecule was produced, which radioactive tracers showed to have only phenylalanine. Uracil was thus shown to code the amino acid, phenylalanine. (b) More efficient means of making and preserving cell-free extracts have been developed along with better methods for preparing synthetic RNAs.

3.24 (a) It is redundant because the same amino acid can be coded by more than one base triplet. Each of the 20 common amino acids is coded by two or more codons. (b) DNA has been found to be the genetic material in virtually all kinds of organisms that have been studied. In some viruses that have no DNA, RNA takes over the function of carrying genetic information. Both RNA and DNA viruses have the same type of code. With the exception of the RNA viruses, DNA is probably the universal genetic material.

3.25 Yanofsky and associates demonstrated from genetic linkage studies and determination of the linear sequence of corresponding amino acid substitutions that the positions of mutational alterations in cistron A correspond in sequence with amino acid substitutions in polypeptide chain A of tryptophan synthetase.

3.26 Colinearity is basic to the mechanism of genetic transcription and genetic translation.

3.27

Requirement	Role
DNA	template
DNA polymerase I	joining free ends
DNA polymerase III	replicating enzyme
polynucleotide ligase	sealing ends for double helix
cell system, deoxyribose,	constituents and energy

necessary bases, phosphates

3.28 Blueprints transcribe into building instructions and translated into structures with boards, bricks and mortar by skilled craftsman may be likened to DNA, mRNA, tRNA activities in assembling amino acids into proteins.

3.29 Khorana demonstrated that the gene could be assembled and thus verified the parts and methods that had been postulated.

CHAPTER 4

4.1 (a) 0; (b) 0; (c) 0; (d) 0; (e) 0; (f) +; (g) +.

4.2 (a) 23; (b) 23; (c) 23; (d) 23; (e) 46; (f) 46; (g) 23; (h) 46.

4.3 (a) 200; (b) 50.

4.4

$$\frac{M}{m} \times \frac{m}{m} \quad \text{P}$$

$\textcircled{M}$ $\textcircled{m}$ $\textcircled{m}$ gametes

$$\frac{M}{m} \qquad \frac{m}{m}$$

F_1 Half of the progeny are expected to have myopia.

4.5 Model in text Fig. 4.4.

4.6 The chromosome mechanism is similar in animals and plants. Division of the cytoplasmic part of the flexible animal cell is accomplished by constriction (cytokinesis) whereas the rigid plant cell forms a partition or cell plate.

4.7 Meiosis includes a pairing (synapsis) of corresponding maternal and paternal chromosomes. In the cell division that follows, the chromosomes that have previously paired, separate. This results in a reduction of chromosome number from $2n$ (diploid) to n (haploid).

4.8 (a) Many plants have male and female parts on the same plant or in the same flower. Unlike animals, plants have a gametophyte stage that consists (in higher plants) of a few cell divisions. (b) The chromosome mechanism is essentially the same in the gamete formation of plants and animals.

4.9 An egg and an endosperm nucleus are developed in the ovule. Two haploid nuclei are introduced by the pollen tube. One male gamete fuses with the egg and the other with the endosperm nucleus. The fertilized egg is the zygote that develops into an embryo. The endosperm forms the nutrient material that supports the developing embryo.

4.10 Model for diagram Fig. 2.11. About $\frac{1}{4}$ of all children would be expected to have only the gene for intestinal polyposis, $\frac{1}{4}$ to have only the gene for Huntington's chorea, $\frac{1}{4}$ to have neither, and about $\frac{1}{4}$ to have both.

4.11 Model in text Fig. 4.4.

4.12 Model in text Fig. 4.4.

4.13 *AaBb*

4.14 Bisexual organisms. Asexual reproduction provides for no genetic variation ex-

cept that of rare mutations. Self-fertilization tends toward homozygosity or pure
lines. Bisexual reproduction in higher
organisms is associated with great hereditary variation through recombination.

4.15 (a) Early primary oöcyte; (b) prophase,
first meiotic division; (c) suspended prophase; (d) first meiotic division is completed just before ovulation of each egg.

4.16 Male Y and female X bearing sperm.

4.17 (a) Female (tetraploid); (b) intersex;
(c) intersex; (d) metamale; (e) female,
(diploid); (f) male (sterile if no Y chromosome is present).

4.18 Female gametes would be (2X2A),
(2XA), (X2A), and (XA). Zygotes and sex
would be 3X3A female (triploid), 3X2A
metafemale, 2X3A intersex, 2X2A female
(diploid), 2XY3A intersex, 2XY2A female,
XY3A metamale, XY2A male.

4.19 (a) male; (b) female; (c) male; (d) female.

4.20 (a) male; (b) female; (c) male; (d) female.

4.21 The single gene (ba) removes the female part of the monoecious plant and
makes the stalk only staminate. Another
gene (ts) transforms the tassel into a
pistillate structure. A plant of the genotype ba ba ts ts would be only pistillate
(female) whereas a plant with the
genotype ba ba ts⁺ ts⁺ would be only
staminate (male).

4.22 (a) 0; (b) 1; (c) 1; (d) 2; (e) 3; (f) 0.

4.23 (a) Female; (b) male; (c) Although genetic determiners for both sexes are
present in the young worms, they become
females when isolated from other worms.
A substance from the female proboscis
stimulates the maleness genes to express
themselves. It can be shown experimentally that an extract made from the
female proboscis will influence young
worms in the direction of maleness.

4.24 (a) $\frac{3}{4}$; (b) $\frac{1}{4}$.

4.25 (a) $\frac{3}{4}$; (b) none.

4.26 (a) 3 hen feathered: 1 cock feathered;
(b) all hen feathered.

4.27 2 rods, 1 hook and 1 rod, 1 knob and
1 rod, 1 hook and 1 knob.

4.28 (a) $\frac{1}{2}$ $X^{bb}X^{bb}$, bobbed females,
$\frac{1}{2}$ $X^{bb}Y^{bb+}$, wild males; (b) $\frac{1}{4}$ $X^{bb+}X^{bb+}$,
wild females, $\frac{1}{4}$ $X^{bb}X^{bb+}$, wild females,
$\frac{1}{4}$ $X^{bb+}Y^{bb}$, wild males, $\frac{1}{4}$ $X^{bb}Y^{bb}$, bobbed
males; (c) $\frac{1}{2}$ $X^{bb+}X^{bb+}$ and $X^{bb}X^{bb+}$, wild females, $\frac{1}{4}$ $X^{bb+}Y^{bb}$, wild males, $\frac{1}{4}$ $X^{bb}Y^{bb}$,
bobbed males; (d) $\frac{1}{4}$ $X^{bb+}X^{bb}$, wild females,
$\frac{1}{4}$ $X^{bb}X^{bb}$, bobbed females, $\frac{1}{2}$ $X^{bb+}Y^{bb+}$ and
$X^{bb}Y^{bb+}$, wild males.

4.29 (a) Because this trait is transmitted as a
maternal effect, all three genotypes could
have dextral or sinistral coiling, depending
on the genes of the mother. (b) The mother
and grandmother might be expected to
determine the coiling characteristics in
their immediate progeny. An ss mother
would produce sinistral progeny. Male
parents have no immediate effect on this
trait.

4.30

$s^+s^+ \times ss$	P
(s^+) (s)	gametes
s^+s	F₁, all coiled to right
$1s^+s^+ : 2s^+s : 1ss$	F₂ genotypes
right right left	Phenotypic results from inbreeding the F₂ snails

Phenotype depends on the mother's genotype. All F₂ snails would coil to the right
because the F₁ mothers were s^+s. Following inbreeding, progeny of ss snails would
coil to the left.

4.31 (a) Dark young, dark adult; (b) light
young, dark adult. (c) The pigment condition is at first influenced by the cells of the
mother. When the hoppers grow older the
condition determined by the genes of the

individual snail replaces the one that originated from the mother (if there is a difference in genotype between mother and young).

4.32 The eyes became lighter as the kynurenine that diffused from the *AA* host into the egg was metabolized and broken down by the *aa* individuals that were unable to manufacture more kynurenine.

4.33 (c) A gynandromorph.

CHAPTER 5

5.1 (a) Prepare a test cross or F_2 (if dealing with self-fertilizing plants) and compare results with those expected from the hypothesis of independent assortment. If they do not fit, linkage may be the next hypothesis. (b) The parental combinations should occur in greater proportion in the progeny. If the parental combinations are not known and only progeny data are available, determine whether the expressions controlled by the two dominants or those controlled by a dominant and recessive are in greater proportion. (c) With most materials, a test cross is easily prepared and the results can be compared with the simple $1:1:1:1$ ratio rather than with the more complex $9:3:3:1$. (d) In self-fertilizing plants it is easier to self the F_1 plants and produce an F_2 than to emasculate the flowers and make a test cross, but the interpretation of the results requires more involved mathematical calculations.

5.2 The one class that is represented by $351/1000$ is out of proportion for independent assortment. At least two of the three gene pairs must be in the same chromosome.

5.3 The three classes given fit the hypothesis of independent assortment, suggesting that the gene pairs are on different chromosome pairs.

5.4 30 percent

5.5 (a) $\dfrac{a^+b^+}{a^+b^+} \times \dfrac{a\ b}{a\ b}$ P

$\underline{(a^+b^+)}$ $\underline{(a\ b)}$ gametes

$\dfrac{a^+b^+}{a\ b}$ F_1

(b) $\underline{(a^+b^+)}$ 40%

$\underline{(a^+b)}$ 10%

$\underline{(a\ b^+)}$ 10%

$\underline{(a\ b)}$ 40%

(c) $\dfrac{a^+b^+}{a\ b}$ 40%

$\dfrac{a^+b}{a\ b}$ 10%

$\dfrac{a\ b^+}{a\ b}$ 10%

$\dfrac{a\ b}{a\ b}$ 40%

(d) Coupling.

5.6 (a) $\dfrac{a^+b}{a\ b^+}$ F_1

(b) $\underline{(a^+b)}$ 40%

$\underline{(a^+b^+)}$ 10%

$\underline{(a\ b)}$ 10%

$\underline{(a\ b^+)}$ 40%

(c) $\dfrac{a^+b}{a\ b}$ 40%

$\dfrac{a^+b^+}{a\ b}$ 10%

$\dfrac{a\ b}{a\ b}$ 10%

$$\overline{\dfrac{a\ b^+}{a\ b}}\qquad 40\%$$

(d) Repulsion.

5.7 The parental gametes would be in the proportion of 30 percent each and the recombinations in the proportion of 20 percent each. The zygotes from the test cross would be in the same proportion as the F_1 gametes.

	(a)	(b)
5.8 (a) Two dominant expressions (a^+-b^+-)	66	51
One dominant and one recessive (a^+-bb)	9	24
Other dominant and other recessive (aab^+-)	9	24
Two recessives $(aabb)$	16	1
	100	100

5.9 (a) No; (b) –;

(c) $\overline{\dfrac{b}{b^+}\ \dfrac{ts}{ts^+}} \times \overline{\dfrac{b}{b}\ \dfrac{ts}{ts}}$

5.10 (a) Yes; (b) 16.7;

(c) $\overline{\dfrac{b^+vg^+}{b\ vg}} \times \overline{\dfrac{b\ vg}{b\ vg}}$

5.11 (a) Yes; (b) 16.3;

(c) $\overline{\dfrac{b^+vg}{b\ vg^+}} \times \overline{\dfrac{b\ vg}{b\ vg}}$

5.12 (a) Yes; (b) 34%;

(c) $\overline{\dfrac{c^+b}{c^+b}} \times \overline{\dfrac{c\ b^+}{c\ b^+}}$

5.13 (a) $\overline{\dfrac{d^+p^+}{d\ p}}$; (b) $\overline{\dfrac{d^+p}{d\ p^+}}$;

(c) $54d^+-p^+-:21d^+-pp:21ddp^+:4ddpp.$

5.14 (a) $\overline{\dfrac{a^+b^+}{a^+b^+}\ \dfrac{c^+d^+}{c^+d^+}} \times \overline{\dfrac{a\ b}{a\ b}\ \dfrac{c\ d}{c\ d}}\qquad$ P

(b) $\overline{\dfrac{a^+b^+}{a\ b}\ \dfrac{c^+d^+}{c\ d}}\qquad\qquad$ F_1

(c)

a^+b^+	c^+d^+	12	a^+b^+	c^+d	8
a^+b^+	$c\ d$	12	a^+b^+	$c\ d^+$	8
$a\ b$	c^+d^+	12	$a\ b$	c^+d	8
$a\ b$	$c\ d$	12	$a\ b$	$c\ d^+$	8
a^+b	c^+d^+	3	a^+b	c^+d	2
a^+b	$c\ d$	3	$a\ b^+$	$c\ d^+$	2
$a\ b^+$	c^+d^+	3	a^+b	$c\ d^+$	2
$a\ b^+$	$c\ d$	3	$a\ b^+$	$c\ d$	2

5.15 (a) Either cataract or polydactyly. The genes would be in repulsion. In the separation at meiosis each gamete would get one or the other. A cross over would be required to produce a gamete with both or neither.

5.16 (a)

$(sr\ e^+)$	:	46%
(sr^+e)	:	46%
(sr^+e^+)	:	4%
$(sr\ e)$	:	4%

(b) striped, gray: 46%
    not striped, ebony: 46%
    not striped, gray: 4%
    striped, ebony: 4%

5.17 (a) Four kinds in proportion of 5%; 45%; 45%; 5%. (Repulsion.) (b) Wild type: 5%; vestigial, red: 45%; long wing, cinnabar: 45%; vestigial, cinnabar: 5%.

5.18 (a) female   male

$$\frac{B}{} \qquad \frac{B}{B^+}$$

$$\times$$

$$\frac{C}{C^+} \qquad \frac{C}{C^+}$$

Note: In this problem the Y chromosome is symbolized as in Drosophila. Chickens may not have a Y chromosome. The mechanics would be the same whether a Y chromosome is present or not.

(b) Summary:

barred, crested males:	6
barred, noncrested males:	2
barred, crested females:	3
nonbarred, crested females:	3
barred, noncrested females:	1
nonbarred, noncrested females:	1

5.19 (a) Neither. Both the dominant and recessive would be expressed with equal frequency in males. (b) The dominant would be expressed more frequently in females.

5.20 (a) If it breeds true in succeeding generations, it is probably hereditary and may be assumed to have arisen through mutation. (b) The criss-cross pattern of inheritance is evidence for sex linkage. (c) Females whose fathers had white eyes could be crossed with white-eyed males. Half of the females would be expected to have white eyes. This would conform to genetic theory based on sex linkage.

5.21

(a) $\dfrac{W}{}$, $\dfrac{W^+}{}$, $\dfrac{W^+}{W^+}$ and $\dfrac{W^+}{W}$, $\dfrac{W}{W}$.

(b) $\dfrac{W^+}{W^+} \times \dfrac{W}{}$ P

$\dfrac{W^+}{}$ $\dfrac{W}{}$ $\longrightarrow$ gametes

$\dfrac{W}{W^+} \times \dfrac{W^+}{}$ $F_1 \times F_1$ (all red-eyed)

$\dfrac{W}{}$ $\dfrac{W^+}{}$ $\dfrac{W^+}{}$ $\longrightarrow$ F_1 gametes

$\dfrac{W}{W^+}$ $\dfrac{W^+}{W^+}$ F_2 females: all red-eyed

$\dfrac{W}{}$ $\dfrac{W^+}{}$ F_2 males: $\frac{1}{2}$ red, $\frac{1}{2}$ white

               females: $\frac{1}{2}$ red, $\frac{1}{2}$ white

(c) (1) $\dfrac{W}{W^+} \times \dfrac{W}{}$ males: $\frac{1}{2}$ red, $\frac{1}{2}$ white

(2) $\dfrac{W}{W^+} \times \dfrac{W^+}{}$ females: all red

               males: $\frac{1}{2}$ red, $\frac{1}{2}$ white

5.22

(1) $\dfrac{rg^+}{}$; (2) $\dfrac{rg^+}{rg}$; (3) $\dfrac{rg}{}\longrightarrow$;

(4) $\dfrac{rg^+}{rg}$; (5) probability is $\frac{63}{64}$ for $\dfrac{rg^+}{rg^+}$

and $\frac{1}{64}$ for $\dfrac{rg^+}{rg}$; (6) $\dfrac{rg^+}{rg}$; (7) $\dfrac{rg}{}\longrightarrow$;

(8) $\dfrac{rg^+}{}\longrightarrow$; (9) $\dfrac{rg^+}{}$.

5.23 Half of their sons would be color blind, and their daughters would be normal.

5.24 No. A son receives his X chromosome from his mother. The father contributes a Y chromosome.

5.25

$\dfrac{rg^+}{rg} \times \dfrac{rg}{}\longrightarrow$ P

$\dfrac{rg^+}{}\quad \dfrac{rg}{}\quad \dfrac{rg}{} \quad \underline{\quad\quad}\longrightarrow$ gametes

$\dfrac{rg^+}{rg}\quad \dfrac{rg}{rg}\quad \dfrac{rg^+}{}\quad \dfrac{rg}{}\longrightarrow$ sons and daughters:
$\frac{1}{2}$ normal
$\frac{1}{2}$ color blind

5.26

(a) $\dfrac{h}{h} \times \dfrac{h^+}{}\longrightarrow$ P

$\dfrac{h}{}\quad \dfrac{h^+}{} \quad \underline{\quad\quad}\longrightarrow$ gametes

$\dfrac{h}{h^+}\quad \dfrac{h}{}\longrightarrow$ F$_1$ normal daughters, hemophiliac sons

(b) $\dfrac{h^+}{h} \times \dfrac{h}{}\longrightarrow$

$\dfrac{h^+}{}\quad \dfrac{h}{}\quad \dfrac{h}{} \quad \underline{\quad\quad}\longrightarrow$ gametes

$\dfrac{h^+}{h}\quad \dfrac{h}{h}\quad \dfrac{h^+}{}\quad \dfrac{h}{}$ F$_1$ daughters:
$\frac{1}{2}$ normal,
$\frac{1}{2}$ hemophiliac
F$_1$ sons:
$\frac{1}{2}$ normal,
$\frac{1}{2}$ hemophiliac

(c) $\dfrac{h^+}{h^+}\quad \dfrac{h}{}\longrightarrow$ P

$\dfrac{h^+}{}\quad \cdot\dfrac{h}{} \quad \underline{\quad\quad}\longrightarrow$ gametes

$\dfrac{h}{h^+}\quad \dfrac{h^+}{}\longrightarrow$ F$_1$ all normal

5.27 $\frac{1}{2}$ chance for each son or $\frac{1}{4}$ for each child ($\frac{1}{2}$ for male X $\frac{1}{2}$ affected).

5.28 (a) $\frac{1}{2}$; (b) $\frac{1}{2}$.

5.29 (a) Daughters 0, sons 1; (b) daughters all Xg; none hairy pinna, color blindness only if mother is a carrier; sons all have hairy pinna; color blindness and Xg only if mother is a carrier.

5.30 Map distances may be converted to probabilities as follows: the chance that crossing over will occur between st and ss is 0.14 and the chance that crossing over will not occur is 0.86; the chance that crossing over will occur between ss and e is 0.12 and the chance that it will not occur is 0.88.

By applying the multiplication theorem, the probabilities for the different combinations expected to be represented in the gametes can be calculated. Only one kind of gamete is produced by the male so the proportion of zygotes will be the same as the proportion of gametes. Phenotypes may be expected in the following percentages:

red eyes, normal bristles, gray body:
$$\frac{(0.12 \times 0.86 \times 100)}{2} = 5.16\%$$

red, normal, ebony:
$$\frac{(0.86 \times 0.88 \times 100)}{2} = 37.84\%$$

red, spineless, gray:
$$\frac{(0.14 \times 0.88 \times 100)}{2} = 6.16\%$$

red, spineless, ebony:
$$\frac{(0.14 \times 0.12 \times 100)}{2} = 0.84\%$$

scarlet, normal, gray:
$$\frac{(0.14 \times 0.12 \times 100)}{2} = 0.84\%$$

scarlet, normal, ebony:
$$\frac{(0.14 \times 0.88 \times 100)}{2} = 6.16\%$$

scarlet, spineless, gray:

$$\frac{(0.86 \times 0.88 \times 100)}{2} = 37.84\%$$

scarlet, spineless, ebony:
$$\frac{(0.12 \times 0.86 \times 100)}{2} = 5.16\%$$

5.31

purple, salmon silk, pigmy:
$$\frac{0.81}{2} = .405 = 40.5\%$$

green, yellow silk, normal:
$$\frac{0.81}{2} = .405 = 40.5\%$$

purple, yellow silk, normal:
$$\frac{0.09}{2} = .045 = 4.5\%$$

green, salmon silk, pigmy:
$$\frac{0.09}{2} = .045 = 4.5\%$$

purple, salmon silk, normal:
$$\frac{0.09}{2} = .045 = 4.5\%$$

green, yellow silk, pigmy:
$$\frac{0.09}{2} = .045 = 4.5\%$$

purple, yellow silk, pigmy:
$$\frac{0.01}{2} = .005 = 0.5\%$$

green, salmon silk, normal:
$$\frac{0.01}{2} = .005 = 0.5\%$$

These predictions are based entirely on probability assuming equal crossing over in all areas along the chromosome. Interference could curtail the double crossover class.

5.32

(Tu is a dominant mutant, Tu^+ is recessive.)

$Tu^+\ j_2gl_3$: $\dfrac{0.893}{2} = 0.446 = 44.6\%$

$Tu\ j_2^+gl_3^+$: $\dfrac{0.893}{2} = 0.446 = 44.6\%$

$Tu\ j_2gl_3$: $\dfrac{0.047}{2} = 0.024 = 2.4\%$

$Tu^+j_2^\pm gl_3^+$: $\dfrac{0.047}{2} = 0.024 = 2.4\%$

$Tu^+j_2 gl_3^+$: $\dfrac{0.057}{2} = 0.029 = 2.9\%$

$Tuj_2^+ gl_3$: $\dfrac{0.057}{2} = 0.029 = 2.9\%$

$Tuj_2 gl_3$: $\dfrac{0.0015}{2} = 0.001 = 0.1\%$

$Tu^+j_2^\pm gl_3$: $\dfrac{0.0015}{2} = 0.001 = 0.1\%$

5.33 The classes with the smallest numbers ($+ + w$ and $y\ ec+$) must be double cross-overs. The gene in these classes that differs from the parentals must be in the center, that is, $+w+$ and $y+ec$. With the sequence established, the single cross-overs can be identified and percentages calculated.

y	w					ec
0	1 1.5 2	3	4	5 5.6 6		

5.34 The smallest class is the double cross-over class and the v locus must be in the center. The double cross-over value must be added to each single cross-over value.

y				v				B
0 5 10 15 20 25 29.4 35 40 45 50.4								

5.35 (a) Yes. The test cross results do not follow the $1:1:1:1:1:1:1:1$ pattern expected from a trihybrid cross under random assortment.
(b) Parental cross: $s\ ss\ e$

(c) $\dfrac{s +^{ss} +^e}{s +^{ss} +^e} \times \dfrac{+^e\ ss\ e}{+^s\ ss\ e}$

Test cross: $\dfrac{s\ +^{ss} +^e}{+^s\ ss\ e} \times \dfrac{s\ ss\ e}{s\ ss\ e}$

(d) 30.3 units

(e) 14.0 units

(f) $\dfrac{18}{.14 \times .163\,(1000)} = 0.79 =$ coefficient of coincidence

5.36 (a)

a^+b	: 45%	$a\ b$	: 5%
a^+b^+	: 5%	$a\ b^+$	: 45%

(b) a: 10%, b: 20%.

5.37 Parental ditype: $2a^+b: 2ab^+$, nonparental ditype: $2a^+b^+: 2ab$.

5.38 Tetratype: $1ab: 1ab^+: 1a^+b: 1a^+b^+$

5.39 All four strands of a tetrad can be identified, phenotypes can be observed in the haploid stage and crossing over can be shown by appropriate crosses to involve all four of the strands.

5.40 (a) A parallel exists between the Mendelian pattern and the chromosome mechanisms in meiosis; sex and other phenotypes were found to be determined by factors in particular chromosomes: nondisjunction and linkage and crossing over studies identify gene with chromosomes. (b) Cytological demonstrations of crossing over. (c) Test crosses provide a simple standard for comparison with results expected from independent assortment. In self-fertilizing plants F_2 crosses are more easily made and tables are used to make the comparisons. Both test crosses and F_2 results are satisfactory. The material involved must be considered in choosing the best method. (d) Numerous experiments have shown that this is true. Chromosome maps are based on this fact. (e) Experiments on Neurospora and Drosophila cited in the text have demonstrated four strand crossing over.

CHAPTER 6

6.1 Cross the two varieties together and observe the results. If monohybrid ratios are obtained in the F_2, the contrasting traits may depend on a single pair of alleles.

6.2 Wild-type alleles can only be postulated when contrasting traits associated with the same locus are discovered. Eventually, biochemical studies may be used for detecting wild-type alleles directly, but thus far only phenotypic comparisons have been used for identifying most loci and determining their positions in chromosomes.

6.3 (a) The same basic symbol is used conventionally for all members of a group of multiple alleles, with superscripts to identify particular alleles. (b) Most series of multiple alleles are associated with gradations in the same phenotype. Some produce quite different end results, such as legs in place of antennae in Drosophila.

6.4 (a) All colored; (b) 3 colored: 1 albino; (c) 3 colored: 1 chinchilla; (d) 1 chinchilla: 1 albino; (e) 3 colored: 1 himalayan; (f) 1 himalayan: 1 albino.

6.5 (a) Yellow and yellow; 2 yellow: 1 agouti light belly ($A^Y A^Y$ is lethal); (b) yellow and agouti light belly; 2 yellow: 1 agouti light belly: 1 black and tan; (c) black and tan and yellow; 2 yellow: 1 black and tan: 1 black; (d) agouti light belly and agouti light belly; all agouti light belly; (e) agouti light belly and yellow; 1 yellow: 1 agouti light belly; (f) agouti and black and tan; 1 agouti: 1

black and tan; (g) black and tan and black; 1 black and tan: 1 black; (h) yellow and agouti; 1 yellow: 1 agouti light belly; (i) yellow and yellow; 2 yellow: 1 agouti.

6.6 (a) 1; (b) 2; (c) 2; (d) 10.

6.7 30

6.8 60

6.9 (a) 28; (b) by crossing fish carrying the different alleles and checking for monohybrid results.

6.10 (a) all AB; (b) 1 A: 1 B; (c) 1 AB: 1 A: 1 B: 1 O-type; (d) 1 A: 1 O-type.

6.11 The man with AB-type blood was not the father of the child with O-type blood.

6.12 The baby with O-type blood was not the daughter of the woman with AB-type blood.

6.13 (a) All children would be heterozygous (Rr) and Rh positive. The mother became immunized during the first pregnancy and the next child was affected. (b) All future children would be expected to be affected.

6.14 Half of the children carried by the rr mother would be Rr (Rh positive) and half would be rr (Rh negative), like the mother. Because the woman was sensitized at the second pregnancy the Rr children would be expected to be erythroblastotic and the rr children, normal.

6.15 (a)

AA	$L^M L^M$	AA	$L^M L^N$	AA	$L^N L^N$
AA^B	$L^M L^M$	AA^B	$L^M L^N$	AA^B	$L^N L^N$
Aa	$L^M L^M$	Aa	$L^M L^N$	Aa	$L^N L^N$
$A^B A^B$	$L^M L^M$	$A^B A^B$	$L^M L^N$	$A^B A^B$	$L^N L^N$
$A^B a$	$L^M L^M$	$A^B a$	$L^M L^N$	$A^B a$	$L^N L^N$
aa	$L^M L^M$	aa	$L^M L^N$	aa	$L^N L^N$

(b) Many possible combinations are available in these and other blood systems that may be useful in checking identity and paternity.

6.16 (a) wild type (b) garnet

6.17 Size, shape, and edge characteristics of plaques are determined by the genetic ability of the virus to lyse the host cells and leave clear areas on the culture plate. If no plaques are formed, the host cells are resistant to the attack of the phage or are lysogenic for the phage and do not undergo lysis.

6.18 When rII mutants are grown on strain K of *E. coli* they do not ordinarily form plaques. Recombination will not occur in strain K. If a mixture of strains K and B is introduced, recombination may occur in B. If clear plaques are formed on the mixture, either mutations or recombinations must have occurred. When the strains introduced to the culture are not mixed and the plaques occur in the order of one per 10^8 phages, the most likely explanation for the change is spontaneous mutation.

6.19 Complementation was detected by introducing more than one rII mutant into the same K culture. Neither mutant alone could produce a plaque but some combinations resulted in large plaques. This led to the discovery that the rII region was composed of two cistrons, A and B, which complemented each other.

6.20 (a) same cistron, (b) different, (c) functional, (d) nonfunctional.

6.21 (a) a_1 and a_2 in same cistron, functionally allelic

$$\frac{a_1 \ a_2}{+ \ +} \quad \text{cis, normal, complementation}$$

$$\frac{a_1 \ +}{+ \ a_2} \quad \text{trans, mutant, noncomplementation}$$

(b) b_1 and b_2 functionally nonallelic, coupling linkage, normal complementation

$$\frac{b_1}{+} \ldots \frac{b_2}{+} \quad \text{defective products 1 and 2}$$
$$\phantom{\frac{b_1}{+}} \quad \text{normal products 1 and 2}$$

Repulsion linkage, normal complementa-

tion

$$\frac{b_1}{+} \ldots \frac{+}{b_2} \quad \begin{array}{l} \text{defective product 1} \\ \text{normal product 2} \\ \text{normal product 1} \\ \text{defective product 2} \end{array}$$

6.22 8–10 area.

6.23 Mixed infections of phage on sensitive bacterial cells have resulted in combinations of markers in the progeny different from those of the parents. During the latent period, when the host DNA breaks down, the virus DNA undergoes a reorganization. Genetic material of the different viruses may recombine and fragments of DNA from the host cell may be incorporated with that of the virus and may be transmitted to the next host cell invaded.

6.24 Recombination resulted in nutritionally competent cells.

6.25 By receipt of an F factor in conjugation.

6.26 Through sexduction.

6.27 When a temperate phage enters a host cell it may control a mutual tolerance between the host and the prophage. The cell is not lysed and the phage DNA becomes incorporated into the genome of the host. Lysogeny is a genetic property of the phage.

6.28 (a) On entering a host cell, the virus loses its protein coat and infectivity. Its DNA becomes incorporated into the genome of the host, replicating synchronously with the host. (b) It takes over the genetic function of the host and directs protein synthesis through the machinery of the host.

6.29

mechanism of recombination	agency for DNA transfer	relative size of units transferred	genetic consequences
transduction	phage	small	donor DNA integrated in host (abortive not incorporated)
sexduction	F factor	varying	donor DNA near F added to recipient chromosome
transformation	direct chemical contact	large	donor DNA replaces recipient DNA at specific sites

CHAPTER 7

7.1 Determine from analysis of the culture medium and the system whether an enzyme is synthesized only in response to an inducer or the rate of enzyme production is decreased when the concentration of certain metabolites in the medium is increased.

7.2 The presence of the raw materials on which enzymes act provides a more efficient control mechanism than feedback from the product of enzymatic activity. Cells are not burdened with storage of unused enzymes, but can activate the process of enzyme production when the raw materials are available.

7.3

Gene	Function	Mechanism
(1) Regulator	Controls operator	Makes repressor which interacts with operator
(2) Operator	Initiates synthesis	Controls structural genes z and y
(3) Structural z	Specifies beta-galactosidase	z mRNA
(4) Structural y	Specifies galactoside permease	y mRNA

7.4 (a) Repressor would be removed and operon would function continuously; (b) structural genes would function continuously; (c) beta galactosidase would not be produced; (d) galactoside permease would not be produced.

7.5 (a) (1) inductive; (2) (5) constitutive (b) (2) (5)

7.6 (a) inductive, (b) constitutive, (c) constitutive, (d) inductive, (e) constitutive

7.7 (a) $\dfrac{i^+o^cz^+y^-}{i^-o^+z^-y^+}$ (b) $\dfrac{i^+o^+z^+y^-}{i^-o^0z^+y^+}$

7.8 (a) o^c maps within the operon but i^- does not; (b) in o^c/o^+ arrangement. o^c is constitutive for its own operon, but i^-/i^+ is inducible; (c) o^c affects genes in cis position but i^- affects genes in trans position as well as cis.

7.9 The system could have developed from a series of tandem duplications of a single ancestral gene. Mutational changes making the system more efficient and therefore favored in selection could have brought the system to its present level of efficiency.

7.10 (a) Polarity could be demonstrated by inducing inactivating mutations in genes near the operator and determining that the activities of genes in the operon sequence but more distant from the operator were inhibited. (b) A characteristic adjustment in enzyme level controlled by genes in the operon series beyond an inhibiting mutation would be evidence of modulation. (c) All enzymatic steps beyond the mutant would be inhibited.

7.11 Clusters of genes functioning as units are more common in bacteria than in higher forms that have more elaborate genetic control systems and more recombination, which would break up clusters.

7.12 Cytoplasmic repressor or activator substances such as those suggested in early developmental studies would fit very well into the operon model that was developed for *E. coli.*

7.13 Negative mechanisms such as the repressor in the lactose operon block the transcription of the operon whereas positive mechanisms such as the activator in the arabinose operon promote activity of the operon.

7.14 In prokaryotic cells that do not store mRNA, protein synthesis is initiated directly by mRNA transcription. When the mRNA transcription is "turned on," translation machinery becomes active and protein enzymes are soon available. When transcription is "turned off," protein synthesis ceases. Eukaryotes apparently have steps between initiation of transcription and subsequent participation of mRNA in protein synthesis.

7.15 The Britten and Davidson model provides for such agents as hormones in initiation of gene activity, action of non-contiguous genes in differentiation, heterodisperse RNA in formation of mRNA, regulatory and protective codons, and repetitive DNA that is transcribed into cell-specific patterns.

7.16 (a) Histone represses gene activity by combining with DNA. The quantity of histones present with respect to that of DNA determines the extent of repression. (b) Histone synthesis is associated with the nucleoli of metabolic cells. (c) Histones become associated with chromosomes as the chromosomes form in preparation for cell division.

7.17 The four hormones listed influence the production of RNA in cells. RNA is essential to protein synthesis.

7.18 Actinomycin-D suppresses the rate of RNA production and also offsets the effect of hormones, thus indicating that hormones stimulate RNA production.

7.19 Some carcinogens apparently derepress genes for RNA synthesis and thus stimulate gene activity and body growth by cell division. Regulating mechanisms do not keep the growth under proper control and tumor formation results.

7.20 When salivary chromosomes are followed in their developmental sequence, puffs that are controlled by a hormone occur in specific regions in a regular pattern. Puffs have been interpreted as regions of particular activity in RNA synthesis and have been related to particular gene loci. The chromosomes are large and easily observed. Staining techniques are available to identify areas of RNA synthesis and the gene sequence along the chromosomes is known and identifiable with phenotypes in the flies.

7.21 (a) These organisms can be cultured readily and investigated in the laboratory. Techniques are available for studies of biochemical pathways. (b) (1) Organisms such as *E. coli* and other bacteria that have a single linkage group, many biochemical mutants, and mostly asexual reproduction, would be most likely to develop clusters of genes that operate together. (2) Repressing effect of histones could be studied in cells such as chicken erythrocytes, which have a large amount of histone. (3) Higher animals such as mammals that have steroid hormones would be best for studies of RNA synthesis control (4) Drosophila and other Diptera have enlarged salivary chromosomes with puffs.

7.22 (a) Genes controlling pigment production are present in all cells of the rabbit. The chemical reaction required for black pigment can proceed at a temperature of 27°C, but it is blocked at higher temperatures. (b) While the rabbits are developing in the body of the mother the temperature is about 33°C, and after birth the main part of the body is maintained at that temperature. Extremities and treated areas in this experiment were cooler as the reaction proceeded.

7.23 (a) Wild-type host tissue apparently can compensate for the chemical blocks induced by *v* and *cn* genes. (b) Two steps in the production of wild-type pigment are involved. Discs with *v* can synthesize a material needed for pigment production when supplemented with a substance from the *cn* host. The *cn* discs, however, had their pigment production blocked at a later stage and required a substance that could not be supplied by the *v* host.

7.24 At present there is no known artificial method for producing and supplying to the cells the necessary enzymes for pigment production. Enzymes that operate in the cell must be built up in the cell under the direction of genes.

7.25 Differentiation provides the necessary cell forms and functions: organization results in appropriate groupings of cells and tissues; and growth, through cell division, determines the ultimate size and shape of the whole organism as well as its various parts.

7.26 The cytoplasm is the location of metabolic activities in the cell. Quantities of metabolic products could represent the sensitive trigger or feedback mechanisms to turn operons on and off. Genes are constant but cytoplasm is undergoing chemical change. Genes or operons may be activated by changes in cytoplasmic substances.

7.27 (a) It has been assumed that environmental agents influence developing phenotypes the same way that genes do. Studies on phenocopies could suggest mechanisms of gene action. (b) They represent one of the few available approaches to physiological genetics. At best they are only suggestive and there is a great distance at present between the phenocopy and an understanding of gene action. (c) Mutations are transmitted in inheritance, whereas phenocopies are not inherited.

7.28 (a) Rumplessness occurs sometimes without any known cause, that is, as an embryological error that interferes with normal development. (b) Additional errors may be induced by upsetting reactions that involve time and space relations in the embryo. (c) Insulin interferes chemically with processes that are ordinarily involved in the production of normal chickens.

7.29 Phenocopies in man are often difficult to distinguish from inherited abnormalities that have delayed action.

7.30 Different gene products have different threshold levels in relation to alteration by environmental conditions. Some are remarkably stable in a wide range of environmental situations. Others are readily modified.

7.31 70%

7.32 (a) Glutamic semialdehyde → ornithine → citrulline → arginine. (b) (1) Citrulline, (2) ornithine.

CHAPTER 8

8.1 (a) transition, (b) transition, (c) transversion, (d) transversion, (e) frameshift, (f) transition.

8.2 (a) ClB or Muller-5 method, (b) attached X-method. (See Figs. 8.14, 8.15)

8.3 Selective media. If the organisms will not grow on the minimal medium, they may be screened in media with different supplements to determine which requirement the mutant is unable to produce.

8.4 Bacteria treated with a mutagen or expected to carry mutations may be introduced to media with particular drugs in appropriate concentrations. Colonies that appear have originated from cells carrying mutations for resistance. X-rays or other known mutagenic agents are usually introduced before the selection for mutations is undertaken. Wild-type bacteria can be introduced to selective media to see if the mutation frequency exceeds that for spontaneous mutations.

8.5 Probably not. Man is larger than a bacterium with more cells and a longer life span. If mutation frequencies are converted to cell generations the rates for human cells and bacterial cells are similar.

8.6 A dominant mutation presumably occurred in the woman in whom the condition was first known.

8.7 The sex-linked gene is carried by mothers and the disease is expressed in half of their sons. Such a disease is difficult to follow in pedigree studies because of the recessive nature of the gene, the tendency for the expression to skip generations in a family line, and the loss of the males who carry the gene. One

explanation for the sporadic occurrence and tendency for the gene to persist is that by mutation new defective genes are constantly being added to the load already present in the population.

8.8 Plants can be propagated vegetatively but no such methods are available for widespread use in animals.

8.9 The sheep with short legs could be mated to unrelated animals with long legs. If the trait is expressed in the first generation, it could be presumed to be inherited and to depend on a dominant gene. On the other hand, if it does not appear in the first generation, F_1 sheep could be crossed back to the shortlegged parent. If the trait is expressed in $\frac{1}{2}$ of the backcross progeny it might be presumed to be inherited as a simple recessive. If two short-legged sheep of different sex could be obtained, they could be mated repeatedly to test the hypothesis of dominance. In the event that the trait is not transmitted to the progeny resulting from these matings, it might be considered to be environmental or dependent on some complex genetic mechanism that could not be identified by the simple test used in the experiments.

8.10 Enzymes may discriminate among different nucleotides being incorporated. Mutator enzymes may utilize a higher proportion of incorrect nucleotides whereas antimutator enzymes may select fewer incorrect bases in DNA replication.

8.11 If both mutators and antimutators operate in the same living system, an optimum mutation rate for a particular organism in a given environment may result from natural selection.

8.12 *Dt* is a mutator gene that induces somatic mutations in developing kernels.

8.13 (a) yes (b) a block would result in the accumulation of phenylalanine and a decrease in the amount of tyrosine which would be expected to result in several different phenotypic expressions.

8.14 These hemoglobins can be distinguished by mobility of molecules in an electric field, and by "fingerprints" showing the amino acid in part 4, position 6 of the beta chain.

8.15

	DNA	m RNA
valine	CAT	GUA
lysine	TTT	AAA

8.16 It depends on an alteration of the hemoglobin molecule.

8.17 Mutations. Transitions, transversions, frameshifts.

8.18 All human chains are basically similar. Differences can be explained by mutations over a period of time. Some 21 homologous sites are recognized on all four human chains and the myoglobin chain.

8.19 Evidence for the evolution of cytochrome C is considerably more extensive than that for hemoglobin. Homologies can be recognized throughout the major part of the phylogenetic tree from yeast to man.

8.20 Irradiate the nonresistant strain and plate the irradiated organisms on a medium containing streptomycin. Those that survive and produce colonies are resistant. They could be transferred to a medium without streptomycin. Those that survive would be of the first type; those that can live with streptomycin but not without it would be the second type.

8.21 3%, 4%, 6%

8.22 Each quantum of energy from the X-rays that is absorbed in a cell has a certain probability of hitting and breaking a chromosome. Hence, the greater the number of quantums of energy or dosage the more likely breaks are to occur. The rate at which this dosage is delivered does not change the probability of each quantum.

8.23 The person receiving a total of 100 r would be expected to have 2X as many mutations as the one receiving 50 r.

8.24 Ultraviolet light produces mispairing alterations mostly in pyrimidines (for example, cytosine to thymine transitions) during the replicating process. Thymine may be altered to cytosine, which pairs with guanine. A reverse mutation may occur when cytosine is changed to thymine, which pairs with adenine. A T-A base pair may thus be changed to a C-G and the reverse mutation may occur from C-G to T-A.

8.25 Nitrous acid brings about a substitution of an OH group for an NH_2 group in a single base giving hypoxanthine. Base pairing properties are altered: GC $\rightleftharpoons$ AT. A directed mutation may thus be accomplished.

8.26 Transitions

8.27 Nitrous acid acts as a mutagen on non-replicating DNA (resting DNA) and produces transitions from A to G or C to T, whereas 5-bromouracil does not affect resting DNA but influences the replicating mechanism resulting in two way transitions GC $\rightleftharpoons$ AT.

8.28 Large numbers of seeds have been given massive X-ray doses and screened to see if induced mutations have improved any of them for particular environments. In one study on oats, a few seeds produced plants with improved disease resistance.

8.29 The immediate effects are surface and internal burning. Irradiation is known to induce leukemia, and in general to shorten the life span, an effect that is difficult to evaluate. Genetic effects are insidious; small doses produce mutations and have a cumulative effect. The number of mutations generally is proportional to total dosage. This is a potential danger to future generations.

8.30 Mutations induced by acridine dyes resulted in inserted and deleted nucleotides that inactivated cistrons by changing the framing pattern. Activity could be restored by compensation of a deletion, by an insertion or by three alterations of one kind.

CHAPTER 9

9.1 The DNA from the sample with 50 percent of GC base pairs probably came from a prokaryote whereas the DNA with the 5 percent of GC pairs probably came from a eukaryote with an abundance of AT repetitive satellite DNA.

9.2 In early development the paternal chromosomes of the male mealy bug become heterochromatic. Only maternal euchromatin is present in the male embryo. In spermatogenesis, only euchromatin products form sperm. In developing females all chromosomes are euchromatic and oögenesis is regular.

9.3 (a) A recessive gene presumed to be carried in heterozygous condition unexpectedly came to expression. Bridges postulated that a section in the homologous chromosome carrying the wild-type allele was missing (that is, there was a chromosome deficiency). On another occasion, Bridges found that a recessive gene presumed to be homozygous did not come to expression. He postulated that a gene acting as a dominant allele must be present elsewhere in the chromosome set (that is, there was a duplication). (b) It was impossible to distinguish microscopically between the structural parts of homologous chromosomes at the time these genetic results were obtained. (c) The discovery of attached-X chromosomes in Drosophila, meiotic configurations in maize, and salivary chromosomes in Drosophila provided tools for cytological verification.

9.4 Salivary preparations in Drosophila provide larger chromosomes to work with than meiotic stages in maize. Chromosome parts can be identified with the aid of Bridges' chromosome maps. Salivary chromosome studies are made from somatic cells, whereas maize studies are made from germ cells.

9.5 Enlarged size, somatic pairing, identifiable bands, and distinguishable anatomical features along the length of the different chromosomes make salivary gland chromosomes especially useful.

9.6 Chromosomes can be spread on a slide, fixed, and stained in a single operation with a single solution. When they are well spread out on the slide the linear sequence of an individual chromosome can be followed. On the other hand, it would be necessary to reconstruct the chromosomes if sections were employed for study.

9.7 The chromonemata duplicate themselves many times but the chromosomes and cells do not divide. Bundles are thus developed. The cross bands represent groups of identical chromomeres.

9.8 Linkage maps are constructed by placing the relative gene positions calculated from crossover data along a line representing a chromosome. Cytological maps are constructed by microscopic observations of chromosomes from actual cell preparations. The salivary chromosome maps prepared by Bridges are linkage maps superimposed on cytological maps.

9.9 When a section of a chromosome carrying a dominant gene becomes deleted, a recessive allele carried in the homologous chromosome may come to expression.

9.10 (a) The extent of a deficiency can be determined by testing the genes on either side of the point known to be in the deficiency to see if they behave as pseudodominants. (b) The determination can be made cytologically by microscopically observing the extent in a suitable chromosome preparation (for example, a salivary chromosome preparation with chromosomes heterozygous for the deficiency).

9.11 (a) This is not an easy determination. New mutants of a particular kind may occur infrequently at a particular locus and produce phenotypic changes like those associated with a deficiency. A mutation from v^+ to v in this case would appear unlikely, but if it did occur, the pair of recessive genes would behave in a regular Mendelian pattern. Test cross or F_2 results could be predicted and checked. In animals from which more complete data are available, deficiencies are usually homozygous lethal, resulting in modified F_2 ratios. If other recessives in the same chromosome, presumed to be heterozygous, came to expression, the case for a deficiency would be strengthened. (b) Cytological determination in mice is also difficult. There are no enlarged salivary or meiotic chromosomes available for study. If a chromosome could be shown to be shorter or structurally different from its homologue the deficiency hypothesis would gain support.

9.12 (a) When crossing over occurs in the area of a paracentric inversion, there is the formation of acentric and dicentric chromosomes that do not separate properly to the poles in division. The gametes carrying crossover chromatids are abnormal and inviable. (b) Crossovers within pericentric inversions result in unbalanced chromosome arrangements that make the crossover gametes or zygotes inviable. (c) Crossing over is reduced to some extent, but the main "suppression" results from inviable gametes or zygotes.

9.13 All flies would be beaded with the genotype Bd^+l/Bd^+l.

9.14 The seedlings would be approximately 1 white: 2 green: 1 yellow; White and yellow seedlings would die and all mature plants would be green.

9.15 (a) autosomal, (b) chromosome III.

9.16 (a) A loop is formed or a buckling occurs in the unpaired normal chromosome segment corresponding in size with the deficiency. (b) A loop similar in appearance to that described for a deficiency is formed. (c) A loop is formed by the pairing chromosomes; one member of the pair is reversed, making it possible for the corresponding segments of the inverted and uninverted parts to pair. (d) A cross is formed by the pairing chromosomes. See Figs. 9.6, 9.7, 9.9, 9.17.

9.17 (a) Altered linkage groups, pollen or ovule sterility, and position effects suggest the presence of translocations. (b) Cross configurations, and rings and chains of chromosomes in the meiotic prophase that can be seen microscopically represent cytological evidence.

9.18 Unbalanced chromosome arrangements occur in gamete formation, making pollen inviable. See Fig. 9.19.

9.19 (a) An inversion has occurred involving section 456. (b) A loop would be formed with paired chromosomes and the elements of the inverted segment would be in reverse order (that is 654).

9.20 (a) segment 34 has been deleted; (b) segment 4 duplicated; (c) segment 876 inverted. See Fig. 9.23 for model.

9.21 See text figure for model.

9.22 If a structural alteration in a chromosome can be inseparably related to a phenotypic change, a position effect would be indicated. The phenotypic effect is caused by the change in position of genetic materials rather than by addition or deletion.

9.23 (a) When a section of euchromatin is moved by structural change to a location in or near heterochromatin, variegated position effects may occur. (b) Apparently genes moved to a different chromosome location behave differently than those remaining in the environment to which they are adjusted.

9.24 (a) The Philadelphia chromosome is a fragment of a No. 22 chromosome that is associated with a particular disease. (b) It is used as a diagnostic factor for chronic granulocytic leukemia.

9.25 (a) 45, XY, -21, (b) 45, XX, $-D$, $-G$, $+t$ (DqGq) (c) 46, XY, -5, -12, $+t$ (5p12p), $+t$ (5q 12q).

9.26 Inversions and translocations act as isolating factors among individuals in populations, thus preventing normal gene exchange and promoting speciation. (See also Chapter 13.)

9.27 Techniques for preparing slides for critical microscopic observation were not effectively applied to human chromosome studies until about 1956. Better sources of human material from surgical procedures and tissue-culture techniques are now available.

9.28 (a) Human chromosomes are small, quite uniform, and by older techniques, difficult to distinguish. They could be separated into seven groups following the Denver conference but enough overlap occurred among members of the same group (partly because of variations in techniques) to make it impossible to consistently identify individual chromosomes. (b) Identification agreed upon at Denver, London and Chicago was based on chromosome length, position of the centromere, and satellites. At the Paris conference in 1971 chromosome banding techniques were shown to be suitable for identifying all human chromosomes.

CHAPTER 10

10.1 (a) 0; (b) 2; (c) 1.

10.2 Nondisjunction of chromosomes in the production of gametes (eggs) seems to be the explanation for most, if not, all trisomy.

10.3 (a) $\dfrac{1}{490,000}$; (b) 60,000; (c) $\dfrac{1}{70}$.

10.4 $\dfrac{1}{2}$

10.5 $\dfrac{1}{3,500,000}$.

10.6 $\dfrac{1}{4,000,000}$; (b) $\dfrac{1}{40,000}$.

10.7 (a) 17 purple: 1 white; (b) 11 purple: 1 white; (c) 15 purple: 3 white reduced to 5 purple: 1 white; (d) 3 purple: 1 white. (e) The mutant trait occurs less frequently than expected in the results of crosses because of the presence of an extra chromosome, usually carrying a wild-type allele.

10.8 (a) $\dfrac{1}{2}$ 2n and $\dfrac{1}{2}$ 2n + 1; (b) 11 normal: 1 eyeless.

10.9 (a) XO is basically female, XXY male, and XXX metafemale. The human Y carries male-determining capacity. (b) Drosophila Y has no influence on sex determination. Melandrium Y carries male determiners.

10.10 Some plants and animals with a particular extra chromosome can be recognized phenotypically. Genes that give trisomic ratios (for example, genes for purple or white in Datura) are located in the particular chromosome that makes the trisome. Monosomics are useful because recessive genes on a chromosome that has no homologue (for example, a monosome) express themselves and can thus be associated with a particular chromosome.

10.11 Tetrasomics and nullisomics apparently occur in nature but they are less viable and usually die before they are detected.

10.12 Monoploids have only one set of genes and therefore no gene segregation. They could be used for experimental work where it is desirable to relate genes with phenotypes or traits. In molds such as Neurospora, monoploids represent the most important part of the life cycle and they are used extensively for genetic studies.

10.13 Polyploid tissues grow through cell division that can occur regularly among polyploid cells. Sexual reproduction of whole animals requires gamete formation, fertilization, and sex determination. Irregularities associated with polyploidy nearly always result in inviability and sterility.

10.14 Polyploidy might occur through a doubling of chromosomes in somatic cells or the failure of the reduction division resulting in unreduced gametes.

10.15 (a) Autopolyploidy is the doubling of a 2n complement resulting in 4 similar genomes. Allopolyploidy occurs through hybridization in which two 2n complements are involved. (b) Because autopolyploidy results in 4 duplicate sets of chromosomes, pairing is irregular; parts of all four similar chromosomes may be paired in different places, thus interfering with normal reduction division. This and more subtle genetic factors make autopolyploids sterile and incapable of perpetuating themselves. On the other hand, allopolyploidy provides for two pairs of chromosome sets. The chromosomes can pair properly and produce gametes. Apparently it has been important in the evolution of many plant groups.

10.16 Tetraploids have an even number of chromosome sets or genomes. If they have arisen through hybridization and the four genomes are represented as two pairs, synapsis, an essential part of meiosis, can be accomplished with regularity. Triploids with three genomes have inherent difficulties in pairing and gamete formation and usually they do not perpetuate themselves through normal sexual reproduction. They may arise through fusion of unreduced (2n) and reduced (n) gametes but usually they do not reproduce triploids unless apomixis or some other deviation from sexual reproduction occurs.

10.17 Poor pairing, the most obvious cause of sterility, is associated with chromosome irregularity. More subtle genetic factors are also involved.

10.18 (a) A valuable criterion available to the taxonomist is chromosome number. It is at least as significant as any well-defined morphological characteristic. (b) Numerical as well as structural differences in chromosomes have been significant in the mechanics of evolution. Analysis and comparison of chromosome numbers within and between taxonomic groups have aided in solving problems of evolution.

10.19 (1) Doubling of chromosomes in the section of the hybrid that grew vigorously. (2) Propagation of cells from polyploid area to reproduce a plant with 2 sets of chromosomes from each parent species.

10.20 (1) A few cells of the hybrid were apparently polyploid with 24 chromosomes. (2) These gave rise to pollen grains with 12 chromosomes. (3) When these pollen grains were used to fertilize species B eggs a few 19-chromosome plants were produced. (4) These produced gametes with different numbers of chromosomes, some with 12. (5) Gametes with 12 chromosomes fused to form 24-chromosome plants with the characteristics of a new polyploid species.

10.21 Colchicine treatment interferes with spindle formation in cell division and some chromosome mechanisms. Irregular numbers of chromosomes are included in the daughter cells. Sometimes the membrane forms around the entire two sets of chromosomes that normally would separate to two cells and the number in the single cell is doubled.

10.22 (a) Colchicine induced polyploidy provides a means of accomplishing in the laboratory a process that has apparently occurred naturally in many important plant groups. Some polyploids have qualities that give them practical advantages. New polyploids are being produced and tested against established commercial varieties of grapes, tomatoes, and other plants in which they have valuable properties. (b) Induced polyploidy has great theoretical significance as a tool for discovering the mechanism that has occurred in the evolution of some plant groups. The origin of modern varieties of cotton, wheat, and other valuable and interesting polyploids presents a great challenge for present and future investigations.

10.23 Polyploidy associated with hybridization could account for the differences among some known strains of cotton and wheat. The processes through which some modern polyploids may have developed in nature have been reconstructed in the laboratory. See text for details.

10.24 Progeny of interspecific crosses are usually sterile. If hybridization is combined with chromosome doubling, fertile hybrids may occur and perpetuate their new chromosome arrangement. They do not cross with the diploids, and, if they are perpetuated in nature, they might well be classified as new species.

10.25 (a) *Triticum spelta* has been reproduced experimentally by doubling the chromosomes of emmer wheat and goat grass and crossing the two polyploids. (b) Raphanobrassica was developed experimentally by crossing the radish and the cabbage. The F_1 progeny were sterile but some unreduced gametes were obtained and the F_2 tetraploid (Raphanobrassica) was produced from them. (c) *Spartina townsendii* apparently occurred in nature from the crossing of *Spartina alterniflora* with *S. stricta* and the doubling of the chromosome number of the hybrid. (d) *Primula kewensis* was produced by the crossing of *Primula floribunda* and *P. verticillata* and the doubling of the chromosome number of the hybrid. (e) The origin of New World cotton has been reconstructed experimentally by crossing Old World cotton with upland cotton and doubling the chromosome number of the hybrid.

10.26 (a) haploid; (b) diploid; (c) trisomic; (d) monosomic; (e) tetrasomic; (f) triploid; (g) tetraploid.

CHAPTER 11

11.1 Maternal inheritance. It is controlled by extrachromosomal factors and not by the genes of the mother.

11.2 (a) The cytoplasm provides the environment in which the genes act. Therefore the mother would be expected to influence secondary actions of certain genes more than the father. (b) Nonchromosomal genes would be carried in the cytoplasm and would be inherited through the maternal line.

11.3 (a) Sex-linked genes are located in the X chromosome and a characteristic crisscross inheritance could be detected from appropriate crosses if sex linkage were involved. (b) Cytoplasmic inheritance would be transmitted through the maternal line because most of the cytoplasm of the zygote comes from the egg. A series of backcrosses could be made from F_1 males and females to the appropriate females and males of the two parent types. If the trait is transmitted repeatedly for several generations from the maternal parent to her progeny and not through the paternal parent, it may be cytoplasmic. (c) If the trait was transmitted from mother to progeny but did not persist in the maternal line, it might be attributed to the influence of the mother's genes on the developing egg or embryo, that is, a maternal effect. In this case nuclear genes would be involved but they would be the genes of the mother rather than those of the individual itself.

11.4 Inherited changes in the proplastids and multiplication of cells in which the changes occurred resulted in sectors showing different color characteristics as in the four-o-clocks. If the plastid characteristics are determined solely by eggs, only female gametes would determine green, pale green, or variegated plants. In examples from other species, a small amount of cytoplasm is carried with sperm which also results in variegated plants.

11.5 Male sterility facilitates crosses involving plants that are ordinarily self-fertilized. Large-scale crossing for obtaining hybrid vigor is accomplished more economically if the plants are male sterile.

11.6 Kappa particles may be microorganisms that have developed an intimate symbiotic relation with paramecia of a particular genotype.

11.7 Maternal inheritance. *O. hookeri* must have yellow plastids and *O. muricata* must have green plastids.

CHAPTER 12

12.1

$AABB \times aabb$	P	
$AaBb$	F_1	
$AaBb \times AaBb$	$F_1 \times F_1$	

$$AA \begin{cases} BB \\ 2Bb \\ bb \end{cases}$$

$$2Aa \begin{cases} BB \\ 2Bb \\ bb \end{cases} \quad F_2$$

$$aa \begin{cases} BB \\ 2Bb \\ bb \end{cases}$$

Summary of F_2: 1/16 red; 4/16 dark; 6/16 medium; 4/16 light; 1/16 white.

12.2 Three pairs of genes were operating in Problem 12.2, and only two pairs in 12.1.

12.3 (a) *Aabbcc* or *aaBbcc* or *aabbCc*; (b) *AaBbcc* or *AabbcC* or *aaBbCc*; (c) *AaBbCc*.

12.4 (a) *aabbcc*; (b) *AAbbcc* or *aaBBcc* or *aabbCC*; (c) *Aabbcc* or *aaBbcc* or *aabbCc*; (d) *AaBbcc* or *AabbCc* or *aaBbCc*; (e) *AaBbCc*.

12.5 (a) *AaBbcc* × *Aabbcc* or any other combination with two capital letter genes in one parent and one (of the same genes) in the other. (b) *AaBbCc* × *AaBbcc* or *AaBbCc* × *AabbCc* or *AaBbCc* × *aaBbCc*; (c) *AaBbCc* × *AaBbCc*.

12.6 *AaBb* genotype, medium or mulatto phenotype.

12.7 1/16 black (*AABB*), 1/16 white (*aabb*) and 14/16 with pigmentation intensity between black and white (4/16 dark, 6/16 intermediate, 4/16 light).

12.8 (a) No. The greatest degree of pigmentation possible would be medium (*AaBb*). (b) If the parents were genetically *aabb* they could not have a baby with the *A* or *B* pigment-producing genes.

12.9 (a) There would be more gradations in pigmentation and more than five classes would be required to classify the phenotypes. Individuals homozygous for all pigment-producing genes or all genes for white would be considerably more infrequent than the model based on two pairs suggests. (b) Better methods of classifying phenotypes, and more data showing the proportions of different color classes in mixed populations will be required to determine more precisely the number of genes involved. The actual results of matings between large numbers of people with only colored ancestry and those with only white ancestry as well as those of matings involving other particular genotypes would be useful in making such a determination.

12.10 (a) 1 ft; (b) Model, Fig. 12.4. Summary of F_2 phenotypes: 1 plant, 6 ft; 4, 5 ft; 6, 4 ft; 4, 3 ft; and 1, 2 ft.

12.11 2012/8 = 251 or 1 extreme in about 256; 4 pairs of genes were involved.

12.12 (a) $\bar{x} = 20.45$; (b) $s = 1.54$; (c) $s_{\bar{x}} = 0.34$.

12.13 (a) The mean measures the magnitude or average of the sample in the units of measurement. (b) The standard deviation measures the variation within the sample. (c) The standard error of the mean measures the reliability of the sample in terms of sample size and variability in population sampled.

12.14 Many. In a sample of 20 the sizes of the parents were not approached.

12.15 (a) $\bar{x} = 10$; (b) $s = 1.3$; (c) $s_{\bar{x}} = 0.29$; (d) $L = \bar{x} \pm 0.82$ where $ts_{\bar{x}} = 0.82$.

12.16 Few, perhaps 2 pairs. One in 20 represented the size range of the small parent and 3 in 20 that of the large parent. One in 16 would be expected if two pairs of equally effective genes were operating.

12.17 (a) $\bar{x} = 11.65$; (b) $s = 1.37$; (c) $s_{\bar{x}} = 0.124$; (d) $L = \bar{x} \pm 0.25$ where $ts_{\bar{x}} = 0.25$.

12.18 (a) $\bar{x} = 56.2$; (b) $s = 1.14$; (c) $s_{\bar{x}} = 0.23$; (d) $L = \bar{x} \pm 0.47$ where $ts_{\bar{x}} = 0.47$.

12.19 (a) The mean was near that of the parent that required a short time for maturity. The small standard deviation shows that there is not much variation, and the small standard error indicates that a sample of 24 plants was sufficiently large to properly sample the population. (b) The end of the curve representing the shorter time required for maturity. (c) The F_2 parent of the sample recorded was probably homozygous.

12.20 (a) $\bar{x} = 73.2$; $s = 2.3$; $s_{\bar{x}} = 0.48$. The mean was slightly larger than that of the original parent with the long time for maturity. The larger standard deviation indicates more variation, probably because of more heterozygosity in parents.

The larger standard error indicates that a sample larger than 24 plants would have better sampled the population. (b) $\bar{x} = 65.25$; $s = 2.72$; $s_{\bar{x}} = 0.56$. The mean is intermediate between those of the original parents (P) and somewhat larger than those of the F_1 and F_2. The parent was taken from the end of the curve between the mean and the end representing long time for maturity. The larger standard deviation indicates more heterozygosity than that found in the other F_3 samples but not as much as shown by the F_2. The larger standard error indicates that a larger sample would be desirable. (c) $\bar{x} = 65.25$; $s = 1.29$; $s_{\bar{x}} = 0.26$. The mean is the same as that of sample (b) but the standard deviation and standard error are smaller, indicating less heterozygosity in the parent and a more adequate sample to represent this population.

12.21 (a) The parameter in question will fall within the ± limits calculated in 90% of similar trials. (b) The population mean (parameter) will fall within ±1.02 cm of the sample mean in about 68% (± one standard deviation) of similar trials. (c) The variance estimates the amount of variation in the population sampled.

12.22 (a) P is the probability that a value as great as or greater than the one calculated will occur in repeated trials. (b) The confidence interval is the + or − distance from the sample mean in which the parameter mean will be expected to occur in a specified proportion of the trials. (c) No hard and fast rule exists for determining whether particular data are significant or not significant. The investigator may decide on the basis of probability whether he will accept or reject a value with a particular probability of occurrence. Arbitrarily, if the probability values are equal to or less than 0.05 the results are usually considered to be "significantly different."

12.23 The population mean (parameter) will fall between 10 cm and 12 cm in 95% of the trials.

12.24 (a) 4 pair, (b) 8 inches, (c) $AABBccdd \times aabbCCDD$ (d) transgressive variation.

12.25 8 pairs

CHAPTER 13

13.1 (a) No, unless they are associated with phenotypes that are favored in selection; (b) mutation, selection, migration, chance (that is, random genetic dirft) and meiotic drive; (c) random mating.

13.2 0.0126 (about 13 per 1000), assuming Hardy-Weinberg equilibrium.

13.3

Group	L^M	L^N
a. Central Amer. Indians	0.78	0.22
b. Amer. Indians	0.39	0.61

c. Assuming a common origin of the two presently isolated American Indian populations and the small size of the populations, the founder principle and random genetic drift could have been major factors in the divergence in allele frequency.

13.4 (a) $L^M = 61\%$ $L^M L^N = 34\%$ $L^N = 5\%$ (b) T 0.455; t 0.545, assuming equilibrium

13.5 (a) TT 0.21; Tt 0.49; tt 0.30; (b) about 2/7; (c) about 5/7; assuming equilibrium.

13.6 A 0.262; A^B 0.0064; a 0.674, assuming equilibrium.

13.7 O-type, 92.16%; A, 5.85%; B, 1.93%, AB, 0.06%, assuming equilibrium.

13.8 (a)

	Males:	$r(a)$	$p(A)$	$q(A^B)$
		0.69	0.25	0.06
	Females:	$r(a)$	$p(A)$	$q(A^B)$
		0.77	0.18	0.05

(assuming equilibrium)

(b) Heterozygous males: 0.46; heterozygous females: 0.37

Homozygous males: 0.54; homozygous females: 0.63.

(c)

r^2	p^2	$2pr$	q^2	$2qr$	$2pq$
0.487	0.062	0.348	0.003	0.074	0.026

13.9

RR	Rr	rr
0.25	0.50	0.25

13.10 $A = \dfrac{0.8 + 2(0.5)}{3} = \dfrac{1.8}{3} = 0.6$

$a = 0.4$

males $= 0.6A; 0.4a$

females $= 0.36A; 48Aa; 0.16a = 0.6A$ and $0.4a$

Generations

	0	1	2	3	4	5	6
males:	0.8	0.5	0.65	0.575	0.6125	0.59375	0.603125 $= A$
females:	0.5	0.65	0.575	0.6125	0.59375	0.603125	0.5984375 $= A$

Therefore equilibrium is approached in the 6th generation but more generations at random mating would be required to stabilize equilibrium at $0.6A$ and $0.4a$.

13.11 $\Delta q \sim -0.0048 =$ change of q per generation. Since s is small, the value is approximate.

13.12 No AA or Aa; 100% aa.

13.13 $-0.00032 =$ change of q per generation.

13.14 $A = 0.09$, $a = 0.91$.

13.15 $A = p$ initial $= 1$

$a = 1 - p = q$ initial $= 0$

$p_1 = p_{\text{initial}} - u_p$

$= 1 - 5.2 \times 10^{-3} = 0.9948$

$p_2 = 0.9907$

$q_2 = 1 - p_2$

q_2 generations $= 0.0093$

$\hat{q} = 0.86$

13.16 0.01

13.17 0.99

13.18 Random fluctuations in gene frequencies due to sampling errors are most effective in very small populations. They become less effective as population size increases. Random genetic drift is a direct result of the sampling process. Since in this process the variance is pq/N and the standard deviation is $\sqrt{pq}\,/N$ (where N is the population size), as N becomes smaller the standard deviation becomes a significant factor.

13.19 The X and Y chromosomes have a differential effect on preferential segregation in meiosis.

13.20 People carrying the gene Si^s for sickle cell anemia are immune to the falciparum malaria parasite and are thus favored in selection in areas where malaria is prevalent.

13.21 Falciparum malaria is a selective agent which takes a heavy toll of Si^A/Si^A people. Most Si^s/Si^s people die in childhood from the sickle cell disease. Those heterozygous Si^A/Si^s are immune to malaria and not seriously affected by the sickle cell disease. To maintain this balance, a high incidence of malaria is required.

$\hat{q}_a = \dfrac{S_A}{S_A + S_a}$ where $S_a =$ selection coefficient of Si^s/Si^s

where $S_A =$ selection coefficient A of Si_A/Si_A

$0.10 = \dfrac{S_A}{S_A + 1}$

$S_A = 0.11$

13.22 $Si^s = 0.10$

$Si^A = 0.90$

$S_a = 1$

$$q_1(Si^s) = \frac{(q - sq^2)}{(1 - sq^2)} = \left(\frac{0.1 - 0.01}{1 - 0.01}\right)$$

$$= \frac{.09}{.99} = .090$$

$$q_2 = \frac{0.09 - 0.0081}{1 - 0.0081} = \frac{0.0819}{0.9919} = 0.0826$$

= frequency of Si^s in two generations.

13.23 To become a biological unit (race or species) of organisms, a population must have its own gene pool. This is accomplished in nature by some type or types of isolation that prevent the exchange of genes with members of other populations.

13.24 The blood is likely that of a black person.

13.25 Antigens listed are common among the natives of New Guinea. S is not found among Australian natives.

13.26 (a) Yes, (b) Japanese

13.27 (a) Wingless insects had a selective advantage on an island where those with wings were in danger of being blown out to sea. (b) Those without wings remained in their protected dishes and produced more wingless flies, whereas those with wings were blown away.

13.28 If two genes are linked the proportions of all types of possible dihybrid gametes depend on the frequency of crossing over. The amount of time needed for the frequency of coupling-type gametes to equal the frequency of repulsion-type gametes is inversely proportional to the linkage distance between the two genes.

13.29 With the introduction of new genes into this isolated gene pool the fitness of the individuals will become drastically reduced. This disruption of the genome and subsequent loss of fitness may lead to extinction of the population.

CHAPTER 14

14.1 (a) Inbreeding as such has no effect on gene frequency. (b) It increases homozygosity and decreased heterozygosity.

14.2 Inbreeding tends to stabilize the type but decreases the vigor of individuals. Outbreeding results in less constancy in the population but in an increase in vigor of individuals.

14.3 They have already lost essentially all of the deleterious recessives from the breeding population through past inbreeding and selection.

14.4 If self-fertilization resulted in a loss of vigor the plant must be cross-fertilized in nature, otherwise the loss would already have been sustained and no further loss would be expected.

14.5 (a) Inbreeding tends to bring recessive genes to expression and thus provides material for natural selection. Outbreeding hides recessives, and natural selection has little effect upon them. (b) Inbreeding develops constancy of type and provides a method of perpetuating desirable combinations. Outbreeding breaks up established combinations but is associated with increased hybrid vigor.

14.6 (a) Pure lines are completely homozygous. Results of various crosses can be predicted and experiments can be readily performed to verify the predictions as Mendel did in his garden pea experiments. (b) All variation within the pure line is environmental.

14.7 The expected pure lines would have the following phenotypes: yellow seed, green pod, inflated pod; yellow seed, green pod, constricted pod; yellow seed, yellow pod, inflated pod; yellow seed, yellow pod, constricted pod; green seed, green pod, inflated pod; green seed, green, pod, constricted pod; green seed, yellow pod, inflated pod; green seed, yellow pod, constricted pod.

14.8 Even in pure lines mutations occasionally would create new heterozygotes.

14.9 About 62 of the 2000 plants would be heterozygous at the end of five generations.

14.10 $1:2:1$.

14.11 The systems are less efficient in order listed (See Fig. 14.3).

14.12 (a) The abnormality apparently arose through mutation and was maintained through continued inbreeding. (b) The losses can be avoided by detecting carriers and eliminating them from the breeding stock.

14.13 (a) Maximum vigor would be transmitted in the seeds from this cross and would be manifest in the F_1 progeny. (b) The seed parent would be vigorous and would produce a good quantity of seed but the quality, with reference to hybrid vigor, would be poor. The F_2 plants would be variable with only a small proportion as vigorous as the F_1 plants. (c) If A, B, C, and D are dominant, this cross should produce as much hybrid vigor as (a). If they are not dominant, the hybrid vigor would be slightly less than that of (a). In any case the parents would be strong hybrid plants on which an abundance of good quality seed could be produced. The F_1 plants raised from the seed would be as vigorous or nearly as vigorous and uniform as those from (a).

14.14 (a) The heterozygous combination would be most efficient for production of vigor. (b) Matings between plants homozygous for A and those homozygous for B

would result in the greatest vigor. (c) A pure breeding strain with increased vigor is not likely because the linked recessives in homozygous condition would offset the advantage of heterosis. If the situation in nature were as simple as that indicated in this problem the recessives, w and l, might be removed by crossing over but in actual populations many recessives for lack of vigor seem to be linked with the dominants postulated to explain hybrid vigor.

14.15 Heterosis probably has great significance for traits with high heritability but it has been difficult to demonstrate. Prospects for the future are good, especially for meat production and beasts of burden.

14.16 (a) Inbred strains of mice have been useful in cancer studies. Strains with high incidences of different kinds of cancer have been developed for experimental purposes over a period of many years. (b) Cattle have been bred for type, and valuable breeds such as the Hereford are now available because of intense inbreeding. (c) Inbred lines are used for the crosses designated to produce hybrid seed.

14.17 Before the inbreeding process was begun, cancer occurred sporadically in mice. It was impossible to follow hereditary patterns in families and to determine whether some types were due to hereditary or environmental factors. Through inbreeding and selecting for various kinds of cancer, strains have been developed with high incidence of leukemia, and mammary, gastric, and lung cancer. This suggests that there is a genetic basis for these types of cancer and that many genes that can be made homozygous through inbreeding are involved.

14.18 (a) Apparently there are many recessive defective genes in the human population. Inbreeding would bring them to expression and would increase the number of abnormal traits in future generations. Recessives for desirable traits also would become homozygous and express themselves. (b) Outbreeding would keep recessives hidden. It would make more difficult the task of tracing defective genes in pedigrees.

14.19 (a) $\dfrac{1}{10,000}$; (b) $\dfrac{1}{16}$; (c) $\dfrac{7.2}{10,000}$; (d) $\dfrac{1}{64}$; (e) $\dfrac{1}{16}$.

14.20 (a) $\dfrac{1}{64}$; (b) $\dfrac{1}{8}$; (c) $\dfrac{1}{8}$.

14.21 (a) 25 inches, (b) 15 inches

14.22

$S_{r_1} = 0.875$
$S_{r_2} = 0.250$

frequency of $r_1 = \hat{q} = \dfrac{S_{r_1}}{S_{r_1} + S_{r_2}}$

$$\hat{q} = \dfrac{0.875}{0.875 + 0.250}$$

$$\hat{q} = 0.778$$

frequency of $r_2 = \hat{p} = \dfrac{S_{r_2}}{S_{r_1} + S_{r_2}}$

$$\hat{p} = \dfrac{0.250}{0.875 + 0.250}$$

$$\hat{p} = 0.222$$

14.23

$$h^2 = \dfrac{V(G)}{V(P)} = \dfrac{250}{600} = 0.417$$

14.24

$$\overline{W}_{m_1 m_1} = \dfrac{600}{300} = 2 \qquad \dfrac{2}{4} = 0.50$$

$$\overline{W}_{m_1 m_2} = \dfrac{1200}{300} = 4 \qquad \dfrac{4}{4} = 1.00$$

$$\overline{W}_{m_2 m_2} = \dfrac{900}{300} = 3 \qquad \dfrac{3}{4} = 0.75$$

$S_{m_1} = 1.00 - 0.50 = 0.50$
$S_{m_2} = 1.00 - 0.75 = 0.25$

$$\hat{q} = \dfrac{0.50}{0.50 + 0.25} = 0.67$$

$$\hat{p} = \dfrac{0.25}{0.50 + 0.25} = 0.33$$

CHAPTER 15

15.1 Animal behavior, the sum total of the animal's responses to its environment, has a hereditary basis. The structural and physiological characteristics on which behavior depends are inherited. Now some behavioral patterns are shown to have a genetic basis, but most of these are also strongly influenced by environmental factors.

15.2 Behavior is more difficult to explain in terms of specific genetic mechanisms than tangible, structural characteristics because behavior traits are complex and the genetic control is intimately interwoven with environmental influences. It is difficult to determine exactly what is inherited, but the general genetic basis as indicated by response to selection is well established.

15.3 Such studies indicate that simple Mendelian inheritance can be applied to behavior traits. They suggest that other behavior patterns that now seem complex and involved with environmental factors may be reduced to common denominators when the mechanics are disentangled.

15.4 Such studies identify differences in the behavior of different species and provide specific areas where hypotheses might be developed and experiments designed to investigate mechanisms that have evolved in these species.

15.5 High and low lines have been selected for preference to alfalfa pollen. A trait that responds to selection and becomes established in a line of bees must have a genetic basis.

15.6 (a) Hormones provide chemical coordination and thus influence the general health and temperament as well as the mating behavior of an animal. (b) Glands that produce hormones develop under genetic control, and hormone production in mature animals can be altered by genetic factors.

15.7 Scott and Fuller showed that genetically determined behavior differences develop under the influence of environmental factors.

15.8 In one case a low threshold of excitability was developed in females, presumably by selection, that resulted in their mating with wild-type as well as yellow males and thus perpetuating the sex-linked gene y.

15.9 Dogs have a great amount of variability, even within breeds, that responds readily to selection.

15.10 The dog is an intelligent and highly adaptive animal. Man and the dog apparently adopted each other at an early period in their known cultural history and have grown up together. Men and dogs work together in many ways for mutual advantage. Dogs live in the same shelters with men, feed on men's leftovers, provide in their own bodies a meat supply for emergencies, and adapt to the requirements for man's companionship.

15.11 In their African environment noisy dogs would call attention to themselves and their masters and either man or nature (in form of carnivores) would tend to remove the conspicuous.

15.12 A large component of behavioral patterns in a population having undergone long periods of selection is genetic. Basic differences in motivation, cultural overlay, hormonal response, and physical variations make it difficult and unsettling to transfer into a population behavioral insights that have not developed in that population.

15.13 A trait with survival value is likely to be predominantly and basically genetic. Selection in nature tends to preserve individuals and populations that have survival values.

15.14 Selection for certain bone structures, and other anatomical features as well as brain and nerve characteristics, sets a genetic pattern that favors either short, fast runs or prolonged endurance, or other favorable performance characteristics. These types of horses have been strongly selected and severely culled in developing them to their present state.

15.15 To be sure that the experiments were properly controlled and environmental factors were ruled out, questions such as the following must be considered: Are diets exactly alike? Do mothers differ in treatment of young? Does early environment provide more diversity for one strain than the other? Are both strains run through the same maze? Does time of day when tested alter results? Are temperature and light and possible disturbing factors controlled?

15.16 Aggression can be decreased in animal and human populations by training, by reducing stimuli that arouse aggressive tendencies, and by avoiding overcrowding.

15.17 Drosophila are better known genetically than mice. All chromosomes have been mapped and numerous genes have been located on chromosomes. It is possible, not only to detect behavior traits, but to determine some basic genetic mechanisms that are involved.

15.18 From the results of reciprocal crosses between geotactic positive and geotactic negative strains of Drosophila it can be shown whether the F_1 and F_2 plotted pattern is that of quantitative inheritance (polygenes) and whether a crisscross pattern (from mother to son, sex linkage) can be detected.

16.1 Experimental methods are not used, families are small, the period between generations is long, and it is difficult to obtain objective data. Modern statistical, cytological, and biochemical methods and renewed interest in human genetics make the prospects for increasing knowledge in this field very good. Much that has been learned from animal and plant experiments can now be applied to human genetics.

16.2 Most genes producing gross abnormalities in man are rare. At the same gene frequency, a dominant will be expressed more often than a recessive. A trait dependent on a rare dominant gene will not skip generations (if it is fully penetrant). Thus, its genetic basis is more readily apparent.

16.3 (a) Pedigree analyses make use of actual family history data. Until recently, most contributions to human genetics were based on this method. (b) Statistical studies were introduced by Galton and have been used extensively. New techniques have made it possible to do a great deal with small amounts of data properly ascertained. Such studies are impersonal. Individuals are lost and only group averages or other statistical units are considered. (c) Studies of twins provide a means for determining the influence of the environment on human traits. Some of the most valuable studies of human genetics have made use of identical twins.

16.4 Human hereditary diseases are incurable in the sense that the genes are already present in the individual and cannot be removed. The symptoms of hereditary diseases can be treated and sometimes prevented or removed. Symptoms of several hereditary diseases respond to environmental treatment and are now being removed, or in some cases avoided. With more knowledge of the abnormalities, more effective treatments of the symptoms will be possible.

16.5 Frequencies of defective genes would be expected to increase and an increasingly higher proportion of the population would be dependent on medical treatment for the symptoms of such diseases.

16.6 The objective of eugenics is to improve the genetic quality of the human population.

16.7 Greater knowledge and application of genetic principles can be valuable for prenatal and postnatal diagnosis of genetic disorders, restoring missing enzymes to cells, and treatment of congenital abnormalities. Some people fear that action could extend beyond that justified by basic knowledge. This could result in monstrosities in nature and misdirection of noble motives as exemplified by some early eugenics movements.

16.8 Moral and ethical considerations are involved when attempts are made to selectively control reproduction in society. Knowledge should precede important action. Although much is known about reproduction, much remains to be learned about human genetics. Questions concerning the types of mankind that should be selected for future generations have not been resolved.

16.9 As the infectious diseases are brought under control, infant mortality is decreased, and people live to an older age, many genetic and constitutional disorders will be encountered. Specialization of physicians and referral systems may be required in coping with tremendous numbers of rare diseases. Since many of these will have a strong environmental component along with a genetic background, much counseling will be required.

16.10 Chromosome anomalies are tangible and can be detected in properly prepared tissue cultures from amniotic fluid. Metabolic disorders require chemical tests which thus far have been difficult to develop and less dependable for diagnosis.

16.11 These diseases are controlled by sex-linked genes passed on from mother to son. They are virtually nonexistent in females but occur in about half the sons of carrier mothers. The sex of the fetus is important for predicting the occurrence of the disease from a particular pregnancy.

16.12 The gene is virtually always transmitted by heterozygous carriers. Sterilization of the relatively few homozygous, diseased individuals would not significantly affect the frequency of the gene in the population. Most people who have the disease are ill in childhood and few survive to the reproductive age. Nature and the disease, control the spread of the gene by homozygous individuals, but the phenotypically normal carriers transmit the gene.

16.13 Cystic fibrosis is associated with widespread abnormalities in mucous membranes and glands. The patient is subject to bacterial infection, mechanical irritation, and digestive disturbance and must be given the greatest care and protection from environmental exposures.

16.14 Genetics and cell biology were not well enough understood in the last century to be applied in medical research or practice. Most of the attention of researchers and physicians was devoted to control of infectious, epidemic diseases. These are now largely understood and quite well controlled. Degenerative and congenital diseases such as cancer, arthritis, nervous disorders, and cardiovascular disease now present the challenge.

INDEX

A

ABO blood system, 189, 190, 445, 453–456, 517, 518
  in American Indians, 453–455
Abortive transduction, 211, 213, 561
Ambystoma jeffersonianum, 361
Acentric chromosome, 319, 541
Acetocarmine smear method, 308
Acridines, 293, 296
Acrocentric, 177, 541
Acrosome, 106, 108, 109, 541
Actinomycin-D, 248, 262
Adaptation, 450, 541
Adenine, 54, 55, 70, 72, 77, 267, 268, 269, 283, 284, 293, 541
Adenosine, 78
Adenosine monophosphate (AMP), 79, 248
Adenosine triphosphate (ATP), 79, 228, 236, 395, 543
Adenylic acid, 245
Adherent ear lobes, 38, 39
Adrenal glands, 131
Adrenal hyperplasia, 521
Adrenaline, 256
Aegilops caudata, 390
  squarrosa, 368
Agglutination, 189
Agglutinin, 189, 541
Agglutinogen, 189, 541
Aging process, 333
Agropyron trachycaulum, 369–371
Alanine, 74, 82
Albinism, 14, 15, 37, 42, 188, 254, 255, 263, 381, 520, 541
  in man, 520, 521
Albino, 37, 263, 380, 381, 520

Alcaptonuria, 255, 256, 541
Alder, J., 495
Alderdice, P. W., 334, 335
Alfalfa, 342, 492
Algae, 4, 157, 386
Allele frequency, factors influencing, 433, 434, 450, 463, 464
Allele frequency for dominance, 429
Allele frequency for intermediate inheritance, 428
Alleles, 9, 10, 12, 17, 47, 116, 117, 124, 169, 187–191, 270, 541
  codominant, 14
  dominant, 12, 21, 269, 480
  functional, 201
  multiple, 187, 188
  recessive, 14, 19, 169
  semidominant, 24
  structural, 201
Allergy, 260
Allium cepa, 99
Allopolyploid, 361–364, 541
Allotetraploid, 366, 541
Alpha chain, 276, 278, 280
Alpha globulins, 519
Alpha hemoglobin, 281
Alpha particle, 290
Altbrünn Monastery, 3
Altenburg, E., 308
Alternation of generations, 111
Alverdes, F., 310
American saltmarsh grass, 364
Ames, B. N., 236, 237
Amido-transferase, 238
Amino acid (AA), 73, 74, 77–84, 228, 232–247, 274–283, 542
  activation of, 79
Amino-acyl-tRNA, 79

Amino group (NH$_2$), 73
Aminopterin, 177
Amniocentesis, 191, 343, 350, 351, 523–525, 542
Amnion, 350, 524, 542
Amniotic fluid, 524, 542
AMP phosphorylase, 524
Amphidiploid, 362–364, 369, 542
Amphidiploids from hybridization, 362–364
Amphiuma means tridactylum Cuvier, 102, 104, 305
Anabolic systems, 235
Anaphase, 97–101, 105, 114, 122, 542
Anderson, E., 108
Anderson, T. F., 214
Androgen, 247, 262, 542
Anemia, 190, 191, 333, 542
Aneuploid segregation in plants, 358, 359, 542
Aneuploidy, 343, 344, 542
Angstroms, 76
Animal breeding, 17, 443, 476–484
Animals, 2, 20, 29, 95, 100, 109, 137, 187, 281
Anlage, 4
Anther, 8, 112, 542
Anthocyanin, 27, 240, 254, 542
Anthoxanthine, 254, 542
Anthranilic synthetase, 251
Antibody, 48, 189, 191, 227, 542
Anticodon, 78, 79, 82, 542
Anticodon loop, 78,
Antigen, 48, 189, 210, 428, 445, 453–456, 542
Antimutator, 272, 273
Antisera, 189, 455, 456
Anucleolate, 77
Aphid, 308
Apical body, 109
Apis mellifera, 493
Apple, Baldwin, 362
   Delicious, 270
   Gravenstein, 362
   McIntosh, 372
Arabinose, 239
Ardrey, R., 506
Arginine, 74, 283
Aristapedia, 187
Artemia salina, 361
Artificial insemination, 16, 133, 530
Ascospore, 160, 385, 542
Ascus, 157, 160, 163, 385, 543
Asexual reproduction, 159, 543
Asexual spores, 286
Ashwell, M., 396
Asparagine, 74
Aspartic acid, 74
Aspergillus, 156
Atavism, 29, 543
Ataxia telangiectasia, 521

Atkins, J. F., 260
Atomic bomb, 300
ATP, *see* Adenosine triphosphate
Attached-X chromosome, 286
Auerbach, C., 293, 301
Autoallopolyploids, 362
Autogamy, 391, 392
Autopolyploid, 361–364, 543
Autoradiograph, 248, 543
Autosomes, 119, 120, 121, 543
Auxotroph, 211, 543
Avena sativa, 342
Avery, A. G., 366
Avery, O. T., 4, 49, 50, 53, 90, 91
Axial, 7, 109
Axial filament, 109
Axolotl, 123
Axoneme, 106, 107, 543

B

Bacillus larvae, 490
Bacillus subtilis, 51
Backcross, 19, 20, 389, 543
Bacteria, 4, 48–51, 64, 198, 208–210, 228, 231, 240, 271, 274, 301, 442
   DNA replication in, 65–57
   irradiation of, 289
   recombination in, 208–210
Bacterial genetics, 189–223
Bacteriophage, 51, 52, 67–69, 86, 196–221, 292, 543
   induced UV mutations in, 292
   lambda, 147, 219–221, 295, 305, 534
   negative strand in, 67
   ΦX174, 67, 70
   positive strand in, 67
   r and r^+, 201, 205, 295–297
   replication in, 67–69
   T1, 198, 199, 216
   T2, 197–199, 204
   T4, 90, 147, 156, 196, 200–207, 268, 273, 295–297
   T4, r11 region, 201, 203, 205–207, 225
   T5, 198
   virulent, 201
   wild type, 292
Bajema, C. V., 457
Balance theory, 121–123
Balanced chromosomal polymorphism, 456
Balanced lethal system, 286, 321, 543
Balanced polymorphism, 452, 543
Balbiani, E. G., 310
Baldness in man, 132, 138
Baltzer, F., 126

G

Painter, T. S., 311
P-aminobenzoic acid, 287
Pancreas, 274
Pangene, 4
Panmictic mating, 426, 452, 555
Panmixis, 426, 452, 555
Pantothenic acid, 286
Para-influenza, 175
Paralithodes camtschatica, 341
Paramecia, extranuclear inheritance, 391, 392
Paramecium, 495
Paramecium aurelia, 384, 392, 393, 497
Paramecium behavior mutants, 497
Parameter, 407, 410, 555
Paranoiacs, 497
Parental, 367, 555
Parental ditype, 163, 165–167
Paris Conference (1971), 333, 345, 515–17, 520
Paromomycin resistance, 386
Parthenogenesis, 124, 308, 361, 555
Partial diploid, 213
Partial gametic coupling, 142
Pascal pyramid, 36
Pasteurella, 222
Patau, K., 351
Patau syndrome, 344
Pathogenic, 221
Pathogens, 373, 555
Patterson, J. T., 335
Pauling, L., 280
Pavan, C., 313
Pawns, 497
Pears, Bartlett, 372
Peas, garden, 3, 7–10, 17–22, 28, 34, 41, 43, 157, 342, 380, 400, 465
  dwarf, 8–10, 20, 21, 30, 41
  tall, 8–10, 20, 21, 30, 41
Pedigree, 37–40, 555
Pedigree analysis, 37–40, 45, 133, 171, 172, 175, 468, 475, 555
Pedigree studies, 14, 15, 37–40, 43, 173, 475, 503
Penetrance, 259, 555
Penicillin, 298
Penicillin resistance, 269, 443
Penicillium, 298
Penrose, L. S., 348
Peptide, 276, 555
Peptide bond, 73, 79, 80, 555
Peptide linkage, 73
Percival, J., 368
Permanent heterozygotes, 322
Petals, 272
Petites, 385, 555
Phage, *see* Bacteriophage
Phage (vector), 208, 211

Pharmacogenetics, 528
Phenocopies, 257, 258, 263, 555
Phenol, 293
Phenotype, 11, 21, 23, 40, 41, 47, 121, 132, 133, 141, 187–189, 192, 195, 210, 252, 254, 258, 259, 267, 269, 274, 286, 297, 318, 327, 333, 342, 475, 556
  mottled, 129, 381
  tortoise-shell, 130
  wild-type, 195
Phenotypic ratio, 12
Phenylalanine, 74, 79, 87, 246, 254, 527, 556
Phenylalanine metabolism in man, 256, 257
Phenylketonuria, 45, 256, 521, 527, 556
Phenylpyruvic acid, 255, 256, 527
Phenylthiocarbamide (PTC), 429
Philadelphia (Ph') chromosome, 331–333
Phillips, D. M., 261
Phosphoproteins, 247
Phosphoribosyl pyrophosphate, 237
Phosphoribosyl-AMP, 236
Phosphoribosyl-ATP, 236
Phosphoribosylformimino-PRAIC, 236
Phosphorus, 51, 52, 87
Phototaxis, 500
Phycomycetes, 495
Phylogenetic tree, 282, 283
Physiological analysis, 206
Pigeons, 41
  Columba livia, 29
  Indian fantails, 29, 30
  tumbler, 477
  wild rock, 29
Pigment, 14, 27, 37, 42, 135, 251, 402
Pine trees, 483
Pirchner, F., 485
Pistil, 112, 123, 125, 556
Pistilate, 125
Pisum sativum, 2, 342
Pituitary, 131
Planococcus citri, 307
Plant fertilization, 115, 120, 125
Plant pigments, 254
Plants, 2, 4, 8, 17, 20, 24, 29, 40, 44, 95, 100, 111–118, 137, 175, 187, 223, 272, 281, 291, 322, 360, 380, 466
  euploidy, 360
  male sterility, 388–391
  polyploidy, 361
  pure lines in, 465
  sex linkage in, 175
Plaques, 200, 220, 225, 295, 556
Plasma, 189
Plasma membrane, 96
Plasmagene, 382, 389, 556

Ribonucleotides, 72
Ribose, 70
Ribosomal RNA (rRNA), 75–77, 169, 234, 245
Ribosomes, 69, 70, 74–80, 110, 200, 558
  cytoplasmic, 76
  E. Coli, 76
Ribothymidine, 78
Rice, 157, 359, 483
Ritossa, F. M., 76
RNA (ribose nucleic acid), 4, 47, 53, 70–88, 92, 110, 128, 245, 267, 293, 440, 558
  activator, 243, 244
  heterodisperse nuclear (HnRNA), 242–247
RNA polymerase, 47, 69, 72, 232, 245, 248, 556
RNA sugars, 70
RNase, 70, 75
Robinson, A. R., 458
Robinson, S. K., 536
Roentgen units, 290, 558
Rolling circle model, 67–69
Romrell, L. J., 107, 261
Rothenbuhler, W. C., 490–492, 506
Rotifers, 97, 495
Roux, W., 4, 100
Rubella, 333, 521
Russell, W. L., 289
Rust, 482
Rutger, J. N., 359
Ryan, F. J., 52, 288, 387
Rye, 111, 114

S

Saccharomyces cerevisiae, 269, 385
Saez, F. A., 135
Sager, R., 52, 288, 386, 387, 397
Salamander, 102, 103, 131, 361
Salmonella, 222
Salmonella phage P$_{22}$, 211
Salmonella typhimurium, 211, 213, 230, 236–238, 262, 269
Santiana, A., 458
Sarabhai, A., 90
Sasaki, M. S., 329
Sawflies, 124
Schull, W. J., 300, 385
Sciara, 308
Scott, J. P., 501, 506
Sea urchin, 108, 399
Sears, E. R., 360, 368, 374
Secale cereale, 111, 114
Second polar body, 102
Secondary oöcyte, 102, 558
Secondary sex characteristics, 131, 247

Secondary spermatocyte, 103, 105, 558
Secretor gene, 518, 558
Seed, 8, 9, 34, 113, 558
Seed coats, 8, 9
Seed parent, 21, 380, 558
Segregation, 8, 9, 17, 30, 40, 115, 162, 380, 400, 426, 558
Segregation distorter (SD), 450
Selection, 2, 434, 441, 449, 469, 475, 558
  differential, 475
  pressure, 449, 558
Selective agent, 443
Selective breeding, 493
Self-fertilization, 465, 466, 558
Self-pollinating plants, 390, 466, 483
Semiclone, 213
Semidominance, 24
Seminiferous tubules, 109
Semisterility, 326, 558
Sendai strain, 175, 536
Sensor genes, 243
Serine, 74, 87, 283
Serology, 48, 49, 518, 558
Sertoli cells, 109
Serum, 189, 517
Serum proteins, 518–520
Sex, 4, 35, 118–135
  chromatin body, 128, 129, 138
  chromosome aneuploidy, 352, 559
  determination, 118–135, 524
  differentiation, 128
  factor, 214–217, 221, 559
  mosaic, 126, 559
  reversals, 131, 559
Sex linkage, 14, 168–170, 559
  in man, 171–173
  incomplete, 169
Sexduction, 218, 225, 559
Sex-influenced dominance, 131, 132, 559
Sex-influenced inheritance, 132
Sex-limited gene expression, 132, 559
Sex-linked alleles, frequency for, 431
Sexual reproduction, 159, 559
Sheep, 132, 138, 271, 477
  Ancon, 271
  Columbia, 477
  Dorset, 132
  Hampshire, 477
  Lincoln, 477
  Merino, 471
  Rambouillet, 477
  Suffolk, 132
Shells, dextral, 133, 134
Shells, sinistral, 133, 134
Shen, M. W., 335